New Cosmic Horizons

New Cosmic Horizons tells the extraordinary story of space-based astronomy since the Second World War.

Starting with the early V2 experiments launched from White Sands, New Mexico, in 1946, this book explores the technological triumphs of space experiments and spacecraft designs and the amazing astronomical results that have been achieved. It also examines the fascinating way in which the changing political imperatives of the United States, Soviet Union and Western Europe have caused their space astronomy programmes to be changed over the years.

This history of astronomy from space is extensively illustrated and unique in its coverage of such a broad range of topics in language accessible to both amateur and professional astronomers, and to other technically minded readers. Where appropriate, parallel developments in ground-based astronomy are outlined to facilitate an understanding of such topics as pulsars, supernovae and the microwave background radiation.

New Cosmic Horizons covers all the major astronomy missions of the first 50 years of space research: the first space rockets, the Soviet Sputnik and American Explorer projects, the subsequent race to the moon, missions to the planets, space-based solar research, and all the wonders of modern astrophysics culminating in the wealth of exciting res

DAVID LEVERING ... n
1963. Six years late ... r-
tium building Ge ... e
European Space A ... e
manager of Mete ... e
Agency. He was ... g
Director at British ... s
responsible, amon ... s
comet, and the P ... e
Telescope. In 198 ... v
called 'Orange', ... h
Aerospace Comm

David Leveringto ... d
astronomy, as we ... t
(Springer 1995 an ... ,
undertaking rese ... -
nomical facilities ... n
the World, 2000 e

Super Cluster?

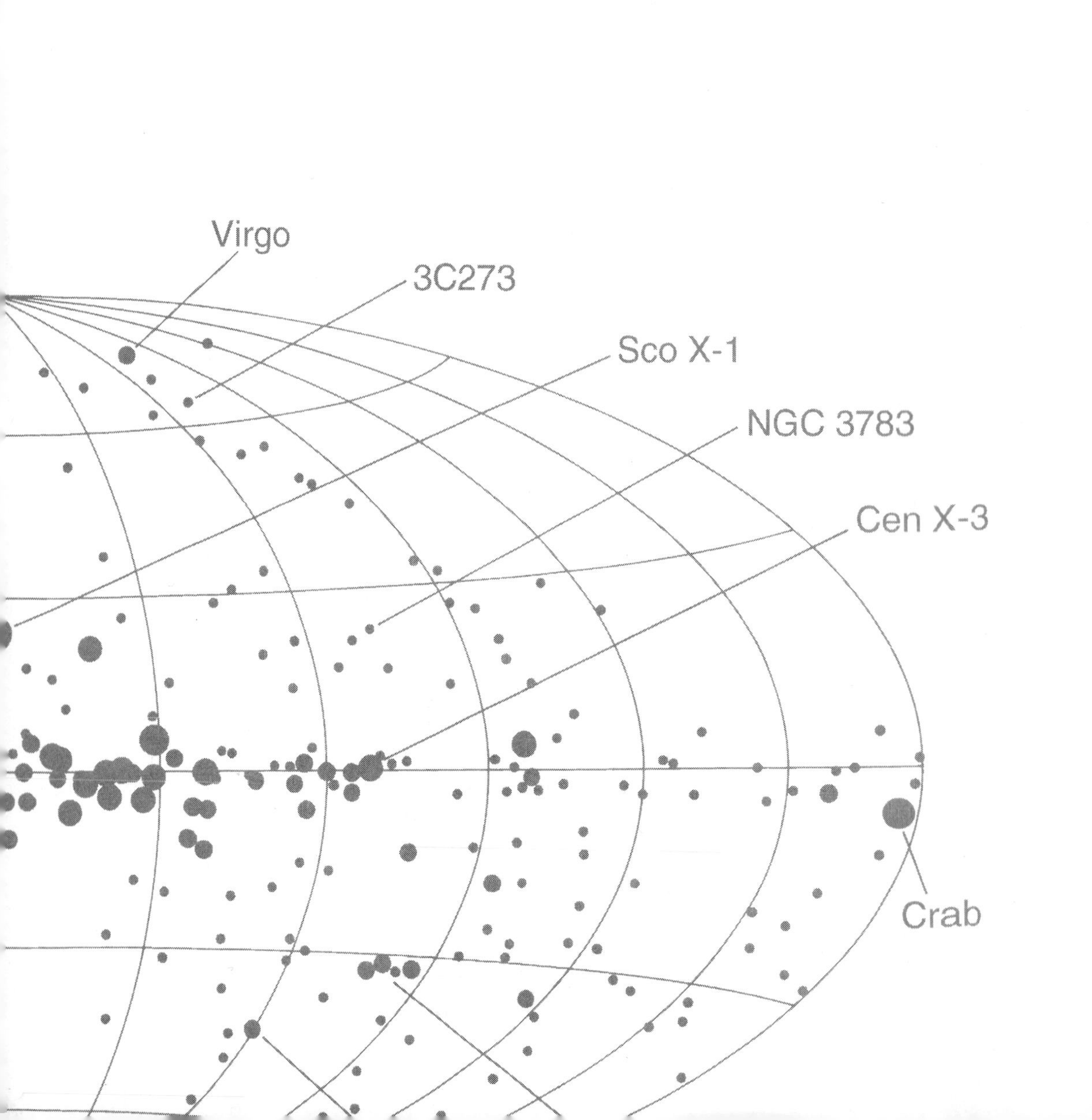
Virgo
3C273
Sco X-1
NGC 3783
Cen X-3
Crab

New **Cosmic** Horizons

Space Astronomy from the V2 to the Hubble Space Telescope

DAVID LEVERINGTON

PUBLISHED BY THE PRESS SYNDICATE OF THE UNIVERSITY OF CAMBRIDGE
The Pitt Building, Trumpington Street, Cambridge, United Kingdom

CAMBRIDGE UNIVERSITY PRESS
The Edinburgh Building, Cambridge CB2 2RU, UK
40 West 20th Street, New York, NY 10011–4211, USA
10 Stamford Road, Oakleigh, VIC 3166, Australia
Ruiz de Alarcón 13, 28014 Madrid, Spain
Dock House, The Waterfront, Cape Town 8001, South Africa

http://www.cambridge.org

First published 2000

Printed in the United Kingdom at J.W. Arrowsmith Ltd, Bristol

Typeface Trump Mediaeval 9.5/13 pt *System* QuarkXPress™ [SE]

A catalogue record for this book is available from the British Library

ISBN 0 521 65137 9 hardback
ISBN 0 521 65833 0 paperback

Contents

Preface

In the ten years or so after the Second World War astronomy was somewhat segmented, in so far as optical astronomers believed that theirs was the true astronomy, and that the new areas of radio and rocket-based astronomy would consume a great deal of money and get largely nowhere. After all, most radio or rocket-based astronomers knew very little about astronomy, as they were engineers or physicists trying to find uses for new types of radio equipment and rockets developed during the war for military purposes. Today, however, astronomy is a much more integrated subject, with research being made using the most appropriate tool for the job, whether it is a spacecraft or a ground-based optical, infrared or radio telescope. So in writing a history of rocket- and spacecraft-based astronomy, one has to decide how much research to include that used ground-based telescopes.

My approach in this book, confronted with this problem, has been to choose those topics or areas where spacecraft or sounding rockets have made a major or significant contribution. I have then explained enough of the work that has been undertaken with other facilities before, during, or after the space-based research both to make that space-based contribution understandable and to put it into context. This has meant that I have ignored those developments in astronomy which have been largely achieved using ground-based telescopes with little or no input from spacecraft.

With a few notable exceptions, such as the 200 inch (5 m) Palomar telescope or the VLA radio telescope, the availability of a particular new ground-based telescope has not had a major effect on the progress of astronomy. That has not been the case with space-based astronomy, however, as here the development of the technology and spacecraft designs has been so rapid, and the number of astronomical spacecraft so relatively few, that each new model has, in itself, yielded significant new results. So the development of spacecraft designs has been fundamental to the development of space-based astronomy, and because of this I have outlined the design, development, and operational phases of the *major* spacecraft before presenting their results. Politics has intervened in the majority of these major spacecraft programmes, so I have also explained the effect of the various political decisions and of funding constraints on these programmes.

Many books in English that discuss spacecraft-based research concentrate heavily on American spacecraft almost to the exclusion of work done elsewhere. This is often unkindly said to be because the authors are American, but the simple fact is that the majority of leading-edge, space-based astronomical research has been undertaken using American spacecraft or spacecraft with a large American contribution. Having said that, however, much work has been done elsewhere and I have tried to redress the transatlantic balance in this book by including, in particular, astronomical research undertaken in both the Soviet and European Space Agency (ESA) programmes. After all, the Soviets contributed much in the early days to lunar and planetary research, and ESA has more recently produced important astronomical spacecraft.

Any spacecraft requires the support in one way or another of hundreds or sometimes thousands of people, and it is quite wrong to talk about spacecraft as being designed by one named individual, as is done sometimes. For example, one medium-sized ESA scientific spacecraft[1] would involve an ESA management team of about twenty or thirty engineers, a spacecraft Prime Contractor and Co- and Sub-contractors totalling about another 400 engineers or so, experimenters in universities across Europe (and sometimes elsewhere) totalling a few dozen scientists, and their contractors who design and build the experiments totalling probably about another 100 people. These are the people directly involved in designing and building a spacecraft. Then there are the people who design and build the launcher, and those who control the spacecraft in orbit. I could go on, but I think the point is made that spacecraft are designed, built, launched and operated by large teams of highly skilled and dedicated engineers and scientists, and so the attribution of spacecraft to one or two named individuals is not correct.

Not only are spacecraft the results of a team effort, but so, generally speaking, are the analysis and interpretation of the results, which is why some scientific papers have a long list of authors that may total a dozen or so people in some cases. The four COBE papers in the *Astrophysical Journal* of 10th January 1994, for example, included 23, 18, 11 and 20 names as authors, so it is quite incorrect in many cases to attribute a discovery or development to one or two named individuals. The usual solution in discussing these developments is to choose the leader or leaders of the team by name, and include the others by the words 'and colleagues' or 'et al', but it is often not possible to tell from the paper who was the leader or leaders,

[1] The detailed management structure for NASA spacecraft programmes is different, but the numbers of people involved is broadly similar for a similar-sized programme.

although secondary sources sometimes name these people. If the leader is not clear, I have named the first-named author of the paper.

It is also often difficult to know what dates to quote. Precedence on discoveries is usually based on dates of first publication, but the recent trend to quote pre-publication dates, submission dates, or dates of electronic publishing is making this more difficult. In astronomy we also need to consider the date when an observation was made, or the date that the significance of the observation was realised, which in exceptional cases can be some years later than that of the observation, particularly with satellite-based research. I have tried my best to use whatever date seems appropriate in each case, but because of the above problems these dates may not always be the same as those mentioned elsewhere. Both may be correct, of course, but based on different criteria. As this book is a general history, rather than a detailed history which would refer to an exhaustive list of primary sources, these differences in date, which usually are no more than one year, are generally not of major significance.

It is not desirable, in my view, to try to write a scientific history that covers discoveries and developments right up to the last possible date, as by the time the book is published (and read) it is inevitably out-of-date. History also requires considering matters in their proper perspective, and developments that seem important today may pale into insignificance later, and vice versa, again arguing against the idea of trying to keep the text fully up-to-date. On the other hand it is important, considering that the subject of this book is the history of relative modern astronomy, to try to keep the book as up-to-date as possible. Sputnik was launched in 1957, and this book was written between 1997 and 1999, so it seemed both practical and sensible to try to cover the main astronomical developments up to about 1996, thus covering about forty years of spacecraft-based research. Furthermore, as this book is a history of astronomy from space, rather than a history of astronomical spacecraft, I have not included details of any spacecraft that were launched after this date, even though they were being designed and built before 1996. I have also deliberately excluded the Mars Pathfinder and Lunar Clementine spacecraft from the text, as they are both recent[2] programmes and mark the start of a new era of planetary and lunar exploration.

It has not been practical nor sensible to try to include details of all lunar, planetary and astronomical observatory spacecraft in this book, so some

---

[2] The first results of the Mars Pathfinder spacecraft were not available until 1997, which is after my nominal cut-off date.

selection has been inevitable. Broadly speaking I have chosen, therefore, to outline those spacecraft or spacecraft programmes that have produced the most significant results. Sometimes I have presented the story from a spacecraft perspective, with sections headed Skylab, Uhuru or COBE, etc., and sometime from an astronomical perspective, with sections on solar flares, pulsars or gamma-ray bursts, etc., depending on circumstances. The Einstein Observatory, for example, has a section of its own as it produced important results in a number of areas, whereas BeppoSAX, which is historically not as important, is only mentioned in the section on gamma-ray bursts where it had a major impact. On the other hand, the X-ray Timing Explorer spacecraft is not mentioned at all, because at the time of writing its pre-1997 results did not seem to be of major significance. Not everyone will agree with my selection of spacecraft and astronomical topics, as any selection is bound to be subjective, but I believe that taken as a whole the selection is fair.

This book is written for the reader with a basic understanding of astronomy, whether as an amateur or professional. I have tried, therefore, to choose a set of units which would be acceptable to all these readers, some of whom may prefer imperial or cgs units, whereas others may prefer SI units where they exist.

Last, but not least, I would like to express my special thanks and appreciation to Richard Baum and Anthony Kinder who read the chapters on the Moon and Planets, Alan Cooper who read the chapter on the Sun, and Roger Emery who read the chapters on astrophysics. They had the kindness and patience to read the text and suggest modifications of both fact and style, to make the book more accurate and readable. To them all I am most grateful, but if there are any errors of fact, misinterpretation or misunderstanding remaining, they are entirely mine.

Chapter 1 | THE SOUNDING ROCKET ERA

1.1 Early Work with the V2

Just before the end of the Second World War, as the Allies were tightening their net on Nazi Germany, the Germans decided to move all 4,000 staff of their V2 rocket research establishment at Peenemünde on the Baltic coast to a safer location in the south of the country. The site chosen was the V2 Mittelwerke factory near Nordhausen in the Harz mountains. The team were not to stay in their new location for very long, however, as almost immediately Wernher von Braun, the technical director, and 500 of his key scientists and engineers were on the move again to a safer refuge. After burying their design archive in an abandoned mine shaft, they moved in April 1945 to an old army camp at Oberammergau in the Bavarian Alps. It was ironic that this small town, the site of a three hundred year old Passion Play, should be the place where some of the team,[1] who had been working on one of the most sophisticated weapons of war, should surrender to the Americans.

Shortly after the move to Oberammergau, the V2 factory at Nordhausen also fell into American hands, giving them six weeks to remove as much hardware as possible before the area was handed over to the Russians as part of the Potsdam agreement. Over a period of nine days, under the supervision of Colonel Holger Toftoy of the US Army Ordnance Department, railway wagons were loaded with 360 tons of V2 parts to be transported to American ships waiting to take them to the USA. The design archive was also located,

[1] Von Braun was not at Oberammergau when the Americans arrived, however, because he had broken his arm and had been moved, via a local hospital, to Oberjoch in the Bavarian Alps. He surrendered there to the Americans with some of his colleagues.

and eventually von Braun and 130 of his key staff followed the V2 hardware and archive across the Atlantic.

Not only was the V2 to play a major rôle in early space research, but von Braun and his team were to be key in the development of American military rockets, and be involved in the design of the Saturn launcher that was to put the first man on the Moon in 1969. Not to be outdone, the Russians also used V2 parts and German V2 engineers in their early military rocket programme, although the Germans were not as prominent in the Russian programme as in the American one. The Russian programme was to result in the launch of the Earth's first artificial satellite, Sputnik 1, in 1957.

When the V2 parts arrived in America in the late summer of 1945 they were sent to the US Army's White Sands Proving Ground, then under construction in the New Mexico desert, whilst the German engineers were located under strict military supervision at the inappropriately named Fort Bliss in Texas. Their job was to assemble and test the V2 rockets (see Figure 1.1) in support of an American project called "Hermes", which was to produce a tactical ballistic missile system. The Joint Chiefs of Staff insisted that the project should be a model of interservice cooperation and Holger Toftoy, who was put in overall charge, also decided to include scientific researchers in the team to undertake fundamental research into the upper atmosphere.

In early 1946 a V2 panel was set up to coordinate this scientific programme on behalf of Army Ordnance. The organisations represented on the panel, some of whom were to figure prominently in the early years of space research, were the Naval Research Laboratory (NRL), the Johns Hopkins Applied Physics Laboratory (APL), General Electric (GE), who was the Hermes contractor, the universities of Harvard, Princeton and Michigan, and the Signal Corps. Ernst Krause of NRL was chairman of the V2 panel, and the NRL V2 team included Edward Hulbert, who had been involved in ionospheric research since the 1920s, Richard Tousey and Herbert Friedman, who were to make significant discoveries over the next few years using rocket-borne experiments, Milton Rosen, who was to become technical director of the first American attempt to launch an artificial satellite (Project Vanguard), and Homer Newell, who was to become director of the Office of Space Sciences in NASA. Other key figures in the V2 programme included James Van Allen (APL), who was to discover the radiation belts around the Earth that now bear his name, and astronomers Fred Whipple of Harvard and Leo Goldberg of Michigan. Most of those actively involved in the V2 scientific programme were very young, being generally in their twenties or early thirties, and so they were ideally suited to cutting edge research.

Figure 1.1 A V2 rocket at the White Sands Proving Ground in New Mexico. The V2 could reach a maximum height of about 180 km. (Courtesy *White Sands Missile Range*.

There was also a smattering of older people, however, which added experience to the team.

As a weapon of war the V2 could carry a 900 kg warhead a range of some 200 to 300 km. As a research vehicle it was expected to carry a similar payload to a height of about 150 to 180 km. The first launch from White Sands took place on 16th April 1946 with a Van Allen Geiger counter to detect cosmic ray primaries as its main scientific instrument, but control of the rocket was lost at an altitude of 5 km before it crashed into the desert. Further experiments launched on 10th and 29th May also failed. The first because, although the V2 was successful, the payload was destroyed on impact, and the second because the telemetry system failed and the rocket exploded.

Further experiments launched in June and July also failed, and then on 10th October 1946 the scientific teams had their first success when Richard Tousey's NRL group photographed the ultraviolet spectrum of the Sun down to a wavelength of about 230 nm for the first time. There was great excitement as first *The Washington Post* and then other American newspapers carried the story, but the resolution of the spectrum was too coarse for anything other than a simple analysis. Over the next two years further NRL and APL spectrographs flew producing higher resolution spectra but, although they showed that the intensity of the solar ultraviolet was less than expected, and provided data on the distribution of atmospheric ozone, their resolution was still not high enough for a detailed analysis of the Fraunhofer lines.[2] Consequently, astronomers began to lose interest in the work.

1.2 Solar X-Rays and the Earth's Ionosphere

The V2 was not the only sounding rocket available in the USA in the mid 1940s. The Jet Propulsion Laboratory (JPL) of the California Institute of Technology (Caltech)[3] had started designing the WAC Corporal sounding rocket in 1944 which, on its first successful flight of 11th October 1945,

[2] Fraunhofer lines are explained in the Glossary near the end of this book.

[3] The first tentative steps in rocket research had started at Caltech in 1936, but in 1944 work started on designing a rocket for military use. For legal reasons, this work could not be undertaken by the university as such, so the rocket team was hived off as a separate centre for guided missile research under the name of the Jet Propulsion Laboratory.

reached an altitude of 72 km. It was very much smaller than the V2, however, with a payload capacity of only 11 kg. This was insufficient for many scientific purposes, and so in 1946 APL started work on an enhanced version called Aerobee,[4] which was initially designed to launch a payload of 70 kg up to an altitude of 110 km. First launched in 1947, the Aerobee largely took over the scientific research programme that had been started with the V2 when the last V2 was launched in 1952. The Aerobee was so successful, in fact, that it was used for scientific research until 1985, when over 1,000 had been launched.

One aim of the early V2 scientific experiments was to understand how the various ionisation regions[5] of the Earth's upper atmosphere were created and maintained. It appeared as though the ionising radiation came from the Sun, as radio soundings in the 1920s and 1930s showed that the Earth's ionosphere exhibited both diurnal and seasonal variations. It could also be disrupted from time to time by solar flares, but the type of solar radiation producing these changes was unclear. Most researchers in the 1940s thought that the different regions of the ionosphere were caused by solar ultraviolet radiation of different wavelengths, but in 1938 Edward Hulbert of NRL and the Norwegian Lars Vegard had independently suggested that the lower ionisation regions may be caused by solar X-rays. This was a revolutionary suggestion at the time, as it implied that the Sun must have regions much hotter than the known 6,000 K temperature of its surface. In 1941, however, the Swedish physicist Bengt Edlén confirmed an earlier suggestion by Walter Grotrian of the Potsdam Astrophysical Observatory that the Sun's corona was very much hotter than its surface. Edlén calculated a minimum coronal temperature of 2 million K, which would be quite sufficient to produce X-rays.

The first observational indication that the Sun may be emitting X-rays was obtained by Robert Burnight of NRL on a V2 flight of 5th August 1948. He used photographic plates covered by thin beryllium windows, which filter out visible and ultraviolet wavelengths, and found that they showed darkening after development, which he attributed to X-rays. Unfortunately, a repeat experiment on 9th December 1948 using beryllium and aluminium windows showed no such darkening. At about the same time Tousey, Watanabe and Purcell of NRL also found indications of solar X-ray emission

---

[4] The name *Aerobee* is a combination of *Aerojet*, the name of the rocket prime contractor, and *Bumblebee*, the APL programme name.

[5] The different regions of the ionosphere reflect radio waves of different wavelengths. (See also Glossary under 'ionosphere'.)

using thermoluminescent material[6] flown on V2 rockets, but it was left to Herbert Friedman, also of NRL, to *prove* that X-rays in the Earth's upper atmosphere came from the Sun.

Friedman's first successful experiment, to prove the solar origin of these X-rays, used photon counters sensitive to three ultraviolet and one X-ray band on a V2 which reached a height of 151 km on 29th September 1949. Those counters sensitive to soft X-rays started to respond at an altitude of about 85 km, and showed a gradual increase in X-ray intensity up to the peak altitude of about 150 km as the rocket traversed the E region of the Earth's ionosphere. Moreover, the response of these counters was higher when they pointed towards the Sun, proving the solar origin of these X-rays. This initial, relatively crude experiment indicated that the intensity of the X-rays may not be high enough to sustain the E region. After a number of rocket failures, Friedman was able to prove that the X-ray intensity *was* sufficient to sustain the E region, however, using a more sophisticated experiment launched on a Viking sounding rocket on 15th December 1952.

Tousey, Watanabe and Purcell's V2 experiments from 1948 to 1950, mentioned above, which used thermoluminescent material, demonstrated that the Lyman-α emission line of hydrogen at about 120 nm dominated the far ultraviolet solar spectrum, but their experiment was not designed to produce a spectrum. Friedman's Lyman-α measurements on V2 and Viking flights in September 1949 and December 1952, respectively, suffered from similar limitations, but both Tousey and Friedman's experiments showed that the intensity of the Lyman-α solar radiation was enough to maintain the Earth's ionospheric D region, which exists at from about 50 to 90 km altitude.

At about the same time, a University of Colorado team led by William Rense had been designing an experiment to measure the ultraviolet solar spectrum, using a grazing incidence spectrograph fitted to an Aerobee sounding rocket. This rocket incorporated a biaxial control system, also designed by the University of Colorado, to keep the spectrograph pointing at the Sun. Unfortunately, the first Aerobee with this control system fitted, which was flown on 12th April 1951, failed when a fuel line broke at an altitude of about 30 km. Further failures followed, but on 11th December 1952 everything finally worked, and the first solar spectrum showing the Lyman-α emission at 121.6 nm was produced. Four years later the Colorado group obtained the first photograph of the Sun in the light of Lyman-α.

[6] Themoluminescent materials absorb high energy radiation which can be later released as visible light by heating.

Over the next few years, Friedman's group continued to study solar X-rays, finding that the majority of solar radiation in the range 0.8 to 2.0 nm was due to the corona, then in 1956 they also found that solar flares produce very high energy X-rays. This NRL group measured the total energy emitted by the Sun in the whole X-ray band to be only about 1 erg/cm^2/s at our distance from the Sun, however, compared with a total solar energy at all wavelengths of 6×10^{10} erg/cm^2/s at the same distance. So, if the Sun was a typical star, the prospect of being able to detect *stellar* X-rays appeared very remote without an enormous improvement in detection efficiency.

Friedman's NRL group also flew ultraviolet sensors at night using Aerobees, and on 17th November 1955 they found a great deal of stellar radiation in the far ultraviolet, although they were unable to locate the sources. During a daylight launch in July 1956 (see Section 10.3), they also measured some hard X-rays that did not appear to come from the Sun, but they were unable to detect these X-rays again on subsequent night-time launches, leading them to conclude that these non-solar X-ray measurements were spurious.

So in the first ten years of using sounding rockets, physicists and engineers had expanded their work from measuring the Earth's atmosphere and ionosphere, to measuring ultraviolet and X-ray emissions from the Sun, to detecting ultraviolet emissions from stars, although the detection systems were not accurate nor sensitive enough to detect ultraviolet emission from individual stars. The sounding rocket work was still very speculative as far as *astronomers* were concerned, however, as most of the work was still being undertaken by physicists and engineers developing new instruments and new measuring techniques to see what research could be carried out using sounding rockets as carriers. In other words, the research was equipment-led rather than astronomy-led.

Although astronomers had been involved from time to time in this early sounding rocket work, they had generally been disillusioned by the technical risks involved and by the relatively poor quality of the results. Institutionally, astronomers were also more used to working alone or in small groups using optical telescopes, and so they generally found it very difficult to work with teams of rocket engineers and physicists located hundreds of kilometres away from their astronomical observatories. What was just as bad, as far as many astronomers were concerned, was that the instruments used in rocket-based research were generally electrical, not optical, so many of the results appeared as electrical signals rather than in the photographic form to which they were used. Interestingly, astronomers also found similar problems in trying to work with radio engineers and physicists who were developing and working with the first radio telescopes during the same period.

Throughout the above period a great deal of work had been undertaken in the USA on developing various new sounding rockets and military missile systems. As mentioned above, the Aerobee sounding rocket was used extensively by scientists after the stock of V2s had been exhausted.[7] Unfortunately the Aerobee's maximum payload capability of 90 kg compared very unfavourably with the 900 kg of the V2. In the meantime, however, a team led by Milton Rosen of NRL had designed a sounding rocket to launch payloads in the intermediate 200 to 500 kg range. Initially called Neptune and then renamed Viking,[8] this rocket had its maiden flight on 3rd May 1949. Unfortunately, the engine shut down early and the rocket only reached about one-third of its expected altitude. Further problems occurred with the second and third Viking launches, and then on 11th May 1950 the fourth firing was successful, lifting a payload of 440 kg to an altitude of 170 km. The launch took place from the surface of a ship (the *Norton Sound*) stationed on the geomagnetic equator. Unfortunately, only a limited amount of experimental data was received.

The Viking was a considerable advance on the V2 as a sounding rocket as it had:

- an aluminium airframe to reduce weight and provide a non-magnetic shroud that would not interfere with cosmic ray and other experiments. (The V2 airframe was made of steel.)
- a servo-controlled, gimballed motor for stabilisation and control, which was more reliable than the graphite vanes used by the V2.
- a design that could accept a wide range of payload weights and still remain stable. (The V2 had to be ballasted to keep it stable).

Although the Viking provided a better environment for experiments than any other large sounding rocket of the period, it was very expensive, costing about $200,000 per vehicle, compared with about $30,000 for a reassembled V2 or about $15,000 for an Aerobee.[9] The costs per kilogram of

---

[7] The last of the 67 V2s to be launched from White Sands lifted off on 19th September 1952. About two-thirds of the V2s were successful; the highest altitude reached by a V2 being 212 km on 22nd August 1951.

[8] The name was changed because the US Navy were developing an aeroplane also called Neptune.

[9] These costs, which excluded the development costs and the costs of launch operations, are only approximate as they depended, amongst other things, on how many rockets were ordered in a batch.

payload for the Viking and Aerobee were a similar order of magnitude, but there was a problem in finding enough scientists who wanted to fly experiments on a similar trajectory at a similar time to fill a Viking payload bay. So the Viking was not really a great success as a sounding rocket, only twelve being launched, of which seven were successful.

1.3 Non-Solar X-Ray Sources

Although it was clear that it would be impossible to detect X-rays from *solar-type* stars with the instruments available in the mid 1950s, as the X-ray emissions would be too faint, advances made in ground-based radio astronomy over the previous few years had shown that there were a number of objects that were bright at radio wavelengths but dim in visible light. Some astronomers reasoned that a similar situation may also apply in other wavebands, so there may be some detectable X-ray sources after all. With this in mind, the Langley Aeronautical Laboratory published a report in May 1958 suggesting that astronomical observations should be made from space in the infrared, ultraviolet, X-ray and γ-ray bands.[10] In October of the same year Sir Harrie Massey published a paper for the Royal Society in the UK making a similar recommendation. Other papers appeared at about this time discussing the possibility of successful observations in these wavebands. To give a maximum chance of success with either spacecraft[11] or sounding rocket experiments, however, the design of X-ray detectors needed to be substantially improved, and that is where Riccardo Giacconi and Bruno Rossi came in.

In 1958 the National Academy of Sciences established a Space Science Board to advise NASA on a space research programme. Three members of the Board, John Simpson (University of Chicago), Leo Goldberg (Harvard University) and Lawrence Aller (University of California) suggested that a survey of the sky should be made in X-rays. A fellow Board member, Bruno Rossi (Massachusetts Institute of Technology, or MIT) was looking for new areas of research at the time and he was readily persuaded that such an X-ray programme would be worth the effort.

Rossi, whilst still working full-time as professor of physics at MIT, had become chairman and main consultant of a small technical company that had just been set up by two of his former MIT students, Martin Annis and

[10] Only radiation in the visible, radio and parts of the infrared wavebands is detectable from the surface of the Earth because the other wavebands are absorbed by the Earth's atmosphere. [11] The first spacecraft had been launched in 1957, see Chapter 2.

George W. Clark. This company, called American Science and Engineering Inc. (AS & E), hoped to be able to develop commercial products on the back of research into nuclear radiation that was being paid for by the Department of Defense (DoD). When NASA started operations in October 1958, AS & E decided that they would also like to get involved in space research and recruited Riccardo Giacconi, a young physicist who had recently come to the USA from Italy, as director of space research. Giacconi's background was in cosmic ray research, where he had been studying high energy nuclear collisions, but he found that work frustrating because primary cosmic rays could only be observed at high altitude, and their observation was particularly difficult. In addition, high energy particle accelerators were beginning to take over the study of high energy nuclear collisions. Then in September 1959 Giacconi met Rossi at a party, where Rossi suggested that Giacconi start a programme in X-ray astronomy. Giacconi needed little persuading and he started to throw all his energies, which were considerable, into this new field.

A few months later Giacconi, Clark and Rossi wrote a seminal paper discussing the type of processes that might generate detectable quantities of non-solar X-rays and the type of astronomical sources in which such processes may occur. They concluded that there may be four types of sources that could possibly be detectable at X-ray wavelengths, namely hot, bright stars, rapidly spinning stars with large magnetic fields, stars that flared, and supernova remnants. The best known supernova remnant was the Crab Nebula, which is the remains of a supernova that had exploded in the year 1054, and in their report Giacconi and his colleagues suggested that this remnant, which was a strong emitter of radio waves, may also emit X-rays.[12] It was still evident, however, that if any of these potential sources were to be detected, they would be relatively weak, so the design of X-ray detectors needed to be substantially improved.

Friedman's group at NRL had used large area Geiger-type counters to detect solar X-rays in the 1950s, but this type of detector suffered from relatively high levels of background noise. Rather than continuing to try to improve his detectors, Friedman's group had basically given up X-ray research in the late 1950s and migrated to the ultraviolet waveband. This was thought to be more promising as ultraviolet sources would be at a lower temperature than X-ray sources and so, presumably, should be more numerous.

Ten years earlier, the German physicist Hans Wolter had shown theoretically that it was possible to build an X-ray microscope using glancing

[12] Interestingly, Herbert Friedman had made the same suggestion in a *Scientific American* article a few months earlier.

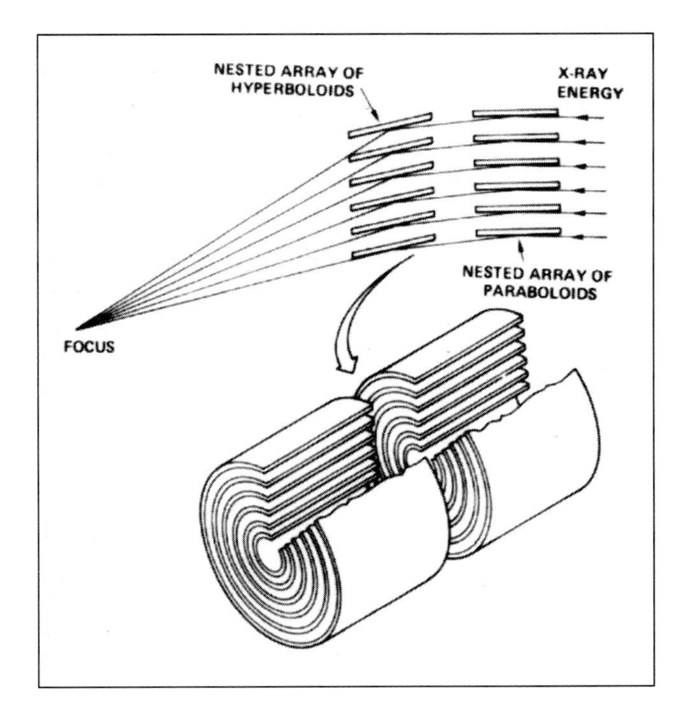

Figure 1.2 The principle of the Giacconi–Rossi design of X-ray telescope. The X-rays are reflected at glancing incidence by two sets of concentric cylinders. The space-facing cylinders (on the right hand side) have a parabolic shape, whilst the other cylinders have a hyperbolic cross-section. (From Golub and Pasachoff 1997, *The Solar Corona*, Cambridge University Press.)

incidence of X-rays off a parabolic and hyperbolic mirror in succession to bring them to a focus. He abandoned the idea of constructing such a microscope, however, as the mirrors were too difficult to construct. Giacconi came across this idea by chance in 1959, and thought that an X-ray telescope could probably be constructed using such glancing-incidence mirrors. Rossi, hearing of Giacconi's idea, suggested that, as the X-rays were being focused by glancing incidence, a number of cylindrical mirrors could be constructed and fitted, one inside the other (see Figure 1.2), to increase the effective collecting area and hence improve the detection rate.

In parallel with trying to construct and launch such an X-ray telescope, which Giacconi recognised could take some years, Giacconi decided to try to improve the detection of X-rays using more conventional detectors. He made a proposal to NASA to detect X-rays from stars using these conventional detectors, but this first proposal was rejected. He then contacted the Air Force Cambridge Research Laboratories (AFCRL) and suggested that he should measure solar X-rays and try to detect X-rays from the Moon, which he surmised may be produced by X-rays or high-energy particles from the Sun impacting on the lunar surface.

The AFCRL accepted his proposal in early 1960, but the first experiment, using a small (1 cm^2) Geiger counter, failed when a problem occurred with the small Nike-Asp sounding rocket. Undeterred, Giacconi modified his next proposal to AFCRL to add the possible detection of non-solar X-rays as an aim. AFCRL accepted his proposal and agreed to fund four Aerobee flights.

An Aerobee allowed Giacconi to launch a much larger experiment than using the smaller Nike-Asp rocket. This new experiment consisted of three 10 cm^2 area Geiger counters, with a scintillation counter behind each. Both X-rays and cosmic rays would trigger the Geiger counters, but only cosmic rays would penetrate to the scintillation counters. So, using this so-called 'anticoincidence' arrangement, the noise generated in the Geiger counters by cosmic rays could be detected by the scintillation counters and be discounted. These X-ray detectors, as a result, were 100 times more sensitive than Friedman's but, even so, it was thought that they would not be nearly sensitive enough to detect non-solar X-rays, but it was worth a try.

The first launch of this new AS & E experiment, which took place on 24th October 1961, was a failure, as the payload door, which allowed the X-ray detectors an unrestricted view of space, failed to open. The second launch attempt, on 18th June 1962, was a resounding success, however, even though one of the three Geiger counters failed. The doors opened on schedule and allowed the experiment to operate for over five minutes. Each time the rapidly spinning rocket took the detectors passed the Scorpio region of the southern sky, in a direction of about 200° (see Figure 1.3), they detected an increase in X-ray intensity. In addition, they found both a much fainter source in a direction of 60°, and a diffuse source of X-ray radiation across the whole sky that they scanned.

Giacconi and his colleagues' initial reaction was that the peak was due to the Moon, but they soon found that the attitude measurements were correct, and the Moon was some 30° away from the position of the large peak. Other possible causes of the two peaks and background radiation were considered, but in the end Giacconi and his colleagues concluded that all the data were real and caused by non-solar X-rays. The results were announced by the research team of Giacconi, Gursky, Paolini and Rossi to an astonished audience at the Third International Symposium on X-ray Analysis held at Stanford University in August 1962. Friedman, who was in the audience, congratulated Giacconi and then went back to NRL to re-start his own non-solar, X-ray astronomy programme, which he had largely abandoned over five years previously.

The third AS & E Aerobee launch on 12th October 1962 was also a success, although Herbert Gursky *et al.* could not detect the intense source

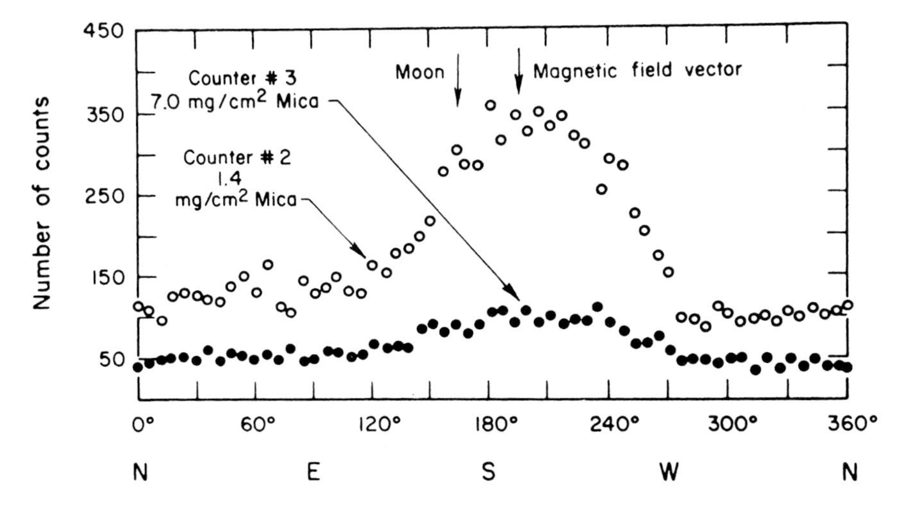

Figure 1.3 The data provided by the Aerobee experiment flown on 18th June 1962, showing a clear peak for counter 2 at about 200° some 30° from the direction of the Moon, with a smaller peak for counter 3. (Counter 1 failed). The measurements of neither counter fell to zero, indicating that there is a general background of X-rays across the area of sky measured. (R. Giacconi *et al.*, *Physical Review Letters* 9 (1962), p. 440, Copyright 1962 by the American Physical Society.)

of X-rays (now called Sco X-1) seen on the previous launch, as Sco X-1 was out of sight below the observational horizon. Two new sources were discovered, however, including one in the region of the Crab Nebula, but the position measurements were not accurate enough to be sure that the latter source was the Crab.

The existence of the X-ray source Sco X-1 was confirmed by C. Stuart Bowyer and Herbert Friedman (NRL) using a sounding rocket experiment launched in April 1963. This confirmation by the respected Friedman group finally convinced many previously sceptical astronomers that non-solar X-ray sources really did exist. The new NRL detector, which had been patented by Bowyer, had the very large window area of 65 cm^2 and was ten times as sensitive as the detector of the AS & E group. In addition, it was able to measure the position of possible sources much more accurately, allowing Bowyer and Friedman to measure the position of Sco X-1 to within about 1°, and to determine that its angular diameter was less than 10 arcmin. The NRL team also detected the diffuse X-ray background radiation and found a source which was clearly associated with the Crab Nebula.

The last of the four AS & E Aerobee launches on 10th June 1963 confirmed the Sco X-1 source and detected yet another source. So after just three successful sounding rocket launches, with a total observing time of only about fifteen minutes, the AS & E group had detected X-rays from Sco X-1, from the region of the Crab Nebula, and from three other unknown sources, together with measuring a diffuse X-ray glow across the whole sky. The NRL group with one launch had confirmed Sco X-1 and the Crab Nebula as sources, and had measured their positions much more accurately than the AS & E group. They had also confirmed the existence of the diffuse X-ray glow.

So detectable non-solar X-rays clearly existed, but how were they produced? In the early 1960s a number of production mechanisms were proposed. Firstly, X-rays could be generated by high-energy synchrotron radiation which is produced when energetic electrons spiral in an intense magnetic field. Or they could be produced when an electron collides with a photon and transfers some of its energy to the photon in a process called the 'inverse Compton effect'. Alternatively, X-rays could be produced in the bremsstrahlung ("braking") process, in which a photon is emitted when an electron is decelerated by passing close to an atomic nucleus. Or maybe they could be generated as part of the black body radiation by a very hot, very diffuse gas like the Sun's corona.

Whilst astronomers were speculating on these possible X-ray processes, a young physicist named Hong-Yee Chiu was investigating possible processes taking place in neutron stars. In this work he calculated that the surface temperature of a neutron star could be about 10^7 K on formation, causing it to emit X-rays of about 0.3 nm wavelength. After about 1,000 years, however, the temperature would reduce to about 10^6 K, producing X-rays of 3 nm wavelength.

Theory showed that a neutron star is produced following the supernova explosion of a heavy star, and one of the best-known supernova remnants, which had recently been shown to be an X-ray emitter, was the Crab Nebula discussed above. Although the Crab Nebula was also known to emit synchrotron radiation in the visible and radio wavebands, Friedman and his colleagues at NRL thought that the electron energies required to produce X-rays by synchrotron radiation would be too high. So maybe, Friedman thought, the source of the Crab's X-rays was a neutron star. If that was the case, the X-rays would appear to be coming from a point source when observed at the distance of the Earth.

In early 1964 Friedman learnt from his radio astronomy colleagues at NRL that the Moon would eclipse the Crab Nebula on 7th July of that year. If Friedman could show that the Crab's X-rays suddenly disappeared as the Moon gradually eclipsed the X-ray source, he would have been the first person to have 'observed' a neutron star. If the Crab's X-rays came from an extended source, however, the cut-off of the X-rays would be gradual.

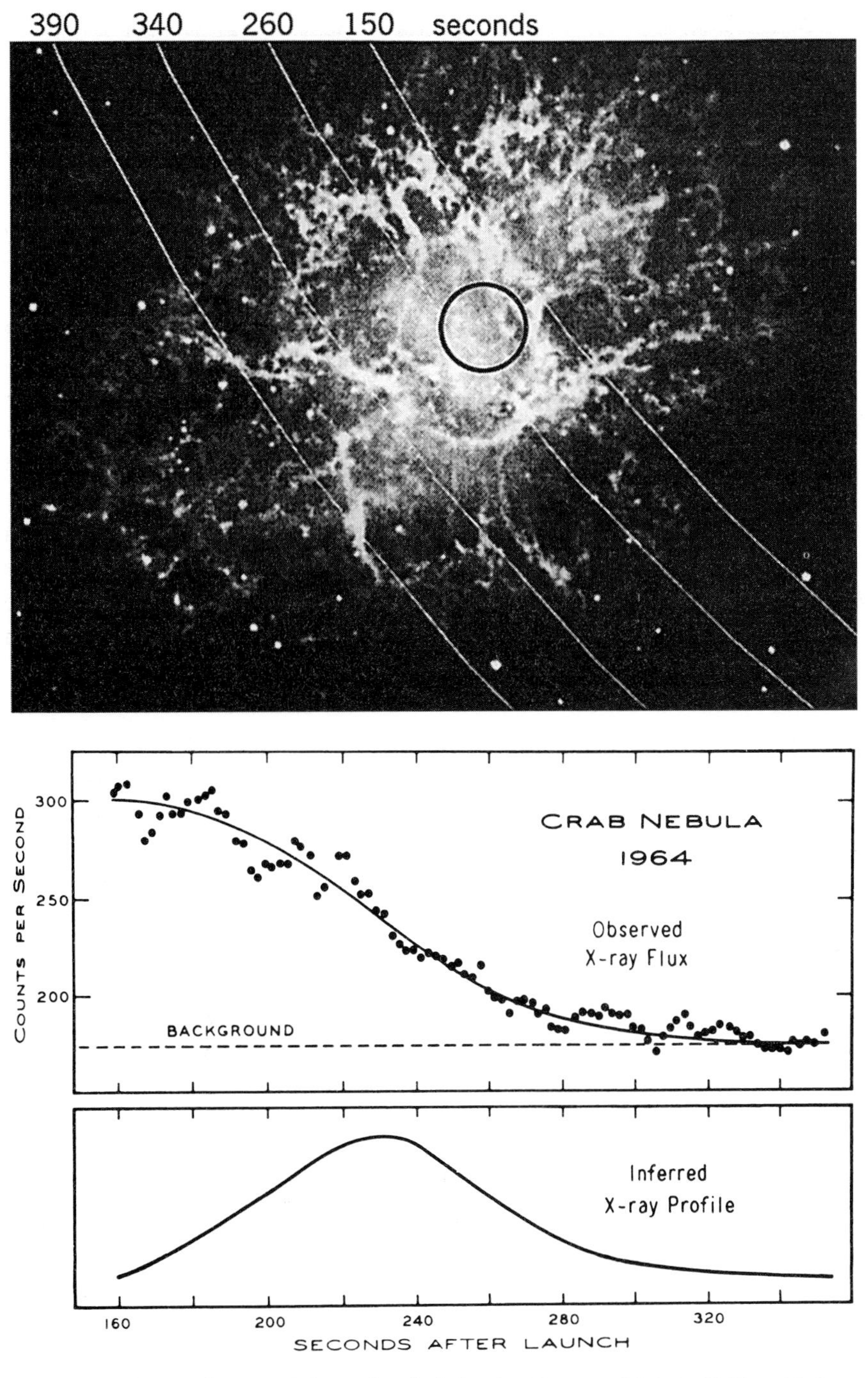

Figure 1.4 The occultation of the Crab nebula by the Moon on 7th July 1964 enabled the size of the Crab X-ray source to be determined using a rocket-borne X-ray detector. The white lines on the photograph indicate the position of the limb of the Moon at the times shown after the rocket was launched. The circle shows the approximate size of the X-ray source derived from the measurements shown below. (Copyright Naval Research Laboratory, Courtesy *Sky & Telescope*.)

The difficulties in undertaking such an eclipse observation from a sounding rocket were severe, as Friedman's group had only five months to design and build the experiment. But just as worrying was the fact that they had to rely on a very unreliable attitude control system to point the X-ray detectors towards the Crab during the eclipse. In addition, there could be no delay in the launch of the Aerobee rocket, even by only a few minutes, or they would miss the eclipse. In the event, all went well, but the experiment showed that the Crab's X-rays did not originate from a point source (see Figure 1.4). In fact, the source had a diameter of about 1 arc minute, which equates to about 1 light year at the Crab's distance from Earth, so it could clearly not be a neutron star. Shortly afterwards, in August and October 1964 the AS & E group were able to measure an approximate spectrum for Sco X-1 and show that it was not that of a black body. So Sco X-1 could not be a neutron star either.[13]

As is evident from the last few pages, the use of sounding rockets for astronomical research did not end with the launch of the first spacecraft, Sputnik 1, in 1957. In fact, sounding rockets are still used for some types of research today, but as the years have progressed, more and more astronomy has been undertaken by spacecraft and less by sounding rockets.

This discussion of the first tentative steps in space-based X-ray astronomy is a convenient place to finish my description of the sounding rocket era, and now return to the start of the spacecraft era some years earlier.

---

[13] The Moon never eclipses Sco X-1 as seen from the Earth, so Friedman's Crab Nebula experiment could not have been repeated for Sco X-1 to establish its diameter.

Chapter 2 | THE START OF THE SPACE RACE

2.1 Sputnik 1

The propaganda coup achieved by the Soviet Union when they launched the world's first artificial satellite, Sputnik 1, on 4th October 1957, was a surprise to both the Russian[1] and American leadership. Neither Khrushchev nor Eisenhower had thought in advance that the launch of the first Earth satellite would be of any great military or political significance. In fact, on the day after the launch the official Soviet newspaper *Pravda* carried only a short factual report, following the low-profile lead given by the Soviet leadership. On the following day, however, *Pravda*, realising that the launch had been a great propaganda success in the West, carried banner headlines announcing the event. In the USA the news of the launch was greeted by the general public with a mixture of surprise and panic, as people began to realise that American cities were now no longer immune from direct nuclear attack by Soviet missiles. Why did Uncle Sam appear to be sleeping? How did the apparently unsophisticated Russians manage to upstage the Americans? What significance did the event have as far as astronomical research was concerned? After all, Sputnik 1 had been launched as part of the USSR's contribution to the International Geophysical Year (IGY), which was a purely scientific venture. To answer all these questions we must return once more to the work of the German V2 team.

---

[1] Although Russia was only a part of the USSR, I will adopt the practice of the time in this book and use the names 'Russia' and the 'USSR' interchangeably.

2.2 The American Programme Pre-Sputnik

It had been evident for some time before the Second World War that it would be necessary to build a multistage rocket in order to launch a spacecraft into Earth orbit, but it was left to von Braun and his team to design the first such multistage rocket, given the code names A-9/A-10.[2] This was in 1943–44, and the military requirements were foremost in the minds of the Germans at that time, so this A-9/A-10 rocket was designed to aim a warhead at New York, a distance of over 5,000 km from a nominal launch site in Portugal. The war came to an end before the rocket design had been completed, however, but such a rocket, had it been developed, could also have orbited a small spacecraft.

A few months after the end of the war General H.H. (Hap) Arnold, commander of the Army Air Force,[3] forecast that the USA would shortly build long-range ballistic missiles with atomic warheads and 'spaceships operating outside the atmosphere'. To assist in the development of these concepts, Arnold funded a group called Project RAND (Research ANd Development) at the Douglas Aircraft Company. In 1946 RAND and the Navy Bureau of Aeronautics independently undertook a preliminary design study of an Earth satellite system, emphasising its military applications, but further work was required before it could be seriously considered as a hardware project. Nevertheless, the RAND report suggested that it would be possible to place a 250 kg spacecraft into a 500 km high orbit within five years at a cost of about $150 million. The Air Force and Navy work continued for another two years, but in 1948 the Research and Development Board of the DoD decided that it was still too early to authorise development and construction of such a system, as the military benefits did not appear to justify the costs. The R & D Board agreed that exploratory work should continue, however, but under the coordination of RAND. Two years later in a clarification of rôles, the Air Force was made responsible for the development of long-range ballistic missiles, which could be used as a spacecraft launcher, if required.

President Eisenhower, as a military leader of many years standing, was permanently worried in the early 1950s about the possibility of a surprise Russian attack on the USA. He lived in dread of another Pearl Harbor, whose memory was burnt on the hearts of all Americans who were old enough to

---

[2] The A-10 was the first stage, and the A-9 was the winged second stage.

[3] The US Army Air Force became the US Air Force under a radical restructuring of the American armed services in 1947.

remember that dreadful event in 1941. Although the Americans had their armed forces stationed in Europe after the war, and so did not need to depend on long-range strategic missiles to attack the Soviet Union, the USSR had no such bases close to the USA, and so the development of long-range strategic missiles was of crucial importance to them. Because of this, Eisenhower felt uncomfortable about persistent intelligence reports in the early 1950s that the USSR was developing ICBMs (Inter-Continental Ballistic Missiles) with nuclear warheads to attack the USA. So in 1954 the president asked a Technological Capabilities Panel (TCP), chaired by James R. Killian, Jr., president of MIT, to examine the threat of a surprise attack on the USA and suggest methods of avoiding or containing it.

The TCP Report "Meeting the Threat of Surprise Attack", which was issued early the following year, had a major effect on American thinking at the highest level, recommending, as it did, an acceleration in the development of medium- and long-range ballistic missiles and the immediate construction of a high altitude U-2 aircraft for reconnaissance. In addition, the report recommended starting a small scientific satellite programme to establish the principle of 'freedom of space' in international law, which could then be used to cover military satellites as well. The U2 aircraft and future military observation satellites were intended to give the USA early warning of any impending attack and ensure that there would be no future Pear Harbor, and the development of long-range ballistic missiles meant that the USA would no longer be dependent on the use of bases in foreign countries to launch a missile attack on the Soviet Union.

Meanwhile, the scientists themselves were gradually warming to the idea of using satellites for scientific research. On 5th April 1950, at a dinner held in James A. Van Allen's house in honour of the renowned Oxford geophysicist Sydney Chapman, discussion turned on how to coordinate the rapidly expanding research into geophysical problems. This led to Lloyd Berkner, an active member of the International Council of Scientific Unions (ICSU), to suggest that a coordinated international programme should be implemented analogous to the International Polar Years of 1882–83 and 1932–33. Further discussions were held with other interested scientists over the next few months, leading in October 1951 to the ICSU agreeing to the establishment of an International Geophysical 'Year' (IGY)[4] to last from 1st July 1957 to 31st December 1958. Three years later, on 4th October 1954, the Special Committee for the IGY suggested that satellites should be launched

[4] It was originally called the Third International Polar Year, but its name was changed to the International Geophysical Year in October 1952.

during the IGY to help explore near-Earth space. Within about a year, both the USA and USSR had agreed to launch such satellites as part of the IGY.

Meanwhile, on 16th March 1955 the US Air Force approved the construction of a military observation satellite under the code-name WS 117L, and in the following month the NRL proposed to the DoD a scientific satellite programme for the IGY, using a modified Viking sounding rocket as the first stage. In parallel, the Army Redstone rocket team under Medaris and Von Braun had suggested launching a small scientific satellite, called Project Orbiter, using a Jupiter IRBM (Intermediate Range Ballistic Missile). A few months later, in August 1955, the Navy's Viking-based, Vanguard programme[5] was selected for launching six ten pound (4.5 kg) scientific satellites, as the American satellite contribution to the IGY, and the US Air Force was told specifically to slow down its military satellite programme to ensure that a scientific satellite was launched first to establish the principle of 'freedom of space'. The Army satellite programme was shelved, as the Americans wanted their scientific spacecraft to be launched by a civilian rather than a military rocket to establish this principle. In addition they did not wish to compromise the early completion of their high-priority IRBM programme by starting a parallel scientific satellite programme using the same Army design team.

So in mid 1957, just before the launch of Sputnik 1, the USA were committed to launching a few very small scientific satellites using a launcher based on a sounding rocket design, with the first launch estimated by the NRL for early 1958. The CIA were also aware in mid 1957 that the Soviet Union might launch their IGY spacecraft first. The Americans did not accelerate their programme, however, as a result of this intelligence information, as President Eisenhower saw little propaganda value in being the first to launch a small scientific satellite to orbit the Earth. In fact Eisenhower's overriding concern at that time was the rapidly escalating cost of the satellite programme, which had increased from $20 million to $110 million over the previous two years.[6] Because of this he cautioned the National Security Council that America might restrict the number of spacecraft actually launched during the IGY if the first one or two launches were successful.

---

[5] The Vanguard programme was originally called the LPR (Long Playing Rocket) project; the letters LPR being chosen as they were memorable as an abbreviation for a 'long playing record'.

[6] This increase came at a time when the budget for guided missiles had increased from $14 million in 1954 to over $1 billion in 1957, putting a strain on the American budget.

2.3 The Soviet Programme

What had happened on the other side of the Atlantic to ensure that the so-called technologically backward USSR was in a position in 1957 to beat America to be the first to launch an Earth-orbiting satellite?

It was generally thought in the West, after the launch of Sputnik 1, that the Russians had won the satellite launching race because they regarded launching a satellite as of major strategic and propaganda importance. So, the theory went, the Russians had gone all-out to be the first in this field, whereas the Americans, seeing little advantage in winning the race, and thinking that the Russians were too technologically backward to succeed anyway, did not put enough effort into the work. In short, it was thought that the Soviets were shrewd and single-minded, whilst the Americans were too laid back and arrogant.

What had been often forgotten in the 1950s was that, whereas the Russians had exploded their first atomic bomb in 1949, some four years after the USA, they had exploded their first experimental hydrogen bomb in 1953, only nine months after the Americans.[7] Although the Russian devices were said to have been cruder in design than their American counterparts, and some of the design details had been stolen from the Americans by atomic spies, nevertheless these successes showed that the Russians were not as far behind America in technology as many liked to pretend. If the Russians could virtually catch up with the Americans in nuclear technology, was it any wonder that, with the same single-minded approach, they could beat the Americans to launch a spacecraft? The logic, post-event, is flawless but, in reality, the Russians showed no more determination to be the first to launch a satellite than the Americans. In fact Khrushchev thought that the Americans would win anyhow, although in both countries the scientific satellite programme was treated as a side-show of no great importance.

The rocket used to launch Sputnik was based on the R-7 ICBM designed by a team led by Sergei Pavlovich Korolev.[8] Although some preliminary design work had been undertaken before final programme approval in both

---

[7] Neither the Russian nor American H-bombs of 1953 and 1952, respectively, were able to be delivered by an aircraft as they were too bulky. They were more 'proof-of-concept' designs. Operational H-bombs were exploded by the USA in 1954 and by the USSR in the following year.

[8] Korolev's name as the 'father of Sputnik' was not released until just after his death in 1966. The Russians were extremely secretive about their whole space programme until Gorbachev's period of 'glasnost' in the 1980s.

cases, both the R-7 programme and that of its American ICBM counterpart (the Atlas) were to be finally approved at about the same time, shortly after the first successful H-bomb tests of both countries. The difference was that the Atlas vehicle underwent a major design change about a year after the basic R-7 design configuration had been frozen, and this resulted in the first successful launch of the R-7[9] being one year ahead of that of the Atlas. As far as the American scientific satellite programme was concerned, however, this delay in the Atlas programme was to have no effect, as Eisenhower was unwilling to use a military vehicle (in this case, the Atlas) to launch a scientific spacecraft. The Russians on the other hand had no such concerns.

Although the R-7 ICBM programme had been approved in the USSR in late 1953, there was, at that stage, no Russian satellite programme. It was not until 26th May 1954 that Korolev formally proposed to Dimitri Ustinov, the Minister of Armaments, that the R-7, which was to be designed to carry a 5.4 ton H-bomb warhead a distance of 7,000 km, could be used as a satellite launcher. Korolev got no reply. Then in October 1954 the Special Committee for the IGY called for the launch of scientific satellites as part of the IGY. The Russians did not reply directly to this request, but on 15th April 1955 Moscow radio announced that 'A permanent Interdepartmental Commission for Interplanetary Communications' had been created under the auspices of the Russian Academy of Sciences, and most people in the West took this to mean that the USSR was about to start a scientific satellite programme as part of the IGY. Three months later, on 29th July 1955, feeling under some pressure because of the Moscow radio announcement, the Americans announced that they would launch some small scientific satellites as part of the IGY.

Prompted by the American announcement, Leonid Sedov, the head of the Soviet delegation to the Sixth Congress of the IAF in Copenhagen, suggested that the Soviets were already embarked on a similar programme, but with larger satellites. But this was not true, as it was not until 30th January 1956 that the USSR Council of Ministers issued a decree authorising such a programme. So, although the Americans did not know it at the time, they had a head start on the Russian scientific satellite programme.

The late start on the Russian satellite programme was made worse by the general lethargy shown by the USSR Academy of Sciences, Russian industry, and the Russian military establishments, who did not want to see their military rocket programme compromised by a scientific satellite pro-

---

[9] The R-7 was the first ICBM in the world to have a successful launch when it reached its target in the USSR on 21st August 1957.

gramme. This resulted in the original spacecraft design being delayed to such an extent that Korolev decided to design a new, much smaller and simpler satellite, called Prostreishiy Sputnik or PS. Korolev only made the suggestion to the State Commission for the R-7 ICBM of launching this small PS satellite, however, instead of the originally intended one ton satellite,[10] after the first successful launch of the R-7 ICBM on 21st August 1957. Even then, with the euphoria of the first successful launch, it took two hard meetings of the State Commission before Korolev's proposal was accepted. The PS satellite was then manufactured in the ridiculously short time of one *month*, and successfully launched as what we now call Sputnik 1 (see Figure 2.1) on 4th October 1957.

Korolev had achieved his main objective with Sputnik 1 of being the first to launch an artificial Earth satellite. Sputnik 1's mass of 85 kg, although modest, was fifty times the mass of the intended American Vanguard spacecraft which had not yet been launched. The military significance of the first Sputnik launch was dramatically eclipsed just one month later when Korolev's team launched the 508 kg Sputnik 2 containing the dog Laika, clearly showing that their rocket was powerful enough, when used as an ICBM, to launch a nuclear warhead against the USA. Pressure was now on the Americans to launch their satellite, and what was to have been a test launch of their new Vanguard rocket was turned into a full-scale satellite launch attempt on 6th December. Unfortunately, the first stage failed, and the rocket exploded on the launch pad in the full glare of publicity.

2.4 **Early Scientific Results**

After the launch of Sputnik 1, von Braun had tried once more to get authorisation to use an Army Jupiter-C rocket to launch a spacecraft, but his request had been rejected. After the launch of Sputnik 2, however, President Eisenhower changed his mind and authorised the Army to launch a scientific satellite as soon as possible. So it was on 1st February 1958[11] that the von Braun team launched the 14 kg Explorer 1 spacecraft at their first attempt. This was followed by the unannounced failure of the USSR to

[10] This satellite, originally called 'Object D', was eventually launched as Sputnik 3 on 15th May 1958.

[11] 03 h 48 m on 1st February Universal Time (UT), or 22 h 48 m on 31st January EST (Eastern Standard Time, i.e. local time).

Figure 2.1 Sputnik 1, the world's first artificial satellite, that was launched by the Soviet Union on 4th October 1957. (Courtesy The British Interplanetary Society, *Spaceflight* magazine.)

launch Sputnik 3 on 3rd February (see Table 2.1), a second Vanguard launch attempt failure on 5th February, and a failure to orbit Explorer 2 on 5th March. Finally on 17th March a spacecraft was successfully orbited by the Vanguard launcher. In the first twelve months of the space age the Russians were to successfully launch three spacecraft out of five attempts (one of the failures being a lunar probe), whereas the Americans were to have three out of five successful Explorer launches, one out of seven successful Vanguard launches, plus one failed lunar probe.

The scientific instrumentation on these early spacecraft was generally very simple and, in the case of Sputnik 1, it consisted of just pressure and temperature sensors. Nonetheless, analysis of the orbit of Sputnik 1 enabled data to be inferred on the density of the Earth's upper atmosphere, but the first major scientific discovery was that made by James Van Allen of the University of Iowa using data from Explorer 1.

When James Van Allen examined the data from his Geiger counters on Explorer 1, he found that they showed an increase in radiation levels with increase in altitude, as expected, as the effect of the Earth's atmosphere reduced. After saturating at an altitude of about 800 km, however, the intensity unexpectedly decreased to virtually zero at higher altitudes. It was difficult to analyse the data, because Explorer 1 did not have a tape recorder, and so the data could only be obtained when the spacecraft was within visibility of a ground station. Because of this, the altitude dependency could not be clearly separated from any latitude/longitude effect. The launch of Explorer 2 was a failure on 5th March 1958, but Van Allen was still trying to make sense of his Explorer 1 measurements when Explorer 3 was launched on 26th March.

The payload of Explorer 3 was almost identical to that of Explorer 1, except for the addition of a small tape recorder. This tape recorder was a real bonus, as it meant that measurements made when the spacecraft was out of visibility of the ground station could be stored and relayed back to Earth when the spacecraft returned to within the ground station's visibility. Putting his Explorer 1 and 3 results together, Van Allen concluded that there was a belt of elementary particles, probably in the form of low energy electrons around the Earth. This announcement, made on 1st May, was a shock to everyone. Further results with Explorer 4, launched on 26th July, confirmed the existence of this so-called Van Allen belt (the inner belt in Figure 2.2).

Sputnik 2 had carried a pair of Geiger–Muller tubes and they could have shown evidence of the inner Van Allen belt before Explorer 1, had the Soviet scientists not had great difficulty in analysing the results. Similarly, Sputnik 3 would also have been able to map the inner Van Allen belt had its tape

Table 2.1. *Spacecraft launch attempts in the first twelve months of the space age*[a]

Spacecraft	Country	Launched[b]	Mass of spacecraft (in kg)	Orbit[c] (in km)	Launcher	Comments
Sputnik 1	**USSR**	**4.10.57**	**85**	**215×940**	**R-7**	**No scientific instruments other than internal pressure and temperature sensors.**
Sputnik 2	**USSR**	**3.11.57**	**508**	**212×1660**	**R-7**	**Carried the dog Laika in a pressurised cabin.**
Vanguard	USA	6.12.57		Failure	Vanguard	Launcher exploded on launch pad.
Explorer 1	**USA**	**1.2.58**	**14**	**360×2540**	**Juno 1**[d]	**Carried Geiger counter to measure cosmic rays (James Van Allen) and two micrometeoroid detectors. Discovered Van Allen Radiation Belt.**
Sputnik	USSR	3.2.58		Failure	R-7	Launcher exploded at 15 km altitude. This was the first launch attempt of Sputnik 3.
Vanguard	USA	5.2.58		Failure	Vanguard	Launch failure.
Explorer 2	USA	5.3.58		Failure	Juno 1	Fourth stage failed to ignite.
Vanguard 1	**USA**	**17.3.58**	**1.5**	**650×3970**	**Vanguard**	**Geodesy spacecraft. Discovered that the Earth is pear-shaped. First use of solar cell power.**
Explorer 3	**USA**	**26.3.58**	**14**	**190×2800**	**Juno 1**	**Similar experiments to Explorer 1 except for the addition of a small tape recorder. Mapped Van Allen radiation belt.**
Vanguard	USA	28.4.58		Failure	Vanguard	Second stage launcher failure.
Sputnik 3	**USSR**	**15.5.58**	**1,327**	**214×1860**	**R-7**	**Carried experiments to measure micrometeorites, density of upper atmosphere,**

						cosmic rays, solar radiation, and charge particles. Tape recorder failed.
Vanguard	USA	27.5.58		Failure	Vanguard	Second stage launcher failure.
Vanguard	USA	26.6.58		Failure	Vanguard	Second stage launcher failure.
Explorer 4	**USA**	**26.7.58**	**18**	**260×2200**	**Juno 1 with new 4th stage**	**Mapped Van Allen belt for 2½ months. Also measured effect on the radiation belt of three small nuclear explosions at about 200 to 500 km altitude on 27th & 28th Aug. and 6th Sept. Showed that electrons produced by these explosions are trapped by the Earth's magnetic field.**
Pioneer 0 Moon probe	USA	17.8.58		Failure	Thor-Able-1	Launcher failed 77 seconds after lift-off. First attempt by either country to send a spacecraft to the Moon. (The USA were hoping to send the spacecraft into orbit around the Moon.)
Explorer 5	USA	24.8.58		Failure	Juno 1	Launch failure
Lunik	USSR	23.9.58		Failure	SL-3[e]	Launch failure. First Russian attempt to send a spacecraft to the Moon.
Vanguard	USA	26.9.58		Failure	Vanguard	Second stage launcher failure.

Notes:

[a] Those spacecraft shown in bold type were successfully launched, the others being failures.

[b] Universal time.

[c] Earth orbit.

[d] Jupiter-C rocket plus fourth stage.

[e] An R-7 with an extra upper stage.

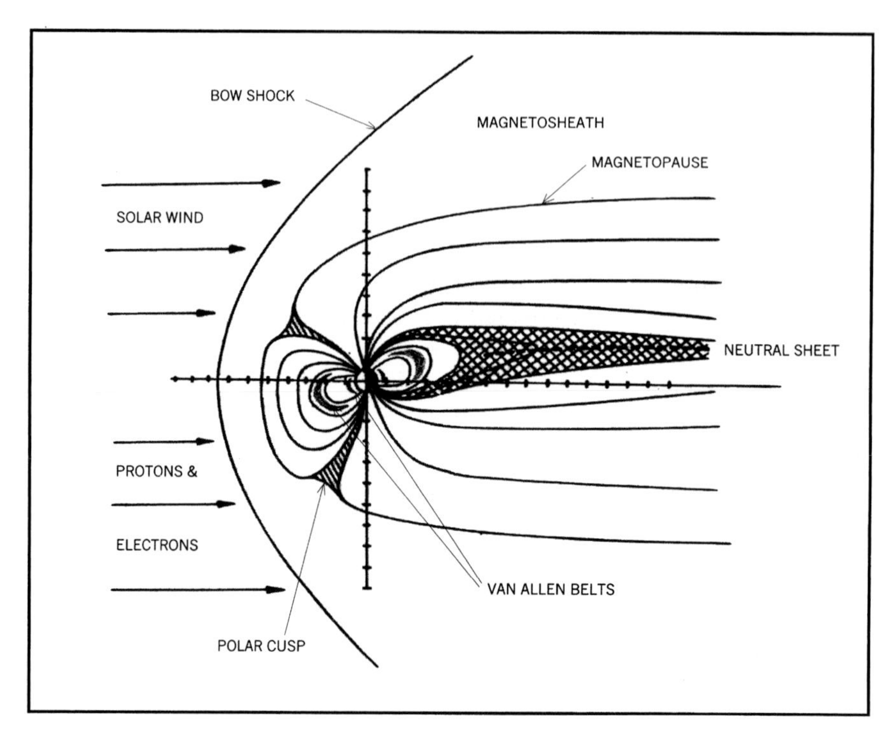

Figure 2.2 The inner and outer Van Allen radiation belts shown within the general configuration of the Earth's magnetosphere, which was determined gradually during the first ten years of the space age. The exact size, structure and composition of these belts varies with solar activity. The inner Van Allen belt is relatively stable and composed mostly of high energy (30 MeV) protons, whereas the outer belt is variable and composed of electrons and low energy protons. The first indications of the tail-like structure of the magnetopause centred around local midnight were obtained by Explorer 10 in 1961 (see Section 10.2) and Explorer 14 in 1968. (The 'tick' marks on both axes are at intervals of two Earth radii)

recorder not failed. It did, nevertheless, show that the inner belt was composed mostly of protons, rather than the low energy electrons suggested by Van Allen.

The one successful Vanguard launching in this first year of the space age also yielded a major discovery when it detected, contrary to expectations, that the Earth is not like a slightly-squashed ball in shape, but is pear-shaped, with the south pole about 20 m nearer to the centre than expected and the north about 20 m further away.

The scientific results outlined above may appear modest when compared with the amount of effort and money expended on these early space satellites, but that is only to be expected of research into any radically new field. Furthermore, the provision of scientific results was not the primary motivation of either the Russian or the American programmes. Once the space race had started, both superpowers became more interested in the prestige involved than in any scientific results, demonstrating, as they saw it, the advantages of their own political systems, which they were then trying to 'export' to countries in the unaligned third world.

The maximum spacecraft mass orbited in the first twelve months was the 1,327 kg of Sputnik 3, compared with the American maximum of 18 kg for Explorer 4, showing that the Russian spacecraft launch vehicles were far more powerful than those used by the Americans, although the latter compensated for this by their much more mass-efficient spacecraft designs. So the Americans could achieve as useful scientific results with their smaller spacecraft as the Russians, but the Americans would still be at a disadvantage in trying to orbit a man, unless they could develop and use substantially more powerful rockets.

2.5 The Formation of NASA

The public outcry in America at the launch of Sputniks 1 and 2, and the apparent lack of concern shown by the Republican President Eisenhower in these events, led the Democratic senator from Texas and Senate Majority Leader, Lyndon B. Johnson, to take centre stage and push for an investigation into the situation. So on 25th November 1957 the Senate set up a subcommittee, chaired by Johnson, to review the whole range of America's defence and space programmes post-Sputnik. It found serious problems of underfunding and disorganisation, and the Senate, based on this analysis, voted on 6th February 1958 to create a Special Committee on Space and Aeronautics to frame legislation for a permanent space agency. Meanwhile, stung into action, President Eisenhower asked his new scientific advisor, James R. Killian, Jr., to review the American position on space matters, *vis-à-vis* the USSR, and advise on a course of action.

In March 1958, the President's Science Advisory Committee, chaired by Killian, recommended that all non-military space work should be controlled by an enlarged National Advisory Committee for Aeronautics (NACA). Many scientists viewed this selection of NACA with misgivings, as it had a reputation for being too cautious and it had little or no experience of managing large contracts. But the selection had the advantage that the space programme could start more quickly if the job was given to an organisation like

NACA that already existed, rather than to a brand new one that would need setting up from scratch. In the meantime the president had created the Advanced Research Projects Agency (ARPA) in the DoD to take temporary control of all space programmes, whilst the structure of the new space organisation was being defined and the appropriate legislation passed.

The key question that Eisenhower and his advisors had to answer was whether the new space agency should be a purely civilian organisation, as suggested by the President's Science Advisory Committee, or whether it should control both the civilian and military space programmes, as in the USSR. Although joint control may well have been best from an overall cost-efficiency point of view, the security problems of having a highly visible civilian space programme integrated with a covert military space programme persuaded Eisenhower that the new agency should control only the civilian programme. The overall military and civilian space programmes would be coordinated at a high level, however, by a National Space Council whose members would include both the head of the National Aeronautics and Space Administration (NASA, as the new civilian agency was to be called) and the Secretary of Defense.

Accordingly, President Eisenhower signed the National Aeronautics and Space Act into law on 29th July 1958, and NASA started operations on 1st October of the same year. NACA, from which NASA had been born, had an annual budget of about $100m, but this was increased to about $300m in the first year of the new agency.

The 8,000 employees of NACA were located mainly at sites that are still familiar to us today, namely the Langley Aeronautical Laboratory, Virginia (established 1917), the Ames Aeronautical Laboratory, near San Francisco (established 1939), and the Lewes Flight Propulsion Laboratory, Cleveland, Ohio (established 1940). In addition to a small HQ staff in Washington, there was also a sounding rocket test centre at Wallops Island, Virginia, and a high-speed flight research centre at Edwards Air Force Base in California. The new Act recognised, however, that these organisations were by no means sufficient to undertake the American civilian space programme, as they excluded, for example, both the NRL Vanguard and the Army Explorer teams. In the event, the Vanguard team joined NASA on 16th November, but the Army team, which had expanded to about 5,500 people to cover other rocket programmes, did not join NASA until 1st July 1960.[12]

[12] This Army team was part of the Army Ballistic Missile Agency located at Huntsville, Alabama. It was renamed the George C. Marshall Space Flight Center when it was transferred to NASA in 1960.

Chapter 3 | INITIAL EXPLORATION OF THE SOLAR SYSTEM

3.1 Early Lunar Missions

Over the last hundred years, the focus of astronomy has gradually changed from being a study of the Sun and Solar System, with some stellar work, to a study today of stars, galaxies, gaseous nebulae, and the like, with lunar and planetary research taking very much of a back seat. In fact, many astronomers have now concluded that lunar and planetary research is better left to geologists, geophysicists and meteorologists, now that spacecraft have landed on the Moon, Mars and Venus, and have visited all the other planets in the Solar System, with the exception of Pluto.[1]

When the Space Age dawned, however, very little was known about the planets as their images, as seen from Earth, had very poor definition because of the distortion of the Earth's atmosphere. In fact, even the composition of their atmospheres was unclear, because it was difficult to interpret their molecular spectra.[2] At that stage they were clearly astronomical objects. However, our knowledge of these bodies, and even that of the Moon, was about to undergo a revolution as the Space Age developed.

Ever since it became obvious in the early 1950s that it was only a matter of time before satellites were launched to orbit the Earth, some of the more progressive rocket engineers were thinking about how to send spacecraft to

---

[1] Today Pluto is often not considered to be a planet at all, but more of a Kuiper belt object, because of its location, orbit and small size.

[2] Molecular spectra are generally difficult to interpret because of their complexity, and this problem is compounded for the planets as astronomers have to deduce which spectral lines are really due to a planet's atmosphere and which have been added by molecules in the Earth's atmosphere (the so-called 'telluric' lines).

the Moon and planets.[3] Korolev, for example, had a detailed plan in 1955 for adding a third stage to the two-stage R-7 rocket to enable an un-manned spacecraft to be sent to the Moon, but he was to have to wait another three years before a Soviet lunar programme was approved.

In America, meanwhile, just three weeks after the launch of Sputnik 1, William Pickering, the director of Caltech's Jet Propulsion Laboratory (JPL) was pushing for the approval of a plan to send an American spacecraft to the Moon. His proposed programme was not approved, however. Nevertheless, on 27th March 1958, only two months after America had launched their first Earth-orbiting satellite, the Secretary of Defense, Neil McElroy, approved America's first lunar programme, consisting of five Pioneer spacecraft. Three of these spacecraft were to be built by STL for the Air Force Ballistic Missile Division (AFBMD) and launched using Thor-Able rockets,[4] and two were to be built by JPL for the Army Ballistic Missile Agency (ABMA) and launched using Juno 2 launchers.[5]

A lunar mission required much more than just a spacecraft and a means of launching it, however, as communications had to be maintained, preferably continuously, with the spacecraft for the two or three days it took to reach the Moon. To do this required a network of large radio receivers to be built around the globe. Some radio telescopes were pressed into service as temporary communications antennae for this American lunar programme, including the new 76 m diameter dish at Jodrell Bank in the UK. In addition, new spacecraft communication facilities were built by the Americans in Hawaii and Singapore. This suite of facilities would enable the Pioneer probes to be tracked and communicated with, but there was a problem in ensuring that the spacecraft would actually reach the Moon, as very small injection errors in velocity and/or direction could have catastrophic consequences for such a mission. For later missions, on-board thrusters would be provided to enable mid-course corrections to be made, but no such facilities were provided for these early American spacecraft. The engineers had to just 'aim and hope'.[6]

---

[3] For convenience I have excluded solar research from this chapter. It is covered in Chapter 10.

[4] Thor-Able rockets had a Thor IRBM first stage, a liquid fuelled Aerojet Able rocket second stage (which had been developed for the Vanguard programme), and a solid propellant third stage.

[5] Juno 2 was the same as Juno 1 (see Table 2.1), except that the Jupiter first stage had longer tanks.

[6] The same situation also prevailed in the USSR for their early lunar probes.

The Air Force part of the American lunar programme was extremely sophisticated for the time, as the engineers were not planning just to send a spacecraft to the vicinity of the Moon, but were actually planning to put it into an orbit around the Moon. In retrospect, this plan was far too ambitious, as the margin for error in achieving the correct orbit was too small, and so it was doomed to failure. But the Air Force had high hopes of success when their first Pioneer spacecraft was launched from Cape Canaveral on 17th August 1958, just five months after the programme had been approved and less than eight months after the launch of Explorer 1. Unfortunately, just 77 seconds after launch, the turbopump failed in the Thor first stage and the rocket exploded at an altitude of 15 km.

Lunar probes can only be efficiently launched from Earth during a short 'launch window' about once per month. Unfortunately for the Americans, the rocket for the next scheduled launch attempt on 14th September was diverted to a military programme, and so the next lunar launch was scheduled for 11th October. Meanwhile, unknown to the West, the USSR had also failed in their first lunar launch attempt on 23rd September 1958 when their SL-3 lunar rocket[7] exploded 92 seconds after launch.

The Americans had much more success with their second Pioneer launch in October 1958 than the first, but there was a 2° injection error and the second stage shut down too early resulting in a spacecraft velocity about 2% too low. As a consequence, the Pioneer 1 spacecraft[8] only reached a distance of about 114,000 km, or about one third of the way to the Moon, before it fell back to Earth. Prior to that time, however, no spacecraft had gone beyond 4,000 km from Earth, and so Pioneer 1 was the first spacecraft to measure the Van Allen radiation belts above this altitude. In addition, Pioneer 1 was used as the first ever communications spacecraft, relaying military messages from one side of the Earth to the other.

The final Air Force Pioneer launch failed in November 1958 (see Table 3.1), and in the following month the first Army attempt, called Pioneer 3, reached an altitude of about 107,000 km before plummeting back to Earth. There were also two more unannounced Russian failures before the Russian Luna 1 spacecraft[9] became the first spacecraft to reach the vicinity of the Moon in January 1959. Luna 1 was not only more successful than the American lunar spacecraft, but it was some ten times the mass of the US Air Force Pioneer spacecraft, and some sixty times the mass of the Army's Pioneers.

---

[7] The SL-3 was an R-7 with an additional upper stage.

[8] The first Pioneer spacecraft was not officially numbered as it did not go into orbit. It is now sometimes called Pioneer 0.

[9] This spacecraft was originally called Lunik 1.

Table 3.1. *The first spacecraft launch attempts to the Moon*[a]

Spacecraft	Country	Launched[b]	Mass of spacecraft (in kg)	Launcher	Comments
Pioneer 0	USA	17.8.58	38	Thor-Able	Thor's turbopump failed 77 seconds after lift-off at an altitude of 15 km.
	USSR	23.9.58		SL-3	Launcher failed 92 seconds after launch.
Pioneer 1	*USA*	*11.10.58*	*38*	*Thor-Able*	*Reached a distance of 113 830 km before falling back to the Earth. Measured radiation belts above 4000 km altitude.*
	USSR	12.10.58		SL-3	Launcher failed 100 seconds after launch.
Pioneer 2	USA	8.11.58	38	Thor-Able	Third stage failed to ignite. Maximum altitude reached 1550 km.
	USSR	4.12.58		SL-3	Launcher failed 245 seconds after launch.
Pioneer 3	*USA*	*6.12.58*	*6*	*Juno 2*	*Reached a distance of 107 260 km before falling back to Earth. Discovered second radiation belt at an altitude of about 20 000 km.*
Luna 1[c]	**USSR**	**2.1.59**	**361**	**SL-3**	**Designed to impact the Moon**[d]**, Luna 1 flew by the Moon at a distance of 5955 km owing to the booster giving the spacecraft a slightly too high velocity. Discovered zone of intense radiation up to 30 000 to 35 000 km altitude. Found that the Moon has no measurable magnetic field. Released 1 kg of sodium at a distance of 120 000 km from Earth which was visible as an 'artificial comet'.**

Pioneer 4	*USA*	*3.3.59*	*6*	*Juno 2*	*Missed Moon by 60000 km instead of the 32200 km planned. Showed that radiation levels beyond Van Allen belts are acceptable for human flight.*
	USSR	18.6.59			Second stage guidance system failed. Launcher blown-up by range safety.
Luna 2	**USSR**	**12.9.59**			**The first spacecraft to impact the Moon. Measured the radiation belts measured by Luna 1 and Pioneer 4. Confirmed that the Moon has no measurable magnetic field.**
Luna 3	**USSR**	**4.10.59**	**279**	**SL-3**	**Took first photographs of Moon's far side. First scientific spacecraft with an attitude control system.**

Notes:

[a] Those spacecraft shown in bold type were successful, those shown in italics were partially successful, and the others were failures.

[b] Universal time.

[c] Sometimes called Lunik 1.

[d] The fact that Luna 1 had been designed to impact the Moon was only disclosed by the Russians in 1961. Prior to then the USSR had implied that it had been intended to fly by the Moon.

The Luna 1 spacecraft was launched on January 2nd 1959, and 34 hours and 400,000 km later it passed within 5,955 km of the Moon to become the first artificial planet of the Sun. The Russians were deliberately vague in their public announcements at the time, as to whether they had intended Luna 1 to impact the Moon or not, and it was not until 1961 that they admitted that the spacecraft had actually been intended to crash land on the Moon. Nevertheless, Luna 1 was a major success in getting so close to the Moon. It was also a well-instrumented spacecraft with cosmic ray and plasma detectors, a magnetometer and two micrometeoroid detectors. Luna 1 showed that there was a zone of intense radiation up to an altitude of about 30,000 to 35,000 km, which significantly reduced at higher altitudes. The spacecraft also found that the Moon's magnetic field was too weak to be measured by the Luna magnetometer. Some astronomers took this weak magnetic field as evidence that the Moon has a solid core, so there can be no dynamo effect like there is in the Earth's molten core producing the Earth's magnetic field. Other astronomers suggested that the Moon's core could still be molten if it rotated very slowly.

Pioneer 4, the second and last of the 6 kg Army lunar probes, was launched two months later but, although it reached the vicinity of the Moon, it missed the Moon by about 60,000 km. Nevertheless, it showed that the particle radiation levels at this distance were low and would not be a safety problem for future human exploration of the Moon.

The Russians produced two more lunar spectaculars in 1959 when in September Luna 2 became the first spacecraft to impact the Moon, and in the following month Luna 3 took the first photograph of the Moon's far side. Both achievements were great PR successes, although the resolution of the lunar images was very poor by modern standards. So ended the first phase of lunar exploration.

What had been achieved?

The twelve launches of lunar probes in the first two years of the space age (see Table 3.1)[10] had taught both the Russian and American engineers a great deal about how to launch a spacecraft away from the Earth's gravity to another body of the Solar System, and how to communicate and control the spacecraft in the process. But the scientific returns were decidedly modest. The configuration of the Van Allen belts had been better understood, the Moon had been found to have no measurable magnetic field, and the far side

[10] Luna 3 was launched on the second anniversary of the launch of Sputnik 1, so strictly speaking Luna 3 should be considered as starting the third year of the space age. For convenience, however, I have included it in the second year in the above discussion.

of the Moon had been found to have fewer mare regions and more craters than the side visible from Earth. In addition, Luna 2 had been the first spacecraft to measure the solar plasma away from the influence of the Earth's magnetic field, which was important in trying to understand what is now called the solar wind and its interaction with the Earth's magnetic field (see Section 10.2).

Between 4th October 1959 and the end of 1962 there were ten further lunar launch attempts, eight American and two Russian, but all were failures. In the meantime the first planetary probes were launched by the USSR, the first of which was launched towards Mars on 10th October 1960.

3.2 The First Missions to Mars and Venus

How did astronomers picture conditions on Mars and Venus before the arrival of the first spacecraft?

Speculation that there was some form of life on Mars had received a boost towards the end of the nineteenth century by the independent observations of Schiaparelli and Lowell of linear markings on the surface. Lowell attributed these linear markings to canals built by Martians, although the observations were disputed by many astronomers at the time. This idea of intelligent Martians fell into disrepute in the early twentieth century as more information became available on the Martian atmosphere and surface.

Mars was seen from observations from Earth to have dark areas which appeared to change with the season and, before the first spacecraft reached Mars in the 1960s, it was thought that these dark areas were probably covered with lichens or moss. So maybe there was some form of life, even if was only in the form of simple vegetation. The reddish coloured areas of Mars, also seen from Earth, were variously thought to consist of limonite sand (a hydrated iron oxide) or felsitic rhyolite, which is an igneous rock. The surface temperature was measured as varying from about −100 °C at dawn to about 10 °C at noon.

White polar caps, whose appearance changed with the season, could be seen from Earth, and, prior to visits by the first spacecraft, these were thought to consist of very thin water ice or a thick layer of hoar frost. The atmospheric surface pressure was estimated as being about 100 millibars, or about 10% of that on Earth, and, although the atmosphere was basically transparent, there appeared to be occasional clouds and dust storms. In 1952 Gerard Kuiper found evidence of carbon dioxide in the atmosphere, but some astronomers found evidence for oxygen and water vapour, whilst others disputed this.

Venus, the other planet to be targeted by spacecraft in the early 1960s, was thought to be a much more inhospitable place than Mars, although the surface conditions on Venus were very difficult to determine as it was permanently covered in cloud. The cloud top temperatures of Venus had been measured in the 1920s as varying from about 60 °C on the day side to about −20 °C on the night side. Because of the permanent cloud cover, however, it was impossible to measure the surface temperature directly, although it was calculated to be in the range of 80 °C to 130 °C. Carbon dioxide had been found at the top of the Venusian atmosphere using spectroscopic analysis, but it was thought that the lower clouds may consist of water vapour or carbonic acid or possibly formaldehyde.

In 1954 Donald Menzel and Fred Whipple of the Harvard College Observatory resurrected the popular nineteenth century view that there were oceans of water on Venus, with clouds of water vapour. This concept was put into question in 1956, however, when analysis of radio emissions from Venus indicated a surface temperature of at least 300 °C. Nevertheless, many astronomers doubted the validity of these temperature measurements, so there was, prior to the arrival of the first spacecraft, still a very big question about the surface conditions of Venus. This question was compounded, in the late 1950s, when Gerard de Vaucouleurs of the Harvard College Observatory deduced a surface atmospheric pressure of the order of five bars (i.e. five times that on Earth). Then Carl Sagan, who at that time was a doctoral student at the University of Chicago, deduced a surface pressure of an incredible 100 bars.[11] So, was Venus hot and humid with oceans of water and clouds of water vapour, or was it so hot that the surface was one great desert with no water to be found anywhere? And what was the surface pressure? Was it really one hundred times that on Earth or was it a much more believable 5 bars?

It was, of course, necessary to guess what the surface conditions were like on Venus in order to design a spacecraft probe to land on the surface and live to tell the tale, even if only for a few minutes. Clearly designing a spacecraft to land on a hard surface at a temperature of 300 °C, in an atmosphere with a surface pressure of 100 bars, would be very different from designing one to land in a water ocean at 80 °C, in an atmospheric surface pressure of 5 bars. In the event the Russians, who were to be the first to land a probe on Venus, eventually designed their spacecraft to survive a landing on a hard surface at a maximum temperature of 80 °C and with a maximum atmos-

[11] As a comparison, this is the pressure about 1,000 m below the surface of the Earth's oceans.

pheric pressure of five bars. In the early 1960s, however, which is the period currently under review, it was enough of a challenge to get a spacecraft to fly-by Venus or Mars, let alone land on the surface of either.

It had been difficult enough receiving signals from spacecraft *en route* to the Moon a mere 400,000 km away, so the prospect of being able to communicate over planetary distances of over 200 million km was a daunting one in 1960. The Russians solved this problem by building a 1,500 ton array of eight 16 m diameter antennae on one enormous rotating frame at Yevpatoriya in the Crimea. They used a liquid helium cooled maser as a detector, and were able to produce signals of 100 kilowatts in the transmitting mode. Because the Russians had only this one receiver, however, they could not communicate with their planetary spacecraft continuously, as they were not always visible from Yevpatoriya. So the Russians had to arrange for their spacecraft to intercept their target planets when the planets were visible from this one ground station. The Americans, in contrast, built a series of antennae dotted around the world in their Deep Space Network, so they had no such limitations.

The first two planetary probes, which were launched to Mars in October 1960 (see Table 3.2), were unsuccessful when the Russian SL-6 Molniya launcher failed. The first launch attempt to Venus, which was also made by the Russians, also failed on 4th February 1961 when the motor that was due to launch the spacecraft from its Earth parking orbit failed to fire. A few days later, however, the Russian Venera 1 (see Figure 3.1) was successfully launched towards Venus from Sputnik 8 in Earth orbit, although contact was lost five days later when Venera 1 was about 1.9 million km from Earth. Three months later, at the request of the Russians, the 76 m Jodrell Bank radio telescope dish was used to try to communicate with the spacecraft as it passed within about 100,000 km of Venus. No signal was detected, but in a second attempt by Jodrell Bank a month later scientists detected a faint signal, although it was not strong enough to enable any data to be decoded.

The American plans for probes to reach Venus had changed over the years. In November 1958 they had planned to build two 169 kg Pioneer spacecraft to be launched by Atlas-Able boosters during the June 1959 Venus launch window.[12] Then in January 1959 the Russians successfully launched Luna 1 to the Moon, after a string of American (and Russian) lunar failures.

---

[12] The launch windows for launches to Venus, which depend on the relative positions of the Earth and Venus, occur at about 19 month intervals (i.e. the synodic period of Venus). They are each approximately 50 days wide.

Table 3.2. *The first spacecraft launch attempts to the planets*[a]

Spacecraft	Country	Launched[b]	Mass of spacecraft (in kg)[c]	To Planet	Launcher	Comments[c]
Korabl 4	USSR	10.10.60		Mars	SL-6[e]	Launcher third stage failed.
Korabl 5	USSR	14.10.60		Mars	SL-6	Launcher third stage failed.
Sputnik 7	USSR	4.2.61		Venus	SL-6	Spacecraft became stranded in Earth orbit when the final rocket stage failed to fire.
Sputnik 8/ Venera 1	USSR	12.2.61	644	Venus	SL-6	The 644 kg Venera 1 probe was launched from Sputnik 8 near the end of the first Earth parking orbit. Venera 1 included a magnetometer, cosmic ray detector, charged particle detector, solar radiation detector, and micrometeorite detector. Contact with Venera 1 was lost five days after launch at a distance of 1.9 million km from Earth. It made measurements of the solar wind.
Mariner 1	USA	22.7.62	204	Venus	Atlas-Agena B	Range safety officer had to destroy the rocket 293 Agena Bseconds after launch after it started to veer off course.
Sputnik 19	USSR	25.8.62	890	Venus	SL-6	Venus probe (an MV-1 lander) stranded in Earth orbit when the final rocket stage failed to fire.
Mariner 2	**USA**	**27.8.62**	**204**	**Venus**	**Atlas-Agena B**	**Mariner 2 flew by Venus at a distance of 34 830 km on 14th December 1962. It confirmed that the solar wind flows outwards from the Sun continuously. The surface temperature of Venus was measured to be 425 °C. Venus was found to have no measurable magnetic field and no radiation belts.**
Sputnik 20	USSR	1.9.62	890	Venus	SL-6	Venus probe (an MV-1 lander) stranded in Earth orbit when the final rocket stage failed to fire.

Sputnik 21	USSR	12.9.62	890	Venus	SL-6	Venus probe (an MV-2 fly-by spacecraft) stranded in Earth orbit when the final rocket stage failed to fire.
Sputnik 22	USSR	24.10.62	890	Mars	SL-6	MV-4 Mars fly-by spacecraft destroyed when final rocket stage exploded in Earth orbit.
Mars 1	USSR	1.11.62	890	Mars	SL-6	Contact was lost with the MV-4 Mars fly-by spacecraft on 21st March 1963 when it was 106 km from Earth and still three months from Mars closest approach. The spacecraft made measurements of a third high altitude radiation belt around the Earth, and encountered the Taurid meteor swarm between 6000 and 40000 km from Earth. It measured the intensity of cosmic rays as twice that measured by the Soviet Luna missions in 1959. The solar magnetic field was measured as varying from 3 and 9 gammas. On 30th Nov. 1962 the spacecraft encountered an intense stream of solar particles which reached a peak of 600 million particles/cm^2s^{-1}.
Sputnik 24	USSR	4.11.62	890	Mars	SL-6	MV-3 Mars lander spacecraft destroyed when final rocket stage exploded in Earth orbit.

Notes:

[a] Those spacecraft shown in bold type were successful, the others were failures.

[b] Universal time.

[c] This is the mass of the spacecraft that was to be sent to the planet, not the mass of the spacecraft that may have been put into Earth parking orbit. This latter was considerably heavier as it included the mass of the rocket motor to enable the planetary spacecraft to escape from that parking orbit. The mass of Sputnik 8 in Earth parking orbit was 6,474 kg, for example, but of this the mass of the spacecraft that was to fly to Venus was just 644 kg.

[d] Those Russian planetary spacecraft that never reached their Earth parking orbit were unknown in the West as the Russians never publicised them, and those that were stranded in their parking orbit were not announced as failed planetary probes. They were simply given a Sputnik designation, suggesting that they had always been intended to be Earth-orbiting spacecraft. The Americans, on the other hand, announced all their planetary missions in advance.

[e] Also called Molniya. It had the same first stage as the SL-3, but a new second and third stage.

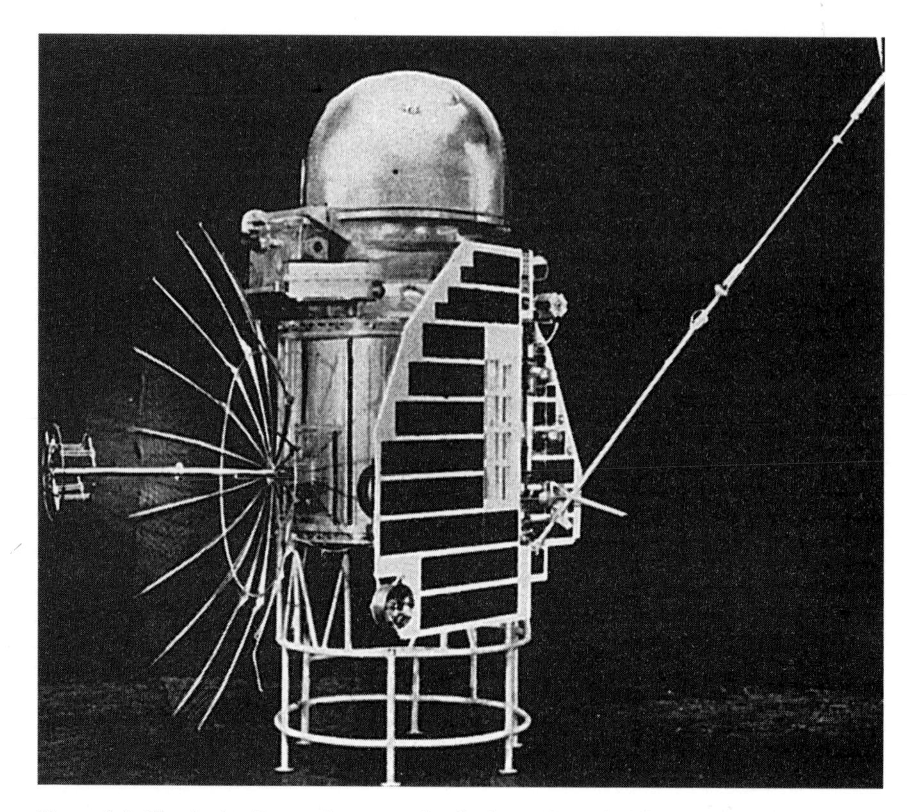

Figure 3.1 The Soviet Venera 1 spacecraft which was launched from Sputnik 8 on 12th February 1961. Note the pressure vessel type of construction, typical of Soviet spacecraft, which allowed the instrumentation to operate in near Earth-like conditions. (Courtesy *Sky & Telescope*.)

So NASA decided to re-designate their so-called 'heavy' 169 kg Pioneers as lunar spacecraft to catch up with the Russians.[13]

Having missed the June 1959 Venus launch window, NASA decided in 1960 to develop a series of spacecraft for launches in the next three Venus launch windows of mid 1962, early 1964 and late 1965. These spacecraft were given the generic name 'Mariner'. Mariner A was to fly-by Venus rather than impact it, Mariner B was to eject a probe to crash-land or soft-land on Venus or Mars, and the smaller Mariner R, which was based on the lunar Ranger spacecraft, was to fly-by Venus. The first two Mariners would weigh

[13] Three of these heavy Pioneers were launched subsequently to the Moon in November 1959, September 1960 and December 1960. In each case the Atlas-Able booster failed.

about 1,000 kg each and would be launched by an Atlas-Centaur, which had not then been developed, whereas Mariner R, being lighter, could be launched by the already-developed Atlas-Agena launcher.

On 30th August 1961 the Mariner A spacecraft was cancelled, as it was then clear that the Centaur booster would not be ready for a 1962 launch, and the Mariner A's mission was transferred to two smaller Mariner R spacecraft which were to be launched to fly-by Venus in 1962. In February 1962 the Venus mission for Mariner B was cancelled, then in April 1962 the Venus mission was reinstated and the Mars mission cancelled, and finally on 14th March 1963 the choice was reversed yet again and the Mariner B spacecraft was reassigned to a Mars landing mission. Finally, Mariner B was cancelled on 6th May 1963 for technical and budgetary reasons.

There was less than one year between the approval of the double Mariner R mission to Venus and its planned launch in the late summer of 1962. Two spacecraft, called Mariners 1 and 2 (see Figure 3.2), were built with seven experiments on each. There was a magnetometer to measure the interplanetary magnetic field during the three month voyage to Venus, a plasma analyser to measure the solar wind, and an ion chamber to measure electrically charged particles in the solar wind. In addition, a particle flux detector would measure cosmic rays and a dust detector would record micrometeorite impacts. During the planetary fly-by phase a scanning microwave radiometer would enable the surface temperature of Venus to be determined, and an infrared radiometer would measure cloud-top temperatures. In the event, the launch of Mariner 1 was a failure when the launcher veered off course and had to be destroyed by the range safety officer. Mariner 2, on the other hand, which was launched on 27th August 1962, flew past Venus on 14th December 1962 at a distance of 35,000 km, after only one course correction manoeuvre.

The Mariner 2 microwave radiometer measured a Venusian surface temperature of an incredible 425 °C,[14] and the surface atmospheric pressure was estimated at about 20 bars. The infrared radiometer showed that the cloud tops were at a height of between 60 and 80 km from the surface, with cloud-top temperatures of from −30 °C to −50 °C. In addition, the magnetometer and particle detectors found that Venus had no measurable magnetic field and that there were no radiation belts around the planet. The interplanetary magnetic field between the Earth and Venus was found to vary randomly

[14] This temperature was estimated assuming a surface emissivity of 90%, as deduced from Earth-based radar measurements. Even if the emissivity was 100%, the surface temperature would still have been about 300 °C.

Figure 3.2 Mariner 2 in its orbital configuration. Most of the electronics was in the sexagonal structure whose temperature was controlled by thermal louvres on its sides. This more open configuration, which was typical of American spacecraft, meant that the electronic units had to operate in vacuum conditions. Mariner 2's communications with Earth were via an omnidirectional antenna, which is the small drum at the very top of the spacecraft, and via the high-gain dish antenna at the bottom. The microwave radiometer antenna is the small dish just above the sexagonal platform. The wing-like panels on either side, which were deployed in orbit, were covered with solar cells to provide electrical power. (Courtesy NASA and NSSDC.)

from 2 to 10 gammas (i.e. about 10,000 times less than that of the Earth) over periods varying from minutes to days, and the velocity of the solar wind was found to vary from 400 to 700 km/s.

Mariner 2 was a big psychological boost to the American space programme as, for the first time, they had managed to upstage the Russians in space by building the first spacecraft to successfully fly-by another planet. The three Soviet attempts to reach Venus during the same 1962 launch window all failed to leave their Earth parking orbit (see Table 3.2). Similar problems plagued two of the three Mars-bound spacecraft also launched by the USSR in 1962, but the second of these three spacecraft, called Mars 1, was successfully launched towards the red planet on 1st November.

Initially all was well with Mars 1 as it sped away from Earth at 3.9 km/s, measuring the intensity of a third high altitude radiation belt around the Earth. It found that the interplanetary magnetic field varied between 3 and 9 gammas (similar to that measured by Mariner 2), and it measured the intensity of cosmic rays, the amount of interplanetary dust, and the strength of the solar wind. Then communications were lost with the spacecraft on 21st March 1963, during a trajectory correction manoeuvre, when Mars 1 was about 106 million km from Earth and approaching the half-way point of its journey.

The Russian attempt to launch a spacecraft to Mars had been affected in October 1962 by very severe Earth-bound problems, however. This was the time of the Cuban missile crisis when the world stood on the brink of nuclear war, and in the middle of October the Mars launch team were told to remove their launcher from the launch pad and replace it with an ICBM targeted at the USA.[15] Korolev protested to the political command in Moscow, who changed their mind as the missile crisis developed. But that was not the end of the problem, because when the Mars spacecraft was finally launched on 24th October it exploded in Earth orbit into a number of large pieces of debris. These pieces were initially thought by the US Ballistic Missile Early Warning System to be part of a possible Soviet missile attack on the USA but, fortunately, a trajectory analysis quickly indicated that this was not the case.

So ended the first phase of planetary exploration, with a total of just one success out of twelve launch attempts by both the USA and USSR, but that one success showed the enormous potential of spacecraft for unravelling the mysteries of the Solar System. Incidentally, the Americans had a success rate of one out of two launches to Venus, compared with the Soviets no successes out of five launches. Only the Russians had tried to reach Mars, but they had no successes out of five launch attempts, although Mars 1 did provide useful data on conditions in interplanetary space before communications had been prematurely lost.

---

[15] At that time the USSR incredibly had only three launch pads for ICBMs, two at Baikonur (one of which had the Mars launcher on it), and one at Plesetsk.

Chapter 4 | LUNAR EXPLORATION

4.1 The American Ranger Programme

It had been clear ever since the launch of the dog Laika in the massive Sputnik 2 in November 1957 that the USSR were intent on putting a man in space. It was equally clear in the early months of the space race that the Americans, with their much less performant spacecraft launchers, would have considerable problems in beating the Russians to such an achievement. Sending a man to the Moon, however, was obviously a much more difficult task, which would require the construction of new vehicles in both the USSR and America, so maybe the Americans could beat the Russians in this. Such was the thinking behind the early American plans to explore the Moon with first un-manned, and then manned lunar spacecraft, but at the beginning no manned programme was approved.

The first American attempts to impact or orbit the Moon, outlined in Chapter 3, showed how difficult it was to reach the Moon, let alone try to impact or orbit it. So the Americans, following those earlier disappointments, decided on a more cautious step-by-step approach, starting with their next series of lunar spacecraft called Ranger. A programme of five Ranger spacecraft was approved in early 1960 for launch in 1961–62 using an Atlas-Agena launcher. Unlike the earlier American lunar probes, the Ranger spacecraft could be adjusted in both orientation and velocity *en route* to the Moon. Two Block I Rangers would loop around the Moon and come back towards Earth, whereas the three Block II spacecraft would include a 44 kg spherical capsule that would be ejected from the main spacecraft just before the latter crash-landed on the Moon. The main Block II spacecraft contained a television camera to take photographs right up to the moment of impact, and a gamma-ray spectrometer to determine the chemical constituents of the lunar surface. The landing capsule,[1] which would be decelerated to a landing speed of about 150 km/h just before impact, contained a seismometer to measure 'moon-quakes' and a temperature probe.

In May 1960, just a few months after the approval of the Ranger programme, the much more sophisticated Surveyor Lunar Orbiter programme was also approved. This new spacecraft, which was much heavier than the 300 kg Rangers, was to have a main spacecraft 'bus' that orbited the Moon, and a smaller spacecraft probe that would soft-land on the Moon, in a much more controlled manner than that of the hard-landing Ranger capsules. Unfortunately, problems with the development of the Centaur upper stage soon reduced the predicted payload of the planned Atlas-Centaur launcher. As a result, the Surveyor spacecraft was split in late 1960 into two autonomous spacecraft, Surveyor A to soft-land on the Moon and Surveyor B, which was to be a photographic orbiter.

The Ranger and Surveyor programmes were designed to help unravel the nature of the Moon for purely scientific purposes, and to obtain data that would enable a manned lunar programme to be undertaken at some unknown date in the future. One question that was of key interest to both astronomers and the engineers designing manned landing probes was the nature of the Moon's surface. Were the lunar maria filled with soft dust some hundreds of metres deep, as Thomas Gold of the Royal Greenwich Observatory suggested in 1955, or was the dust only a few centimetres thick? If Gold was correct it would be virtually impossible to land a spacecraft, manned or un-manned, in the mare areas of the Moon, whilst landing in mountainous areas would not be allowed for a manned spacecraft because of safety considerations.

To astronomers who were still trying to produce a satisfactory theory of the origin of the Moon, data on the age and type of material on the lunar surface was essential. There was also the question about whether the Moon still had a molten interior and, if so, how large was it and what was it made of? Were the lunar craters caused by meteorite impacts or by volcanic activity? Even though the Moon was only 400,000 km away, we knew surprisingly little about it.

The Ranger programme soon ran into trouble. Just six months after programme approval, financial constraints on NASA resulted in one hundred engineers working on the Ranger programme being laid off. Then in October 1960 the Russians tweaked the Americans' tails when they tried to launch

[1] Interestingly, the spherical landing capsule of this 'high-tech' spacecraft was made of balsa-wood to try to cushion the impact on landing.

two spacecraft to Mars, followed in February 1961 with two Venus probes. Although none of these probes was successful, it was clear that the Russians were still in advance of the Americans. Then on 12th April 1961 the Russians landed the final blow of this round when they launched the first man, Yuri Gagarin, into space. He was returned safely to Earth, to a heroes welcome, one orbit later.

The Americans, meanwhile, had elected the young, dynamic John F. Kennedy to the presidency. Kennedy was determined to show America's technological and military supremacy to the outside world. Stung by the launch of Gagarin, Kennedy asked his vice president, Lyndon B. Johnson, to undertake an urgent review of where America stood in space exploration compared with the Soviets. In particular, he asked Johnson on 20th April 1961 to recommend which high profile space programme the Americans should commit to, in which they could beat the Russians. On 5th May the USA launched its first man, Alan B. Shepard, on a suborbital flight. On 8th May Johnson's report was sent to the White House, and two days later it was accepted by President Kennedy. The result, which was announced to the nation and the world on 25th May by the president, committed America to "achieving the goal, before this decade is out, of landing a man on the Moon and returning him safely to Earth".

Kennedy's commitment of America to the Apollo lunar programme, which is what the manned lunar programme was called, meant that all lunar programmes would be realigned to concentrate on providing information for the Apollo programme, at the expense of scientific research. So, although money was then poured into the Ranger and Surveyor programmes, their objectives were redefined with the manned Apollo programme in mind. A third series of Rangers, the Block IIIs, designed to undertake high-resolution photography of the Moon,[2] as the spacecraft crashed onto the surface in a kamikaze mission, was added to the programme in August 1961. This was followed by approval of five similar Block IV spacecraft in October 1962, and by the approval of six Block V hard-landing spacecraft in March 1963.

The Ranger programme started badly with the two Block I and three Block IIs all failing; the Block Is in 1961 because of problems with the Atlas-Agena launcher, and the Block IIs in 1962 because of problems with the spacecraft. This catastrophic series of events was made the subject of a formal enquiry, which quickly concluded, in December 1962, that the heat

[2] This was their only mission. They had no ejectable capsule like the Block II Rangers.

sterilisation[3] of the Block II spacecraft was responsible for most of the spacecraft problems by damaging the electronics. This was made worse by the lack of adequate redundant systems in the spacecraft design. As a result the Block III programme was delayed[4] by one year whilst design changes were incorporated.

JPL were having problems managing all the new lunar programmes, but this was eased when in December 1962 budgetary problems, partly caused by the redesign of the Block III Rangers and the addition of the five Block IVs two months earlier, caused NASA to cancel the Surveyor B orbiters. Then in July 1963 the Block IV Ranger programme was cancelled, and the money saved was used to support a new high-resolution Lunar Orbiter spacecraft designed to investigate possible landing sites for the manned Apollo spacecraft.

So in the course of just nine months from October 1962 to July 1963, the Block IV Ranger programme had been approved and cancelled, the Surveyor B orbiter spacecraft had been cancelled, and the Block V Ranger programme and the Lunar Orbiter spacecraft programme had both been approved. Five months later, in December 1963, the Block V Ranger programme was also cancelled.

These rapid changes made it look as though NASA did not know what they were doing and were being profligate with American taxpayers' money. But the programme changes were caused by the rapidly evolving Apollo programme. They were typical of any crash programme, where concepts are bound to change rapidly and where the programme has to be changed according to development problems with the various spacecraft and launch vehicles. In such circumstances, it is far better to cancel programmes that are no longer needed, rather than try to plough on to the bitter end and waste even more money. The NASA management had to be flexible and, because of the importance of the prize, President Kennedy and Congress were prepared to be flexible also.

The first Block III Ranger spacecraft, namely Ranger 6, was launched from Cape Canaveral on 30th January 1964 but, although the launcher worked perfectly, the spacecraft hit the Moon blind as the television cameras

---

[3] The heat sterilisation followed a decision by NASA in 1959 that all spacecraft that might impact any celestial body must be sterilised to avoid contamination by Earth-based organisms. The sterilisation used with the Block II Rangers involved cooking the spacecraft for 24 hours at 125 °C!

[4] The planned launch date of the first Block III Ranger spacecraft was one month after the date of publication of the Board report.

failed to switch themselves on. A Failure Review Board was immediately convened by NASA.[5] It quickly concluded that the fault was caused by an umbilical connector which had been shorted by burning vented fuel from the Atlas first stage just two minutes into the flight. Six months later, following modifications to correct for the Ranger 6 failure, the next Ranger was launched and, to everyone's great relief, this seventh in the series became the first successful Ranger spacecraft. It impacted the Moon only fifteen kilometres from its target point on 31st July 1964, taking its last image at an altitude of just 500 m, showing craters as small as one metre in diameter. Importantly the photographs indicated that the surface dust was not very deep and could probably support the Apollo lander, although it was going to be difficult to find a landing site free of craters. Two more Rangers followed, both successfully, before the Block III Ranger programme and the Ranger programme as a whole was completed in March 1965. (Images from the last Ranger spacecraft are shown in Figure 4.1).

So during the period from August 1961 to March 1965 the Americans had launched nine Ranger spacecraft, the first six of which were failures and the last three successes. Over the same period the Russians had an even worse success rate as all nine of their lunar spacecraft, seven of which had soft-landing modules, were failures, mainly due to problems with their new SL-6 launcher. In fact, only one of these nine Russian spacecraft, Luna 4, left Earth orbit, and that missed the Moon by 8,000 km.

The Ranger programme had cost about $270m at 1965 prices (or about $1.4bn at 1998 rates) but what had it achieved? Technologically and managerially it had achieved a great deal, showing how a relatively immature design and management structure could be modified and developed to produce a reliable spacecraft. As far as understanding the Moon was concerned, it was also a valuable step on the way. The surface dust was found to be very shallow, and a number of lunar features became easier to understand and interpret, but it could not throw any light on the age of the Moon, nor the composition of its surface rocks, as all the Block II Ranger spacecraft had failed. For that a soft-lander was required, and that is where Surveyor came in.

---

[5] Congress also carried out a failure review under Joseph Karth, chairman of the Subcommittee on Space Science and Applications, which concentrated on management and organisational aspects. This unearthed a number of problems at JPL, together with problems between NASA and JPL at both an organisational and contractual level. Eventually these were resolved, but only after JPL had been forced to become more responsive to NASA's overall management responsibility.

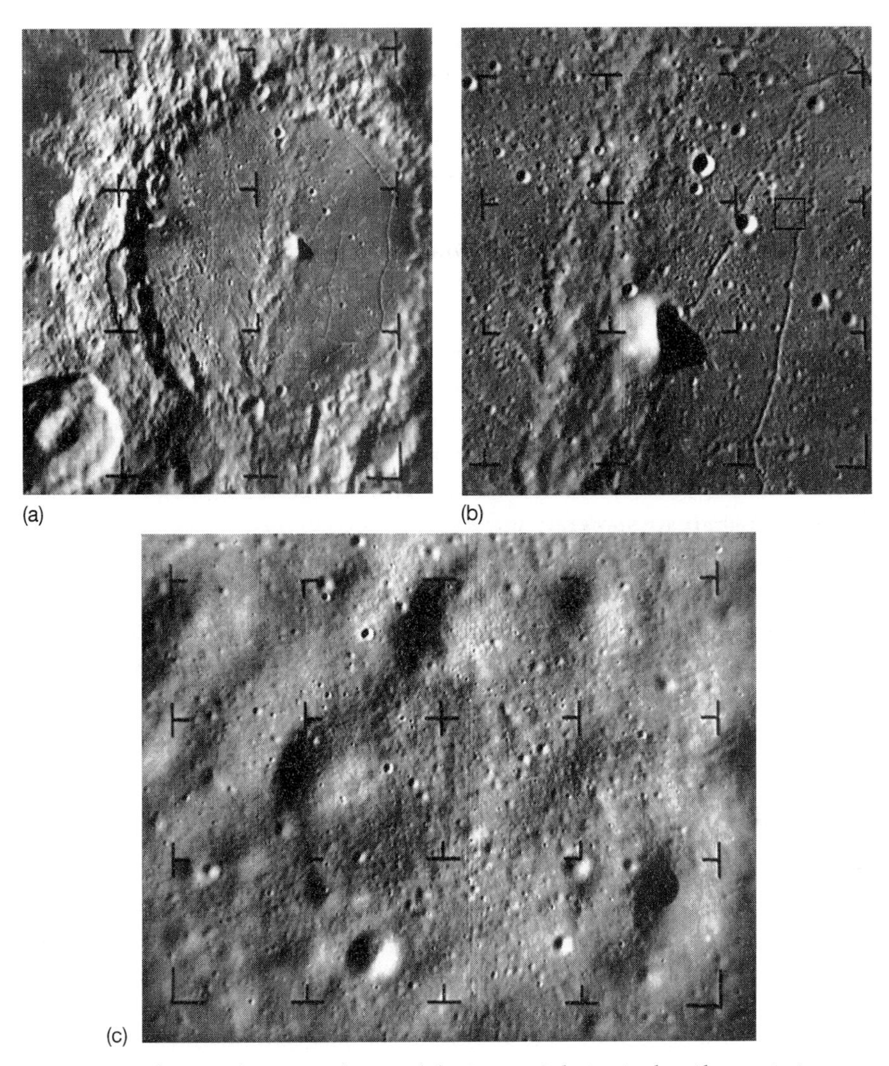

Figures 4.1 Photographs returned to Earth by Ranger 9 during its kamikaze mission to the Moon in March 1965. (a) (which shows an area about 130 km across) is part of a Ranger image of the crater Alphonsus taken 170 seconds before impact. (b) (about 45 km across) shows the area near the central peak about 54 seconds before impact, and (c) (about 3 km square) was taken just 3 seconds before impact. The area covered by (c) is that shown by the small square near the top right hand side of (b).

The smallest crater shown in (c) is about 10 m diameter, whereas on the last image received from Ranger 9 the smallest craters were only about 1 m diameter. (Courtesy NASA and NSSDC; Ranger 9 images A035, A060 and A070.)

In retrospect, what Ranger did was to change the Moon for ever from an astronomical object that had to be understood from afar, into a geophysical one that could be studied close-up. Surveyor and Apollo were to complete the change.

4.2 Korolev and the Russian Programme

The period of just over a year between the launch of the last Ranger spacecraft and that of the first Surveyor was to see the launch of the first successful Russian lunar spacecraft since Luna 3 almost six years previously. Ironically, the first of these Russian lunar spacecraft, Zond 3, had originally been intended as a Mars probe, but problems with the launcher had caused Zond 3 to miss the 1964 Mars launch window. Instead it was redesignated as a test spacecraft and was launched in July 1965 to take photographs of the Moon as it flew by at a closest approach distance of 9,200 km. Its 25 photographs of parts of the far side of the Moon had a maximum resolution of 3 km. They showed chains of craters, which are unusual on the near side, and also revealed a more mountainous landscape than the near side.

The next two Russian lunar spacecraft were failures. Both were attempts to soft-land a capsule of the Moon, but the retro-rocket of Luna 7 fired too early and that on Lunar 8 fired too late, causing both spacecraft to crash-land on the Moon. The next spacecraft in the series, Luna 9, was the first to successfully soft-land a capsule on the Moon, however. But its success was somewhat muted as it took place just two weeks after the sudden death at the age of 59 of the father of the Russian space programme, Chief Engineer Sergei Pavlovich Korolev, on 14th January 1966.

Korolev and his relatively small team of engineers had been working under intense pressure for more than ten years when, in December 1965, they had been given responsibility for the Russian manned lunar orbiting programme.[6] In the same month Korolev was diagnosed as suffering from a bleeding internal polyp, which required a relatively straightforward surgical operation scheduled for January 1966. Unfortunately, Korolev haemorrhaged on the operating table and died shortly afterwards.

---

[6] Korolev had originally been responsible for the N-1 rocket programme that was to land a man on the Moon, whilst Vladimir Chelomei had been responsible for the parallel Proton rocket programme that was to orbit a man around the Moon. Korolev was also given responsibility for the latter programme on 15th December 1965.

The Soviet public had never been told the name of the man who had masterminded their space programme, which had included the first ever spacecraft, Sputnik 1, and the launch of the first man to orbit the Earth, Yuri Gagarin. Korolev's approximate equivalent in America, Von Braun, had become something of a folk hero in the USA, and yet Korolev's name was not even known in the USSR until it was revealed two days after his death by Pravda in an extensive obituary. On the following day the remains of this previously unknown hero was given the honour of a lying-in-state in the Hall of Columns, with an honour guard that included such top-level politicians as Brezhnev,[7] Mikoyan, Podgorny and Andropov, the cosmonauts Yuri Gagarin, Gherman Titov and Valentina Tereschkova, leading academicians and many more important Russians. Then two weeks later Luna 9 successfully achieved the mission that Korolev had worked on for so long, the first un-manned soft-landing on the Moon.

Luna 9 (see Figure 4.2) was launched on 31st January 1966 into an Earth parking orbit, from which it was launched towards the Moon near the end of the first orbit. A mid-course correction was made on the following day, and 75 km above the lunar surface the retro-rocket was fired. Then, one second before landing, the 100 kg capsule which contained the scientific instruments was ejected from the lander, coming to rest on the inner slope of a 25 km diameter crater in the Ocean of Storms (Oceanus Procellarum). It took four minutes for the capsule's stabilisation petals to open and the communications antennae to deploy before contact was made with Earth. This confirmed that the capsule from Luna 9 had survived the landing and was the first man-made object to transmit signals from the surface of another world.

The capsule contained just two 'instruments', a panoramic TV camera and a radiation detector. Three panoramic photographs were transmitted, between each of which the lander settled slightly on the lunar soil, fortuitously allowing stereoscopic imaging. Much to the Russians' annoyance, however, the images had been received by Jodrell Bank in the UK and distributed to the world's press before the Russians were prepared to release them. Although Bernard Lovell of Jodrell Bank protested his innocence, he did not repeat this early release with later Russian probes.

[7] At that time Brezhnev was the First Secretary of the Communist Party and the most powerful man in the USSR, Mikoyan was the previous President and Podgorny was the current President of the USSR. Andropov was to succeed Brezhnev in 1982. Yuri Gagarin and Gherman Titov were the first and second men in space, and Valentina Tereschkova was the first woman in space.

Figure 4.2 Luna 9, the first spacecraft to transmit signals from the surface of another world, which landed on the Moon on 2nd February 1966. The landing capsule is under the dome-shaped thermal cover at the top right. (Courtesy NASA.)

As it turned out the Jodrell Bank photographs exaggerated the vertical scale by a factor of 2.5, as the Russians had not provided information for correctly scaled images to be produced. As a result these British photographs were somewhat misleading, and it was not until the official Russian photographs were released a few days later that satisfactory interpretations could be undertaken. The surface appeared to have a rough texture like pumice or slag of volcanic origin, and to be covered with rocks of various sizes. There seemed to be no dust, thus confirming the conclusions from the Ranger images that the surface appeared to be strong enough to support a manned lander. The radiation detector also showed that the radiation level was safe for astronauts.

The next Russian lunar launch, designated Cosmos 111, failed to leave its Earth parking orbit on 1st March 1966, but the subsequent launch of Luna 10 at the end of the month was a resounding success, with the spacecraft successfully entering lunar orbit on 3rd April. The suite of instruments on this, the first spacecraft to enter lunar orbit, included a three-axis magnetometer, a gamma-ray spectrometer, and micrometeorite, cosmic-ray, infrared, X-ray and solar wind detectors. It had no cameras as it was a purely scientific spacecraft, unlike the American Lunar Orbiter which was planned to be a photo-reconnaissance spacecraft to locate suitable places for manned landings.

The Luna 10 magnetometer detected a weak magnetic field of about 0.1% that of the Earth, but the magnitude of the field did not change with altitude, implying that it was caused by the Moon's distortion of the interplanetary magnetic field, rather than being generated by the Moon itself. Radio occultation measurements, made as the spacecraft just disappeared or just reappeared from behind the Moon, showed that the Moon had no detectable atmosphere.

4.3 Surveyor, Orbiter and Apollo

Two months later the Americans were ready to launch their first Surveyor soft-landing spacecraft after severe problems with both the Centaur launcher upper stage and the spacecraft had apparently been resolved.

Surveyor 1 had a mass of 995 kg at launch[8] but, as in the case of Luna 9, the bulk of this mass was taken up by the solid-fuelled retrorocket. Unlike Luna 9, however, the instrumentation payload was retained with the lander and not jettisoned just before touchdown, so the whole of the Surveyor lander was designed to soft-land on the Moon, not just an instrumentation capsule. Because of problems with the Centaur, the allowable spacecraft mass was lower than originally anticipated, and problems with the spacecraft itself resulted in the payload of Surveyors 1 and 2 consisting of just a television camera, rather than the full suite of scientific instruments originally intended. There were other instruments on the spacecraft, however, designed to monitor its engineering performance, that could also provide scientific information about the Moon. In particular, strain gauges on the landing legs could provide information on the rigidity of the lunar surface, and the radar altimeter used in landing could provide information on the radar reflectivity of the surface.

Much to everyone's surprise and relief Surveyor 1 worked almost perfectly, soft-landing on the Moon on 2nd June 1966. All the Pioneer spacecraft had failed to reach the Moon, and the first six Ranger spacecraft had also been failures, but Surveyor had worked first time. This was some reward for the 700% cost overspend of the programme. The images received showed that Surveyor had landed on a dark, level area of the Ocean of Storms

---

[8] In comparison, Luna 9 had a mass of 1,580 kg at launch.

near a number of craters up to about 100 m or so in diameter. The strain gauges on the legs indicated that the surface had a slight 'give' like uncompacted terrestrial soil.

Next up was the first American Lunar Orbiter, which was launched by an Atlas-Agena in August 1966. This photo-reconnaissance spacecraft had a sophisticated 68 kg photographic payload that, unusually for the Americans, used film rather than television cameras for imaging. This film system, in which the film was scanned[9] on board after development, was used because of the very high resolution required. Unfortunately, the motion-compensation[10] system failed on Lunar Orbiter 1, so only the medium-resolution images were successful. Nevertheless, the image quality was excellent, allowing a number of possible Apollo landing sites to be identified.

Although Lunar Orbiter was not a scientific spacecraft, tracking its orbit precisely yielded a significant scientific discovery. Luna 10 had shown that the Moon was pear-shaped, with an elongation facing away from Earth, but Lunar Orbiter 1 showed that there were mass concentrations or 'mascons' under the Moon's maria. Five Lunar Orbiters in all were flown between August 1966 and August 1967, all of them achieving lunar orbit and returning most of the planned images to enable Apollo site selections to be made. In addition, Lunar Orbiter 5 virtually completed the high-resolution imaging of the far side of the Moon.

In parallel the Surveyors were also largely successful, with just two failures, namely those of Surveyors 2 and 4, in the programme of seven spacecraft. Surveyor 3 had a soil scoop added to the payload which showed that the lunar soil in the Ocean of Storms where it landed was soft and clumpy. Both Surveyors 5 and 6 included an alpha-particle scattering experiment which found that the chemical composition of the lunar soil in their landing areas of the Sea of Tranquillity (Mare Tranquillitatis) and the Central Bay (of the Sinus Medii) was similar to that of terrestrial basalt, which is produced by volcanic activity.

Surveyor 7, the last Surveyor,[11] was launched on 7th January 1968. It landed on the ejecta thrown over the lunar highlands when a meteorite impacted the Moon to form the crater Tycho. This spacecraft had both a soil

---

[9] This system scanned the negative with a 0.005 mm diameter light beam at about 300 lines/mm. [10] To compensate for spacecraft movement during exposure.

[11] The more scientifically oriented, heavier Surveyors 8, 9 and 10 were cancelled in December 1966 to allow more effort (and money) to be focused on the manned Apollo missions.

scoop and alpha scattering experiment. The scoop dug a number of trenches up to 15 cm deep, and weighed a rock by measuring the current needed to operate its scoop arm. This enabled the density of the rock to be estimated at about 2.8 g/cm^3.[12] The alpha scattering experiment found that the iron content of the soil was lower than in the maria areas analysed by Surveyors 5 and 6. Amazingly, no further un-manned American probes have soft-landed on the Moon since Surveyor 7 thirty years ago.

Whilst the Americans were launching five (out of seven) successful Surveyor landers and five (out of five) successful Orbiters, the Russians were not idle, but their achievements were not as impressive with one (out of one) successful lander and two (out of four) successful orbiters. The Russian lander (Luna 13) had an explosively driven probe that found that the local lunar soil had the load-bearing characteristics of average terrestrial soil. It also had a soil density meter, using a gamma-ray source, which measured the average density to a depth of 15 cm of about 0.8 g/cm^3, showing that the soil was not very compacted. Luna 13 also took panoramic photographs like the American Surveyor landers, although the Russian photographs were of lower resolution. The two successful Soviet orbiters (Luna 12 and 14) had a scientific payload similar to Luna 10, but with a photographic module added. The quality of the images was not as good as that of the American orbiters, however.

Surveyor 7 had been launched in January 1968 and Luna 14 had been launched three months later. Both were the last of the series, and at the end of that year the Americans launched Apollo 8, the first manned flight around the Moon, and the following month the Russians made the first attempt at an automatic sample return mission using their new Proton launcher. The Russian attempt failed, as did their next five lunar missions, whilst the Americans had one more successful manned lunar orbiter (Apollo 10) and two successful manned lunar landers (Apollos 11 and 12), before they had their one and only Apollo mission failure (Apollo 13)[13] in April 1970.

Apollos 8 and 10[14] were pathfinder missions for the first manned lunar landing and achieved little in the way of scientific results. When Neil Armstrong and Buzz Aldrin landed on the Moon on 20th July 1969 in the Sea of Tranquillity (Mare Tranquillitatis), however (see Figure 4.3), they collected 22 kg of Moon rocks which they brought back to Earth. They also left

[12] The average density of the Moon as a whole is 3.3 g/cm^3.

[13] Fortunately, all three crew members of Apollo 13 returned safely to Earth, although at one stage in the mission this looked highly unlikely.

[14] Apollo 9 was an Earth-orbiting manned mission.

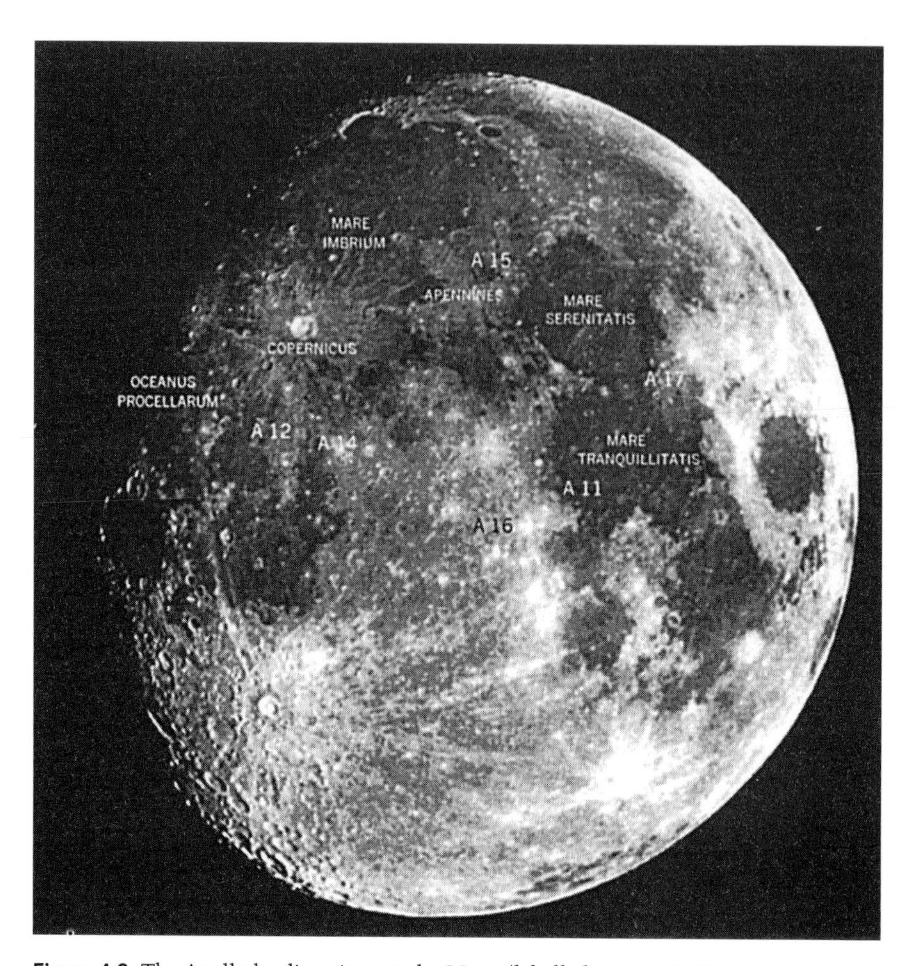

Figure 4.3 The Apollo landing sites on the Moon (labelled A11 to A17).

a seismometer (see Figure 4.4) and laser reflectors behind on the Moon; the seismometer to measure moonquakes and the reflectors to measure the Earth–Moon distance with high precision. The lunar material that they brought back was found to be generally basalt, which is volcanic rock, but some of the samples also contained spherical glazed particles that were thought to have come from impact craters.

Prior to the launch of space probes it was unclear as to whether the solid lunar surface was still relatively thin, with virtually all the lunar features (craters, mountains and maria) being caused by volcanic activity, or whether the Moon was completely solid and had remained unchanged for most of its 4½ billion year lifetime, except for the occasional impact of meteorites that

Figure 4.4 Buzz Aldrin photographed by Neil Armstrong near to the passive seismic experiment that they set up on the Moon on Apollo 11. The lunar module is seen in the background. The surface in this mare area of the Moon is remarkably flat. (Courtesy NASA; AS11-40-5948.)

had produced the maria and craters. In the latter cold Moon theory, it was thought that the energy of the largest impacts would have been sufficient to melt the rocks locally and form large pools of molten rock which would then have cooled to form the flat maria. The images from Luna 9 in January 1966 gave the impression that the surface of the Ocean of Storms, where it landed, was of volcanic origin, but it was not until Surveyor 5 in September of the following year that the chemical composition of the lunar soil was measured for the first time. This probe, which landed in the Sea of Tranquillity, showed that the soil there appeared to be of volcanic origin, but it was not until the first rock samples were brought back to Earth by the Apollo 11 astronauts that the volcanic origin of the maria was firmly established.[15] So the Moon had once been geologically active, with a molten core. The Apollo 11 samples also gave a date for the age of the Tranquillity lavas of 3.65 billion

15 Although chemical analysis undertaken by the Surveyor probes could determine what chemicals were in the lunar soil, it was only the microscopic analysis of samples returned to Earth that could confirm the true nature of the rocks

Figure 4.5 Alan Bean removing the television camera from Surveyor 3 to take back to Earth for examination of micrometeorite damage. The lunar module that brought him to the Moon is seen on the horizon. (Courtesy NASA; AS12-48-7133.)

years, or about 1 billion years after the Moon was thought to have been formed. Interestingly, there was no trace of water, even in the interstices of the rocks, nor was there any organic material, so the Moon appeared, from these limited samples at least,[16] to have had no opportunity to support even the most basic forms of life.

Apollo 12 landed in the Ocean of Storms (Oceanus Procellarum), only two hundred metres from Surveyor 3 (see Figure 4.5). The rocks returned from Apollo 12 were generally similar to those of Apollo 11, but they were about 400 million years younger, showing that the maria did not all form at the same time.

The next successful mission, Apollo 14, was the first manned mission to the lunar uplands. It landed at Fra Mauro, a region of hills and craters on the edge of the Mare Imbrium, with the astronauts being asked, on one of

[16] No traces of water or organic material were found in any of the samples returned by any subsequent Apollo (or Luna) missions either.

the two moon-walks, to climb to the rim of the 350 m diameter Cone crater. This was to find samples of rocks that had been thrown out of the crater when it was formed. Impact rocks called breccias were found, and the samples that the astronauts collected indicated that the impact[17] that had created the 1,140 km diameter Imbrium basin, and the adjacent Fra Mauro hills, had occurred about 3.85 billion years ago. Rocks made of so-called KREEPy[18] lavas were also found of the type that had first been recognised in small quantities in the Apollo 12 samples. These lavas appeared to have been produced by partial melting of the primitive crust.

Six months after the launch of Apollo 14, Apollo 15 landed near the 1.5 km wide, 400 m deep Hadley Rille at the foot of the lunar Apennines (see Figure 4.6). Mission Control hoped that the astronauts would find samples here of the original lunar crust in the form of anorthosite, which would have risen to the surface of the Moon whilst it was still molten. To assist the astronauts in their exploration, a motorised vehicle called the Lunar Rover was provided for the first time (see Figure 4.7). During their exploration the astronauts came across an unusual white rock which, once cleared of dust, was found to consist almost entirely of crystallised plagioclase, which is the main constituent of anorthosite. They had discovered what came to be known as the Genesis Rock, which was found to be about 4.5 billion years old, or only about 100 million years younger than the Moon itself. They had found an example of the original lunar crust. Other rock samples confirmed the date of 3.85 billion years, deduced from Apollo 14 samples, for the Imbrium impact that had created the Apennines and the Fra Mauro hills. The astronauts also discovered a few examples of green-coloured rock, the colour of which was later found to be due to very small green-coloured glass spheres. These had apparently been brought to the lunar surface from hundreds of kilometres down by fire fountains. On the surface of the beads were volatile elements, like zinc, lead, sulphur and chlorine, which were the remains of the gases that had caused the eruptions. Surprisingly, these samples of green-coloured rock were only analysed after Apollo 17 had returned with samples of orange-coloured rock, which were also found to have been produced by fire fountains.

So the history of the Moon was gradually being pieced together in late 1971 as the Apollo 16 mission was being planned. The Moon had apparently

---

[17] G. K. Gilbert had been the first person, as long ago as 1893, to suggest that the Imbrium basin was an impact structure.

[18] So called as they are rich in potassium (K), rare Earth elements (REE), and phosphorous (P).

Figure 4.6 The Hadley Rille as seen by the Apollo orbiter. The landing site for Apollo 15 is indicated by the letter A. The mountains are part of the lunar Apennines (see also Figure 4.3) which rise to a height of over 15,000 ft (4,500 m), and the flat area to the upper left is the edge of the Imbrium basin. (Courtesy NASA; AS 510)

formed about 4.6 billion years ago and, within about 100 million years, a crust had formed of light anorthosite rock as the magma ocean had cooled, with the heavier iron- and magnesium-rich materials sinking. The ALSEP packages[19] left on the Moon by the various Apollo missions had shown that

[19] All the Apollo landing missions left behind experiments on the Moon, varying from the two instruments left by Apollo 11 (see Figure 4.4) to much more sophisticated sets left by subsequent missions. These experiments, which had an autonomous power supply and antenna for communicating with the Earth, continued to operate for some time after the astronauts had left the Moon. They were known as the ALSEP packages.

Figure 4.7 James Irwin and the Lunar Rover of Apollo 15 as photographed by David Scott. The inverted umbrella-shaped antenna on the right allowed the astronauts to communicate directly with Earth when they were out of sight of the lunar module. (Courtesy NASA; AS15-86-11602.)

the loose lunar soil, or regolith, was about 2 to 8 metres thick over the maria, and some 15 metres or more over the uplands. The regolith covered the crust, which was now some 60 km thick, and below that was the mantle of iron- and magnesium-rich rocks which had been, about 4 billion years ago, the source of the maria (volcanic) basalts when the mantle was still molten. The numerous craters seemed to have been formed by meteorite impacts, although the existence of some volcanic craters could not be ruled out. After Apollo 15 geologists wanted to find samples of volcanic rock from places other than the maria, and the shape and relatively light colour of the Cayley plains, in the Descartes highlands, suggested to them that they were made of rhyolite, which is volcanic rock of higher silica content than basalt.[20] So the Descartes highlands was the target chosen for Apollo 16.

Apollo 16 was launched on 16th April 1972 on what was becoming, to the general public, yet another routine mission. Although the man in the street was losing interest in these manned lunar missions, however, the Apollo scientists were not, as each mission continued to provide surprises.

[20] Rhyolite is also lighter coloured and more viscous than basalt.

For example, most of the rocks that the Apollo 16 astronauts found on the Cayley plains were breccia or impact rocks, with a few anorthosite remnants of the ancient crust. There were no examples of the predicted rhyolite, so at least the area of the Descartes highlands where the astronauts had landed was not volcanic.

So far no geologists had been to the Moon. NASA had relied, instead, on astronauts who had received special geological training for their missions. That was about to change with Apollo 17, however, which had the geologist Harrison Schmitt on board. The area selected for this last Apollo landing was the Taurus-Littrow valley on the edge of the Sea of Serenity (Mare Serenitatis), chosen because of its unusual very dark surface with what looked like very small volcanic craters nearby. In the event, the most interesting discovery was made on the edge of a 100 m diameter *impact* crater where Schmitt discovered a layer of orange soil under the surface dust. This spectacular find proved to be made of very small beads of glass, 3.7 billion years old, with a very high titanium content. The material had originated deep inside the Moon, but it had been brought to the surface by a fire fountain, produced by the impact that had made the crater 19 million years ago. In addition, the black colour of the valley proved to have been due to volcanic glass generated in a similar way — the different colour being due to a different mineral content.

So ended the Apollo programme, as the remaining Apollo missions had been cancelled to save money. Was it worth the $25 billion in real-year dollars (or about $120 billion in 1998 dollars) that the Americans had spent on it? Could the same or more have been achieved more cheaply by automatic sample return missions? As a purely scientific enterprise the Apollo programme was of limited usefulness, but Apollo was clearly not a purely scientific programme. Its main purpose was, instead, to prove to the world that the USA was the most powerful nation on Earth that could win any technological race that it put its mind (and money) to. It achieved that purpose in the most visible way with spectacular images of men exploring the Moon. But by the time of Apollo 17 the public in America and elsewhere had lost interest in the enterprise and America was playing to almost an empty house.

It is interesting to speculate that if the USA had been the first to launch a satellite, and had been the first to put a man in space, there would not have been the public imperative behind the Apollo programme and, as you read these words, Man may not yet have put foot on the Moon. But it would be naive in the extreme to believe that the $120 billion saved (1998 dollars) would have been spent on more cost-effective un-manned lunar or other astronomy missions. It is much more likely that the money would have been spent on American social or military programmes.[21]

4.4 The Soviet Response

Meanwhile, whilst America was undertaking its Apollo programme, the Russians had a programme based on the new N-1 rocket to put a cosmonaut on the Moon. Unfortunately, development problems with this launcher caused its first (un-manned) launch to be delayed until February 1969, only five months before Armstrong and Aldrin landed on the Moon. In the event, the first launch of the N-1 failed, and the programme was eventually cancelled in 1974 after all four development launches had failed.

The un-manned Soviet exploration of the Moon was taken over in 1969 by a series of spacecraft launched by the Proton launcher, but the first six lunar missions using this new rocket were also a failure. In fact, the first successful Proton-based lunar mission was Luna 16, which was launched over a year after Apollo 11 had landed the first men on the Moon. Luna 16 was an automatic sample return mission, but it returned only 101 g of material, compared with the 22 kg returned by Apollo 11. In addition, Luna 16 could only sample the surface near the spacecraft, whereas Apollo 11 had men who could walk some distance from their lander and choose which rocks to collect. Two months after the launch of Luna 16 the Russians launched Luna 17, which solved part of this problem. On board was the first self-propelled wheeled vehicle to travel across the lunar surface. This strange-looking contraption, called, Lunokhod (see Figure 4.8), undertook automatic soil analysis some distance from the lander, but the soil could not be returned to Earth as the Lunokhod vehicle had replaced the lunar ascent stage of the lander.

Over the period 1971–1976 the Russians launched eight further spacecraft in the un-manned Luna series; one was destroyed during launch (and so was not officially given a Luna number), two were orbiters, one contained an up-dated Lunokhod vehicle, and four were sample return missions. Only two of the latter returned with samples, and these only weighed a total of 200 g.

---

[21] America was involved in the Vietnam war at the time of the manned lunar missions, and was also experiencing a period of great social unrest. These areas could easily have used any surplus money available. In fact the missions after Apollo 17 were cancelled to provide money for precisely these areas.

Figure 4.8 A model of the Russian lunar rover, Lunokhod 1, which landed on the Moon, shown mounted on its landing module. There were two ramps on the lander in case either was blocked or damaged. Lunokhod 1 was driven by mission control in the Soviet Union, and over ten months it covered a distance of about ten kilometres. (Soviet Lunar Programme, Courtesy NASA.)

The samples of lunar surface material returned by Luna 16 from the Sea of Fertility (Mare Foecunditatis) were of basaltic composition, similar to the samples returned by Apollo's 11 and 12, whereas those returned by Luna 20 from a highland region had a high concentration of anorthosite material like Apollos 16 and 17. Lunokhod 2 (carried on Luna 21) found residual magnetism in lunar rocks and Luna 24, which landed near a 10 km diameter crater in the Sea of Crises (Mare Crisium), found evidence of an ejecta blanket caused by a meteorite impact about 3 billion years ago.

Although the results from these later Russian lunar probes almost pale into insignificance compared with the scientific results from Apollo, the Russian programme cost a great deal less. In terms of cost per kilogram of material returned, however, the 380 kg of hand-picked material returned in toto by the Apollo astronauts was undoubtedly much less expensive than the 0.3 kg returned by the Luna missions.

Chapter 5 | MARS AND VENUS: EARLY RESULTS

5.1 Mariner 4 to Mars

America's brave new world had been badly battered by the events at Dallas, Texas, on 22nd November 1963 when President Kennedy was assassinated. The man who had decided to put an American on the Moon was dead and America was in a state of deep shock. The event shook the world at large also, and even Nikita Khrushchev, Kennedy's enemy of just one year before at the time of the Cuban crisis, visited the American embassy in Moscow to sign the book of condolence.

In 1960 Kennedy, during the presidential election campaign, had committed himself to the concept of the 'New Frontier' in America, which he tried to develop during his short presidency. Part of this concept involved extending the final frontier in space, not just to the Moon but to the planets also.

The American lunar programme had a momentum all of its own in November 1963, and no-one would have dreamt of cancelling it in the aftermath of the president's assassination,[1] as Kennedy had backed it heavily and it was generally agreed that the programme should be completed as a legacy to his memory if nothing else. Planetary exploration was another matter, however, as the Americans had only taken the first halting steps along that path by the end of 1963. Fortunately for the space programme, the man who

---

[1] The Cold War had thawed a little in 1963 after the Cuban missile crisis of the previous year, but with America being gradually sucked into Vietnam, Congress was looking at ways of saving money. As a result, in 1963, before the president's assassination, Congress had cut $500m from NASA's budget and questioned the usefulness of Apollo. These questions stopped abruptly when Kennedy was killed.

took over from Kennedy, his vice-president Lyndon Johnson, was an even bigger supporter of the American space programme than Kennedy. In particular, Johnson had been chairman of the National Space Council, and he had been the person that Kennedy had turned to in 1961 for his advice on how to counter the effects of the Russian launch of Yuri Gagarin, the first man in space.

So as 1964 dawned the un-manned American planetary programme was still on track. In 1962 America had launched two spacecraft to Venus, one of which (Mariner 2) had been a great success (see Section 3.2), but they had so far made no attempts to reach Mars. Therefore, encouraged by the success of Mariner 2, NASA decided to pass up the opportunity of a Venus launch in 1964 and make their first attempt to send a spacecraft to Mars instead. As in the case of Venus, two identical Mariner spacecraft, Mariners 3 and 4, were built for the attempt. In addition a third spacecraft, Mariner 5, was built as a potential back-up, in case of problems with both of the other two.

Mariner 3 was launched by an Atlas-Agena on 5th November 1964, but problems with the newly designed shroud[2], which protected the spacecraft during launch, meant that the spacecraft's solar arrays could not be deployed. So the only power available to the spacecraft was that provided by the on-board batteries, and when they ran down eight hours after launch all communications were lost.

It was now a race against time to determine precisely what had gone wrong, make the appropriate modifications, and launch the second spacecraft, Mariner 4, before the launch window closed one month later on 4th December. In an amazing effort of dedication and round-the-clock working a new shroud was built in magnesium,[3] rather than the original fibre-glass, tested and installed on the waiting launcher. Then, just 23 days after the Mariner 3 failure, Mariner 4 was on its way to Mars following a perfect launch. A few days later, a course correction motor put the spacecraft on a trajectory to fly-by Mars at a closest approach distance of about 10,000 km. Mariner 4 was, in fact, the first spacecraft to carry a course correction motor

---

[2] This shroud was made of a fibre-glass skin over a honeycomb core. It is thought that the skin separated from the core, due to a mixture of thermal and aerodynamic forces during launch, and the skin then jammed the shroud so that it could not be ejected after launch.

[3] The new shroud was 23 kg heavier than the previous one. To compensate for this, the ground-controlled rocket destruction system, which was controlled by the range safety officer, was replaced by an automatic system. This also had to be designed, manufactured and tested in the ridiculously short time available.

Figure 5.1 Mariner 4, the first spacecraft to successfully fly-by Mars. The small f/8 Cassegrain telescope, which was designed to photograph the Martian surface, is at the base of the spacecraft underneath the octagonal main structure. It is on the opposite side of the spacecraft to the dish-shaped, high-gain antenna which was designed to communicate with Earth during the fly-by phase. The circular device in the centre of the central sexagonal face is part of the propulsion unit, with the thermal control louvres on either side. The deployed solar arrays were each about 2 metres long. (Courtesy NASA/JPL/Caltech.)

that could be fired twice if necessary, but the first correction was so accurate that the second firing was not required.

Although Mariner 4 had the same generic name as its illustrious predecessor, Mariner 2, its design was completely different (see Figure 5.1). This is because Mars is much further from the Sun than Venus, the target of Mariner 2. This resulted in Mariner 4 receiving only 22% of the sunlight per unit surface area when near its target compared with the Venus probe. So the surface of Mariner 4 would be much colder, and would need much larger solar arrays to generate the power required by the spacecraft experiments and subsystems. In addition Mariner 4 had to operate for 9 months, instead of the 4 months of Mariner 2, and so had to be more reliable, and it also had to communicate over 240 million kilometres, rather than the 80 million kilometres of Mariner 2.

Mariner 4 had on board a number of instruments to analyse the interplanetary space environment *en route* to Mars, but the main instrument was a 30 cm focal length, f/8 Cassegrain telescope to take images of the Martian surface as the spacecraft flew by the planet. The images were to be stored on a tape recorder and transmitted in slow-time later. In addition, a magnetometer would measure the intensity of the Martian magnetic field, and three Geiger–Müller counters and a solid state detector would detect if Mars had any Van Allen radiation belts. Finally, after closest approach, as the spacecraft passed behind the planet as seen from Earth, the intensity of the radio signals would be monitored to estimate the density of the Martian atmosphere. This so-called radio occultation experiment received a good deal of discussion before it was approved, as it involved sending the spacecraft behind Mars before the images stored on the tape recorder could be transmitted back to Earth. So, if NASA could not re-establish communications with the spacecraft after the planetary occultation, all the images would be lost.

Most spacecraft have idiosyncrasies of some sort and Mariner 4 was no exception, as its star sensor (used for attitude measurement) often lost lock on Canopus (its target star), and one of the Geiger counters failed, but these problems apart the spacecraft behaved perfectly *en route* to Mars. When Mariner 4 was 17,000 km from the planet, it began its 22 image picture-taking sequence of Mars, which continued on 14th July 1965 until the spacecraft was about 11,900 km from the surface.[4] The images received were stored on the tape recorder and successfully transmitted back to Earth after planetary occultation at the incredibly slow rate of 8.3 bps,[5] requiring about 8½ hours to transmit each image. The results caused a sensation.

At first everything seemed normal as the images were slowly received, with astronomers, physicists and engineers peering at the first relatively indistinct images, trying to make out surface features, but with only limited success. Then image number 7 was received (see Figure 5.2) and what it showed stopped everyone in their tracks, as craters began to appear. More craters were seen on images 8 to 14, but then the contrast of the images suddenly got worse and little more could be seen on the remainder.

Whenever surprises occur, there is always a review to see if anyone had predicted what had been found. In this case, craters had apparently been seen on Mars by Edward Barnard in 1892 using the Lick 36″ (91 cm) refractor, and by John Mellish in 1917 using the Yerkes 40″ (102 cm) refractor. Their existence had been predicted by D.L. Cyr in 1944, Clyde Tombaugh (the discoverer of Pluto), Ernst Öpik and, most recently, Fred Whipple of the Smithsonian Astrophysical Observatory. Nevertheless, the discovery of craters on Mars

---

[4] The closest approach of 9,800 km occurred 18 minutes after the last image was taken, but by then the camera was pointing at space beyond the planet's limb.

[5] bps stands for 'bits per second'.

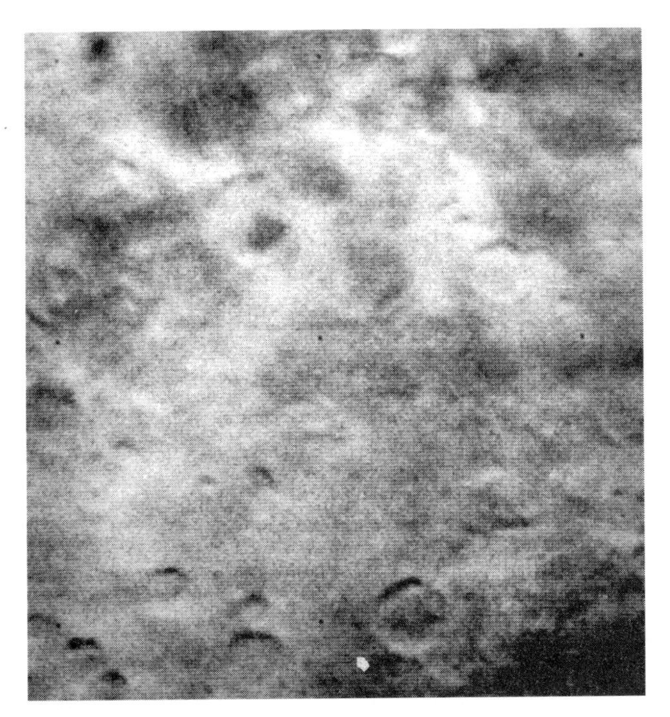

Figure 5.2 The historic image number 7 from Mariner 4, covering an area of 260×310 kilometres, which was the first to show craters on Mars. (Courtesy NASA and NSSDC; Mariner 4, frame 07B.)

was a major surprise to the astronomical community as a whole, which was not expecting such a discovery. The atmospheric pressure measured by the radio occultation experiment was also a surprise as it turned out to be only about 5 millibars at ground level, rather than the 10–80 millibars predicted,[6] and compared with the 1,000 millibars at the surface of the Earth. The Mariner images showed no evidence that water had ever existed on Mars, and water could not exist there now with such a low atmospheric pressure. The atmosphere was also so thin that it would not shield the surface from solar ultraviolet radiation which would have killed any micro-organisms, even if there had been water to sustain them. So conditions were highly unfavourable for life on Mars now, and there probably was never any form of life in the past, but without actually landing on Mars and examining and analysing the surface rocks, astronomers could not be absolutely sure. That would have to wait for a later mission.

[6] In 1963 Lewis Kaplan, Guido Münch and Hyron Spinrad of JPL had estimated the surface atmospheric pressure on Mars to be about 10–40 millibars, compared with the previously generally accepted estimate of around 100 millibars. Then in the following year a NASA sponsored Summer Study widened the range to 10–80 millibars.

Mariner 4 also found no measurable magnetic field on Mars, indicating that it had no molten magnetic core.[7] The apparent lack of volcanoes also indicated that the layer of rock near the surface was not molten either. Finally, Mariner found no Van Allen radiation belts (which is consistent with there being no measurable magnetic field), so charged particles emitted by the Sun would impact the surface virtually unimpeded, again being a threat to life.

Although the Mariner 4 data indicated that Mars was unlikely to have supported life in the past, some astronomers pointed out that Mariner had imaged only about 1% of the surface at a resolution of only three kilometres. The other 99% of the planet may look different, and evidence of the previous existence of water, for example, may be seen when the resolution was significantly improved. But the prevailing view following Mariner 4 was probably best summed up by the Mariner 4 imaging team, chaired by Robert Leighton of Caltech, who concluded, and I quote:[8]

- Reasoning by analogy with the Moon, much of the heavily cratered surface of Mars must be very ancient – perhaps 2 to 5 billion years old.
- The remarkable state of preservation of such an ancient surface leads us to the inference that no atmosphere significantly denser than the present very thin one has characterised the planet since that surface was formed.
- Similarly, it is difficult to believe that free water in quantities sufficient to form streams or to fill oceans could have existed anywhere on Mars since that time. The presence of such amounts of water (and consequent atmosphere) would have caused severe erosion over the entire surface.

5.2 Russian Disasters

The Russian planetary spacecraft of 1961 and 1962 had had a disastrous time (see Section 3.2) with four of the spacecraft being stranded in their parking orbit around the Earth, and two others being destroyed in that orbit by explosions of the launcher fourth stage. Clearly there were design problems with the SL-6 launcher, and Korolev and his team spent 1963 making a series of modifications. They then made four attempts to launch a spacecraft to Venus in 1964, and one attempt was made to reach Mars.

---

[7] The dynamo effect in the molten magnetic core of the Earth causes its magnetic field. So the lack of a magnetic field on Mars indicated that either its core was not molten or, if it was molten, that it was composed of non-magnetic material or that its internal circulation was very slow.

[8] See NASA News Release 65-249 of 29th July 1965.

The first three Venus launch attempts were failures, with the first two failing even to reach their Earth parking orbit, and with the third being stranded in its parking orbit. Only the fourth spacecraft, Zond 1, was successfully launched towards Venus, but communications with it were lost some 14 million kilometres from Earth after an attempted course correction manoeuvre. The Mars probe, Zond 2, was also launched successfully, but one of its solar panels failed and communications were lost when the spacecraft was still three months away from Mars. So the Americans had not only beaten the Russians to Venus in 1962 with Mariner 2, they had also beaten them to Mars in 1965. The last three American Ranger probes to the Moon had also been successful in 1964/65 (see Section 4.1) so the Americans, at last, seemed to have the measure of the Russians in space.

The Americans did not launch a spacecraft to Venus during the 1964 launch window and they also passed up the opportunity to visit Venus during the 1965 window. After all, they had had a successful mission to Venus with Mariner 2, and most attention at NASA was on the manned lunar programme in the mid 1960s. The Russians had only managed to launch two of their nine Venus-bound spacecraft from their parking orbit around the Earth up to 1965, and neither of these had reached Venus in operational condition. So in 1965 the Russians made four more attempts to reach Venus.

The first Soviet launch was that of Venera 2 on 12th November 1965, which was planned to undertake visible light photography and ultraviolet and infrared spectroscopy during a fly-by of Venus. The SL-6 launch was perfect and the launch trajectory to Venus was so accurate that no course correction manoeuvre was required. Three days later, Venera 3 was launched, so the Russians had two spacecraft *en route* to Venus at the same time. This success with their SL-6 did not last, however, and the other two launches on 23rd and 26th November were failures. The first because of an explosion in Earth parking orbit and the second when the initial launch failed.

Unfortunately, communications were lost with Venera 2 just before it was due to start its photographic mission as it closed in on Venus. It was suspected that the cause of the failure was overheating problems, but there was still Venera 3 following just behind.

The exact mission of Venera 3 is a mystery even today. The Russian Academician Leonid Sedov said early on in the flight that it was a twin of Venera 2 and was to take photographs as it flew by Venus on the opposite side to Venera 2. The Soviet news agency TASS, on the other hand, said that Venera 3 included a landing capsule which replaced the photographic module of Venera 2. TASS also said that communications were lost with Venera 3 just before the lander was due to separate from the main spacecraft, and that Venera 3 struck the surface at 9.56 a.m. Moscow time on 1st March 1966. For some time it was thought in the West that Venera 3 was the first spacecraft to have impacted Venus, but later information showed that communications with Venera 3 had been lost some three months earlier than stated by the Russians, so whether the spacecraft had hit Venus or not can only be a matter of surmise. Whatever happened to Venera 3, the Russians had not yet managed to get any data from Venus in spite of thirteen launch attempts.

5.3 Venera 4

The next Venus launch window saw the launch of Venera 4, which was to become the first successful Russian planetary probe. Venera 4 had a completely new thermal control system, to avoid the overheating problems of the type experienced by Veneras 2 and 3, and a new course correction motor that could be fired twice. The one ton spacecraft carried the usual type of sensors to measure conditions *en route* to Venus, and had a 383 kg landing capsule that was designed to float in the event that the Venusian surface was liquid. The landing capsule was also provided with a glass-fibre/polymer heat shield and a double parachute system to enable it to soft-land on the surface. Power was provided by a battery designed to last for about 100 minutes.

Venera 4 was launched on 12th June 1967 and a course correction burn was satisfactorily achieved seven weeks later. The spacecraft reached the vicinity of Venus on 18th October and detached its landing capsule whilst it was still about 45,000 km from the planet. The probe hit the top of the atmosphere at a velocity of about 24,000 mph (10.7 km/s) and was slowed by the heat shield to about 700 mph (300 m/s) when the first parachute was deployed. The main parachute opened a little later, eventually slowing the descent rate to only 7 mph (3 m/s), but contact was lost 100 minutes after the probe had first entered the atmosphere, when the battery was almost completely discharged.

For some months the Soviets maintained that the Venera 4 landing probe had stopped transmitting when it reached the surface, but later analysis showed that the probe's altimeter was at fault and the probe had failed some distance above the surface. It is thought that this failure was either because of the very high temperature of the Venusian atmosphere, or because the probe was crushed by the very high atmospheric pressure. In either case, the Venera 4 landing probe was the first man-made object known to have landed on any planet, even though it was no longer in working order.

The first measurements as the main spacecraft had approached Venus showed that the magnetic field of Venus was very weak, and that it was not very much stronger than the interplanetary field near Venus. As a result the bow shock, where the solar wind is deflected by the Venusian magnetic field, was found to be only about 500 km above the planet's surface, compared with 20,000 km for the Earth. Data transmission from the probe had stopped at an atmospheric pressure of 20 bar and a temperature of 270 °C, and the atmosphere was found to be composed of about 98% carbon dioxide, with the remainder being water vapour[9] and oxygen. This was a surprise as in 1960 Hyron Spinrad of JPL had reanalysed Adams' and Dunham's spectrograms of 1932, in which carbon dioxide had first been detected, and had concluded that the atmosphere was composed of about 96% nitrogen and only 4% carbon dioxide, and Joseph Chamberlain of the Kitt Peak National Observatory had even suggested that the carbon dioxide figure should be less than 4%.

5.4 Mariner 5 to Venus

The Americans launched Mariner 5, their first spacecraft to Venus since 1962, two days after Venera 4. Mariner 5 was intended to fly-by Venus at an altitude of about 4,000 km and use a radio occultation experiment to determine the pressure profile of its atmosphere. Separate experiments would enable its surface temperature and magnetic field to be measured.

Mariner 5 arrived at Venus just one day after Venera 4 and confirmed that the planet's magnetic field was very weak, and that the bow shock was only a few hundred kilometres above the surface. It also showed that the surface temperature was at least 430 °C, which is enough to melt lead, with a surface pressure in the range of 75 to 100 bar.

For some time there was concern about the apparent disagreement between the surface temperature and the surface-level atmospheric pressure of Venus deduced from the Venera 4 landing probe (270 °C and 20 bar) and from Mariner 5 (430 °C and 75 to 100 bar). It was noted, however, that the

[9] The water vapour diagnosis was later found to be erroneous.

two sets of measurements could be reconciled if the last data received from the Venera 4 probe was not from the surface but from an altitude of about 27 km, and the Russians then admitted that their altimeter reading could be in error. The final proof that Venera 4 had not reached the surface was provided when Earth-based radar measurements of the diameter of Venus were compared with the distance that the Venera 4 probe had been from the centre of the planet when it had stopped transmitting. The result was consistent with the altitude of 27 km deduced from the comparison of the Venera 4 probe and Mariner 5 data.

Although the results of Mariner 2, Venera 4 and Mariner 5 differed in some detail, by 1968 a general idea of the Venusian environment had been deduced although, as yet, there had been no attempts to photograph the planetary surface as it was permanently enveloped in cloud. Venus was shown to be a hellish place, with a surface temperature of at least 430 °C and with a surface atmospheric pressure of about 75 to 100 bar. So the pre-war concept of Venus as being a twin of the Earth, but a little warmer, because it was closer to the Sun, was completely wrong, and the surface temperature estimates made from radio measurements made in 1956 and the atmospheric pressure estimates made by Carl Sagan in 1961, which were disputed by many astronomers at the time, were correct. In addition, the Venusian atmosphere was found to consist mostly of carbon dioxide, and the greenhouse effect produced by this gas was the reason for the very high temperatures. Venus was also found to have a very weak magnetic field with, consequently, no Van Allen radiation belts and a bow shock only about 500 km above the surface.

5.5 Veneras 5 and 6

The Russians had originally been sceptical of the estimates of very high surface temperatures and atmospheric pressures on Venus, but they now accepted these and started to design landers to withstand these extreme conditions. They were anxious to continue with their Venus probes, however, before such new spacecraft could be produced, and so as an interim measure the two Venus probes that they launched in 1969 were modified Venera 4 designs. The pressure rating of these spacecraft, called Veneras 5 and 6, was increased from the 20 bar for Venera 4 to 27 bar, and the parachute area was reduced to allow for a more rapid descent through the atmosphere, thus subjecting the spacecraft to the very high temperatures for as short a time as possible.

Venera 5 was launched on 5th January 1969 and its twin, Venera 6, followed from the same launch pad just 5 days later. Incredibly, Venera 6 was launched in a snow storm, which would normally have stopped operations, but the Russians wanted to use the same launch pad for two manned spacecraft shortly afterwards. Both Venera spacecraft behaved perfectly, arriving at Venus on 16th and 17th May 1969. Venera 5 stopped transmitting at an altitude of 24 km, when the atmospheric pressure and temperature were 27 bar and 320 °C, and Venera 6 met a similar fate at about the same altitude. Both spacecraft confirmed the scientific results of Venera 4.

5.6 American Funding Problems

The Americans had successfully launched Mariner 2 to Venus in 1962, Mariner 4 to Mars in 1964 and Mariner 5 to Venus in 1967. Each of these spacecraft weighed about ¼ ton and was launched by an Atlas-Agena rocket. In 1960 NASA had also produced tentative designs of a one ton spacecraft to visit Venus and Mars,[10] but it had had to be cancelled in 1963 because of development problems with the intended Atlas-Centaur launch vehicle. Even more sophisticated designs were also considered by NASA in the 1960s for spacecraft to visit Venus and Mars, and in order to understand these developments we need to go back to 1961.

In 1961 NASA had been given carte blanche to put a man on the Moon, and money was suddenly available in substantial amounts for that and other space programmes. True most of the money was for the manned programme, but NASA felt invigorated by the political and public support at the time, and in 1962 they started to examine the outline design of a three ton spacecraft called Voyager[11] to be sent to Venus and Mars in 1967 and 1969, respectively, by a Saturn 1B rocket. In the following year the study was extended to include a two ton spacecraft to be launched by a Titan IIIC and a 27 ton spacecraft to be launched by a Saturn 5! It soon became clear as the study work progressed, however, that an intermediate step was necessary before flying any of these large planetary spacecraft, and so an Advanced Mariner spacecraft was approved in August 1964 for a launch in 1969, which would then be followed by a three ton Voyager spacecraft to Mars in both 1971 and 1973.

---

[10] Mariners A and B, see Section 3.2.

[11] This should not be confused with the Voyager spacecraft that was sent to the outer planets in 1977, which was entirely different.

No sooner had the Advanced Mariner programme been approved than it was cancelled because of short-term budgetary pressure. At the end of 1964 the plan was changed yet again to now send two 3 ton Voyager spacecraft to Mars in 1971 followed by two in 1973, all using Saturn 1B-Centaur launchers. Each of these spacecraft was to consist of an orbiter and a sophisticated lander, which would sample the atmosphere and surface of Mars for any signs of organic or inorganic activity as indicators of life.

In July and August 1965, however, the Mariner 4 results became available that showed that the Martian atmosphere was thinner than expected. This meant that the Voyager lander would have to be redesigned to provide larger parachutes and retrorockets, as aerobraking would not be as efficient as previous envisaged. This increased the weight of the spacecraft and meant that it could no longer be launched on a Saturn 1B-Centaur. In October 1965, therefore, it was decided to launch the Voyager spacecraft in pairs on a Saturn 5 in 1971 and 1973.

Many of the scientists in NASA viewed the decision to use a Saturn 5 with horror, as it was far too large to launch Voyager-sized spacecraft to Mars, even in pairs. There was a suspicion that the decision, which was taken at NASA HQ, was part of an undisclosed plan to find a use for Saturn 5 rockets after the Apollo lunar landings had been completed. Unfortunately, the choice of Saturn 5 meant than an option of launching one Voyager to Mars per launch window, if budgets got tight in the future, was no longer a realistic possibility.

But there was also a more urgent problem that surfaced at the end of 1965 when the Bureau of the Budget cut NASA's total 1967 budget request from $5.6 billion to $5.0 billion before it went to Capitol Hill. This was at a time when the priority manned lunar programme was asking for more money, and so the planetary programme took the largest financial cuts. So in December 1965 the Saturn 5 launches to Mars were delayed until 1973 and 1975. This meant, unfortunately, that there would be no American spacecraft visiting Mars until 1973, which was over seven years away. In addition, there was no approved Venus spacecraft programme at that time. To overcome these problems, NASA decided in December 1965 to modify the Mariner 5 spacecraft that had been built as a spare for Mariner 4 and send it to fly-by Venus in 1967, and shortly afterwards it was decided to build two new Mariner spacecraft, Mariners 6 and 7, each weighing just under half a ton, to fly-by Mars in 1969.

The American space programme had been placed under severe financial pressure in the second half of the 1960s as, like other Federal agencies, the NASA budget was cut to help to pay for the escalating Vietnam war and the results of ethnic unrest in American cities. NASA was also under considerable pressure in 1966, in particular, to transfer substantial amounts of money from its unmanned to its manned space programmes as the latter hit further trouble. James Webb, the head of NASA, was largely successful, however, in resisting this pressure to transfer money. But the message from Congress and the Bureau of the Budget that money was very tight did not seem to have reached the lower levels in NASA, as by September 1966 they were discussing using not just Saturn 5 launchers for Mars probes in 1973 and 1975, but also using the same expensive launchers for additional probes to Mars in 1977 and 1979! Although landers would fly on each of the four Saturn 5 missions, the first automatic biological laboratory looking for signs of life would be delayed from 1973 to 1977.

But the situation was completely different in 1966 from 1961 when President Kennedy had directed NASA to land a man on the Moon by the end of the decade. America was no longer behind the Soviet Union in space activities and everyone, including the Russians (although they would not publicly admit it) was sure that America would be the first to put a man on the Moon. So the political imperative behind the American space programme was becoming diluted by 1966, and it would largely disappear when the Apollo programme reached its goal a few years later, unless political circumstances changed substantially in the meantime. NASA would then have to justify every programme thoroughly, instead of getting them approved on the coat-tails of Apollo, which is what had happened in the early 1960s.

So what appeared to be a perfectly logical plan in 1966, of using rockets that had already been developed for the Apollo programme to send spacecraft to Mars, would stand or fall purely on whether the political and public interest was there to justify the large amounts of money involved. America had become more insular and uncertain in its power and rôle in the world by 1966, however, as the Vietnam war continued to damage its prestige. Thousands of Americans were dying fighting a largely untrained, ill-equipped peasant army using all the most sophisticated weaponry at America's command,[12] and America was losing. Serious violence had also broken out in American cities. Now was not the time to spend large sums of money on some extravagant PR exercise, which is what the American space programme now appeared to be to Congress and the man in the street.

NASA gave in on the Saturn 5 based Voyager programme only very slowly, however. On 17th May 1967 Voyager study contracts were awarded to Martin Marietta and McDonnell Douglas, but in July 1967 Congress

---

[12] Excluding nuclear weapons.

refused[13] to allow NASA extra money to cover the Voyager work, and in August the programme was finally cancelled.[14] So when Mariner 5 flew by Venus in October 1967 NASA had just one funded planetary programme left, that of Mariners 6 and 7 that were due to be launched to Mars in 1969.

5.7 **Mariners 6 and 7**

Mariners 6 and 7 were superficially similar to Mariner 4, the last American probe to visit Mars, in that they had the similar octagonal shape with four solar array paddles, but they were, in fact, much more sophisticated. In particular, the new Mariners each had two television cameras, one with a 508 mm focal length lens and the other with a 52 mm wide-angle lens, instead of the single 305 mm focal length camera carried by Mariner 4. In addition, the detector system on Mariners 6 and 7 provided sixteen times as many pixels per image compared with their predecessor, so the images should be much sharper. The images and data would also be transmitted to Earth at up to 2,000 times faster than the 8.3 bps rate of Mariner 4 as, not only was Mars closer to Earth than during the previous Mariner encounter, but the ground station had been improved to receive fainter signals and the spacecraft transmitter power had also been increased.

In addition to the imaging payload, the new Mariners had an infrared radiometer, an infrared spectrometer and an ultraviolet spectrometer, none of which had flown on Mariner 4, to study the atmosphere and polar caps. Mariner 6 was to concentrate on the equatorial regions of Mars and Mariner 7 on the region of the south polar cap.[15]

---

[13] Relations between Congress and NASA were somewhat strained after the disastrous Apollo 7 fire in January 1967, in which three astronauts were killed, as this brought home to everyone the risks and expense of putting men into space. So Congress were nervous that the Saturn 5 launched Voyager probes could be part of a NASA plan to develop a programme to put men on Mars, which would be exorbitantly expensive. Congressional suspicions were apparently confirmed when the Manned Spacecraft Center in Houston in August 1967 sent out a request for proposals for studies of manned missions to Mars and Venus. Although this request for proposals and the Voyager programme were not actually linked, they were linked in the mind of Congress and so the Voyager funding was cancelled.

[14] At the time of this cancellation, the total estimated costs of the 1973 and 1975 Voyager missions had risen to about $2.3 billion.

[15] The north polar cap was almost absent during the encounter period of Mariners 6 and 7, which was during the northern summer.

The design of the first stage of the Atlas-Centaur booster, to be used to launch Mariners 6 and 7, was so optimised that the propellant tanks had to be pressurised, otherwise they would collapse as their skin was so thin. Unfortunately, about ten days prior to the launch of Mariner 6 an electrical fault with its launcher caused one of the tanks to become depressurised on the launch pad, and the outer skin of the Atlas buckled under its own weight. The tank was quickly repressurised to prevent a complete collapse, but the rocket's outer skin had been damaged beyond repair. So Mariner 6 and its Centaur upper stage were transferred to the Mariner 7 launcher that was being prepared on a nearby pad, and a new Atlas was obtained for Mariner 7.

Mariner 6, which was launched from the Cape on 24th February 1969, was the first planetary spacecraft not to enter an Earth parking orbit before setting out on its planetary-intercept trajectory. Instead, it was launched directly into that trajectory by the launcher. Just over one month later Mariner 7 was also on its way to Mars. On the same day, 27th March, the Russians launched the first of their two Mars probes in this 1969 launch window, and 2½ weeks later they launched their second spacecraft. Unfortunately, neither Russian spacecraft reached their Earth parking orbit.

Everything went well with Mariner 6 until, as the spacecraft neared Mars, debris released when the television scan platform was unlocked attracted the attention of the star sensor. The sensor duly followed this bright object, instead of its target star Canopus, and the ground controllers had to override the sensor and direct it back to its correct target. To be on the safe side, it was then decided to instruct the spacecraft to ignore the star sensor during its critical planetary encounter phase.

There were no more problems with Mariner 6 as it flew over Mars' equatorial regions on 31st July 1969 at an altitude of 3,429 km. The infrared radiometer measured surface temperatures at the equator ranging from 15 °C during the day to −73 °C at night, with a south polar cap temperature of about −125 °C. Similar temperatures had been estimated using Earth-bound radiometers, and the cratered terrain imaged by the television cameras was now no longer a surprise. The craters were much easier to see than in the Mariner 4 images, however, and it was clear that the larger craters were shallower than those on the Moon with fewer central peaks, although the smaller craters had steeper sides and did resemble lunar craters. Mariner 6 also imaged a new type of terrain, which was almost craterless with short jumbled ridges and depressions, unlike anything seen on the Moon. The surface atmospheric pressure was measured as about 6 millibars, the same as that measured by Mariner 4, and the atmosphere was found to be almost

100% carbon dioxide, which was a disappointment to those people hoping to find conditions suitable for life.

Meanwhile, whilst most of the NASA Deep Space Network was involved with the fly-by of Mariner 6, Mariner 7 hit a problem. For some reason it had suddenly started tumbling in orbit, and its radio beam was sweeping past the Earth in a regular manner, rather than being beamed straight at the Earth by the spacecraft directional antenna. A strong signal was sent from Earth directing the spacecraft to switch to its omnidirectional antenna, and a weak but steady signal was received as a result. This enabled the ground controllers to reconfigure the spacecraft, stop the tumbling motion and resume near-normal operations. Fortunately, although there were some problems with subsystem units on board the spacecraft, the faults were not catastrophic and the experiment units were still intact. Apparently, one of the 18 battery cells had exploded and given the spacecraft a push sideways. Electrolyte from the exploded cell had then shorted out some of the electrical subsystems. A little later another cell exploded, but again the results were not catastrophic.

Mariner 7 flew by Mars, just one week after Mariner 6, at an altitude almost exactly the same as its predecessor, but in a trajectory that allowed better imaging of the south polar region. It again showed a cratered surface, except for the area called the Hellas plain which had no discernible craters, so this area was probably younger than the other areas observed. Mariner 7 confirmed that the atmosphere was almost 100% carbon dioxide, with clouds of dust and carbon dioxide crystals, and the ground-level atmospheric pressure was confirmed as being about 6 millibars. The south polar ice cap seemed to be made almost completely of frozen carbon dioxide.

5.8 **Summary**

So the 1960s ended with the first substantial results from spacecraft sent to Mars and Venus. As a result most people had given up the idea of life on Mars, unless it was in a very simple form or not based on carbon and oxygen, and Venus looked extremely unlikely as a candidate for life of any description, unless it was in the cooler upper atmosphere. The lack of an ozone layer around either planet added to the problems that living organisms would have to overcome.

The main difference between Mars as understood at the end of 1969, and Mars of before the space age, was that it was now seen to have a highly cratered and apparently dead surface, rather than one possibly covered by

Table 5.1. *Planetary Launch Attempt Statistics for 1964 to 1969*

	Mars		Venus	
	American	Soviet	American	Soviet
Number of launches	4	3	1	12
Successful missions	3	0	1	3
Successful spacecraft	Mariners 4, 6 & 7		Mariner 5	Veneras 4, 5 & 6

lichens or moss. The ground-level atmospheric pressure was only about 0.5% of that of Earth, instead of 10%, and the disputes on the atmospheric constituents had been largely clarified by the discovery that it is almost 100% carbon dioxide, with clouds of dust and carbon dioxide crystals.

As far as Venus was concerned, the question of whether the surface-level atmospheric temperature and pressure were 300 °C and 100 bar or 80 °C and 5 bar had been largely answered to the benefit of the first figures, although the actual surface temperature seemed to be at least 400 °C. The dispute on the constitution of the atmosphere had been resolved, as it had been found to be almost completely carbon dioxide, although the constituents of the dense clouds were still unknown.

This chapter covers planetary spacecraft launched from 1964 to 1969. Over this period, there were seven spacecraft launched towards Mars and thirteen towards Venus, as shown in Table 5.1. The American success rate was clearly better than that of the Russians, although the Russians' statistics for Venus had improved over the period, with three of their last four missions being successful. Two thirds of the Russian planetary failures were due to continuing problems with their SL-6 launcher.

Chapter 6 | MARS AND VENUS: THE MIDDLE PERIOD

6.1 American Plans for the Early 1970s

The cancellation of NASA's extensive Mars Voyager exploration programme in August 1967 left NASA with no approved planetary spacecraft after Mariner 6 and 7 (see Section 5.7), which were due to be launched in 1969. The early 1970s were also due to see the end of the Apollo series of manned lunar spacecraft, so what was to be done post-Apollo as far as Solar System exploration was concerned?

In the mid 1960s the Space Science Board (SSB) of the National Academy of Sciences and the President's Science Advisory Committee (PSAC) had both recommended that planetary exploration should be the major post-Apollo goal, initially with automatic probes, to be followed eventually by manned spacecraft. But the cancellation of the Mars Voyager programme in 1967 showed that Congress was not yet ready to support major expenditure on planetary missions post-Apollo, although limited expenditure on unmanned planetary probes would probably be supported. As a near-term goal, the PSAC put spacecraft missions to Mars and Venus as of equal priority, but they also said that missions to Mercury and Jupiter were of almost the same importance as these. The SSB's priorities were almost the same, although they favoured Mars as the prime target.

With these priorities in mind, the Planetary Missions Technology Steering Committee at NASA Langley asked JPL and others in September 1967 to suggest a programme of planetary exploration for the early 1970s. Within a month JPL had suggested a 1971 Mars orbiter, two Venus spacecraft in 1972 and a Titan III-launched Mars orbiter/probe to be launched in 1973. Langley themselves suggested a Titan-launched Mars orbiter in 1971, followed by a Venus orbiter/probe in 1972. Then on 8th November the NASA

Administrator, James Webb, in replying to a question from Senator Clinton Anderson, proposed a planetary exploration programme consisting of:

- Two modified Mariner spacecraft to be launched by Atlas-Centaurs as Mars orbiters in 1971.
- A Titan IIIC-launched mission to Venus in 1972.
- A Titan IIIC-launched mission to Mercury via Venus in 1973.
- Two Titan IIIC Mars orbiter/probe spacecraft to be launched in 1973.
- Two Saturn 5-launched Voyager-size orbiter/landers to be launched in 1975.

So Webb himself was trying to reinstate the Voyager programme, although with a launch delay of two years and with smaller near-term missions.

The period at the end of 1967 saw extensive discussions by the various presidential and NASA advisory committees on a near-term planetary programme, resulting in President Johnson endorsing a new Mars exploration programme in January 1968. It started with two 1.5 ton orbiter/hard lander spacecraft to be launched in 1973 by a launcher based on Titan III, to be followed in subsequent launch windows by orbiter/soft lander missions. By the end of the year, further work had hardened-up the plan and the new NASA Administrator, Thomas Paine, was able to approve the Viking Mars orbiter/lander programme using dual Titan IIID-Centaur launches in 1973. The orbiter would be based on the Mariner design, and the lander, which would be released from the orbiter when the latter was in Mars orbit, was designed for a soft rather than a hard landing. The orbiter would be used to look for suitable landing sites, and act as a telecommunications relay between the lander and Earth when the lander was on Mars, as well as undertaking scientific investigations of its own.

Whilst outline designs were being developed for this Mars Viking spacecraft for a 1973 launch, plans were also being produced for a Mariner orbiter programme for the 1971 Mars launch window, and in August 1968 Mariners 8 and 9 were approved for a 1971 launch. These new Mariners were basically Mariner 6-size spacecraft with a large retrorocket added[1] to enable the spacecraft to enter into orbit around the planet. Mariner 8 was to be put into a polar orbit to map the whole of Mars (as it turned beneath the spacecraft), and Mariner 9 was to fly in an orbit inclined at an angle of 50° to the equator to examine changes in appearance of the equatorial and mid-latitude regions with time.

---

[1] These 998 kg Mariner 8/9 spacecraft had almost half of their mass taken up with fuel for the retrorocket.

So by the time that Mariners 6 and 7 were launched towards Mars in early 1969, NASA had two orbiters approved to fly to Mars in the 1971 launch window, and two Viking orbiter/soft-lander spacecraft for 1973, but there were still no approved spacecraft for Venus.

Both the SSB and the PSAC had been pushing NASA to include Mercury and Jupiter missions in its planning for the 1970s. It so happened that between 1970 and 1973 it would be possible to reach Mercury via Venus, because of the relative alignment of these two planets and the Earth at that time. This would not only enable a spacecraft to visit two planets in one mission but, using Venus to propel the spacecraft towards Mercury, in a so-called 'sling-shot' manoeuvre, would enable a smaller launch vehicle to be used. This Venus/Mercury mission was approved in December 1969 for a spacecraft launch in 1973. Then in February 1970 it was realised that the spacecraft could be made to fly-by Mercury on three separate occasions if its orbit was arranged so that it took precisely twice as long to orbit the Sun as did Mercury, returning to the planet every two Mercurian years. Unfortunately, this orbital arrangement meant that the spacecraft would image the same illuminated half of the planet on each pass, because the planet rotates exactly three times every two Mercurian years. Nevertheless, this was too good an opportunity to miss, and this fly-by of Venus and triple fly-by of Mercury became the Mariner 10 mission.

These were the American plans as the first planetary spacecraft of the 1970s was launched, not by America but by the USSR, on 17th August 1970.

6.2 Veneras 7 and 8

The Russians had accepted after Venera 4 and Mariner 5 that the surface atmospheric pressure and temperature on Venus were about 75 to 100 bar and at least 430 °C. But they had only been able to make relatively simple modifications to Veneras 5 and 6 (see Section 5.5), because there was not enough time before the 1969 Venus launch window to do otherwise. The Russians had another 19 months before the 1970 Venus launch window, however, and this enabled them to make major modifications to Venera 7 (see Figure 6.1) to try to ensure that it would still operate for a short period of time on the surface of Venus. To do this, the Venera 7 landing module was strengthened to withstand a pressure of 180 bar and its interior was cooled to a temperature of −8 °C by cold gas injection before it was separated from the main spacecraft (called the spacecraft 'bus'). A new parachute design enabled the parachute to be partially deployed initially, to limit the shock

Figure 6.1 Venera 7, whose landing module, which is the spherical object at the bottom of the spacecraft, was the first man-made object to return data after landing on another planet. (Courtesy NASA and NSSDC; Venera 7, 70-060A.)

on opening to 25*g*, and to allow the landing module to descend rapidly through the hot atmosphere thus limiting the heating time for the capsule. Lower in the atmosphere, the increased heat would melt the nylon cord which had stopped the parachute from opening completely, and the fully opened parachute would then allow a slow descent.

Venera 7[2] worked perfectly during its 4 month voyage to Venus, where it arrived on 15th December 1970 on the *night side* of the planet. Unlike previous Russian landers, which separated from their spacecraft bus about an hour before they hit the top of the Venusian atmosphere, the lander of Venera 7 stayed attached to the cooling system of the bus until the whole spacecraft was in the upper reaches of the atmosphere. After the lander was released, its parachute partially opened as planned at a velocity of about 200 m/s (450 mph) and an altitude of 60 km, and ten minutes later the parachute fully opened, reducing the descent velocity from 27 to 19 m/s (61 to 43 mph).

[2] The Russians also launched a second spacecraft, which became known as Cosmos 359, but this was stranded in Earth orbit by a malfunctioning rocket escape stage.

So far everything had worked as planned, but sixteen minutes after full parachute deployment the lander suddenly accelerated and crashed onto the surface at 16 m/s (37 mph). At first it was thought that the lander had been destroyed, but analysis of what appeared to be radio noise picked up by the tracking antenna on Earth later revealed a very faint signal. Apparently the lander was still working, although its antenna was not now pointed at Earth. Data transmission lasted from the surface for 23 minutes before the lander succumbed to the high ambient temperature.

Venera 7 confirmed earlier spacecraft findings that the atmosphere was about 97% carbon dioxide, and the ground-level temperature was measured *in situ* for the first time as 475 °C, which was also in agreement with previous estimates. Unfortunately, the telemetry system malfunctioned and did not transmit pressure data from the surface, but extrapolating from data transmitted earlier in the descent gave an estimated ground-level atmospheric pressure of about 90 bar.

The data received from Venera 7 did little more than confirm that received from previous spacecraft, so it did not significantly advance our knowledge of Venus. But for the first time a spacecraft had landed on another planet and lived to tell the tale, if only just. That was its real significance and the Russians were justifiably proud of their achievement.

The Russians' experience at the next Venus launch window in March 1972 was similar to that in 1970, as the first spacecraft to be launched, Venera 8, reached Venus, but its twin, launched four days later, did not leave its Earth transfer orbit owing to a problem with its rocket escape stage.

Venera 8 successfully soft-landed on the *daylight side* of Venus on 22nd July 1972, measuring a surface temperature of 470 °C at a pressure of 90 bar so, comparing these results with those from Venera 7, the surface conditions were seen to be the same, within error, on both the daylight and night-time sides of the planet. The 500 kg lander contained a number of new instruments compared with its predecessors, one of which was to measure ammonia concentrations in the atmosphere whose presence had been predicted theoretically and detected from Earth. The concentrations measured of from 0.01 to 0.1% at an altitude of about 40 km were consistent with expectations. The lander also measured a wind speed as it descended of from 100 m/s (225 mph) at 48 km altitude to 1 m/s (2 mph) below 10 km. The probe had landed relatively close to the terminator, and this very low surface-level wind speed was consistent with surface temperatures that did not vary greatly between the day and night sides of the planet. A light meter showed that only 1.5 % of the Sun's illumination

Figure 6.2 Mariner 10, the first spacecraft to fly-by both Venus and Mercury. The twin television cameras are seen at the top, looking like a pair of eyes, and the sun-shade is the umbrella-like structure deployed at the bottom. The magnetometers are packages on the right-hand boom, placed there to be as far away from the spacecraft as possible, to avoid spacecraft-generated interference upsetting their measurements. The device to the left of the sun-shade is the steerable high-gain antenna. (Courtesy NASA/JPL/Caltech.)

reached the ground where the probe landed. Measurement of γ-rays showed that the upper atmosphere shielded the planet from high energy cosmic rays, and γ-ray spectroscopy indicated that the surface at the landing site was probably composed of volcanic rock. This finding was important as it suggested that Venus had once been hot enough for the heavier elements to fall towards the centre of the planet, leaving the lighter elements on top, producing what is called a 'differentiated structure' like that of the Earth and Moon.

The mission of Venera 8 completed this phase of the Soviet exploration of Venus, and to make significant further progress the Russians realised that they would need a more sophisticated lander. As a result they decided to forgo the next Venus launch window (in 1973) and concentrate on producing new Venera designs. The Americans, on the other hand, decided to send the 500 kg Mariner 10 spacecraft (see Figure 6.2) to Mercury via Venus during the 1973 launch window.

6.3 Mariner 10

The main problem in designing a spacecraft to fly-by Mercury was the extra heat that it would be subjected to so close to the Sun.[3] In the case of Mariner 10 this problem was solved by placing most of the scientific instruments (including the cameras) on the shadow side of the spacecraft, and using a sun-shade, louvered side-panels and insulation blankets, together with reflective coatings on the Sun-facing side.

The main instruments on board Mariner 10 were twin television cameras, each of which could use either a 1,500 mm effective focal length Cassegrain telescope or a 62 mm wide-angle lens. An eight-position filter wheel allowed imaging in various wavebands in the near ultraviolet and visible. Although photographs of Venus taken from Earth showed very little structure in its clouds, it was known that the best images were produced at ultraviolet wavelengths. Unfortunately, all but the longest ultraviolet wavelengths are absorbed by the Earth's atmosphere, and it was hoped that photographs taken further into the ultraviolet than possible from Earth may yield more detailed cloud images.

The television cameras on Mariner 10 were mounted on a moveable scan platform to enable them to follow the planet during each fly-by, but most of the instruments were fixed to the spacecraft and could only observe Mercury if Mariner flew-by the planet on the dark side. In the event the first and third fly-bys were on the dark side but the second pass was on the daylight side so the magnetometer, infrared radiometer and charged particle telescope could not be operated satisfactorily during this second fly-by. So the second fly-by was basically limited to imaging.

Mariner 10 had more than its fair share of faults in space, the first of which occurred when the heaters on the TV cameras failed only a few days into the mission. A series of other faults occurred *en route* to Venus and it was beginning to look as though the spacecraft would be lucky to reach Venus in working order, let alone undertake three successful fly-bys of Mercury. Mariner 10 was, in fact, to succeed far better in its mission than anyone thought possible as it approached Venus.

Mariner 10 sent back to Earth over 4,000 images of Venus over the period February 5th to 13th 1974, showing chevron-shaped cloud patterns that

---

[3] Mercury is, on average, only 0.38 AU from the Sun (where 1 AU is the Earth's distance from the Sun). Consequently, the incident heat per unit area on the spacecraft is $1/0.38^2 \approx 7$ times that of an Earth-orbiting spacecraft.

rotated around Venus once every 4 days at about 400 km/h (250 mph). These cloud patterns had first been photographed by Charles Boyer, a French amateur astronomer, in 1957, and he had observed that they rotated around Venus once every 4 days. The spacecraft images were much more detailed, however, enabling the first analysis to be undertaken of the atmospheric dynamics of Venus.

After its intercept with Venus, Mariner 10 was on its way to Mercury, which it began to image from as far away as 5.4 million kilometres. At this distance, surface features could hardly be resolved, but as it closed in hundreds of craters could be seen. Superficially, these craters looked very much like those on the Moon, but on closer examination they were seen to be shallower. This is because of the higher surface gravity on Mercury compared with the Moon. There was no evidence of wind erosion of Mercury's craters, indicating that Mercury has not had an appreciable atmosphere for most of its life.

The most obvious large-scale feature on Mercury was found to be the 1,300 km diameter Caloris Basin, which is ringed by mountains about 2,000 to 3,000 m (6,500 to 10,000 ft) high. The floor of this large basin is unlike anything on the Moon or Mars as it appears to have been filled with material that has subsequently been broken up into both concentric and radial ridges and grooves. On the other side of Mercury, at the antipodes of the Caloris basin, is an area of peculiar rippled terrain,[4] which the Mariner 10 imaging team called the 'weird terrain'. In 1975 Schultz and Gault showed that this weird terrain was probably produced when the seismic energy of the impact that caused the Caloris basin had been focused by the planetary core onto this antipodal region.

Because Mercury is much smaller than the Earth, it would have cooled much faster even though it is closer to the Sun. So it was generally thought that Mercury would now be solid throughout and, because of this, it could have no dynamo like the Earth to produce a magnetic field. It was a big surprise, therefore, when Mariner found that Mercury had a magnetosphere[5] and measured a magnetic field 1.2% times that of the Earth. This doesn't sound much, but it is quite large for a planet whose mass is only 5% that of the Earth, and the size of the magnetic field implies that Mercury has a

[4] Similar terrain exists on the Moon at the antipodes of the Mare Imbrium and Mare Orientale.

[5] Although Mercury has a magnetosphere, Mercury's magnetic field is too weak to produce radiation belts.

metallic core containing about two-thirds of the planet's mass. Mercury's slow axial rotation and lack of atmosphere[6] means that the surface facing the Sun is very hot, but that on the other side just before dawn is very cold, and the temperatures measured of 430 °C (max.) and −173 °C (min.) gave the largest temperature range of any planet in the Solar System.

Mariner 10 passed within 703 km of the surface of Mercury on its first pass on 29th March 1974, some 48,000 km above Mercury on its second pass about six months later, and only 327 km from Mercury on its third and final fly-by on 16th March 1975. Data from the last intercept confirmed the value of Mercury's magnetic field[7] and enabled surface features to be resolved at a maximum resolution of only 100 m.

Mariner 10 was the most successful planetary probe launched up to that time and, even now, it is still the only spacecraft to have visited Mercury. It was the first spacecraft to use a gravity-assist trajectory,[8] the first to visit two planets and the first to visit the same planet more than once.

6.4 Mariner 9, Mars 2 and 3, and the Great Martian Dust Storm of 1971

Over the period from 1970 to 1974 when Veneras 7 and 8 and Mariner 10 were examining Venus, both the Americans and Russians were also attempting to send spacecraft to Mars, five of which were launched in the 1971 launch window. Unlike the American spacecraft, which were designed to orbit Mars, the Russian spacecraft included landing modules (see Figure 6.3).

The first spacecraft to lift off was the American Mariner 8, which was launched on 9th May, but it was destroyed when the range safety officer had to blow up the launch vehicle's Centaur upper stage when it started to veer off course. On the following day the first Russian attempt also failed, owing

[6] Mercury's atmosphere was found by Mariner 10 to have a pressure of only about 10^{-13} bar. Mariner 10 also found that it was composed mainly of atomic helium and hydrogen, and so it appeared to be nothing more than a local concentration of the solar wind caused by Mercury's gravity.

[7] The magnetic field detected during the first Mariner 10 intercept could possibly have been due to an interaction between Mercury and the solar wind, rather than be a magnetic field intrinsic to Mercury itself. The data received during the third fly-by showed that the field was intrinsic to the planet, however.

[8] Using the gravity of one planet to increase its speed and send it on to another in a sling-shot manoeuvre.

Figure 6.3 Mars 3, the lander of which landed on Mars during the great dust storm of 1971. The aeroshell and lander are the conical part at the top. (Courtesy IKI, Moscow.)

to a computer programming error on the Proton launcher, but the next two Russian spacecraft, Mars 2 and 3, both successfully reached Mars. They were overtaken *en route* by the second American spacecraft, Mariner 9.

Originally, Mariner 9 was planned to study temporal changes in the Martian atmosphere and on its surface in equatorial and low latitude regions, and so it was originally intended to orbit Mars at an angle of 50° to the equator. Mariner 8, on the other hand, was planned to map 70% of the surface, and so was expected to orbit Mars in a polar orbit. These two missions were combined into the Mariner 9 mission when its sister spacecraft was lost. In particular, Mariner 9, which was launched on 30th May 1971, had the angle of its orbital plane increased to 65° to increase the amount of Martian surface covered.

The design of Mariner 9 was based on that of Mariners 6 and 7, with a large retrorocket added to enable it to enter Mars orbit. Because Mariner 7 had had a problem with its silver zinc battery, that on Mariner 9 was changed to a nickel cadmium battery, and improvements to other subsystems were also incorporated. The scientific payload consisted of a television imaging system with both a 508 mm focal length f/2.4 lens capable of a one kilometre maximum surface resolution, and a 50 mm f/4 wide-angle lens fitted with a filter wheel for both colour and polarisation work. An infrared radiometer was included to map the surface temperature of Mars, and ultraviolet and infrared spectrometers were to analyse its atmosphere.

Four months after the launch of Mars 2 and 3 and Mariner 9, a large planet-wide dust storm developed. Whilst this was unfortunate for the Mariner 9 orbiter, it was to prove disastrous for the Mars 2 and 3 landers, as the dust storm was still raging when they reached Mars on 27th November and 2nd December, respectively. Unfortunately, there was no way that the Russian landing missions could be delayed as their landers were to enter directly into the atmosphere from their interplanetary orbits.[9] Although the Mariner 9 orbiter had also to go into orbit as planned, the start of its main mission could be delayed, however, until the dust storm blew itself out.

Four and a half hours before arriving at Mars the Mars 2 lander separated from the main spacecraft, which then fired its retrorocket to enable it to enter Mars orbit. The lander, which entered the atmosphere at a velocity of about 6,000 m/s (13,400 mph), was first decelerated by an aeroshield, then by a parachute system, and finally by a retrorocket for the last 30 m. The probe was designed to withstand a landing shock of about 1,000*g* but, in spite of this, no signals were received from the surface, possibly because the spacecraft was damaged by either high cross-winds and/or the dust. Five days later the probe from Mars 3 also landed on Mars but transmitted test data for only 20 seconds before falling silent. Unfortunately, the atmospheric data that Mars 2 and 3 had measured during their descent had been stored on board, ready to be relayed to Earth once they had landed. So their failure on the surface meant that all of this atmospheric data was lost.

Meanwhile, Mariner 9 had arrived at Mars on 14th November 1971. It was placed into an initial orbit that was trimmed two days later to be a 1,400 × 17,100 km orbit inclined at about 65°. This would allow imaging of the same piece of ground every seventeen days, to enable study of transient features such as the seasonal colour changes seen from Earth and the seasonal advance and retreat of the polar caps. Because of the dust storm, the initial surface images received from Mariner 9 showed limited surface detail but, whilst waiting for the storm to blow itself out, the spacecraft took the first detailed images of the two small Martian satellites, Phobos[10] (see Figure 6.4)

---

[9] The Russians did not have enough launcher power to put the whole of the Mars 2 and 3 spacecraft into Martian orbit and then release the lander probes. Had they been able to do this, they would have been able to delay the release of the landers.

[10] Relatively crude photographs had been taken of Phobos by Mariner 7 two years earlier but no surface detail had been visible.

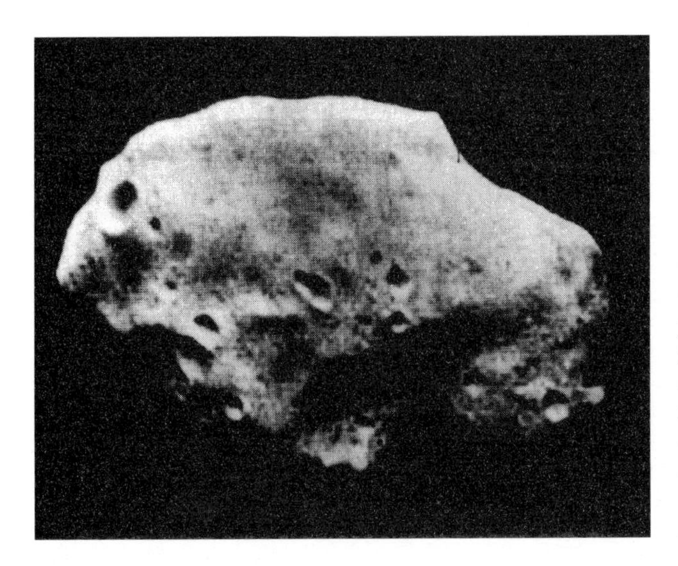

Figure 6.4 Phobos, one of the two small Martian satellites, photographed whilst Mariner 9 was waiting for the dust storm on Mars to abate. This image, taken from a distance of 5,800 km, shows craters as small as 300 m in diameter. (Courtesy NASA and NSSDC; Mariner 9, MTVS 4109-9.)

and Deimos, showing them to be irregularly shaped and covered in craters. Although the Mars imaging mission had to be put largely on hold during the dust storm, other experiments produced useful results. In particular, the infrared spectrometer found water vapour over the Martian south polar cap. This discovery created great excitement, even though the previous Mariners had shown that the Martian surface was now apparently dry and cratered. If there was some water vapour in the atmosphere now, maybe there had been water on the surface billions of years ago.

The dust storm was seen to be blowing itself out over the next few weeks and, as the dust settled, glimpses of craters were seen.[11] Routine imaging started on 2nd January 1972 and the excitement level reached fever pitch as a new Mars was slowly revealed. This new Mars was not the disappointing, crater-strewn Mars revealed by the previous Mariners, but an exciting planet with massive volcanoes and a giant canyon system. The largest volcano, Olympus Mons (see Figure 6.5), was 25,000 metres (80,000 ft) high (Everest is only 9,000 metres above sea level) and 500 kilometres (300 miles) in diameter, which is far larger than any volcano on Earth.[12] What is even more remarkable is that this massive volcano is on a planet only half the diameter of the Earth. The canyon system (see Figure 6.6) imaged by Mariner 9, now called the

[11] During this time the best images were of the south polar cap and the tops of the high ground that appeared above the settling dust.

[12] Mauna Loa in Hawaii is the largest volcano on Earth, rising 9,000 metres above the sea floor and being 120 kilometres in diameter at its base.

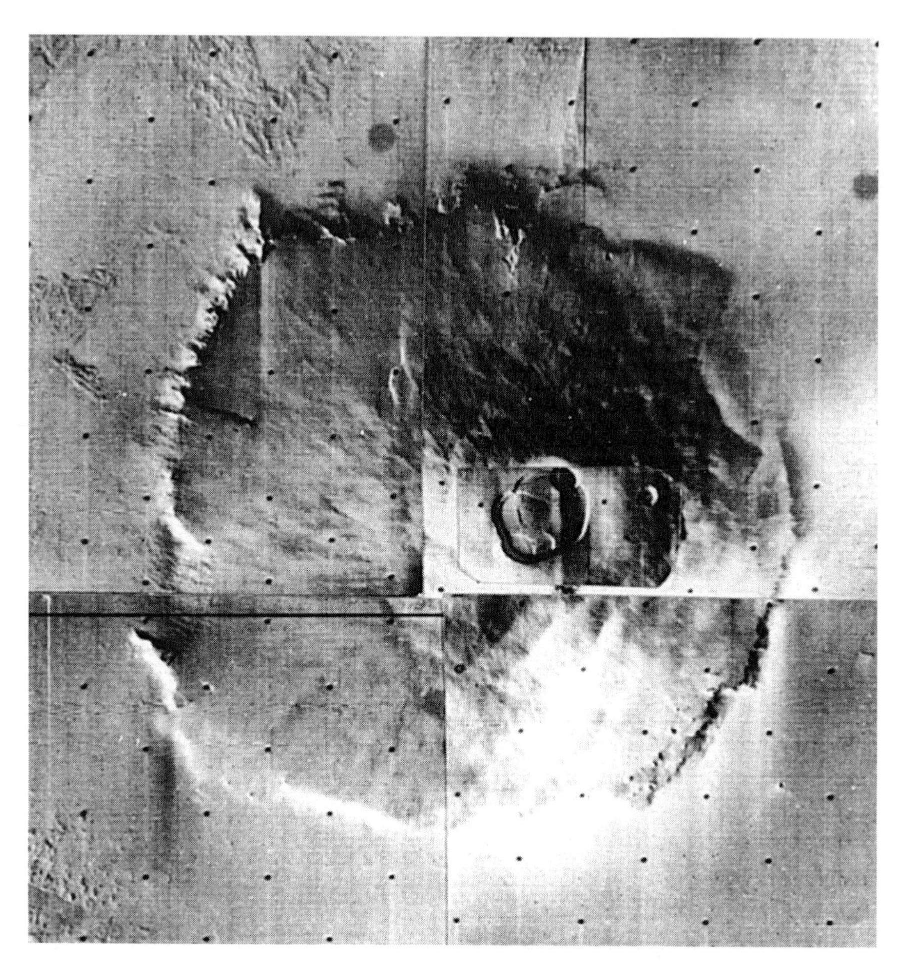

Figure 6.5 The huge volcano Olympus Mons as imaged by Mariner 9 in January 1972. The 80 km diameter caldera is clearly seen just right of centre. (Courtesy National Space Science Data Center, World Data Center-A for Rockets and Satellites, NASA; Principal Investigator, the late Dr Harold Masursky; P-12834.)

Valles Marineris,[13] is also vast, being 4,000 kilometres (2,500 miles) long, up to 200 kilometres (120 miles) wide, and up to 6,000 metres (20,000 ft) deep.[14]

The discovery of such large volcanoes and of a vast canyon system indicates that there has been no plate tectonics on Mars. With no plate

---

[13] After the Mariner spacecraft that discovered it.

[14] The Grand Canyon in the USA is only 150 kilometres long, and at most 2,000 metres deep.

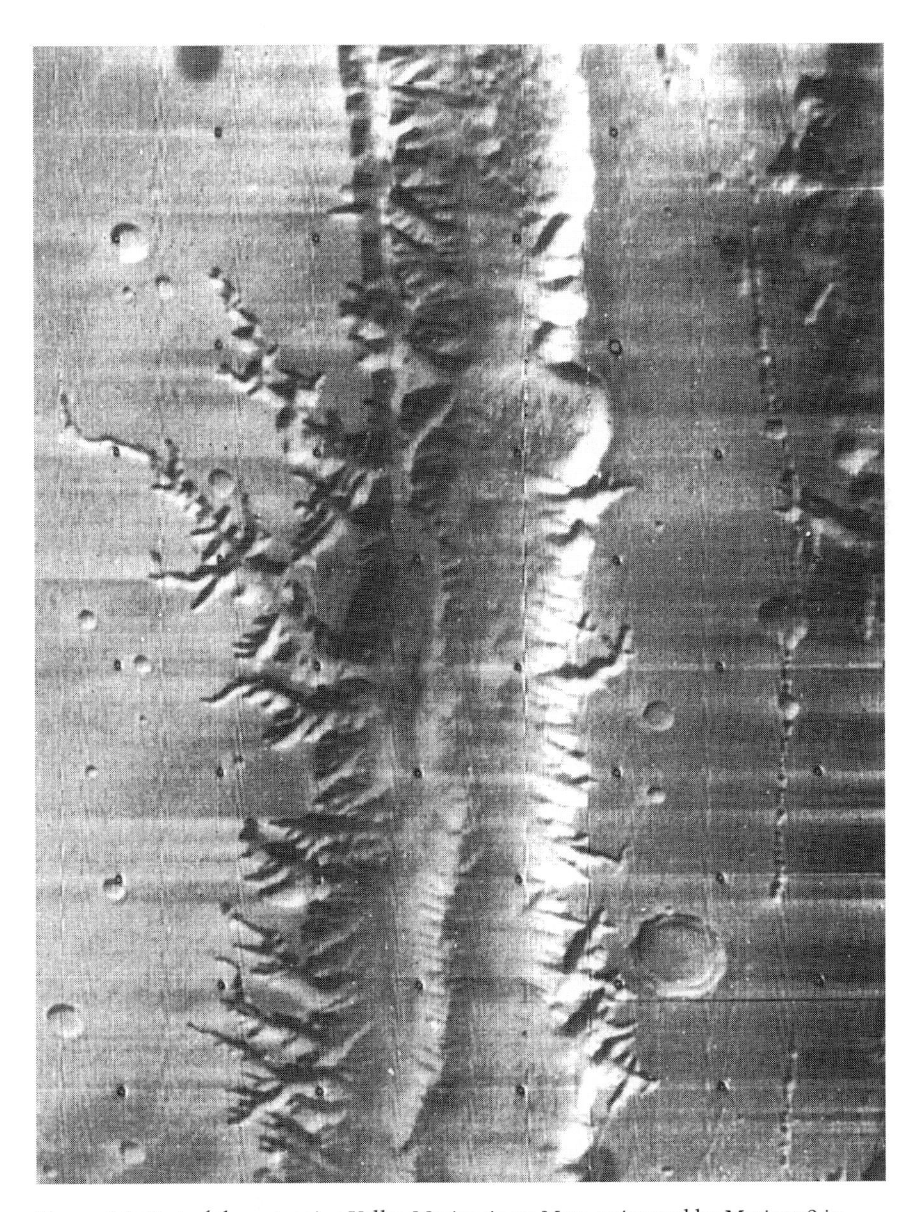

Figure 6.6 Part of the extensive Valles Marineris on Mars as imaged by Mariner 9 in January 1972. The image covers an area of about 380×480 km. (Courtesy National Space Science Data Center, World Data Center-A for Rockets and Satellites, NASA; Principal Investigator, the late Dr Harold Masursky; P-12732.)

movement, hot spots in the underlying mantle would have continued to eject lava through the same place in the crust year after year to build up these enormous volcanoes. On Earth this does not happen as the crust is in motion relative to the mantle. It also appears as though the Valles Marineris could possibly be where two plates started to move apart but were stopped as Mars cooled quickly, shortly after the crust was formed.[15]

As the images from Mariner 9 continued to be received on Earth, astronomers started to wonder, since Mars looked much more interesting than before and had now been found to have some water vapour in its atmosphere, whether there may be signs that water had existed on the surface of Mars in the past. Imagine their delight, therefore, when sinuous channels were discovered in an area called Chryse. Because of the low atmospheric pressure it was clear that water could not exist in liquid form on the surface of Mars today, but it could well have been there in the past when the atmosphere was probably denser.

As has been mentioned above, parts of Mars had been seen from Earth to change colour with the season. This cause of this had long puzzled astronomers, some of whom had interpreted it as signs of plant life on the surface. Unfortunately, Mariner 9 was able to show that the colour change was due to nothing more than seasonal winds alternately covering and uncovering the darker substrate with lighter dust.

The Russian Mars 2 orbiter entered its orbit around Mars two weeks after Mariner 9, followed by the Mars 3 orbiter a few days later. Unfortunately, the camera systems of both Russian orbiters could not be reprogrammed, so they photographed the surface of Mars whilst the dust storm was raging. As a result the images were virtually useless. The Russians had much more success with their other orbiter instruments, however, which generally produced complementary data to that received from Mariner 9.

Mars 2 and 3 measured a very weak Martian magnetic field of 30 gammas, which is not much above that of the interplanetary magnetic field, and is about three orders of magnitude less than that of the Earth. Their infrared detectors, operating at a wavelength of 1.4 microns,[16] showed that, although there is water vapour in the Martian atmosphere, it is so tenuous that, if it were all to be precipitated as water on the surface, it would be less than 1 mm deep. The surface of Mars was found to cool quickly after sunset suggesting that it was covered with dust or sand, although the darker areas, which were warmer than the other areas during

---

[15] Mars would cool much more quickly than the Earth because it is only half the Earth's diameter. [16] 1 micron is 10^{-3} mm.

the day, generally cooled more slowly after dusk indicating that they were made of rock.

The Russian orbiters found that the dust particles in the dust storm were about 60% silicon, in agreement with Mariner 9 results, with sizes generally in the range from 1 to 10 microns. Although the largest particles precipitated out of the atmosphere quickly once the winds had died down, the smaller particles took months to reach the surface. During the dust storm the surface cooled, as the dust obscured the Sun, by about 20 to 30 °C, but the atmosphere warmed up as the dust absorbed the solar energy.

In summary, Mariner 9 had shown us a much more interesting Martian surface than previously seen, with some water vapour in the atmosphere, which Mars 2 and 3 had confirmed. There was no liquid water on the surface today, however, but there were signs in some of the Mariner images that water may have flowed in significant amounts in the past.

If significant amounts of water had been present earlier in the history of Mars under a more substantial atmosphere, then maybe life could have been present. As the planet gradually lost its atmosphere, however, and the water evaporated, damaging solar radiation would probably have killed off all but the most resistant forms of life, but could such life still exist today? Maybe it could do so shielded from the worst of the solar radiation under rocks or beneath the surface. After all, it is possible that there may be water still trapped beneath the surface. All of this may sound improbable, but our exploration of the universe has regularly shown us surprises, and in some cases things have been found which at the time had been thought to be impossible. Finding life on another planet would have such important psychological and religious consequences that we should leave no stone unturned (literally!) to find such evidence. This was the prevailing view in NASA in the early 1970s.

Clearly, a lander would be required to analyse samples of the Martian surface to see if there was any evidence of biological activity, and this is what was intended with the American Viking probes. But before the Americans could launch their Viking spacecraft, the Russians had four more attempts to visit Mars.

6.5 Mars 4 to 7

Because Mars was further away during the 1973 launch window than during 1971, it was not possible for the Russians to have their normal-sized orbiter and lander launched together in 1973 by their Proton launch vehicle. So the first two spacecraft, Mars 4 and 5, were orbiters and the second two, Mars 6 and 7, were fly-by spacecraft with landing modules. All four of these spacecraft were successfully launched over a three-week period in July and August 1973.

Mars 4's retrorocket developed a leak, so it could not be put into orbit around the planet. Mars 5 did go into orbit around Mars, however, in February 1974, and its camera and that on Mars 4, during its un-planned fly-by, imaged an equatorial region that showed more of the sinuous channels first imaged by Mariner 9. A comparison between the atmospheric results from Mars 5 and those from Mars 3 two years earlier showed that there was four times as much water vapour in the atmosphere in 1974 than during the dust storm of 1971. Contact was lost with the Mars 6 lander when the landing retrorocket failed, and a fault on Mars 7 caused the landing capsule to miss the planet completely. All in all, the results from this armada of four Russian spacecraft were very disappointing, and this put paid to any Russian attempts to visit Mars during the next fifteen years.

6.6 **Viking**

When the American Viking programme was approved in 1968 it was planned to launch the first two spacecraft in 1973, and a third in 1975. Each of the spacecraft was to go into orbit around Mars and then, when suitable landing sites had been found, each would release a module to land on the surface. Unfortunately, Viking fell foul of a major reduction in NASA's budget at the end of 1969, which was part of a general tightening of the federal budget to pay for the continuing Vietnam conflict, extra social security benefits and the like. At the same time the estimated cost of the Viking programme (excluding launch vehicles) had increased from $364m in March 1969 to $606m in August of the same year. As a result, at the end of 1969 the first launch was delayed until 1975 to reduce the annual costs. Then in September 1974 the third flight unit was cancelled to reduce the total programme costs.

The decision to delay the launch until 1975 was taken by NASA after the spacecraft contractor had already started design work. Although fewer people would now be required to work on the contract, because the rate of progress could be reduced, many of those that were required would have to work on the programme for longer. So although the annual costs could be reduced, the total costs would be higher. In June 1970, for example, it was estimated that the two-year launch delay would increase total spacecraft costs from $609m to $750m. Unfortunately, the alternative strategy of stopping work for two years was not really an option, as it would have been impossible to reassemble the Viking design team after they had been reassigned to different contracts.

The main purpose of the Viking lander missions was to search for life, so the landing sites chosen for the two spacecraft were those where there was a maximum possibility of finding water. The sites had to be low lying as, not only were they more likely to have water, but the greater atmospheric pressure at low-lying sites would make the landers' 16 metre diameter parachute more efficient. In addition the landing sites should be flat with no evidence of high winds, but here the mission planners had a problem. The best images received from Mariner 9 had a maximum surface resolution of only about 100 metres, and this was not sufficient to detect anything other than the most obvious surface irregularities. So the site selection was something of a gamble, and it was recognised that the chosen landing sites may have to be changed once the images had been analysed from the Viking orbiters. With this proviso, the landing site chosen for Viking 1 was in the Chryse region, where sinuous channels had been discovered by Mariner 9 in an otherwise relatively flat region, and that for Viking 2 was in the Cydonia region near the southern fringe of the north polar hood, where it was hoped to find evidence of liquid water. Both were very low lying sites.

The Viking spacecraft consisted of a 900 kg orbiter with 1,430 kg of propellant, and a 610 kg landing module with 70 kg of propellant, plus an aeroshell, bioshield, etc., giving a total launch mass of 3,530 kg. The orbiter had a twin television camera system to image the surface at a maximum resolution of 35 metres, a water detector, and an infrared instrument to map the surface temperature. Initially, the television cameras and water detectors would try to locate a smooth, moist surface on which to land the landing module and, when that part of the mission had been successfully accomplished, the television system would produce high-resolution images of the whole planet.

The lander was contained in a double-skinned lens-shaped cover attached to the underside of the orbiter (see Figure 6.7). The outer skin was a 'bioshield', designed to protect the lander from biological contamination during its time on the launcher and during its passage through the Earth's atmosphere, and the inner skin was an aeroshell, designed to protect the lander from high temperatures during its initial passage through the Martian atmosphere. Once in space, the part of the bioshield at the base of the lander would be ejected. NASA did not want to contaminate Mars with Earth-based organisms as, apart from anything else, it could make the biological results from the Mars lander very difficult to interpret.

Figure 6.7 The Viking spacecraft showing the lens-shaped bioshield at the bottom, which contained the lander, the orbiter in the centre, and the retro-rocket at the top. The tip to tip distance along the solar arrays was 9.75 m (about 32 ft). (Courtesy NASA/JPL/Caltech.)

The three-legged lander (see Figure 6.8) had twin, high-resolution cameras with a nodding mirror to scan the surface from 40° above to 60° below the horizontal. The images, which each took a minimum of two minutes to build up, could be either monochrome or taken in red, green, blue or infrared light. A scoop would sample the soil and deposit it in the various soil analysers on board, and a meteorological assembly would measure atmospheric temperature, pressure and wind velocity. Mass spectrometers would chemically analyse atmospheric and soil samples, and a seismometer would measure Mars-quakes. The mass spectrometer would be able to detect live or dead organic material in soil samples, and an X-ray fluorescence spectrometer would also measure the concentrations of inorganic material in the soil.

The most sophisticated instrument package on the lander was, however, a miniature biological laboratory to test the soil for signs of elementary forms of life. This laboratory package consisted of three different experiments, called the pyrolitic release, labelled release and gas-exchange release experiments.

In the pyrolitic release experiment, a soil sample would be heated using a lamp that simulated Martian sunlight, minus the ultraviolet, in an atmos-

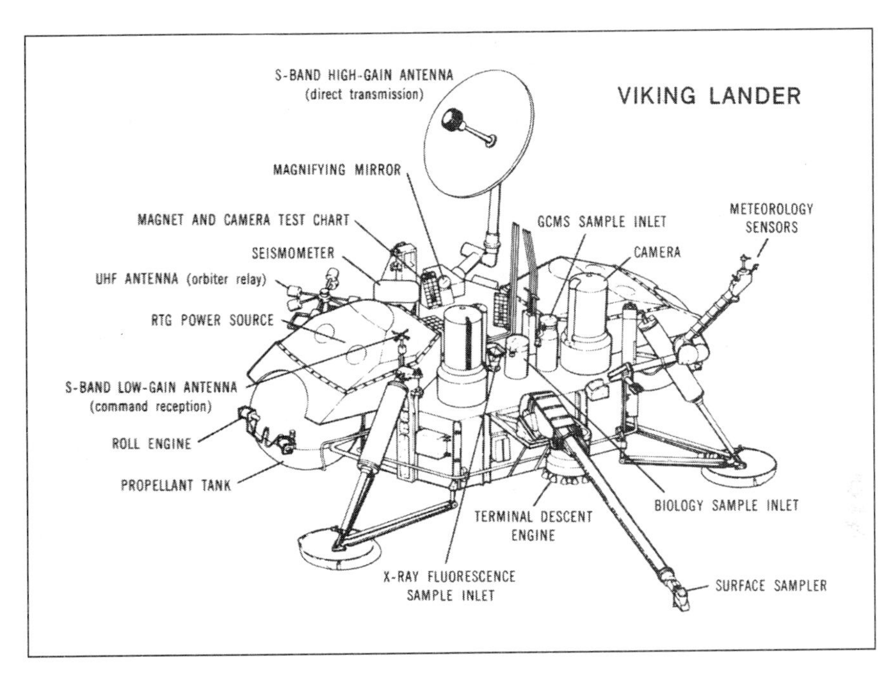

Figure 6.8 A schematic diagram of the Viking lander as it would appear on Mars. An isotope power source (labelled 'RTG power source') produced power instead of the usual solar arrays, as solar arrays would have been inoperative during the 12 hour Martian night. The soil scoop, which was used to fill the various biological experiments, is seen in the right foreground. (Courtesy NASA and *Sky & Telescope*.)

phere of carbon dioxide containing a small amount of radioactive carbon-14. If organisms ingested the carbon dioxide, they would also take in the carbon-14 tracer. After five days the chamber would be flushed with an inert gas, and then the soil/organism mix would be heated up to a temperature of about 625 °C to see if carbon-14 was released. If it was, this would be taken as an indication that there had been living organisms in the soil.

In the labelled release experiment, water and nutrients labelled with (i.e. containing) carbon-14 would be fed to the soil sample. It was then expected that any organisms present would consume the nutrients and release gases labelled with carbon-14 which would be detectable.

Finally, in the gas-exchange release experiment unlabelled nutrients would be fed to a soil sample in a humid carbon dioxide based atmosphere. Detectors would then search for hydrogen, nitrogen, oxygen and methane produced by the organisms.

The launch of Viking 1 was delayed on 11th August 1975, the first day of the launch window, because of a faulty valve on the launcher, and two days later engineers also found that the Viking orbiter batteries had been accidentally discharged. Viking 1 was then replaced on its launcher by the second Viking spacecraft, to allow analysis of the battery problem to take place, and on 20th August this replacement Viking, re-designated Viking 1, was successfully launched towards Mars by a Titan IIIE-Centaur launch vehicle. Viking 2's lift-off was further delayed from 1st September to allow a fault to be corrected with the Viking orbiter's high-gain antenna. Then on 9th September, the last day of its nominal window, it was successfully launched. Just five minutes after lift-off, the launch site was subjected to an intense rainstorm that, had it occurred just a few minutes earlier, would have caused a further launch postponement.

Viking 1 arrived at Mars on 19th June 1976 and decelerated into a 1,500 × 50,300 km orbit with a period of 42.4 hours and an inclination of 37°. Two days later the apoapsis[17] was lowered to 32,800 km to reduce the orbital period to 24.6 hours, to match the rotation period of Mars, with the periapsis[18] being over the chosen landing site in the Chryse region. As a result, the orbiter would have its closest approach to Mars over the chosen landing site at the same time each day. Images of that landing site at 19.5° N, 34° W were first returned on 22nd June and, much to the surprise of the mission scientists, it seemed to be more uneven than they had thought.[19] Their chosen landing site was seen to be on the floor of what looked like a river channel which, although scientifically interesting, would be a very risky place to land. This was a pity, as it was hoped to land Viking 1 on Mars on 4th July 1976, the bicentennial anniversary of American independence, but this was no longer possible. A survey of more images enabled a new landing site to be chosen at 22.5° N, 47.5° W, which was on the western side of the Chryse region, 2.7 kilometres below the datum representing Mars' average surface elevation.

At 08.51 UT[20] on 20th July the Viking 1 lander separated from the orbiter, leaving the upper part of the bioshield attached to the orbiter and,

---

[17] The furthest distance of the spacecraft from Mars.

[18] The nearest distance of the spacecraft from Mars.

[19] A detailed comparison of the Mariner 9 and Viking orbiter images showed that, not only were the latter of higher resolution, but, surprisingly, the atmosphere even during the second half of 1972 had not really settled after the large dust storm observed by Mariner 9 over six months earlier. The residual atmospheric dust had masked some features, particularly large lava flows, in the Mariner 9 images that were now clear from Viking.

[20] All times are the times of receiving the signals on Earth. The true times on Mars were some 19 minutes earlier, because it took radio waves that time to travel from Mars to Earth.

just over an hour later, the lander fired its de-orbiting thrusters. The pitch angle for entry was adjusted at 11.54, and by 12.10 UT the aeroshell, whose surface had reached a temperature of 1,500 °C, had reduced the lander's velocity from about 16,000 km/h (10,000 mph) to 1,000 km/h (600 mph) at a height of about 5,900 m (19,000 ft). The parachute was then opened, the aeroshell base jettisoned, and the lander's legs deployed, and one minute later, at an altitude of 1,400 m (4,600 ft), the parachute and aeroshell top were jettisoned and the retrorocket fired. The final descent to the surface was very slow to minimise the disturbance to the landing area, and the lander touched down on Mars at 12.12 UT at a final velocity of just 2.4 m/s (8 ft/s), only 28 km (17 miles) from its target. Mariners 6 and 7 had shown that the atmosphere was almost 100% carbon dioxide, but during Viking 1's entry sequence its mass spectrometer, whilst confirming this result, also detected small amounts of nitrogen, argon, carbon monoxide, oxygen and nitric oxide in the upper atmosphere. The discovery of nitrogen was a great fillip to those exobiologists hoping to find evidence of life, as it was believed that nitrogen was an essential ingredient in any environment in which life had evolved.

Twenty-five seconds after landing, the first ever image from the surface of Mars began to be transmitted, and over the next few months images of the red, rock-strewn, sandy surface and pink sky were received and analysed on Earth (see Figure 6.9). On the first day the atmospheric temperature varied from −86 °C at dawn to −33 °C in the afternoon, the pressure was 7.6 millibars, and the wind velocity gusted up to 51 km/h (32 mph).

Inevitably there were a few problems with the lander. The seismometer, which had been locked in position until after the landing, failed to unlock, and the scoop also failed to operate properly at first, but this latter problem was soon solved.

The soil was analysed by the X-ray fluorescence spectrometer to be iron-rich clay with 21% silicon and 13% iron as its major constituents, and the atmosphere was found to consist of 95.3% carbon dioxide, 2.7% nitrogen, 1.6% argon and 0.13% oxygen. Interestingly, both the soil and the very fine dust were found to have an abundance of magnetic particles, possibly because these consisted of a form of iron oxide called magnetite. The mass spectrometer found no evidence of organic matter in the surface sample tested, indicating that there was no life in it, but the biological experiments produced confusing results.

The first results from the biological experiments came from the gas-exchange experiment (GEX), where a large amount of oxygen was measured as soon as the surface sample was humidified. Then large amounts of

Figure 6.9 Mars as seen by the Viking 1 lander showing small wind-blown sand dunes and a plethora of rocks. The large boulder, nicknamed 'Big Joe', at the left of centre is about 1 metre high and is covered with red soil. Although there is no water on Mars, the rocks show clear evidence of weathering, due mainly to the wind. (Courtesy NASA and NSSDC.)

radioactive carbon dioxide were produced in the labelled release experiment (LR) when the soil sample was moistened with nutrients. The results of the GEX and LR experiments were so clear and startling, particularly when compared with the lack of organics found by the mass spectrometer, that the experimenters began to doubt whether they were really measuring the effects of life, and chemical reactions were suggested as a cause of the results. The problem was that, when the soil sample in the LR experiment was heated up to 170 °C to kill off any life and then re-tested, no radioactive gases were detected. So, either living organisms had been killed off, or the heat had broken up the chemical constituents so that they no longer reacted. A similar effect was also found with the pyrolitic release experiment (PR), which gave positive results with normal soil, but showed no activity after the soil had been heated. Later results of the PR experiment were not as clear-cut in showing this effect, however.

At face value the GEX, LR and PR experiments all showed that there was life on Mars, yet that was inconsistent with the mass spectrometer results that showed that there was no organic material in the soil samples. It was possible that a small number of living organisms could have activated the biology experiments but have not been sufficiently numerous to be detected by the mass spectrometer, because the mass spectrometer was not that sensitive. For every living organism on Earth, however, there are thousands of dead ones, and there should have been enough dead organisms on Mars to have been detected by the mass spectrometer. Maybe there was life on Mars but the organisms were highly efficient cannibals, leaving few dead organisms, or maybe there was no life and the GEX, LR and PR results were all caused by chemical reactions. Either was possible.

Viking 2 arrived at Mars on 7th August, but imaging of its proposed landing site showed that it was too rough for a safe landing. After a number of orbit manoeuvres a new site was found in Utopia Planitia at 47.9° N, 225.8° W, some 3 km below the datum level. This was about 7,000 kilometres from the Viking 1 lander and about 4° further north than originally planned. A more northerly position was chosen because it was hoped that the soil would have more moisture, being that much closer to the north polar cap.

By 3rd September everything was set for the Viking 2 landing, but when the lander separated from the orbiter, a problem with the power supply to the orbiter gyros caused its attitude to drift, resulting in the orbiter's high-gain antenna losing its lock on the Earth. This meant that the data to be measured by the lander during its descent, which should have been relayed to Earth in real time via the orbiter's high-gain antenna, would be lost. So ground controllers instructed the orbiter, via its omnidirectional antenna, to record the information for later play-back to Earth. The landing sequence had already been pre-programmed, because of the 38 minute round-trip communications delay with Earth (because of the finite speed of radio waves), but the loss of real-time communications meant that the ground controllers were not receiving their 'comfort signals' from the lander showing that every thing was proceeding normally during the landing sequence. In the event everything worked fine and the lander landed only 10 kilometres (6 miles) from its chosen site. Subsequently, full communications were re-established with the orbiter, and the recorded data taken during landing was successfully transmitted to Earth.

The images sent back to Earth by the Viking 2 lander showed a similar red, rock-strewn surface to that seen by Viking 1. This was something of a surprise as the scientists had anticipated seeing gently rolling sand dunes with more dust and less rocks than for the Viking 1 site. The eagerly awaited results of the biology experiments on Viking 2 were as inconclusive as those of Viking 1. Even today, in fact, the reasons for these strange results from the biology experiments are not clear, although most scientists think that they were produced by chemical reactions rather than living organisms.

Viking 2 initially measured a minimum temperature of −87 °C during the northern summer, but during winter the minimum temperature (at dawn) fell to −118 °C. Frost was first observed on the surface in September 1977 when the temperature was −97 °C.

Over their lifetimes, the Viking 1 and 2 orbiters together produced over 50,000 images of the Martian surface, imaging virtually 100% of it to at least 300 metres resolution, with 2% being imaged at about 25 metres resolution. Later in the mission the periapsis of both the Viking orbiters was reduced to 300 km, allowing selected areas to be imaged at a resolution of just 8 metres.

The Viking images showed clear signs of flash floods in the Chryse region (see Figure 6.10), and many landslides, some of which seem to have occurred in saturated soil. Dendritic or branching drainage features were also seen resembling terrestrial river systems. Temperature and water vapour sensors on the orbiters showed that the north polar cap is made of water ice in the northern summer at a temperature of about −65 °C. It is then covered by carbon dioxide ice in the northern winter. In the case of the southern cap, however, the carbon dioxide ice does not completely melt in the southern summer. There was also evidence of extensive permafrost regions, with the permafrost around the polar caps possibly extending down to a depth of several kilometres in places.

Mariner 6 had shown that the south polar cap temperature was about −125 °C, which is the equilibrium temperature of carbon dioxide at the south polar atmospheric pressure of about 6 millibars. The infrared thermal mapper on the Viking orbiter indicated a south polar temperature as low as −139 °C, however, which was difficult to explain unless there was a higher concentration of nitrogen and argon at the south pole than elsewhere on Mars. Nitrogen and argon would not condense at this temperature, and so they could easily provide the extra cooling, but it was not clear if there were higher concentrations of these gases than normal at the south polar region. So scientists could only speculate.

Viking also measured much larger daily temperature ranges than previous spacecraft. In some places, for example, the temperature went from a daily maximum of 4 °C to a minimum of −133 °C, which was difficult to explain. The daily thermal profiles of other areas did not match predictions either, falling more rapidly than expected in the late afternoon, before slowing down to match the predicted temperatures during the night.

The water vapour in the Martian atmosphere was found to be highest in low-lying regions and more water vapour was found during the summer than during the winter, when it was presumably frozen out of the atmosphere. In regions of rough terrain there were marked daily fluctuations in the amount

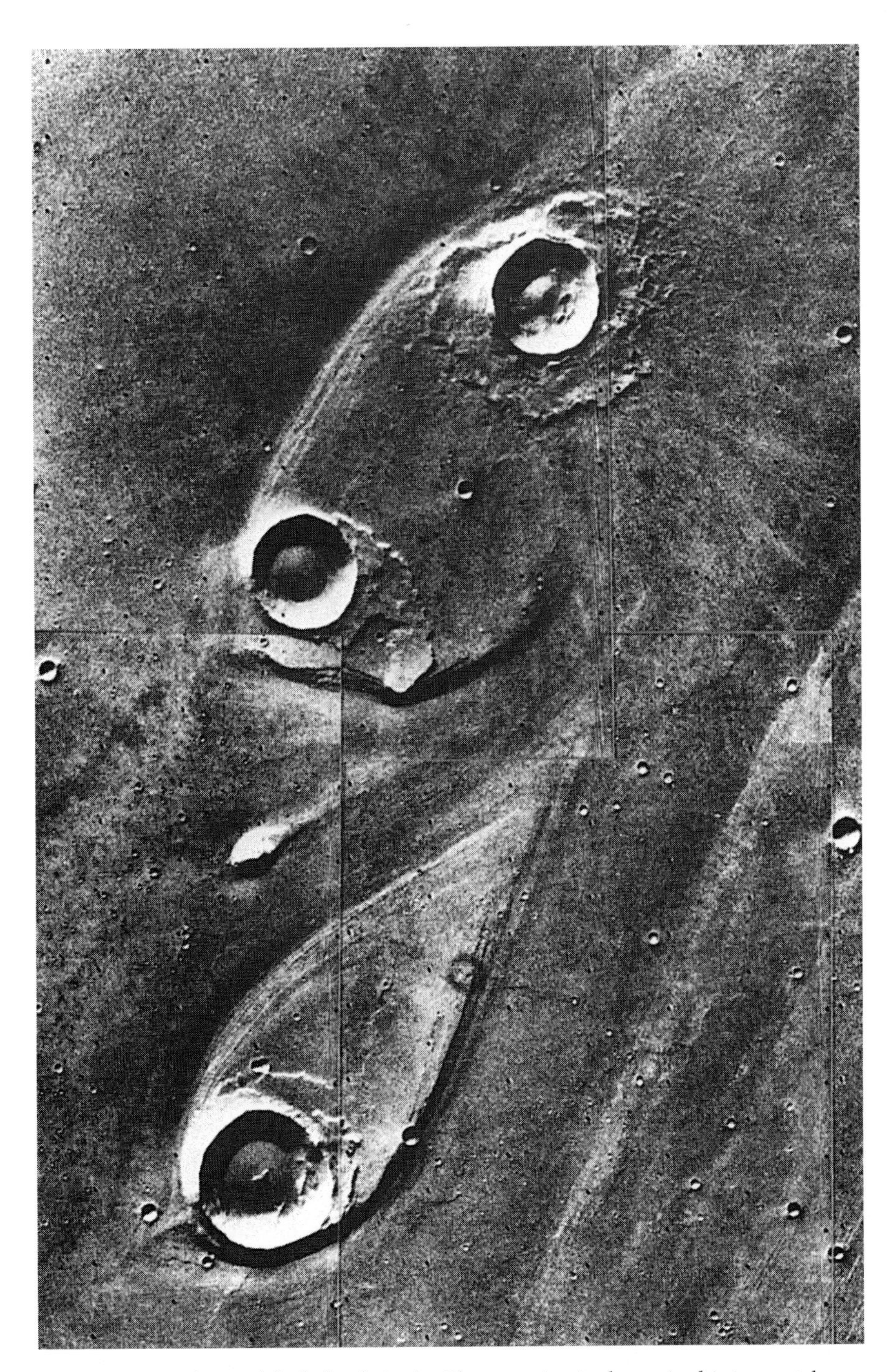

Figure 6.10 Evidence of flash floods in the Chryse region is shown in this image taken by the Viking 1 orbiter. (Courtesy National Space Science Data Center, World Data Center-A for Rockets and Satellites, NASA; Experiment Team Leader, Dr Michael H. Carr; 76.H.480.)

of water vapour, possibly due to changing wind patterns. The atmosphere above the residual northern polar cap was found to be saturated with water vapour in the northern summer. Although the amount of water vapour varied on any particular part of the surface with time, the total amount in the Martian atmosphere stayed constant over the course of a Martian year.

The Viking 2 orbiter failed in July 1978 after about two years in orbit, but the Viking 1 orbiter continued to return images for another two years. The landers lasted even longer, with Lander 2 being shut down in April 1980, and Lander 1 failing in November 1982 when an incorrect command sent by a ground controller accidentally shut it down.

All-in-all the Viking missions were a spectacular success, with the landers, in particular, lasting over ten times longer than their specified lifetimes of 90 days. This highly successful programme clearly showed what could be done, and what could not be done, with un-manned planetary probes. Although the Viking programme cost an insignificant amount compared with the cost of a manned mission to Mars, it still cost a massive $2.5 billion in 1984 dollars (or $4.0 billion in 1998 dollars).

6.7 Veneras 9 and 10

Two months before the Americans launched their first Viking to Mars, the Russians had launched two new spacecraft, called Veneras 9 and 10, to Venus. It may be remembered that the Russians had passed up the opportunity to launch Venera probes in the 1973 launch window, preferring instead to concentrate on producing a brand new spacecraft design for the 1975 launch window. Veneras 9 and 10 were their new designs.

Venera 9 at 4,936 kg was very much heavier and more sophisticated than any previous spacecraft sent to Venus. It consisted of a spacecraft bus, based on the design of the Mars 4 spacecraft, surmounted by a 2.4 metre diameter aeroshell inside which was the lander. After releasing the lander, the main spacecraft would orbit Venus in a highly eccentric orbit and initially act as a relay for the lander, because the landing site was on the daylight side of Venus, which at that time was on the far side of the planet as seen from the Earth. After acting as a relay, the orbiter would undertake extensive observations of the Venusian cloud system using a panoramic camera. In addition, an infrared radiometer on the orbiter would measure cloud-top temperatures, an infrared spectrometer would analyse atmospheric gases, and a photopolarimeter would measure the intensity and polarisation of sunlight scattered by the cloud particles, enabling an estimate to be made of their sizes.

The spherical aeroshell was designed to protect the lander during its initial deceleration in the dense Venusian atmosphere from a velocity of 23,900 mph (10.7 km/s) to 560 mph (0.25 km/s). The aeroshell would then be discarded at an altitude of 40 miles (64 km), and the 1.0 metre diameter lander would be further retarded by two sets of parachutes and by an aerobraking disc which surrounded the spherical lander like the brim of a hat. Finally a crushable base below the lander would cushion the landing.

Venera 9 was launched to Venus by a Proton vehicle on 8th June 1975, and six days later it was followed by its twin, Venera 10. The refrigeration system on the Venera 9 orbiter cooled the inside of the lander shortly before releasing it on 20th October as it approached Venus, and two days later the lander entered the Venusian atmosphere. At the same time, the orbiter undertook a braking manoeuvre to enter into Venus orbit, the first such spacecraft to do so, and the orbit was finally adjusted to become a 1,510 × 112,200 km orbit with a period of 24 hours. The lander, meanwhile, had touched down about 2,500 metres above the mean Venusian surface level on what is now known to be the eastern slope of the shield volcano Rhea Mons. Data was received on Earth via the orbiter for 53 minutes until the orbiter disappeared below the lander's horizon. Three days later the Venera 10 lander also touched down about 2,200 km from Venera 9 at the foot of another shield volcano now called Theia Mons. Data was received on Earth for 65 minutes until the Venera 10 orbiter also disappeared below the lander's horizon.

The atmospheric data received from the landers, as they descended through the atmosphere, showed no great surprises, and generally confirmed the data received from Venera 8, although the new data was more detailed and accurate. In particular the vertical profile of the various cloud layers was more clearly defined, with three layers being detected centred on heights of 64, 55 and 51 kilometres This compares with the Earth where most of the clouds (and most of the atmosphere) are below 10 kilometres altitude. The clouds on Venus were found to be relatively transparent and their opacity as seen from Earth was found to be due to their great depth. At ground level the measured atmospheric temperatures and pressures were now no longer a surprise at 460 °C and 90 bar (Venera 9) and 465 °C and 92 bar (Venera 10).

On the surface the γ-ray spectrometers measured the chemical composition of the rocks, which appeared to be similar to slowly cooled basalt, confirming the volcanic origin deduced from the Venera 8 data. But the most excitement was generated by the receipt of a single monochromatic photograph of the surface from each of the landers.

The Russians had been concerned that there may not be enough sunlight percolating through the thick Venusian clouds to enable clear photographs to be taken of the surface, so they provided floodlights to illuminate the scene. In the event, although the floodlights were used, they created no obvious shadows so they had not been really required.

It had generally been thought that the surface of Venus would probably look like a sandy desert produced by billions of years of wind[21] and heat erosion, but the photographs taken by the Venera 9 and 10 landers showed a very different scene consisting of numerous rocks with little or no sand. Detailed analysis indicated that the surface shown in the photographs is relatively young and probably still active.

In the meantime, whilst the Russian scientists were analysing the landers' data and photographs, the orbiters were continuing to transmit their own scientific data to Earth. This showed that the cloud particles were definitely not water droplets, although they were liquid droplets of some sort and, somewhat bizarrely, the cloud-top temperatures on the night side of the planet were found to be higher than those on the day side. The Venera 10 radar altimeter showed that the surface of Venus was relatively flat, varying by only a few kilometres in elevation from a perfectly smooth surface.

Some years earlier the Venera 5 landing module had detected what was thought to be lightning on its descent through the atmosphere, but as no such detection was observed with subsequent probes the data was considered to be erroneous. But both the Venera 9 and 10 landers detected radio signals during their descent, which were tentatively attributed to lightning, and a spectrometer on board the Venera 9 orbiter also detected what appeared to be very intense lightning over a seventy second period on 29th October. Whether these were true lightning observations or not is still not clear, however.

It had been known since the seventeenth century that, when the phase of Venus is new (i.e. when only a thin crescent is seen to be illuminated from Earth), the dark part of the disc sometimes appears to be slightly illuminated by what is called the 'ashen light'. The existence of this ashen light was disputed for many years, but it eventually became an accepted intermittent phenomenon. It was thought that it is caused by some sort of auroral discharge, or by the scattering of sunlight in the Venusian upper atmosphere producing an extensive twilight. In 1967 an ultraviolet photometer on Mariner 5 had detected a faint ultraviolet glow on the night side of the planet which was thought by many to be the ashen light. Then the Venera 9 and 10

[21] Although the wind velocity at the surface of Venus is very low, the density of the atmosphere is very high, so wind can still cause significant erosion.

orbiters showed that there was a night-side airglow caused by molecular oxygen in the upper reaches of the Venusian atmosphere.

6.8 **Pioneer-Venus**

The last American mission to Venus had been Mariner 10 launched in 1973, but this was really a mission to Mercury that happened to fly by Venus *en route*. Prior to that the last American mission to Venus had been Mariner 5 in 1967. But it was not until 1978 that the USA was ready once more to visit Venus with what was called the Pioneer-Venus spacecraft.

The origins of the Pioneer-Venus missions can be traced back to August 1968 when the Space Science Board of the National Academy of Sciences recommended that a multiple drop-sonde or 'multiprobe' mission be sent to Venus in 1975. Work continued at a relatively low level on this concept, until in early 1972 a Pioneer-Venus Science Steering Group was set up by NASA to consider the various options for Venus missions in the late 1970s. The Group recommended that two identical multiprobe spacecraft, each carrying one large and three small probes, be launched in 1976/77, followed by an orbiter in 1978 and a third probe in 1980. By August 1972 this programme had been reduced to one 380 kg multiprobe spacecraft to be launched by a Delta in 1976/77 and one orbiter to be launched in 1978.

The problem with the August 1972 programme was that it would involve using a Delta launcher which, although highly reliable for Earth-orbiting missions, had never been used for planetary missions. In addition, a Delta could only launch a relatively small multiprobe spacecraft and this would involve considerable costs in miniaturising spacecraft units, as each of the three small probes could only carry 1.5 kg of scientific equipment. The obvious solution was to use a larger launcher that had already been used for planetary missions, so avoiding the costs of miniaturisation, provided this solution could be justified from a total programme cost point of view. By June 1973 NASA had settled on this approach, replacing the Delta with the more performant Atlas-Centaur, which had been already been used for planetary (and lunar) missions.

In August 1974 Congress approved the programme of launching both the multiprobe and orbiter spacecraft on separate Atlas-Centaurs in 1978, within a cost ceiling of $250 million. Hughes Aircraft were then awarded the Spacecraft hardware contract later that year. So the 1978 Venus programme seemed to be on course with all the necessary programme approvals in place, and with the hardware contract awarded, but the American budgetary system has a strange quirk. Although a programme may be approved by Congress, its annual funds still need voting every year, and in June 1975 the House of Representatives voted to cut $48 million from the Pioneer-Venus' 1976 budget, which would have effectively killed the project. The scientific community were incensed that such a valuable programme could be cancelled when over $50 million had already been spent or committed, and they started an extensive lobbying campaign. As a result the money was restored by Congress.

Pioneer-Venus 1, otherwise known as the Pioneer Venus Orbiter or PVO, had a mass of 553 kg, including a 179 kg retrorocket which was designed to put the spacecraft into a 24 hour orbit around Venus, so as to synchronise it with tracking stations on Earth. The key scientific instrument on board the PVO was a radar that would 'see' through the clouds and provide the first detailed map of the surface of Venus,[22] with a maximum surface resolution of 20 kilometres and an altitude resolution of 100 metres.

The 875 kg Pioneer-Venus 2, otherwise known as the Pioneer Venus Multiprobe, consisted of a carrier bus plus four atmospheric probes, none of which was designed to survive surface impact. There was one 316 kg probe and three identical 90 kg probes, all four being designed to study the atmosphere from a height of 200 kilometres to the surface. The probes were to be targeted to enter the atmosphere at radically different locations, so that their results could be compared to give some idea of the atmospheric dynamics.

The large probe consisted of an ejectable conical heat shield that was used for the initial atmospheric deceleration, and a spherical titanium pressure vessel containing the scientific instruments. The pressure vessel would deploy a parachute at an altitude of about 50 km, once the heat shield had been ejected, to further slow its descent. The three small spherical probes, on the other hand, retained their conical heat shields and had no parachutes. Their 3.5 kg of instruments consisted of a net-flux radiometer to measure radiant energy (heat and sunlight), a nephelometer to examine cloud structure, and instruments to measure temperature, pressure and wind speed. Quite a collection in such a small mass. The large probe had a much more sophisticated set of experiments, including a mass spectrometer and gas chromatograph to measure the composition of the atmosphere, as well as instruments like those in the small probes.

---

[22] Although a radar map had already been produced using the 305 metre diameter, Earth-based, Arecibo dish, this map was relatively crude and limited only to a small part of the Venusian surface.

In many ways the best trajectory from Earth for a spacecraft designed to orbit another planet is one that requires the minimum velocity[23] on leaving Earth as, when the spacecraft reaches the planet, there is less braking required to inject the spacecraft into planetary orbit and hence a relatively small retrorocket can be used. For a spacecraft that is to enter the planet's atmosphere directly, however, the requirements are not as stringent, provided the probe can withstand the extra heat generated by a higher velocity. For these reasons the Pioneer Venus Orbiter (PVO) was launched towards Venus some two and a half months before the Multiprobe, although they arrived at the planet within days of each other.

Lift-off for the PVO took place on 20th May 1978 followed by that of the multiprobe spacecraft on 8th August. The PVO's on-board computer turned out to be over-sensitive to the radiation environment of space but, apart from this, both spacecraft performed nominally *en route* to Venus. On 16th November, when the multiprobe spacecraft was still 11 million kilometres from Venus, the large probe was released, and four days later the multiprobe spacecraft was spun up to 48 rpm and the three small probes released. This ingenious manoeuvre was a simple but effective way of ensuring that the three small probes all impacted Venus at widely different points.[24]

On 4th December 1978 the PVO retrorocket was fired to place the PVO into orbit around Venus, which was adjusted over the next two weeks to be a $150 \times 66{,}900$ km orbit inclined at 105° with a period of 24.13 hours. Imaging of Venus' clouds took place around apoapsis, when the whole planetary disc was visible, whereas radar mapping of the surface took place around periapsis to give maximum surface resolution.

Five days after the PVO started orbiting Venus, the four probes from the multiprobe spacecraft entered the atmosphere at four minute intervals, each taking about fifty minutes to reach the surface. These probes confirmed the theoretically based predictions made in 1973 by the Americans Godfrey Sill, Andrew Young and Louise Young that the Venusian clouds were composed mainly of corrosive sulphuric acid. So finally the true nature of the beautiful

[23] This minimum velocity consideration for planetary orbiters is of more importance for relatively close planets like Venus than for more distant ones, where transit time can be more important.

[24] Clearly if they had been ejected one after another every few minutes with the same velocity from a non-spinning spacecraft, they would have followed each other into the Venusian atmosphere at virtually the same location, as the rotation rate of Venus is very slow.

evening star as seen from Earth had been revealed. Not only was its surface hot enough to melt lead, under an atmospheric pressure of ninety times that of Earth, but its clouds were made of sulphuric acid. Maybe its name should be changed from Venus to Hades.[25]

Over the next two years the PVO radar mapped 93% of the surface with a resolution of at least 30 km and with a height resolution of at least 200 m. The result showed a relatively flat surface, with 65% being within ±500 m of the mean level, although there were exceptions like the 10,800 m (35,000 ft) high Maxwell Montes[26] and the 2,900 m (9,500 ft) deep Diana Chasma. Two large elevated plateaux were found (see Figure 6.11), namely Ishtar Terra (named after the Babylonian goddess of love) in the far north, and Aphrodite Terra (named after the Greek fertility goddess) just south of the equator. Ishtar Terra was found to have in its western half a 2,500 km plain, called Lakshmi Planum, that rises about 4,000 m (13,500 ft) above the rolling lowland plains. In many ways Lakshmi Planum, which was found to be bordered on two sides by towering mountain ranges including Maxwell Montes, resembles the Tibetan Plateau, which is bordered by the Himalayas including Mount Everest.

Infrared images from the PVO showed a circumpolar collar[27] of very cold air, about 2,500 km from the pole, 1,000 km across and 10 km deep. The collar, which is at an altitude of about 70 km, is just above the clouds and is about 30 °C colder than the atmosphere on either side. Surprisingly, at an altitude of 85 km the PVO found that the Venusian atmosphere is warmer at the poles than at the equator. The PVO also measured a slow reduction in sulphur dioxide in the atmosphere with time, which was attributed by some scientists to a major volcanic eruption, although there was no other evidence of such an eruption.

6.9 Veneras 11 and 12

The two American spacecraft were not alone in their trip to Venus in 1978, as on 9th and 14th September the USSR launched Veneras 11 and 12.

---

[25] The Greek idea of Hades (hell) was of a dark place. Gedenna would probably be a better name as it is the fiery hell of Judaism.

[26] The Maxwell formation had been detected by radar from Earth (and called simply 'Maxwell') but its nature as a chain of high mountains was first resolved by the PVO.

[27] A similar dark collar had been detected previously in the ultraviolet by Mariner 10 and by some ground-based observers.

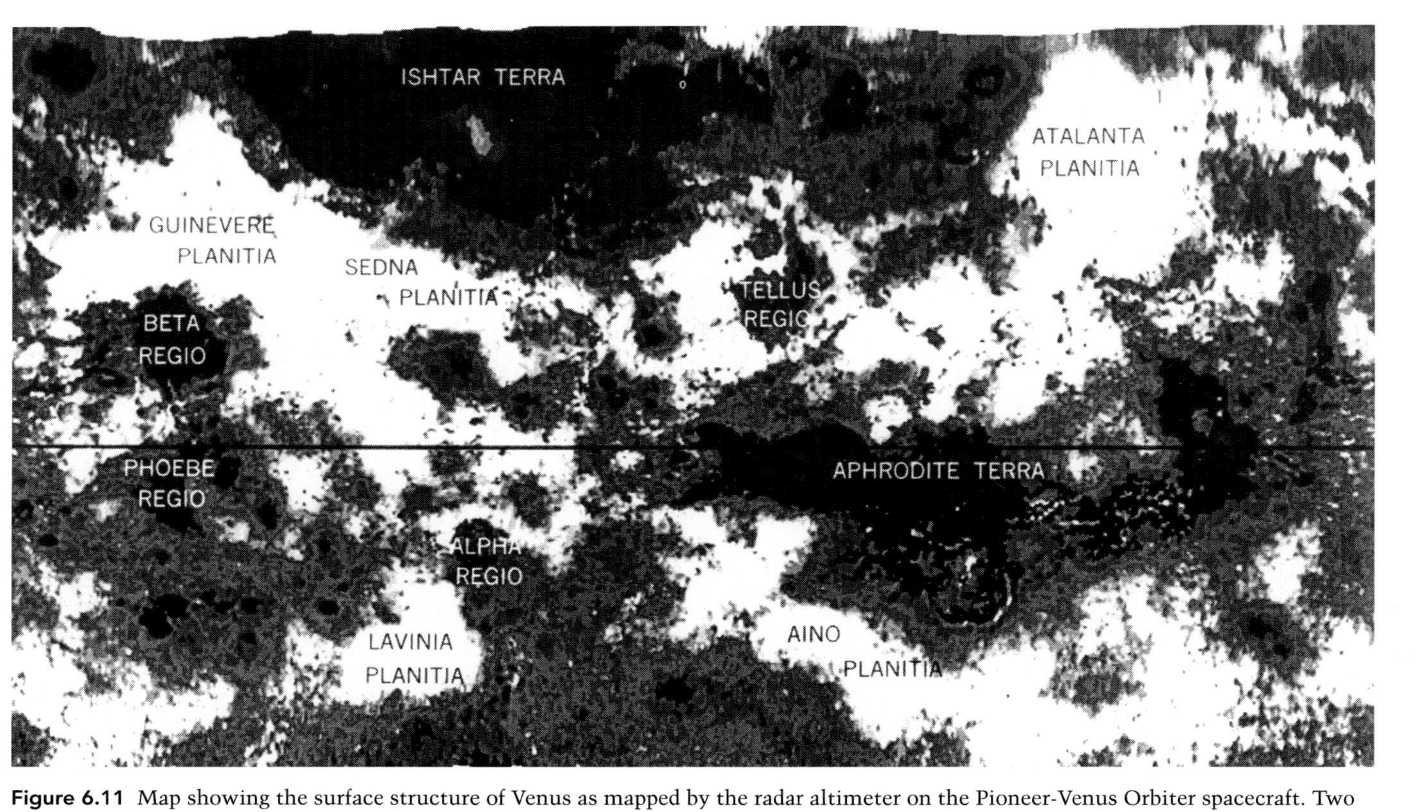

Figure 6.11 Map showing the surface structure of Venus as mapped by the radar altimeter on the Pioneer-Venus Orbiter spacecraft. Two large elevated plateaux called Ishtar Terra in the north, and Aphrodite Terra centred just south of the equator (shown by the line) are clearly seen. The lighter coloured areas (called planitia) are low-lying areas. The map covers an area from about 75° north to about 65° south. (Courtesy MIT, U.S. Geological Survey, and NASA.)

Superficially, these new Venera spacecraft were similar to Veneras 9 and 10 but, because the orbital geometry was not as favourable this time, their masses had had to be reduced by about 500 kg or 10%. As a result, the orbiter part of each spacecraft was replaced by a fly-by spacecraft, although both of the new Veneras still contained a landing module.

The new Russian landers contained a broader range of scientific instruments than their illustrious predecessors, including a camera system. Unfortunately, a design fault meant that the camera covers could not be ejected on either lander when required and so no images were returned from the surface. The soil analyser also failed on both landers.

Much of the data returned by these landers confirmed the information previously transmitted by the Venera 9 and 10 landers, but there were also new results. The droplets in the upper-level clouds were found to be composed mostly of chlorine, although those in the other clouds were confirmed as being largely sulphuric acid. Heat balance measurements also indicated that the heat that Venus radiates into space comes mostly from the upper-level clouds. Sulphuric acid aerosols were found which form during daylight and disappear at night, and which inhibit heat flow into space during daylight hours from lower levels in the atmosphere. These aerosols were the reason why the Veneras 9 and 10 had found that the cloud-top temperatures on the night side of Venus are higher than those on the day side

The relative proportions of argon 36 and argon 40 were recognised in the 1970s as key measurements in understanding the formation of planetary atmospheres. This is because argon 36 was thought to have come from the original solar nebula from which the planets formed, whereas argon 40 is produced on Earth by the radioactive decay of potassium 40 deep inside the Earth, being brought to the surface by volcanoes or through fissures caused by tectonic motion. On Earth there is much more argon 40 compared with argon 36 because of this volcanism and tectonic motion. It was, therefore, highly significant that the Venera 11 and 12 landers found that the ratio of argon 36 to argon 40 in their atmospheric samples was close to unity. This relatively high level of argon 36 on Venus was thought to indicate that its atmosphere was generally formed from the original solar nebula. The fact that Venus also has a significant amount of argon 40, however, indicated that there must also have been some planetary outgassing produced by volcanism or tectonic motion.

In absolute terms there is more argon 36 on Venus than on the Earth which, in turn, has more than on Mars. This is contrary to what had been expected before the launch of Veneras 11 and 12, as it was thought that the solar wind would have been strong enough to strip the inner planets of their original atmospheres, so the above Venus/Earth/Mars sequence should have been the reverse. Not for the first time had astronomers, trying to understand the formation of the Solar System, had to tear up their theory and start again.

Chapter 7 | VENUS, MARS AND COMETARY SPACECRAFT POST-1980

7.1 International Collaboration

The early period of space exploration was a time of competition between America and the USSR, but this changed once Neil Armstrong had set foot on the Moon in 1969, having decisively beaten the Russians to this key psychological target. In the same year President Nixon took over the American presidency and immediately started withdrawing troops from Vietnam, and in 1971 he announced that he would visit China in the following year. This was a time when the Cold War started to thaw, and Nixon was keen to undertake any high-profile projects that would aid détente, with the USSR in particular. One such 'political' project was the manned Apollo–Soyuz Test Project, which was developed ostensibly to test compatible docking systems between Russian and American manned spacecraft. This was in case of emergency, to enable one country to rescue the other's astronauts. Some less well known areas of cooperation were also undertaken in this period, including that between the USA, the USSR, ESRO[1] and Japan on a world-wide, geosynchronous meteorological satellite system, and the first tentative steps at collaboration between the USA and the USSR on Solar System exploration.

The first significant cooperation between the USSR and the USA on planetary missions took place whilst the Venera 11/12 and Pioneer Venus spacecraft were all *en route* to Venus at the same time in 1978. In this case, in-flight data were exchanged between both countries to help the missions

[1] The European Space Research Organisation, which was one of the two forerunners of the European Space Agency, ESA. The other was ELDO, the European Launcher Development Organisation.

of both pairs of spacecraft. Unfortunately, plans for further cooperation had to be reduced following the Soviet invasion of Afghanistan the following year. Nevertheless, the Americans provided the Russians with Pioneer-Venus radar data to assist in their planning of the next Venus-bound spacecraft, Veneras 13 and 14, which were due to be launched in 1981. As a result, the Russians moved the planned landing point of Venera 13 to an area where the rolling plains were thought to be made of granite, leaving Venera 14 to investigate a lowland area, some 900 kilometres away, that was thought to be flooded with lava.

7.2 **Veneras 13 to 16**

Both Veneras 13 and 14 consisted of a fly-by spacecraft and a landing module. The landers each contained a significant number of new instruments compared with the Venera 11/12 landers, including a drilling and analysis system to analyse surface material, a penetrometer to measure the load-bearing characteristics of the planetary surface and a new panoramic camera system with higher contrast, a smaller pixel size, and a colour capability. Soil analysis of the drilled sample was undertaken by an X-ray fluorescence spectrometer, which was much more sensitive than the previously used γ-ray spectrometer.

The Venera 13 spacecraft was launched on 30th October 1981 and its sister spacecraft, Venera 14, followed five days later. Venera 13 ejected its landing capsule on 27th February 1982, which touched down on Venus two days later. It survived for a record 127 minutes on the surface, communicating with Earth via the fly-by spacecraft. The Venera 14 lander followed exactly the same sequence four days later.

En route to the surface, the gas chromatograph on board the landers detected both carbonyl sulphide (COS) and hydrogen sulphide (H_2S) in the Venusian atmosphere. The discovery of COS was particularly important as it provided a key to understanding the formation of the atmosphere. More refined argon and water vapour data were also provided by the landers, compared with their Venera 11/12 predecessors, and the three cloud layers previously observed were also detected down to an altitude of 49 km (Venera 13) and 47.5 km (Venera 14) where the atmosphere became relatively transparent.

The images transmitted of the Venera 13 landing site (see Figure 7.1) showed a sandy landscape littered with sharp rocks, with the dust that was blown onto the bottom of the lander moving between successive images. The site looked similar to those of Veneras 9 and 10, whereas the Venera 14

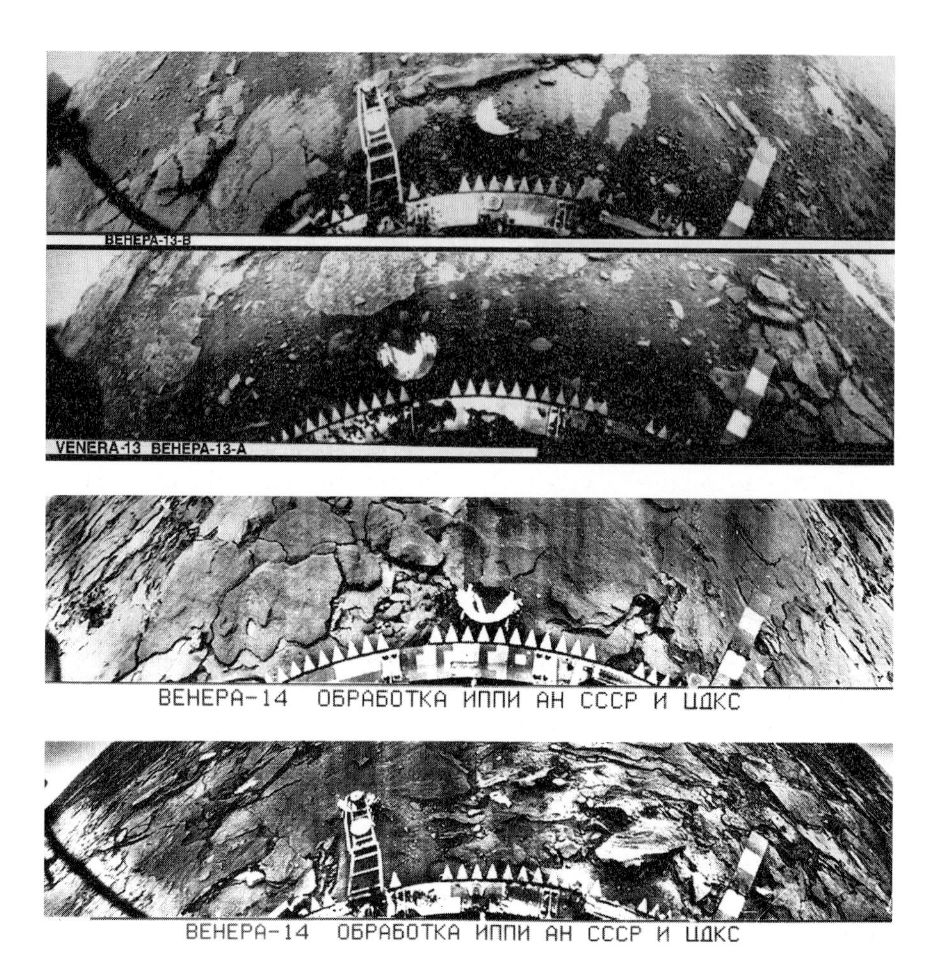

Figure 7.1 Images of the surface of Venus transmitted by the Venera 13 lander (top pair of images) and Venera 14 lander (bottom pair). The images are distorted because the cameras (one on each side of the landers) panned in an arc-like motion. The sawtooth structure at the bottom of the lander is the landing ring, and the crescent-shaped objects on the surface are the camera covers. Both landing sites look remarkably flat, with more dust and pebbles at the Venera 13 site, and large flat expanses of rock at the Venera 14 site. (Courtesy NASA and NSSDC; Venera 13 Lander, YG 06847 and Venera 14 Lander, YG 06848.)

site (see Figure 7.1 also) looked different with much less fine grained material and with rocks of a more rounded shape, indicating that they were older. Both of the new sites showed layering of rocks but the evidence was clearer for the Venera 14 site.

Just half a minute after touch down, the Venera 13 lander started drilling the surface to collect a soil sample. The 1 cm^3 sample was then passed through a complex series of airlocks to gradually reduce the pressure and temperature from 90 bar and 460 °C outside to 0.05 bar and 30 °C inside. The sample's composition was then determined by X-ray fluorescence. The Venera 14 lander followed the same procedure.

Geologists concluded that the material in both of the Venera locations is basalt. That in the rolling plains region of Venera 13 appears to be like leucitic basalt on the Earth, which is rich in potassium and is often found on the slopes of volcanoes, and that in the lowland region of Venera 14 appears to be like terrestrial tholeiitic basalt, which is found on the Earth's ocean floor.

The Americans had produced very useful data on the overall surface topography of Venus using a simple radar altimeter on the PVO (Pioneer-Venus Orbiter) in 1979, but since then the performance of the Earth-based Arecibo radio telescope had been substantially improved to produce higher-resolution radar images than the PVO. The PVO had not imaged the polar regions, however, and these regions could not be satisfactorily resolved from Earth, so there was a gap in our knowledge here. In 1983 the Russians rectified this for the north polar region with the launch of two radar-carrying spacecraft, Veneras 15 and 16.

The radar system carried on the new Russian spacecraft was much more sophisticated than the simple altimeter carried by the PVO. The Russian device was a Synthetic Aperture Radar or SAR of the type first launched by the Americans in 1978 on their Earth observation Seasat spacecraft. With a radar altimeter the distance of a small area of the planet (and hence its altitude) is determined by the time delay between the transmitted and returned signal from and to the spacecraft. So with one pulse emitted by the spacecraft in a given direction the altitude of just one area of ground is determined. In a SAR, however, an *image* can be reconstituted of an area of ground from the signals returned from just one pulse, as both the time and frequency of the returned signals are recorded and analysed. The problem with SARs, however, is that they require a large spacecraft antenna, they produce an enormous amount of data which has to be transmitted to Earth, and they use a fair amount of power. Because of this the new Russian Veneras did not carry a lander, replacing their normal landing module with the SAR. This radar had a surface resolution of about 2 kilometres, which was about ten times better than that of the PVO radar altimeter system.

The SAR antenna was a 1.4×6 metre planar structure which, like the solar array, was folded during launch. Data were to be transmitted to Earth every 24 hours at a rate of 100 kbps, which was 30 times faster than for the previous Veneras, and the solar arrays were doubled in size to provide the extra power required by the SAR. Extra fuel was also provided for the retro-rocket as the whole of the spacecraft had to be slowed down to go into orbit around Venus. In addition to the SAR, Veneras 15 and 16 carried a radar altimeter that had a height resolution of 50 metres (compared with 100 metres for the PVO), an infrared spectrometer, a radio occultation experiment, and instruments to measure conditions in space *en route* to Venus.

Venera 15 was launched on 2nd June 1983, with its sister spacecraft being launched five days later. At arrival at Venus, both spacecraft were put into a 1,000×65,000 km orbit around the planet, inclined at 87° to the equator, with a period of 24 hours. The imaging was produced in one 150×9,000 km strip each day centred on the orbital periapsis of 60° N, which meant that imaging was limited to the northern hemisphere down to a latitude of about 18°. The two spacecraft operated in tandem, with one spacecraft recording data whilst the other was transmitting and vice versa. The orbital path of each spacecraft moved 1.5° eastwards each day, as Venus rotated beneath them, so a complete 360° longitudinal coverage for each spacecraft took 8 months, which was their nominal lifetime.

The images returned by Veneras 15 and 16 showed many ridges and narrow valleys which appear to have been formed by the horizontal motion of the crust. Several ten to forty metre diameter craters were found in Lakshmi Planum,[2] but few craters were seen in the lowland regions, suggesting that they may be young lava plains. In fact crater counts in some lowland regions gave an age as young[3] as about 600 million years. The 100 kilometre diameter summit caldera or impact crater on Maxwell Montes was clearly seen, together with two unusual circular features called Anahit Corona and Pomona Corona, which were thought to have been formed by a mixture of both volcanic and tectonic processes. These two large corona structures and thirty smaller ones are unique to Venus as far as we know. A large area of intersecting ridges and grooves was found to the east and north-west of Maxwell Montes, in the upland region of Ishtar Terra. At first, this type of terrain, where the intersecting ridges and grooves produced both diagonal and random patterns, was called 'parquet',[4] but this term was thought not to be scientific enough and the term 'tessera' (Greek for tile) is now used instead.

---

[2] Lakshmi Planum and Maxwell Montes are in Ishtar Terra, see Section 6.8 above.

[3] Young compared with the age of Venus of about 4.6 billion years.

[4] The first examples of this type of surface produced diagonal patters reminiscent of parquet flooring. Later on the Veneras also found areas of more random patterns.

7.3 Planning The Halley Encounters

The next Venus-bound spacecraft were also Russian, but they were also to fly-by Halley's Comet, which was due to swing by the Sun in 1986, 76 years after its previous visit. The Russian spacecraft called Vega[5] 1 and 2 were part of a major international effort to visit Halley, which included two Japanese spacecraft called Suisei and Sakigake,[6] a European spacecraft called Giotto[7] and, from a distance of 28 million kilometres, the American International Cometary Explorer or ICE spacecraft (see Figure 7.2). We must now retrace our steps, however, to understand why America did not have a real Halley fly-by spacecraft, and the origin of this international collaboration between the USSR, Japan, Europe and the USA.

American plans to intercept Halley's comet were, unfortunately, formulated in the mid 1970s, at a time of peak funding for the Space Shuttle, which was being plagued by programme delays and cost overruns. NASA was also seeking approval during the same period for the Hubble Space Telescope, the Galileo mission to Jupiter and the Ulysses mission to investigate the Sun,[8] and was still spending a significant amount of money on the Mars Viking mission and the outer planets Voyager mission. So it was not the best time to push for another expensive Solar System programme, although it could be argued that there never is a 'best time' to try to get a new programme approved. Not satisfied with a relatively straightforward Halley fly-by mission, however, astronomers pushed for a spacecraft that would fly with the comet for a number of months. This complex mission required an early programme decision and, when this was not forthcoming, an alternative mission was proposed of a Halley's comet fly-by in 1986 *en route* to the Tempel 2 comet in 1988. As the spacecraft flew by Halley it was suggested that it could release a small probe to fly closer to the comet. NASA proposed in 1978 that this small probe could be provided by ESA (the European Space Agency). But at that time ESA was examining alternative options.

---

[5] 'Vega' is a combination of 'Venera' and 'Gallie', the Russian for Halley.

[6] Suisei is the Japanese for 'comet' and Sakigake is the Japanese for 'pioneer'.

[7] Named after the Florentine painter Giotto di Bondone, who in 1303 painted a comet, thought to be Halley's comet, on a fresco in the chapel of the Scrovegni in Padua, Italy.

[8] At that time these were called the Large Space Telescope, the Jupiter Orbiter and Probe, and the Out-of-Ecliptic missions, respectively. The Jupiter Orbiter and Probe mission was approved in 1976, and the Large Space Telescope and Out-of-Ecliptic missions were both approved the following year. Subsequently, the American Out-of-Ecliptic spacecraft was cancelled to save money.

Figure 7.2 Four of the spacecraft that flew by Halley's comet in March 1986. (a) The European Giotto, (b) the Japanese Sakigake, (c) the Russian Vega 2 and (d) the American ICE spacecraft. The drawings are not to the same scale. (Courtesy ESA, ISAS (Japan), IKI (Russia), and NASA.)

In the 1970s BAC[9] had led a European consortium to build two spacecraft for ESA called GEOS to study the solar wind and the Earth's magnetosphere. After the launch of the second GEOS in 1978, it was suggested by Dave Link of BAC that the prototype spacecraft could be refurbished and used to visit a comet, exploring the solar wind on the way. The timing of this suggestion was very appropriate as ESA had just started to examine alternative strategies for a fly-by mission to Halley's comet, including par-

[9] The British Aircraft Corporation, now part of British Aerospace (BAe).

ticipating in the mission suggested by NASA. Eventually, foot-dragging by NASA led ESA to decide to 'go it alone', and then the proposed NASA cometary mission was cancelled in 1981 owing to lack of money.

Deprived of their preferred option, NASA looked at alternative, less expensive ways of visiting a comet and eventually decided to use the ISEE-3[10] spacecraft, which was already in space, to fly-by the comet Giacobini–Zinner, followed by Halley's comet. Re-named ICE, ISEE-3 was planned to fly-by Giacobini–Zinner six months before the Russian, Japanese and European spacecraft intercepted Halley's comet in March 1986. It looked as though the Americans, having been stopped from joining the Halley armada, had decided that they must still be the first to intercept a comet, even if their spacecraft was not designed for such a purpose. Whatever the reason, on 11th September 1985 ICE flew through the tail of Giacobini–Zinner about 7,800 km from the nucleus.

ICE did not have a camera or spectrometers to analyse the cometary dust, as it was not designed as a cometary fly-by spacecraft. Nevertheless, it was able to measure the electromagnetic environment of Giacobini–Zinner and to observe the interaction between it and the solar wind. As the experiments were not optimised for the cometary fly-by, however, this caused considerable difficulties when it came to analysing their results. In particular, the only experiment capable of measuring the composition of the plasma tail was that of Keith Ogilvie of NASA, but this experiment only measured the atomic mass of the ions, so not allowing an unambiguous detection of the type of ions present. In addition it took Ogilvie's experiment 20 minutes to step through its sampling sequence, which was about the entire time spent by the spacecraft inside the comet's plasma tail.

The most accepted theory of comets, at the time of the first spacecraft intercepts, had been proposed by Fred Whipple of the Harvard College Observatory in 1950. In this so-called 'dirty snowball' theory, Whipple proposed that the cometary nucleus is a relatively loose mixture of ice and dust. As the nucleus begins to approach the Sun it starts to heat up, and when its temperature exceeds the sublimation temperature of ice, the ice starts to evaporate and forms the coma or head of the comet. The water molecules are then ionised to form an ion tail. The evaporation of the ice releases the entrapped dust that also enters the head of the comet and forms a separate dust tail. It was of some interest, therefore, when the ICE spacecraft detected many water ions in the tail of the comet Giacobini–Zinner. No clear bow

---

[10] ISEE stands for the 'International Sun–Earth Explorer'. The ISEE-2 spacecraft was an ESA spacecraft, and a number of experiments on the ISEE-3 spacecraft had been provided by non-American organisations, hence the use of the word 'international'.

Table 7.1. *Achieved fly-by distances for the Halley spacecraft*

Spacecraft	Agency	Date in 1986	Fly-by distance from nucleus (km)
Vega 1	Intercosmos (USSR)	6th March	8,890
Suisei	ISAS (Japan)	8th March	151,000
Vega 2	Intercosmos (USSR)	9th March	8,030
Sakigake	ISAS (Japan)	11th March	6,990,000
Giotto	ESA (Europe)	14th March	596
ICE	NASA (USA)	25th March	28,100,000

shock[11] could be detected by ICE, but the turbulence produced by the comet in the solar wind was found to extend further from the nucleus than expected.

International collaboration in the 1970s between the USA, USSR, ESA and Japan on a world-wide meteorological satellite system was followed by similar collaboration between the same countries or organisations for the Halley spacecraft intercept missions of 1986. This informal Halley collaboration was coordinated by the so-called Inter-Agency Consultative Group (IACG), consisting of representatives from Intercosmos in the USSR,[12] ESA, ISAS[13] of Japan and NASA. It first met in Padua, Italy, in 1981, where Giotto had painted a comet, thought to be Halley's comet, on a chapel fresco. Thereafter, the Group met once a year in various locations.

One of the problems discussed at the IACG was how to improve the targeting of the ESA Giotto spacecraft, which it was hoped to send to within something like 1,000 km of the Halley nucleus (see Table 7.1). Unfortunately, it was estimated that the position of the cometary nucleus in space would only be known to an accuracy of about 1,000 km, even after an extensive set of Earth-based observations. Fortunately, the two Soviet Vega spacecraft were due to fly-by Halley a few days before Giotto at distances of about 10,000 km and photograph its nucleus. The Russian tracking system was not able, however, to locate the Vega spacecraft accurately enough in space to give ESA

[11] It was anticipated that comets would behave like planets in creating a bow shock in the solar wind.

[12] Intercosmos was, strictly speaking, an Eastern European organisation, but it was naturally dominated by the USSR.

[13] The Institute of Space and Aeronautical Science.

a good enough position fix on the nucleus, so ESA asked NASA to help with their Deep Space Network, which could achieve the required accuracy. This so-called pathfinder concept was one outcome of the IACG discussions. There were many others at a more detailed level.

The launch constraints of the Halley spacecraft were unique, as never before had a spacecraft to be launched within a relatively narrow launch window[14] and then wait for 76 years for the next opportunity if it missed that window. This thought focused the minds of the various people involved in the various Halley fly-by programmes, and helped to curtail the natural desire of engineers and scientists to improve the designs of spacecraft subsystems and units which were 'adequate for purpose', even if they were not optimal.

Halley's comet has an orbit inclined at 18° to the ecliptic and the most favourable places for a spacecraft to intercept it, from a propulsion point of view, are the two points where that orbit intercepts the ecliptic. This was to occur on 27th November 1985, when the comet was 1.85 AU from the Sun, and on 10th March 1986 when it was 0.89 AU from the Sun. The scientists preferred to view Halley when it was closest to the Sun, as that is when it would be most active, and so all the agencies decided to aim for an encounter around 10th March 1986.

Halley's comet orbits the Sun in a retrograde direction, that is it goes around the Sun in the opposite direction to the Earth and the other planets. Because of this, the closure velocity of the spacecraft with Halley's comet would be much higher than for a planetary intercept, as it would intercept the comet almost head-on.[15] In fact, the spacecraft intercept velocities were all about 70 km/s for Halley (see Figure 7.3), or about 160,000 mph, and this would be the velocity with which any dust particles emitted by the comet would strike the spacecraft.[16] Giotto would be the most vulnerable spacecraft as it was to pass closest to Halley's nucleus.

At first sight the problem of designing a spacecraft to survive its passage through a dust cloud at 70 km/s may not seem too difficult, until it is realised that the kinetic energy of a 0.1 g dust particle travelling at this velocity is the same as that of a small family car travelling at about 100 km/h (60 mph). The solution adopted for Giotto, to protect it from these

---

[14] The Giotto launch window was 22 days wide.

[15] For energy efficiency reasons spacecraft are launched from Earth to orbit the Sun in the same direction as the Earth.

[16] To put this velocity into perspective, it would only take about one minute to cross the United States at this velocity.

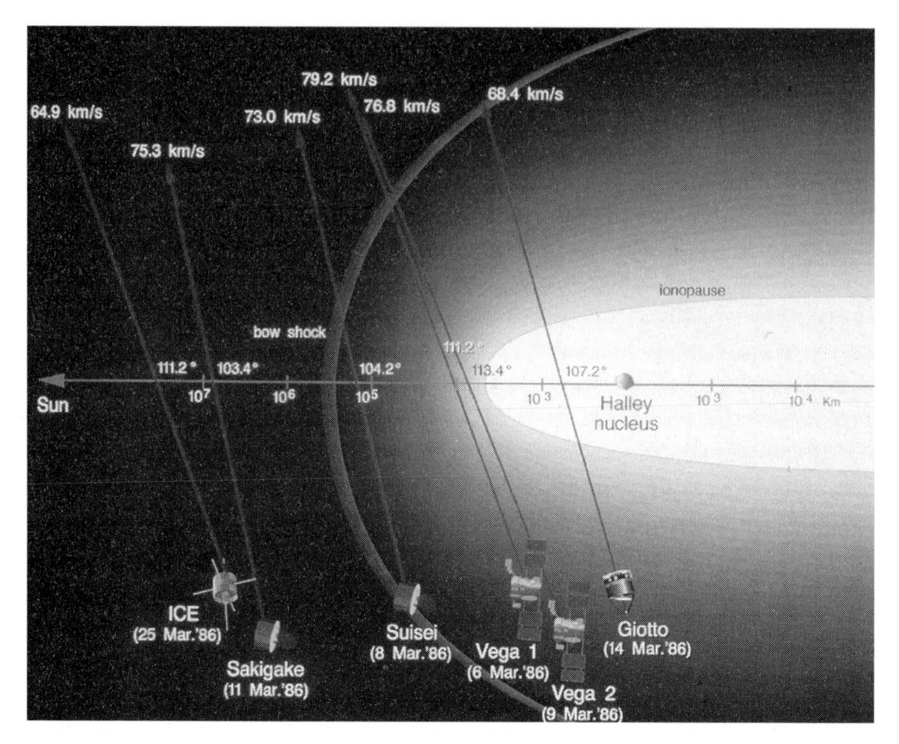

Figure 7.3 The trajectories and velocities of the spacecraft that flew by Halley's comet in March 1986 plotted relative to the comet. Note that the distance scale is logarithmic. The actual closest approach fly-by distances are given in Table 7.1. (Courtesy ESA.)

high-velocity dust impacts, was to use a two-layer bumper shield designed on the Cobham armour principle. In this the first layer of the bumper shield vaporises the dust particles, and the second layer protects the spacecraft from the resultant vapour, or from any particles that manage to penetrate the outer layer. Spacecraft engineers like to test everything in a simulated environment, if at all possible, to ensure that it will work in space. But in the case of the Giotto bumper shield this was not possible as there was no way that a dust particle could be fired at a bumper shield at 70 km/h in vacuum. So the engineers had no alternative but to test the shield at the highest velocity possible and then just kept their fingers crossed that it would work during the Halley intercept.

It was recognised before launch that Giotto may not survive its headlong rush past the Halley nucleus because of the high-velocity dust impacts, so Giotto was advertised as a kamikaze spacecraft that was designed to get as close as possible to the nucleus, although it may not survive the experience. This could be because the dust environment was much more severe than anticipated, or because there may be a problem with the design of the bumper shield that had not been evident during its limited pre-launch tests. Alternatively, a relatively large dust particle could hit the spinning spacecraft off-axis, and start it oscillating or nutating so much that the high-gain communications antenna would be no longer pointing at the Earth. Its beam would, instead, sweep back and forth like the beam from a lighthouse, illuminating the Earth for only a very short period during each oscillation. Even if there were no problem with the spacecraft as a whole, some of the experiments, such as the camera, would be especially vulnerable to dust, as they needed a direct view of the cometary nucleus, and so could only be partially protected by the dust shield.

In 1969 observations with the American OAO-2 (Orbiting Astronomical Observatory) spacecraft had shown that the comet Tago–Sato–Kosaka was surrounded by a spherical cloud of neutral hydrogen over a million kilometres in diameter. Similar clouds were found around Bennett's comet in 1970 and Kohoutek's comet in 1973. The main mission of the Japanese Suisei spacecraft was, therefore, to observe the expected hydrogen cloud around Halley's comet with an ultraviolet imager. Sakigake, on the other hand, which would not fly any closer to Halley than about 7 million kilometres, was mainly designed to examine the solar wind. These Japanese spacecraft only weighed about 140 kg each, compared with the 960 kg for Giotto or 4 tons each for the Vega spacecraft, so their missions were comparatively modest. Nevertheless, as they were the first Japanese spacecraft to visit interplanetary space, the Japanese had to build both a new launcher and a new tracking station to support their missions.

As long ago as 1967 the French scientist Jacques Blamont had suggested a joint French–Soviet Venus balloon expedition, in which a Russian spacecraft would deploy a French balloon and associated experiment package to float in the Venusian atmosphere and measure its characteristics. Since then a number of low-key discussions had taken place between French and Russian scientists on this potential mission, and this eventually led to an agreement to fly a French balloon payload on the next Russian Venera landers after the 1983-launched Veneras 15 and 16. Then in 1980 Blamont pointed out to the Russians that they could use Venus' gravity to redirect the 1984-launched Venera spacecraft to fly-by Halley's comet. The Russians leapt at the idea, but, unfortunately for Blamont, decided to replace the French balloon system with a smaller Russian-made balloon system based on the French design. Thus were born the Russian Vega spacecraft that were to fly by Venus *en route* to Halley's comet.

7.4 Vegas 1 and 2 at Venus

Vega 1 was launched on 15th December 1984, followed six days later by Vega 2. Two days before its closest approach to Venus in June 1985, Vega 1 ejected its lander and balloon payload cocooned together inside the spherical aeroshell. Its first parachute was deployed at an altitude of 64 km after the aeroshell had been jettisoned, and then seconds later the balloon canister separated from the lander at a height of 61 km. At 55 km the balloon canister's parachute opened, and at 54 km the inflation of the 3.5 m diameter radio-transparent balloon was started and its parachute released. The teflon-coated plastic balloon, which was inflated with 2 kg of helium, settled at an altitude of 54 km where the atmospheric pressure is a reasonable 0.54 bar and the temperature an equally reasonable 32 °C. The 6.4 kg instrumental payload, which was coated with a special finish to protect it from the sulphuric acid in the Venusian atmosphere, was suspended below the balloon. Its 4.5 watt transmitter communicated directly with the Earth in 270 second bursts every half hour. In order to receive the signals and track the balloon a total of 20 antennas was used around the world, including NASA's Deep Space Network and a new 70 m diameter Russian dish. This system enabled the balloon's horizontal position on Venus to be determined to within 10 km and its horizontal velocity to within 3 km/h.

Balloon 1 was deployed at a latitude of 7° N at about local midnight, and during its 47 hours of operation it returned valuable data about the atmospheric circulation in the most active cloud layer of Venus. Its instruments included a nephelometer to measure cloud particles, two temperature sensors, an anemometer to measure vertical wind speeds, and a light detector to detect lightning.

The horizontal wind velocity turned out to be an average of 240 km/h (150 mph), with downdraft gusts of up to 12 km/h (8 mph), which were much stronger than expected. The light detector did not detect any unambiguous evidence of lightning, and the nephelometer found no clear areas in the clouds. The balloon drifted over one quarter of the way around the planet in its two days of operation before the battery failed.

Two days later, Balloon 2 was released near local midnight 7° south of the equator, and operated for exactly the same time as Balloon 1. The horizontal velocity was the same as for Balloon 1, but for the first 20 hours the downdrafts for Balloon 2 were very light. After 33 hours, however, Balloon 2 came into a very turbulent area after it crossed over a 5 km high mountain peak, and this turbulence continued for another 2,000 km. Meteorologists were very surprised that a 5 km peak could have such an effect on the atmosphere at an altitude ten times as high as the mountain and for such a great horizontal distance.

Neither landers contained cameras because the timing of the missions dictated night-time landings, but both carried a drilling system and X-ray fluorescence spectrometer to analyse the surface material. During its descent through the atmosphere, Lander 1 found only two cloud layers, not the three previously observed, and found that the total cloud layer extended from about 60 to 35 km above the surface, rather than stopping at the altitude of about 48 km previously observed. Sulphur, chlorine and possibly phosphorous were found in the clouds, which contained an average of 1 milligram of sulphuric acid per cubic metre of atmosphere.

Unfortunately, the drilling system failed on Lander 1, so it was not possible to carry out the planned detailed analysis of the surface material. There were no such problems with Lander 2, however, which touched down successfully in the eastern part of Aphrodite Terra, between lowland plains to the north-west and mountains in the east. This was a new type of terrain as the previous landings had been in the lowland or rolling plain areas of Venus. The X-ray fluorescence spectrometer analysed the surface material excavated by the drilling mechanism, and found that it was similar in constituents to the material in the highland areas of the Earth and Moon. These results, when taken alongside those of previous landers, enabled the first tentative steps to be taken in deducing how the Venusian surface and crust had been formed and modified with time. The theory could only be tentative, however, because of the limited number of Venusian sites visited.

7.5 The Halley Intercepts

Whilst the Vega 1 and 2 balloons and landers were returning their scientific data, the main Vega 1 and 2 spacecraft had had their orbits modified by their swing-by of Venus to enable them to fly-by Halley's comet. As mentioned earlier, not only were the Vegas to undertake their own measurements of the comet, they were also crucial in determining the precise location of its nucleus, so that the European Giotto spacecraft could get as close as possible during its fly-by a few days later. Both the Vega spacecraft, like the other spacecraft in the Halley fleet, were due to fly-by Halley on the sunward side (see Figure 7.3 earlier).

At 6.20 am Moscow time on 6th March 1986, Vega 1 crossed the bow shock of Halley's comet, where the solar wind is deflected around the comet's ionosphere, about 1.1 million kilometres from its nucleus. Then at a distance of 320,000 km, about an hour before closest approach, and a little earlier than expected, Vega 1 detected its first impact by dust particles. As the spacecraft drew ever closer, the concentration of ions, molecules and dust particles continued to rise, and the maximum size of the dust particles was found to increase from 10^{-13} g to 10^{-6} g. In parallel, the camera was taking images of the inner region of the comet to detect the nucleus. The images were too indistinct to be sure, but most of the experimenters viewing them in real-time concluded that the nucleus appeared to be about 4 km in diameter. This was the first time that the nucleus of a comet had been imaged. After encounter Vega 1 was still operational, although the dust impacts had reduced the electrical power from its solar arrays by about 50%.

The nucleus apparently imaged by Vega 1 was about the size expected, but it seemed much darker than predicted, and the infrared spectrometer measured an unexpectedly high surface temperature of about 330 K. This is about 100 K greater than the sublimation temperature of water ice in space, so the water ice in the nucleus must sublimate below this dark insulating surface.

Vega 2 intercepted Halley's comet three days after Vega 1. All went well until 32 minutes before closest approach, when the automatic system for pointing the camera at the brightest part of the comet, which was assumed to be the cometary nucleus, failed. Ground controllers instructed the spacecraft to turn to its back-up system, but Vega 2 was so far from Earth that it took eight minutes for the instructions to Vega 2 to be received by the spacecraft, and a further eight minutes for the signals to come back from the spacecraft to Earth indicating that all was well. If there had been a further problem, there would not have been time before closest approach to analyse and fix it, but fortunately there was not.

The nucleus imaged by Vega 2 looked quite different from the roughly circular object imaged by Vega 1, as it now appeared to be larger and elongated with a bright spot near either end. Later analysis showed that the two bright spots were due to two dust jets pointing almost exactly at the spacecraft, and the main reason for the difference in size and shape detected by the two Vega spacecraft was because the potato-shaped cometary nucleus had rotated from being end-on for the Vega 1 intercept to being side-on for Vega 2. The size was estimated to be 14×7.5 km but this was later updated to be $16 \times 8 \times 8$ km when the Vega and Giotto data were combined.

The Japanese Suisei spacecraft had had its closest approach with Halley 53 hours after Vega 1 and before Vega 2. For the previous four months its ultraviolet telescope had been observing the enormous hydrogen corona that extended some 10 million kilometres from the comet, and found that it tended to brighten at a frequency of 52.9 hours. This was similar to the rotation period of the nucleus deduced in 1985 from enhanced copies of the photographs taken during Halley's last visit to the vicinity of the Sun in 1910.

Unfortunately, it was not possible to operate both the ultraviolet telescope and the plasma instrument on Suisei at the same time so, as the ultraviolet telescope was still operating when Suisei crossed the bow shock, the spacecraft could not transmit plasma data at that time. Such data were transmitted, however, from about half an hour before closest approach to the nucleus until well after the spacecraft had passed the bow shock on its retreat from the comet. This showed that the 400 km/s solar wind of interplanetary space was reduced by about 50% on crossing the bow shock, reaching a minimum of 54 km/s at closest approach of the spacecraft to the nucleus.

At around closest approach, Suisei was hit by two relatively large particles that changed its spin axis orientation by 0.7°. This was very worrying for the Giotto experimenters, as a 1.0° change in the orientation of Giotto's high-gain antenna could break the communications link with Earth and cause the scientific data to be lost. Suisei's closest approach was 151,000 km from the nucleus, but Giotto was expected to approach to about 600 km, so Giotto should experience a much worse dust environment than Suisei. On the other hand the Vega 1 spacecraft, and later Vega 2, which had flown by at an intermediate distance, had not seen such large impacts, so maybe Giotto would be all right. Only time would tell.

The second Japanese spacecraft, Sakigake, passed some 7 million kilometres from Halley three days after Suisei, but even at this enormous distance the effect of the comet was felt. Not only was Sakigake within the hydrogen corona, but the plasma wave intensity was found to increase up to the period of closest approach and then decrease, showing that the plasma waves were connected with the comet. This effect was due to neutral atoms or molecules which are ejected by the nucleus travelling large distances before they are ionised. These so-called pick-up ions then induce the long period plasma waves in the solar wind.

Giotto, the European probe that was to get closest to the nucleus, had been launched some eight months before these events on 2nd July 1985. The data from the Russian spacecraft at their closest approaches in March 1986 had allowed a more accurate trajectory to be calculated for the cometary nucleus. But it had not been decided what the closest approach distance should be for Giotto. Some scientists wanted the closest possible approach and suggested that the spacecraft should be aimed directly at the nucleus, whereas others wanted a fly-by at a distance of about 1,000 to 2,000 km to give the spacecraft a good chance of surviving the encounter and so obtain experiment data as it retreated from the comet. Uwe Keller of the Max-Planck-Institut für Aeronomie pointed out that their camera would not be able to follow the nucleus fast enough if the spacecraft flew by any closer than 500 km. So Roger Bonnet, ESA's Director of Science, who had to adjudicate, decided that Giotto should be targeted to miss the nucleus by 500 km plus the targeting error of 40 km, to give an aiming fly-by distance of 540 km. In the event Giotto achieved a closest approach distance of 596 km.

The Americans and Russians were relatively used to sending spacecraft to fly-by or land on the Moon or planets, and so they were familiar with the tension involved, the type of problems that occur and the instant diagnosis required. Not that much can be done in the final stages, in any case, as the spacecraft is on automatic control because of the long time that it takes signals, even travelling at the speed of light, to go to and from the spacecraft. But each mission is different and the Vega 1 encounter was the first one that the Russians had had to handle in the full glare of publicity. For the Europeans, however, the night of the Giotto encounter was a unique experience, as ESA had never attempted such an intercept before. The main European team was at the ESA ground control centre in Darmstadt (near Frankfurt), but there were many other engineers and scientists spread across Europe, many of whom had worked on this project full time for years at universities and in industry.

At about 7.30 pm on 13th March 1986 Giotto crossed Halley's bow shock, 1.3 million kilometres from the nucleus. A little later the first images started to be received, and then the first dust particles struck the spacecraft whilst it was still some 290,000 km from the nucleus. Beginning at 11.00 p.m., about an hour before closest approach, the images started to appear more frequently, but they were very difficult to interpret as they were presented as coloured isophotes in which each colour represents a different intensity level.[17] At about 8,000 km from the nucleus the first dust particle

---

[17] Mrs Thatcher, the British Prime Minister at the time, was watching the live BBC television programme that, like the other European channels, also carried these images. Because they were so incomprehensible, however, she concluded that the Giotto spacecraft programme (with a UK Prime Contractor!) was a waste of money. An ESA public relations own goal. In fact these images were equally incomprehensible to the scientists themselves, and they rapidly chose the standard black and white alternatives when they were given the choice. Unfortunately for the European television audience, they had no choice and had to continue to stare at the incomprehensible psychedelic images.

penetrated the outer skin of the bumper shield, but not the inner skin. Then about one minute before closest approach Giotto crossed the ionopause 4,700 km from the nucleus, the only spacecraft to do so, where the interplanetary field fell to zero. Thirty-three seconds before closest approach one experiment instrument was damaged by dust impact, at 21 seconds two other instruments failed, and 12 seconds later the camera failed. The last complete image was transmitted at a distance of 1,930 km from the nucleus. Then 7.6 seconds before closest approach the spacecraft signal suddenly stopped, and it looked as if Giotto had been finally destroyed by the dust impacts.

It became quickly apparent, however, that the spacecraft signal was intermittent, so Giotto was oscillating or nutating owing to an off-axis impact by a large dust particle. Unfortunately, some of the experts on the BBC programme that was covering the intercept live concluded that the spacecraft had probably been destroyed, because they were not aware that an intermittent signal had been detected. What was worse, their reaction was by no means unique, and a number of television stations across Europe quickly terminated their transmissions, leaving their audience believing that Giotto was a failure. It was a public relations disaster which could have been avoided with better planning between ESA and the media. Fortunately, after a few minutes news filtered through to the BBC, who were still on air, that Giotto was nutating. Half an hour later the on-board nutation dampers had stabilised the spacecraft sufficiently for it to send back full telemetry data as it receded from the nucleus, wounded but not silenced.

The Giotto mission to Halley's comet was a complete scientific and engineering success. The spacecraft and all the experiments had worked normally until seconds before closest approach when some of the experiments had been silenced by dust impacts. The mission had been advertised as a kamikaze mission, with any post-intercept data being treated as a bonus, and some such data had been received.

Giotto had found that inside the ionopause, which was about 4,700 km from the nucleus, there was a stream of neutral molecules and cold ions flowing away from the nucleus at about 1 km/h. These neutral molecules were the agent that dragged dust particles away with them to form the coma or head of the comet. The cometary nucleus was seen to be an irregular potato-shaped mass (see Figure 7.4) covered with bumps and hollows, one of which clearly looked like a crater. Bright jets could be seen streaming towards the Sun in both the Vega and Giotto images. At $16 \times 8 \times 8$ km the nucleus was somewhat larger than some had expected, but the Vegas and Giotto found that it was also darker than predicted with an albedo or

Figure 7.4 The nucleus of Halley's comet as imaged by the Giotto camera. Two streams of ejecta are clearly seen pointing towards the Sun off the left hand side of the image. The irregular shape of the nucleus is also clear, and on the original image a number of craters can be seen on the left hand surface of the nucleus. (Courtesy MPAE, Dr H.U. Keller and ESA; 88.03.015-004.)

reflectivity of only about 4%. This makes it one of the darkest objects in the Solar System, being blacker than coal.

Halley's inner coma was found by the Vegas and Giotto to consist mainly of water, with smaller amounts of carbon monoxide, carbon dioxide, methane, ammonia and polymerised formaldehyde. It was thought that the polymerised formaldehyde may be the reason why the nucleus is so dark. The Vegas and Giotto found that the dust particles consisted of carbon, hydrogen, oxygen and nitrogen (the so-called CHON elements) and simple compounds of these elements, or of mineral-forming elements such as silicon, calcium, iron and sodium. At the time of the spacecraft encounters the water vapour production rates were about 40 tons/s (for both Vegas 1 and 2) and 15 tons/s (for Giotto) and the dust production rates were about 10 tons/s (Vega 1),[18] 5 tons/s (Vega 2) and 3 tons/s (Giotto), but even at these rates there is still enough material in the cometary nucleus to allow about 1,000 more orbits of the Sun.

One problem that all spacecraft agencies have is deciding when to retire a spacecraft. Some spacecraft solve the problem themselves by failing before the end of their nominal lifetime, but a large number are still operational at that time. It seems obvious that if an agency, or more accurately the taxpayers, have spent $100 million or more on a spacecraft mission lasting say two years, spending another $10 million or so to extend that mission by a further two years say would be money well spent. The problem is that the data returned in the second two years may be a great deal less useful than that returned in the first two years, sometimes because the spacecraft has partially failed. In addition, the agency has to ask its political masters for more money as a matter of some urgency, which usually causes difficulty because of the inertia in the financial approval system. The possibility of extending spacecraft missions is usually limited to observatory-type spacecraft that are continuously observing the cosmos, rather than spacecraft which fly-by their target(s) and then continue into interplanetary space with no further targets in view. But this problem was also visited on ESA when it was realised that Giotto could be retargeted to fly-by another comet after its successful Halley intercept.

It had been anticipated that Giotto would not survive its fly-by of Halley, but now that it had done so, the question arose of what to do with it. Orbital specialists had already calculated what other comets Giotto could visit, in case Giotto survived the Halley intercept, and had suggested the

---

[18] This relatively high figure for Vega 1 compared with Vega 2 helps to explain why it was more difficult to see the nucleus in the Vega 1 images than the Vega 2 images.

comet Grigg–Skjellerup, which it could fly-by on 10th July 1992, via an Earth fly-by on 2nd July 1990. It would take some time to come to any decision on whether to target this comet, however, so to buy time, Giotto's thrusters were operated shortly after its Halley intercept to bring it back to vicinity of the Earth on 2nd July 1990. If by then a decision had been made to fly-by Grigg–Skjellerup, then this could be achieved by using the spacecraft's thrusters when Giotto was in the vicinity of the Earth. So Giotto was put into hibernation whilst the European politicians tried to find the $12 million to fund the Grigg–Skjellerup mission.

Only 60% of the experiments were still operational after the battering that the spacecraft had taken during its Halley fly-by, with the camera being one of those that was inoperative. Whilst the money was still being sought, Giotto passed within 23,000 km of Earth on 2nd July 1990, after six orbits of the Sun, and it was then retargeted to Grigg–Skjellerup. Then the mission was finally approved in June 1991 for the Grigg–Skjellerup fly-by in July 1992.

Halley's comet visits the Sun every 76 years and it has an impressive tail during its period in the inner Solar System. Grigg–Skjellerup, on the other hand, orbits the Sun every five years and never gets further from the Sun than the orbit of Jupiter. Because of this its nucleus has suffered much more erosion than the relatively pristine Halley nucleus, and it is now much less active than Halley. So it was anticipated that the nucleus of Grigg–Skjellerup would be much smaller than the $16 \times 8 \times 8$ km nucleus of Halley, and that it would emit much less gas and dust and have less of an effect on the solar wind than the larger comet.

The position of the Grigg–Skjellerup nucleus in space would only be known to about 1,000 km at the time of the Giotto fly-by, and so there was no possibility of a carefully targeted fly-by of a few hundred kilometres, as in the case of the Halley fly-by. It was decided, therefore, to aim Giotto directly at the estimated position of the nucleus, on the basis that the chances of hitting it were very low and this would produce the closest fly-by distance.

The first evidence of the comet was found about 600,000 km from the nucleus, where pick-up ions were detected that were caused by contamination of the solar wind by cometary gas. Then about 25,000 km from the nucleus, the magnetometer and electron analyser found a remarkable series of waves in the magnetism of the solar wind, with a wavelength of about 1,000 km, which were much stronger and more clearly defined than those at Halley's comet. Just before the bow shock, water, carbon monoxide and other ions were detected, and then about 17,000 km from the nucleus Giotto passed through the bow shock itself. This compared with the 1.3 million km distance of Halley's bow shock from its nucleus, indicating that Grigg–Skjellerup was emitting only about 1% of the gas that Halley did during the Giotto fly-by. Although the camera was no longer operative, and so the nucleus of Grigg–Skjellerup could not be imaged, analysis of the results from the operational experiments indicated that Giotto had passed within about 200 km of the nucleus, even closer than to Halley.

Because of the relative position of the comet, the Earth and the spacecraft, Giotto could not fly past Grigg–Skjellerup bumper shield first. This not only provided Giotto with an increased risk of being written off by the dust, but it also meant that the Giotto dust impact detectors would not be operating optimally. Grigg–Skjellerup orbits the Sun in the same direction as does the Earth, and so the relative velocity of Giotto and the comet was a relatively 'slow' 14 km/s, however, which meant that each dust particle would have less energy on impact, and so would be more unlikely to be detected. As Grigg–Skjellerup was also emitting less dust than Halley, a smaller number of impacts were expected than the 12,000 recorded during the Halley fly-by. It was, nevertheless, something of a surprise when only three dust impacts were recorded by the dust detectors during the whole encounter phase with Grigg–Skjellerup. The three particles were relatively large, however, and confirmed the result from the Halley intercept that the ratio of large particles to smaller ones is higher than originally expected, indicating that comets have a higher ratio of dust to gas than originally thought. So rather than cometary nuclei being 'dirty snowballs' they are more like 'icy mudballs'.

Even though there was less dust, and the impact velocity was lower than with Halley, Giotto was not properly protected by the bumper shield and the spacecraft had already been damaged during its Halley fly-by. It was something of a surprise, therefore, when it was found that Giotto had come through this second cometary fly-by unscathed. It was suggested, at first, that Giotto should be sent past another comet, but this idea was eventually dropped because of the diminishing scientific returns such a mission would achieve and the difficulty of raising yet more money.

7.6 Magellan

In June 1978 NASA had launched an Earth observation spacecraft called Seasat to measure sea surface conditions using, amongst other instruments, a Synthetic Aperture Radar or SAR. In the same year NASA had also sent the Pioneer Venus Orbiter (PVO, see Section 6.8) to Venus to provide the first maps of its surface using a simple radar altimeter. Then in November 1978 NASA invited proposals from scientists to fly experiments on a Venus orbiter spacecraft which would also include a SAR similar to that flown on Seasat. To be launched by the Space Shuttle in 1984, this Venus-Orbiting Imaging Radar (VOIR) spacecraft would go into polar orbit around Venus in May 1985. Its surface resolution would be about 0.4 km, compared with the 20 km resolution of the PVO.[19]

In November 1980 President Carter gave the go-ahead for the VOIR spacecraft mission and, although the subsequent Reagan administration was initially in favour of the programme, its high cost at a time of an expensive over-run on the shuttle programme caused the VOIR mission to be cancelled. Not to be thwarted, NASA now started to look at a cheaper alternative spacecraft using as many existing designs and hardware as possible, and descoping the spacecraft by eliminating all experiment packages except the SAR and radar altimeter. A proposed high-resolution SAR mode for the VOIR spacecraft, which was to have been used on selective targets, was also eliminated.

The largest cost savings, however, were achieved by abandoning the idea of placing the spacecraft in a 300 km altitude circular orbit around Venus and accepting a highly elliptical orbit instead. Unfortunately, this not only made image processing more complicated, but it reduced the surface definition achievable except around periapsis. To save more money this new Venus Radar Mapper spacecraft, as it was then called, used the spare antenna built for the Voyager Jupiter spacecraft (see later) to act as both the SAR antenna and the high-gain antenna for communications with Earth. Further money was saved by using designs from the Galileo Jupiter mission, including that of the on-board computer, tape recorder and power subsystem.

With these modifications, initial funding of the Venus Radar Mapper programme was approved by the Office of Management and Budget in October 1983, for a launch on a Space Shuttle in April 1988. Re-named Magellan, its development was going well until, like many other American space programmes, it was severely hit by the Challenger Space Shuttle disaster of January 1986. In the case of Magellan this provided a double problem, as not only was the launch date delayed, along with all other shuttle launches, but it was also decided that the cryogenic, liquid-fuelled Centaur rocket, which was to boost Magellan from Earth orbit, was now too

---

[19] This compares with the 2 km resolution of the Venera 15 and 16 spacecraft that were to reach Venus in 1983.

dangerous to be carried in the shuttle cargo bay. The obvious solution was to launch Magellan with a Titan 3-Centaur rocket, similar to the one used to launch the Viking Mars spacecraft in 1975. But as part of the deal to get the Shuttle programme approved in the early 1970s, NASA had agreed to phase out the use of expendable launchers and launch all of their spacecraft on the shuttle, and in 1977 they decided that the Shuttle development programme was sufficiently advanced that they would buy no more expendable launchers. This decision was criticised at the time as, not only were the Americans putting all their eggs in one basket, but it was beginning to look as though a shuttle launch would be more expensive than that of an expendable rocket. These criticisms were ignored.

NASA were not prepared, in the immediate aftermath of the Challenger disaster, to reconsider this decision to launch all their spacecraft on the Shuttle, as this was the first complete Shuttle launch failure. It would have been an admission that the original decision had been wrong if it was reversed following its first real test. After all, tragic though the loss of a manned shuttle was, it was unreasonable to assume that the shuttle programme could have run for decades without a major problem occurring, no matter how careful NASA were. The solution adopted for Magellan was to continue to use the Shuttle, but to replace the Centaur by the safer solid-fuelled Inertial Upper Stage (IUS) rocket to launch the spacecraft from Earth orbit. Unfortunately, this IUS was not powerful enough to allow Magellan to follow the fast-track, six month, Centaur trajectory to Venus. It would, instead, take a more-leisurely fifteen months. There was a big advantage to this, however, as it meant that the velocity of approach of the spacecraft to Venus would be lower than in the Centaur case, and so a smaller retrorocket could be used.

Eventually on 4th May 1989 Magellan (see Figure 7.5) was launched from Earth aboard the Space Shuttle Atlantis, the first planetary spacecraft to be launched by the Shuttle, and the first American planetary probe to be launched for ten years. Three hours after lift-off, the 22 ton IUS/spacecraft assembly[20] was gently released by the shuttle. Then one hour later, after the spacecraft solar arrays had been deployed and the shuttle had backed away for safety reasons, the IUS launched Magellan on its way to Venus.

Magellan arrived at Venus on 10th August 1990 and, when on the far side of the planet, the retrorocket was fired to put the spacecraft into orbit. Over the next six days small corrections were made to put Magellan into a 258 × 8,460 km orbit, inclined at 85.5°, with a period of 3 h 15 min. Everything

[20] The IUS weighed a little over 18 tons of this.

Figure 7.5 The Magellan spacecraft, mounted on top of the Inertial Upper Stage, during their deployment from the Space Shuttle Atlantis. Magellan's main antenna at the top doubles as both the SAR antenna and the high-gain antenna for communications with Earth, and the long horn-shaped antenna to its left is part of the radar altimeter. (Courtesy NASA; P-34252.)

went well, until on 16th August contact was lost by mission control at JPL, but 14 hours later a faint signal was picked up by the Deep Space Network (DSN). Fortunately, by this date designers in the West[21] had adopted a fail-safe approach to spacecraft design, in which a spacecraft will put itself into a safe operating mode if it loses contact with Earth, pointing its low gain antenna at Earth[22] and awaiting further commands from ground control. This saved Magellan, as by increasing the ground transmitter power from 18 to 350 kilowatts NASA was able to communicate with the spacecraft and regain control. Five days later exactly the same thing happened, apparently due to a problem with its attitude control computer, and it took until 12th September for everything to be returned to normal. Three days later the first mapping cycle started, which lasted a full Venusian year of 243 Earth days, during which the planet rotated once below the polar orbiting spacecraft.

The SAR antenna operated for 37 minutes per orbit, storing data on an on-board tape recorder. The spacecraft was then slewed so that the SAR antenna was pointing at Earth and half of the stored data transmitted to Earth via this antenna for 57 minutes. After a short breakage in transmission for updating the spacecraft attitude control system, the remaining tape recorded data was transmitted to Earth for another 57 minutes. Then the recording/transmitting cycle started again on the next orbit. Mapping data was recorded from the north pole to 54° South (when the tape recorder was full) on one orbit, and from 72° North to 68° South on the next orbit, with the pattern repeating on alternate orbits.

One Venusian year and 1,852 orbits later the Magellan spacecraft had transmitted to Earth over 3,000 gigabits of data, which was more data than all the previous NASA planetary spacecraft combined. Magellan had covered 84% of the surface of Venus by the end of the first mapping cycle on 15th May 1991, with the SAR antenna looking to the left of its orbit around the planet. NASA then reconfigured the SAR to look to the right of its orbit and started the second mapping cycle, concentrating on those areas of the planet, mostly in the south polar region, that had not been covered by the first cycle. A third cycle was started on 24th January 1992, with the SAR returned to its left-looking mode. This cycle began with a stereo mapping of

---

[21] Although spacecraft designers in NASA and ESA, and their contractors, had adopted this fail-safe approach, those in the USSR had not. Had they done so the USSR may well not have lost the Phobos 1 spacecraft (see later).

[22] In the case of Magellan, the spacecraft was also designed to scan space, by slowly coning its spin axis, to help find the Earth. (Coning is the motion that the spin axis of a child's spinning top follows when it is spinning slowly).

Maxwell Montes and part of Aphrodite Terra. A fourth cycle, starting on 15th September, was designed to map the detailed gravitational field[23] in the equatorial region of Venus, looking for evidence of mass concentrations below the surface.

Initial funding for Magellan had been limited to the first mapping cycle, and so NASA began to have more and more funding problems as the spacecraft continued to exceed its design goal of operating for just this first cycle. It seemed wrong to switch off a $500 million spacecraft (plus $500 million for the shuttle/IUS launch) after just 243 days when it was still working perfectly, so NASA raided other budgets to keep it going. But by the end of 1992 this funding had been virtually exhausted. There was just enough money to start a fifth cycle in May 1993, but the NASA staffing level had to be drastically reduced to keep the mission running up to the end of the year when the money would finally run out. It looked for a time as if this would be the end of the mission, but eventually NASA managed to find the funds to continue work until October 1994 when the mission was finally terminated.

It was important to get as close as possible to the surface to obtain accurate gravitational results, so for the fourth cycle the periapsis of Magellan's orbit had been reduced from an altitude of 258 km to 185 km, but this only allowed accurate gravitational mapping of the regions around periapsis at about 10° North. To achieve a better planet-wide set of data it would also be necessary to reduce the altitude of the apoapsis appreciably below its value of 8,460 km. To do this required far more fuel than was left on the spacecraft, so the mission engineers decided to use aerobraking instead. This manoeuvre, in which the spacecraft with all its appendages deployed would use the air resistance of the Venusian atmosphere to gradually slow it down, was highly risky. But this late in the mission, where all new data were looked upon as a bonus, it was considered a risk worth taking.[24]

So on 25th May 1993 the periapsis was reduced to just 140 km, and then repeated aerobraking in the upper atmosphere over the next 70 days successfully reduced the apoapsis to 540 km. The periapsis was then increased to 197 km, to prevent further aerobraking, and the fifth cycle was started on 3rd August 1993, concentrating on measuring the gravitational field in mid and high latitudes, to complement the equatorial data obtained during cycle 4.

---

[23] By observing the effect of the varying gravitational force on the spacecraft's orbit.

[24] The cynics might also have added that if it had failed it would have solved NASA's problem of trying to find the money to operate it subsequently.

The images returned by Magellan, with a resolution of about 120 metres, were a revelation. In addition to these SAR images, a small radar altimeter on board also determined the surface topography to within a height resolution of 10 metres.[25]

The surface of Venus revealed by Magellan is older than the average age of the Earth's surface, as one would expect as there are no oceans, rivers or rain on Venus to modify its surface, but crater counts showed that it is appreciably younger than the surface of the Moon, Mars or Mercury. Most of the surface of Venus appears to be about 400 million years old, with no features older than about 1 billion years. Although Magellan found thousands of volcanoes, none seemed to be active at the time of observation.

Most of the 900 impact craters imaged by Magellan were fresh-looking, suggesting that weathering[26] on Venus is a slow process. The largest crater imaged was a highly modified, double-ringed crater called Mead of about 275 km diameter. The lack of any larger craters, and the relative paucity of other large double-ringed craters, was due to the relative youthfulness of the planet's surface. Interestingly, no craters were found of less than about 3 kilometres in diameter because the small meteorites, which would normally have produced these small craters, would have been burnt up by the very thick atmosphere. Marks were seen on the surface, possibly caused by the hot gases hitting the surface from such burnt-up meteorites. Craters of between 3 and 20 km in diameter were found to have uneven floors with signs of rubble on them (see Figure 7.6), indicating that the meteorites that caused them were breaking up just before impact. Craters in the 20 to 50 km range were found to be generally pristine in appearance (see Figure 7.7), with a central peak and smooth floor of solidified lava. Some of these craters were found to have an asymmetry in their ejecta blankets, unlike similar sized craters on the Moon and Mercury. In the case of the Moon and Mercury, when a meteorite hits the surface in all but the most glancing impacts, the ejecta is emitted equally in all directions, as there is no atmosphere to prevent it. In the case of Venus, however, when a meteorite strikes the surface at an angle, the wake[27] of the meteorite in the atmosphere prevents the ejecta

[25] These surface and height resolutions compare with 20 km and 100 m for the PVO and 2 km and 50 m for Veneras 15 and 16.

[26] The weathering on Venus is probably caused mostly by the wind, aided by the high atmospheric pressure and density and the high surface temperature.

[27] Because the atmosphere on Venus is so thick, it responds to impacts more like water would on the Earth.

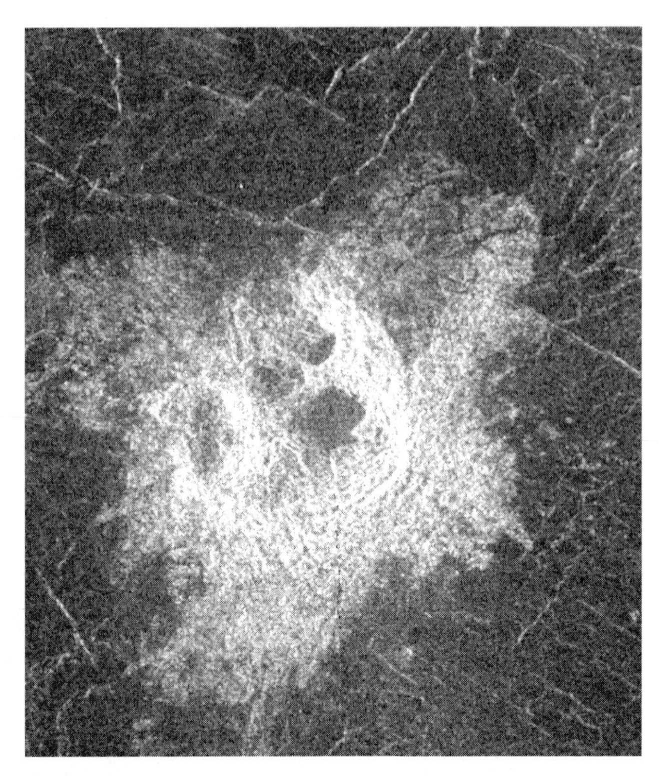

Figure 7.6 This distorted 14 km crater, which is actually four separate craters in contact, was probably caused by a meteorite that broke up just before impact. (Courtesy NASA/JPL/Caltech; PIA 00476.)

being deposited in a back-scattering direction, thus producing the asymmetry observed in the ejecta blanket.

The plains[28] of Venus, which cover about 80% of its surface, have thousands of individual shield volcanoes scattered over them. Otherwise, the surface of the plains is either smooth, reticulated, gridded or lobate in appearance. The reticulated plains have irregular grooves, the gridded plains have regular grooves, and the lobate plains are where various lava flows meet. All of the four types of plain were originally produced by lava flows, which in the case of the reticulated and gridded plains have since been modified. So the surface of Venus is relatively young and of volcanic origin. It was something of a surprise, therefore, when Magellan's gravitational measurements, made later on in its mission, indicated that the rigid lithosphere of Venus appeared to be at least 30 km thick.

The solidified lava flows imaged by Magellan are very interesting, often showing characteristics not seen on Earth. In some places the lava had

[28] The plains described in this section also include the lowland areas of Venus.

Figure 7.7 The three large impact craters in this image range in diameter from 37 to 50 km. They each have clear almost circular rims, distinct central peaks and radar dark, lava-covered floors. (Courtesy National Space Science Data Center, World Data Center-A for Rockets and Satellites, NASA; Experiment Principal Investigator, Dr Gordon H. Pettengill, *The Magellan Project*; P-36711.)

obviously been extremely fluid, running down slopes as shallow as 1°, whilst in other places it had been very viscous. In many ways the fluid lava behaved like water on Earth cutting channels in previous lava deposits to produce features that look more like rivers of water than lava flows. The very viscous lava has produced pancake-like volcanic domes about 30 to 60 km in diameter and about 100 to 750 m high (see Figure 7.8), some of which pre-date faults which have now split them, and some of which cover faults, obviously post-dating them. These pancake domes are not always connected with faults, however, often appearing on relatively smooth surfaces.

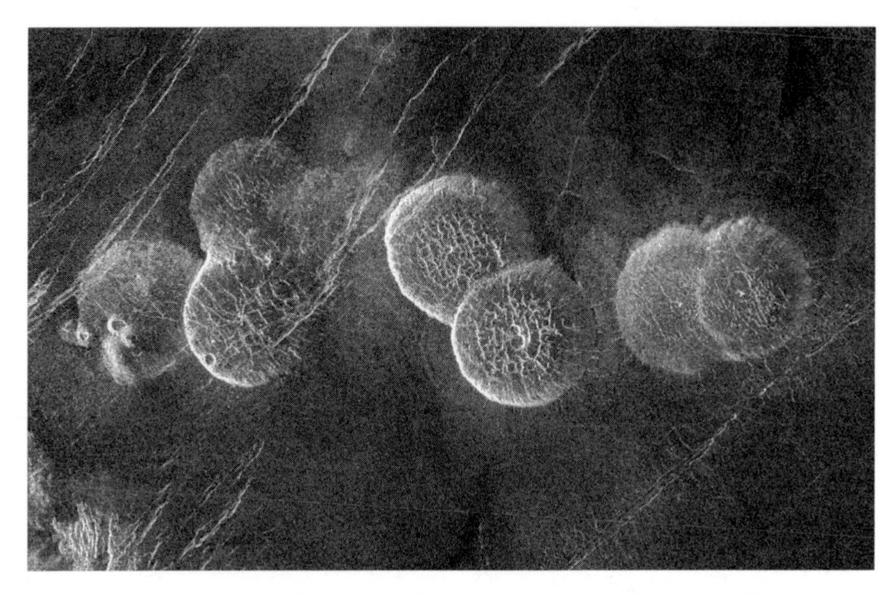

Figure 7.8 These 750 m high, pancake-like, volcanic domes are made of highly viscous lava which has oozed onto the surface. (Courtesy NASA/JPL/Caltech; PIA 00215.)

Venus shows clear signs of past tectonic activity in the highland regions. The deformational (tectonic) features show the results of both compressional and extensional forces. Rifting of the crust has occurred to produce chasmas and abundant faulting in the Aphrodite Terra and Beta Regio highlands, whereas compressional forces have produced the mountain belts of Ishtar Terra. It is unclear whether there is any limited resurfacing proceeding today due to local tectonic activity or not. There is no global network of faults like that on the Earth, however, showing that there has been no obvious *plate* tectonic activity over the last few hundred million years represented by the present surface of the planet.

So ends this general overview of Venus as seen by Magellan, the last Venus probe of the twentieth century.

7.7 **The Phobos Spacecraft**

The success of the American Viking missions to Mars in the mid 1970s, and the continuing problems that had plagued the Russian Mars missions in the early 1970s, caused a hiatus in the exploration of the planet. In the event the next two spacecraft to be launched towards Mars were not launched until

1988, and their main mission was not to observe Mars but Phobos, one of its two small satellites.

Mars' 27×21×19 km irregularly-shaped satellite Phobos (see Figure 6.4 earlier) had been shown by earlier Mars spacecraft to be covered in craters, and in some ways it looked a very uninteresting place to examine in any detail. But it was thought that Phobos and Mars' other satellite Deimos were probably captured asteroids and, as such, were probably bodies left over from the formation of the planets. This was why Phobos was of interest.

Encouraged by the success of their Venera spacecraft and by the international collaboration with the Vega Halley probes, the Russians decided to build two Phobos spacecraft and to invite foreign scientists to provide some of the experiments. Originally, the two spacecraft were to be identical, but mass constraints finally dictated that their designs should differ. Both spacecraft were designed to fly-by Phobos at an altitude of only 50 metres and deploy landers when cruising past at about 2 to 5 m/s (4 to 11 mph).

Phobos 1 was launched on 7th July 1988 and its sister spacecraft Phobos 2 was launched a few days later. Initially all went well, and then on 31st August a faulty command was sent to Phobos 1 which caused it to tumble slowly. Had the design had the fail-safe provisions of Magellan this would not have been a problem, but the Russians had not adopted such a design philosophy and communications with the spacecraft were permanently lost.

All hope now rested on Phobos 2, which successfully reached the vicinity of Mars on 29th January 1989, firing its retrorocket to put it into a 850× 79,750 km orbit around the planet. Phobos 2 imaged Mars over the next few weeks, and on 18th February its orbit was circularised at 9,670 km altitude and the main propulsion unit jettisoned. Imaging of Phobos started on 21st February from a distance of 860 km, and then the orbit of the spacecraft was gradually modified ready for the intercept on 9th April. Unfortunately, near the end of March contact was lost during an attitude manoeuvre.

The loss of both of the Phobos spacecraft was a devastating blow to the Russians, although some interesting data had been produced. For example, *en route* to Mars an X-ray telescope on Phobos 1 had taken 140 high-quality images of the Sun and over 100 cosmic gamma ray bursts had been detected by Phobos 2. Once at Mars the Termoskan scanning infrared radiometer on Phobos 2, which was cooled with liquid nitrogen, made four wide-field images of the equatorial region of Mars in the red to near infrared band from 0.6 to 0.95 microns and in the infrared band from 8.5 to 12 microns. Their resolution varied from 300 metres for one scan to over 1 kilometre for the other three. A comparison between these red and infrared images produced a much clearer understanding of the surface geological formations. In addition, the shadow of the satellite Phobos was seen on some of the infrared images, and this enabled the thermal inertia of parts of the surface to be determined. The Phobos 2 infrared spectrometer (ISM), which operated over the range from 0.76 to 3.16 microns, enabled the column mass of carbon dioxide in the Martian atmosphere to be determined, and this allowed the height of the various surface features to be estimated. Using this technique the summit caldera of Pavonis Mons was estimated to be about 6 km deep.

In addition to data on Mars, Phobos 2 took thirty-seven images of Phobos at resolutions of up to 40 m, during which about 80% of the surface was seen at one time or another. This included some parts not previously imaged by the Viking orbiters. Before the Phobos 2 mission it was thought that Phobos was probably similar in composition to a carbonaceous chrondrite meteorite, but colour photographs taken by Phobos 2 showed that the surface colour was not as homogenous as previously thought, so reopening the question of its composition. The density of Phobos was measured as 1.95 g/cm^3, slightly less than anticipated and low enough to imply that there may be some ice mixed in its core. It was, therefore, a surprise to find that its surface temperature reached 27 °C in sunlight which, according to the ISM spectrometer, produced a bone-dry surface. So Phobos 2 had produced some interesting data about Phobos but, because its mission was so cruelly cut short, it had raised more questions than it had answered.

7.8 Mars Observer

America's next spacecraft to visit Mars after Viking was originally to have been launched in 1984, but the plan was never implemented because of budgetary cuts in the astronomy programme to help pay for the overspend in the Space Shuttle programme. In May 1983, however, the Solar System Exploration Committee reconsidered possible missions to Mars and recommended that the next mission should be devoted to geoscience and climatology. This was originally conceived of as a two spacecraft mission, but it was eventually reduced to one spacecraft to save money. The design of the Mars Observer spacecraft, as it was called, was to be based on RCA's Satcom communications spacecraft, and was to use units from both civilian and military weather spacecraft to save more money. It would be a polar orbiting spacecraft with no lander.

As with Magellan and other spacecraft of the time, the Mars Observer programme was badly hit by the 1986 Challenger disaster, which caused an indefinite delay in the launch of all future spacecraft, pending the results of the various disaster enquiry boards.

Before the Challenger disaster, Mars Observer had been earmarked for a shuttle launch in 1990 using a solid-propellant Transfer Orbit Stage (TOS) to boost it from Earth orbit. After the disaster, NASA initially delayed the launch to 1992. Fortunately the use of the solid-propellant TOS motor did not raise the hazard problems associated with Magellan's liquid-fuelled Centaur, but a two year launch delay would prove very costly in itself. In January 1987, therefore, Bill Nelson, chairman of the House of Representatives' Space Science and Applications Subcommittee requested NASA to launch the Mars Observer in 1990. James Fletcher, the NASA Administrator, refused and retained the 1992 launch date because he was concerned at the log-jam of launches that would occur once the shuttle was operational again. The argument was far from over, however, as the scientists were still lobbying hard for a 1990 date, pointing out that this could be achieved using an expendable Titan 3 launcher, thus decoupling the launch from the shuttle log-jam. Eventually, NASA gave way and agreed to a Titan 3 launch to reduce the shuttle manifest problem but, although a 1990 launch was initially targeted, this was later changed to 1992 because of other problems. All this vacillation, and the changes in launcher type and launch date, took their toll of the budget for what was originally intended to be a low cost mission; the final bill being a massive $1.2 billion (in 1998 dollars).

The Mars Observer was designed to measure the atmosphere and surface of Mars from a circular 390 km altitude polar orbit using a twin camera imaging system, a laser altimeter, a magnetometer, and various spectrometers and radiometers. Its launch was initially delayed by a hurricane warning after the spacecraft had already been mated to the launcher, but its Titan 3 finally lifted off successfully on 25th September 1992.

The first problem was when the TOS rocket stage telemetry failed during the burn to eject the spacecraft from Earth orbit, and no one knew if the rocket had exploded or not. Happily, the TOS had worked perfectly (apart from its telemetry system), but every now and again during the Mars Observer's eleven month voyage to Mars the spacecraft lost orientation. Then, just before the spacecraft was due to be put into orbit around Mars, communications were completely and irrevocably lost.

It has never been clear exactly what caused the loss of the Mars Observer spacecraft. The plan on 21st August, the day that it was lost, was to turn off the spacecraft's transmitters,[29] pressurise the fuel tanks in preparation for

[29] This was to protect the travelling wave tubes in the transmitters from the physical shock caused by opening the pyrotechnic control valves in the fuel pressurisation system.

the orbit insertion burn, and then the transmitters would turn themselves back on when the pressurisation had been completed. No communications were received from the spacecraft after the transmitters were turned off. At first it was thought that the spacecraft had lost orientation again, as it had on its journey to Mars, but it failed to respond to commands from Earth to transmit via its wide-beam antenna. The spacecraft was apparently completely dead, and the Failure Review Board after much analysis concluded that the most likely cause was a leak in the fuel system which had caused the spacecraft to explode.

I will now return to the 1960s to discuss the spacecraft missions to Jupiter and the other large outer planets.

Chapter 8 | EARLY MISSIONS TO THE OUTER PLANETS

8.1 Programmatics and Finance

The mid 1960s were an interesting time for NASA, buoyed up as they were by the success of their Mariner 2 spacecraft to Venus in 1962, and that of Mariner 4 to Mars in 1965. NASA's last three Ranger spacecraft to the Moon had been successful, and their manned programme seemed to be on course for a lunar landing in 1968 or 1969. There was a feeling at NASA that they could do no wrong (provided there was enough money to cover programme slippages). It was in this climate that NASA mission scientists started to look at possible missions to Jupiter and the other gas giants, namely Saturn, Uranus and Neptune.

Although the mid 1960s were a period of economic stringency (see Section 5.6), NASA engineers were looking long term and they assumed that this period of financial restrictions would be only temporary. The President's Science Advisory Committee were keen on sending a spacecraft to Jupiter (see Section 6.1) so what could be done?

In 1963 NASA had contracted TRW to build a series of Pioneer spacecraft (eventually numbered 6 to 9) to investigate the solar wind, and it eventually became clear that modified versions of these spacecraft could be used for Jupiter intercept missions in the early 1970s. Detailed design studies were then undertaken by TRW, and in February 1969 they received a contract to build two spacecraft to fly-by Jupiter. The first of these, Pioneer 10, was to be launched in 1972, followed by Pioneer 11,[1] which was to be launched 13 months later at the next Jupiter launch window.

In the meantime, NASA had investigated the even more interesting

[1] Pioneers 10 and 11 were called Pioneers F and G before launch.

possibility of one spacecraft launched during the period 1976–80 to fly past Jupiter, Saturn, Uranus and Neptune sequentially. Conceived in 1964 by Michael Minovitch of UCLA, this potential mission would use the gravity of one planet to accelerate and redirect the spacecraft on to the next, in the same sort of 'sling-shot' manoeuvre that was to be used by Mariner 10 to fly by Venus and Mercury in the early 1970s (see Section 6.1). This 'Grand Tour' mission, as it became called, attracted much attention as the outer planets were only suitably aligned once every 175 years, so this was an opportunity that could not be missed. Not only did this sling-shot trajectory allow one spacecraft to visit four planets, thus saving a considerable amount of money, but the mission time from Earth to Neptune, for example, was reduced from about 30 years to about 12 years.

As the Grand Tour mission was investigated further, however, it was found that the spacecraft would have to fly uncomfortably close to Saturn's rings. It was not worth the risk of flying through the rings themselves, because of a possible collision with the small particles of which they were composed, so there were two possibilities. Either fly between the planet and the inner edge of the inner ring, or fly outside the outer edge of the outer ring. The first solution would call for a very accurate trajectory, and it was not certain whether the gap between the inner edge of the rings and the planet was empty, but the second solution would add three years on the flight time to Neptune. Whilst these alternative were being considered, another solution was proposed that would both avoid the Saturn ring problem and also enable a spacecraft to fly past Pluto. This involved building two spacecraft; one to fly past Jupiter, Uranus and Neptune, and the second to fly past Jupiter, Saturn and Pluto.

It was now almost 'open house' on suggestions for missions to the outer planets, including a possible Jupiter entry probe mission, to measure Jupiter's atmosphere *in situ*, and a Jupiter orbiter. But what did the planetary scientists really want, as they could not have everything? NASA's Lunar and Planetary Missions Board were asked by NASA in 1968 for their views, and they in turn set up a Panel on the Outer Planets chaired by James Van Allen to consider the alternatives and make recommendations.

Van Allen's panel recommended that the first priority for the next decade should be the four-planet Grand Tour mission, with separate orbiters for each of the four major planets (starting with Jupiter) being a possible alternative. NASA then formed an Outer Planets Working Group to consider and develop this recommendation, and by May 1969 the working group had developed an ambitious programme that seemed to be based on the idea that unlimited money was available. Once more NASA, or at least some of its senior staff, did not seem to have learnt the lessons of a few years before, when NASA's ambitious Mars and Venus missions had been severely curtailed through lack of money (see Chapters 5 and 6).

The proposed programme of the Outer Planets Working Group consisted of (years quoted are launch dates):

- 1972 & 1973 Pioneer 10 and 11 fly-bys of Jupiter (already approved)
- 1974 Test flight of the spacecraft to be used in the 1977 multi-planet mission. The spacecraft in this test flight would fly to Jupiter and then swing out of the ecliptic (the first spacecraft to do so)
- 1976–78 Jupiter orbiter
- 1977 Jupiter, Saturn, Pluto fly-by
- 1978–80 Jupiter atmospheric multiprobe
- 1978–81 Saturn orbiter and/or multiprobe
- 1979 Jupiter, Uranus, Neptune fly-by

Further discussions followed, and then in September 1969 NASA requested new-start funding for the 1974 test flight and for the 1977 and 1979 three-planet Grand Tour missions. Money was still tight in 1969, however, and it rapidly became clear that finance would not be available in the short term for any of these missions, so as an interim measure NASA decided to delay the test flight from 1974 to 1975. In March 1970 President Nixon enthusiastically endorsed the Grand Tour concept, but continuing budgetary problems caused the test flight to be deleted, with the launch of the Jupiter, Saturn, Pluto spacecraft being brought forward one year to partially compensate.

Unfortunately, a dispute had broken out in August 1970 within the scientific community while NASA were trying to juggle with their outer planet programme. Some scientists were adamant that the opportunity for the Grand Tour should not be missed, and that the orbiter and entry-probe missions could be delayed until the following decade. Others felt that the Grand Tour missions were too risky, requiring a newly designed, sophisticated spacecraft called TOPS[2] to operate for ten years or more with a completely untried power generation system, and they suggested that it would be best to explore Jupiter with entry probes and orbiters as the priority mission. In January 1971 NASA announced that it was keeping open the option of a 1976/77 Jupiter orbiter mission and/or a 1976/77 Jupiter,

[2] TOPS stood for 'Thermoelectric Outer Planet Spacecraft', so-called because it relied on a thermoelectric power generator as its power source.

Saturn, Pluto Grand Tour mission as the first stage of their outer planet programme.

In the meantime NASA began to review the complexity of its TOPS spacecraft design, with its self-test and repair (STAR) computer which was being designed to allow the spacecraft to operate largely independently of ground control.[3] As an alternative they considered simpler designs based on the Mariner interplanetary spacecraft or the smaller Pioneer spacecraft. Whilst NASA were doing this, in June 1971 the US Senate cut the start-up funds for these so-called Outer Planet Missions from the $30 million requested by NASA to $10 million. Although this was later revised to $20 million by a congressional committee, it spelt the beginning of the end for the sophisticated 660 kg TOPS spacecraft.

At first TOPS refused to die, and during the summer of 1971 various competing options were considered, including a programme of four TOPS multiplanet Grand Tour missions costing an estimated $750 million, and a programme where two of these Grand Tour spacecraft were replaced by two TOPS Jupiter orbiters in a programme costing about $925 million.[4] Programmes including Mariner- and Pioneer-class orbiters were also costed. Once more, however, as in the previous year the scientific community was split on whether to push for a Grand Tour or orbiter programme. Finance and the anticipated public interest were about to make the decision for them, however.

The crunch came during discussions of the FY1973[5] NASA budget in 1971. Initially, James Fletcher, the new NASA Administrator, was willing to request $100 million for FY1973 for the Outer Planet Missions, but this was eventually reduced in September 1971 to $29 million for four Grand Tour missions costing a total of $750 million. The Office of Management and Budget (OMB) still wanted substantial cuts in the total FY1973 NASA budget, however, so Fletcher was forced to suggest cancelling Apollo 16 and 17. He also pointed out that if the 1973 level of funding were to be repeated in subsequent years, NASA would not be able to afford either the space

---

[3] This was to make the spacecraft as autonomous as possible, because of the long round-trip time for signals to travel between the spacecraft and ground control. Even at the speed of light it would take 90 minutes for such signals to travel to Earth from Jupiter, and back, and for the other planets the round-trip time would run from three hours for Saturn to many hours for the more distant planets.

[4] The extra cost of the orbiters was due to the fact that a new retrorocket would have to be developed to put the spacecraft into orbit around Jupiter.

[5] FY stands for 'Financial Year' or 'Fiscal Year'.

shuttle or Grand Tour programmes. It was at this stage that President Nixon stepped into the fray, prompted by Caspar Weinberger, the deputy director of the OMB.

Unknown to both NASA and the OMB staff who were negotiating the NASA FY1973 budget, Caspar Weinberger thought that the cutting of the NASA budget had gone too far. In August 1971 he suggested to President Nixon that money could be saved from other government programmes to allow NASA an annual budget of about $500 million more than the figure then stipulated by the OMB. Nixon agreed, partly to avoid unemployment in critical states during the election year of 1972, and the Space Shuttle programme was approved in January 1972. The four spacecraft TOPS mission was replaced, however, by two Mariner-class Jupiter–Saturn fly-by spacecraft at a total cost of about $250 million, to be launched in 1977. These two spacecraft were eventually known as Voyagers 1 and 2.

8.2 **Pioneers 10 and 11**

We left Pioneers 10 and 11 with the decision to award contracts in 1969 for two spacecraft to be launched in 1972 and 1973. These two spacecraft were to be subjected to completely new hazards *en route* to Jupiter, and Pioneer 11, which was to use Jupiter's gravity to swing it on to Saturn, was also to fly through the plane of Saturn's rings.

In order to get to Jupiter both the Pioneer 10 and 11 spacecraft were going to have to spend about six months crossing the asteroid belt. This would be risky as the asteroid belt contained an indeterminate number of small asteroids, meteorites and very small particles, but the risk could not be quantified and it had to be accepted, as there was no other practical way to reach Jupiter. Jupiter was also known to possess a very active magnetosphere, but its structure and the density of the high energy particles within it were unknown. It was known, however, that solid state devices carried by all spacecraft could be badly affected by high energy particles, so this was the second hazard of indeterminate size. Finally, the structure and extent of Saturn's rings were only known approximately and the particles in these rings would provide another indeterminate hazard to Pioneer 11.

Very little could be done in designing the Pioneers to help to protect them from the above hazards, apart from choosing components that were relatively insensitive to charged particles and shielding those that were. So the main difference between the design of Pioneers 10 and 11 and those of previous interplanetary spacecraft was in the area of power generation. Jupiter and Saturn are about 5.2 and 9.5 times as far away from the Sun as is the Earth, so the amount of energy falling on each square metre of spacecraft solar array near Jupiter and Saturn would only be about 3.6% and 1.1% of that in the vicinity of the Earth. Because of this, solar arrays were not a practical proposition for these Pioneer spacecraft. So to provide power each spacecraft carried four radioisotope thermoelectric generators, mounted in pairs at the end of two 3 metre booms to minimise interference with on-board experiments.

The 260 kg Pioneer 10 spacecraft was launched towards Jupiter at about the middle of its launch window on 3rd March 1972 by an Atlas-Centaur launcher, augmented by a solid-fuel final stage. This gave the spacecraft a velocity of 32,600 mph (14.5 km/s),[6] which enabled it to pass the orbit of Mars in May 1972 and to reach Jupiter on 4th December 1973 after a journey of just 21 months. Its passage through the asteroid belt caused no problems. Meanwhile, on 6th April 1973 Pioneer 11 was also launched towards Jupiter, with three orbital options depending on the results of Pioneer 10. Pioneer 11 could either fly over Jupiter's equator (like Pioneer 10), or near Jupiter's poles, or it could use Jupiter's gravity to send it to fly past Saturn about six years after launch.

Pioneer 10 first detected high energy electrons from Jupiter at about 300 Jupiter radii (R_J) from the planet. Pioneer then passed through Jupiter's bow shock (see Figure 8.1), where the solar wind interacts with Jupiter's magnetic field, on 26th November 1973, whilst still about 7.7 million km, or about 108 R_J, from the planet. The discovery of high energy electrons from Jupiter in front of the bow shock was a surprise, as this meant that these electrons had escaped from the planet's magnetosphere and had somehow crossed the shock wave. The distance of the bow shock from Jupiter was also greater than expected, indicating that Jupiter's radiation environment may be more active than anticipated and so be more likely to damage the spacecraft than originally thought.

On crossing the bow shock, the magnetic field measured by Pioneer 10 increased from 0.5 to 1.5 gammas, the temperature of the solar wind increased dramatically and its velocity was reduced by a factor of two. On the following day the magnetic field suddenly increased to about 5 gammas when the spacecraft crossed Jupiter's magnetopause into its magnetosphere,

---

[6] This was the highest launch velocity of any spacecraft up to that time. To put this into perspective, it reached the orbital distance of the Moon only 11 hours after launch, compared with 3 days for the Apollo missions.

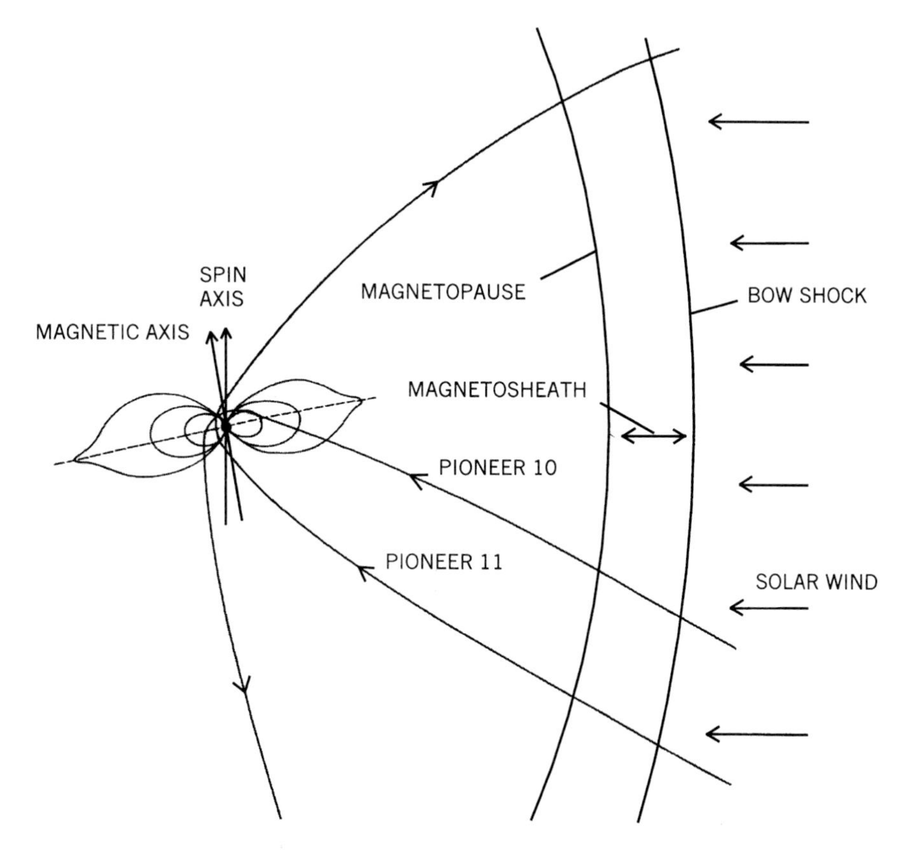

Figure 8.1 The trajectories of Pioneers 10 and 11 are shown as they passed through Jupiter's magnetosphere (the region to the left of the magnetopause). The magnetopause is the boundary where the pressure of the solar wind and that of Jupiter's magnetosphere are in equilibrium. Jupiter's field is seen to be dipolar near to the planet, but to have a stronger component on the magnetic equator further out.

a region protected from the solar wind and dominated by the planet's magnetic field. Then on 30th November, much to everyone's surprise, the magnetic field suddenly decreased back to its interplanetary level and the solar wind was detected once more. Evidently an increased pressure of the solar wind had pushed the magnetopause closer to Jupiter and it and the associated bow shock had overtaken the spacecraft, but eventually the spacecraft crossed the bow shock and magnetopause once more. As Pioneer 10 flew past Jupiter it found that its radiation belts, where protons and high energy electrons are trapped in the planet's magnetic field, are about 10,000 times as intense as the Earth's Van Allen radiation belts.

Later analysis of the data recorded during Pioneer 10's fly-by showed that Jupiter's magnetic field seems to consist of two regions. The first was a dipole field, with a strength of about 4 gauss at Jupiter's cloud tops,[7] that extended to about 20 R_J. Then there was a stronger field, outside the first, which was found to be more confined to Jupiter's equatorial plane, because of the centrifugal forced caused by Jupiter's rapid rotation.[8] The axis of the dipole field was found to be inclined at about 10° to Jupiter's spin axis, with the magnetic centre being displaced slightly from the geometrical centre of the planet. The radiation belts were found to be most intense inside the 20 R_J boundary of the dipole field, and concentrated towards the planet's equatorial plane.

When the high energy electrons were found in front of Jupiter's bow shock, records of early Earth-orbiting spacecraft were re-examined, and they showed an increase in the background level of cosmic ray electrons about every 13 months, which is the same as Jupiter's synodic period. Thus, some of Jupiter's high energy electrons even reach the vicinity of the Earth.

Before Pioneer 10's mission to Jupiter it had been thought that at least some of the four Galilean satellites would be within Jupiter's radiation belts, and astronomers wondered what effect their presence would have. In the event, Pioneer 10 found that all of these satellites were within the belts, and Io and Europa, and to a lesser extent Ganymede and Callisto, tended to sweep up both protons and electrons from the belts. In addition, Io was found to have its own ionosphere, and a cloud of hydrogen atoms was found stretching in an arc along Io's orbit, with Io near its mid-point.

Imaging of Jupiter and its four Galilean satellites started about one month before closest approach, and on 1st December, when Pioneer 10 was still about 2.8 million km or 40 R_J from Jupiter, the resolution of the images had already reached the best achievable from the Earth. About 80 images with resolutions up to six times better than that were returned during the encounter. On 3rd December Pioneer passed within 1.4 million km of Callisto, and 90 minutes later it passed within 450,000 km of Jupiter's largest satellite, Ganymede. Their surface temperatures were found to be about −163 °C (or 110 K) and −125 °C (or 148 K), respectively, but the resolution of the images, which was about 400 km for Ganymede, was insufficient to show any clear surface features. Later in the encounter, Pioneer passed within 320,000 km of Europa and 350,000 km of Io, but the close-up

---

[7] This is about ten times the field strength at the surface of the Earth

[8] Although the diameter of Jupiter is eleven times that of the Earth, it rotates on its axis in only 10 hours.

images of Io were lost when the spacecraft imaging system was temporarily switched off. This was due to spurious commands produced on-board the spacecraft by Jupiter's severe particle environment. Pioneer then flew only 130,000 km (or 1.8 R_J) above Jupiter's cloud tops at about 82,000 mph (37 km/s), before passing first behind Io and then behind Jupiter, before continuing on its way out of the Solar System, passing in 1987 the average distance of Pluto from the Sun. Pioneer 10 had survived the harsh environment close to Jupiter remarkably well,[9] giving spacecraft engineers and scientists confidence that Pioneer 11 should also be in good condition after its close encounter with Jupiter one year later.

Pioneer 10 found other interesting features. Its ultraviolet spectrometer detected helium in Jupiter's atmosphere for the first time, and showed that Jupiter's atmosphere is about 99% hydrogen and helium. This, like the ratio of helium to hydrogen measured on Jupiter, is similar to that of the Sun,[10] thus showing that Jupiter's atmosphere has apparently evolved relatively little since the planet was formed from the solar nebula. Jupiter's brightness temperature of 128 K measured by Pioneer's infrared radiometer was very close to that measured from Earth, confirming that Jupiter emits about twice as much energy as it receives from the Sun. The cloud tops of Jupiter's Great Red Spot (GRS) were found to be colder than those of the adjacent South Tropical Zone, so the GRS was thought to be an enormous high pressure hurricane rising about 8 km above its surrounding area. The scale of the GRS is quite unlike anything on Earth, however, as it has a length three times the diameter of the Earth, and it has existed for over a century at least. The temperature at the top of the bright zones in Jupiter's atmosphere were found to be about 9 °C lower than that of the adjacent dark belts, indicating that the bright zones are regions of rising gas which are about 19 km higher than the belts where the gas eventually descends.

After evaluating the data from Pioneer 10, NASA announced on 19th March 1974 that it would adjust Pioneer 11's intercept trajectory with Jupiter to allow it to fly past Saturn on 1st September 1979. The new trajectory would send Pioneer 11 just 43,000 km (or 0.6 R_J) above Jupiter's cloud tops as it traversed Jupiter flying from the south to north polar regions on

---

[9] The total radiation exposure of Pioneer 10 was about 1,000 times the lethal dose for a human being. In spite of this the spacecraft had only relatively minor damage, with the failure of a few transistors and a minor darkening of exposed optics.

[10] Jupiter's helium to hydrogen mass ratio deduced from Pioneer 10 measurements was 0.28 ± 0.16, compared with the solar value which was thought at that time to be 0.22. This latter value was later revised to 0.28 ± 0.01.

3rd December 1974. Jupiter's gravity would then increase the spacecraft's velocity to about 108,000 mph (48 km/s) as it swung past the planet.

The Jupiter intercept trajectory was a compromise between many competing requirements, the main one being that Pioneer 11 should be able to fly on to Saturn. This required Pioneer 11 to fly closer to Jupiter than Pioneer 10, which would inevitably subject Pioneer 11 to a harsher electromagnetic and particle environment. Fortunately, however, astronomers wanted Pioneer 11 to investigate Jupiter's polar regions, and this meant that the spacecraft could approach Jupiter outside its equatorial plane, and fly-by the planet in a south–north trajectory. This would take it through the equatorially flattened radiation belts relatively quickly, thus subjecting it to this harsh environment for as short a time as possible.

The helium vector magnetometer on Pioneer 10 had been saturated for a time near Jupiter, so a flux-gate magnetometer, capable of measuring much stronger magnetic fields, had been added to Pioneer 11. This new magnetometer should be able to measure the harsher environment expected to be found closer to Jupiter by Pioneer 11.

Pioneer 10 had flown past Jupiter in the same direction as the planet's spin, but its successor was due to fly past the planet counter to this spin direction, so that it could measure the magnetic field and radiation belts around the whole of Jupiter's circumference as the planet rotated. This required a modified motion-compensation system on Pioneer 11 to that flown on Pioneer 10 to eliminate smear of the planetary images. Even so, Pioneer 11's movement for about six hours around its closest approach to Jupiter was still too fast to allow any imaging for that period.

Pioneer 11 first crossed Jupiter's bow shock at about 6.9 million km from the planet on 27th November 1974, and over the next two days it crossed the bow shock and magnetopause three times as the magnetosphere was successively compressed and expanded in response to the varying pressure of the solar wind.[11] The spacecraft then took twenty-five images of Jupiter during its final 52 hour encounter phase, together with one image of Io, Ganymede and Callisto, but, as with Pioneer 10, the images of the satellites had insufficient resolution to show any surface detail. As expected, Pioneer 11's magnetometers measured a higher maximum magnetic field of 1.2 gauss near closest approach than the 0.2 gauss estimated using Pioneer 10, because Pioneer 11 had passed much closer to Jupiter than its predecessor. Similarly the peak counting rate of high energy protons was measured to be

---

[11] The position of the magnetopause varied from about 100 R_J to 50 R_J from Jupiter during the Pioneer 10 and 11 fly-bys.

forty times higher using Pioneer 11 than with Pioneer 10. Finally, Pioneer 11 found that Jupiter's magnetic field very close to the planet was not the dipolar field measured by the earlier Pioneer slightly further out, but was much more complex in structure, varying in intensity from about 3 to 14 gauss at the cloud tops.

In 1951 W.H. Ramsey of Manchester University (England) had suggested that Jupiter consists mostly of hydrogen, the majority of which is so highly compressed that it behaves as a metal. This basic concept was confirmed by the magnetospheric and other data returned by the two Pioneers, although the detailed structure was somewhat different to that proposed by Ramsey. It was now thought that Jupiter probably consists of a 25,000 km diameter rocky core, composed mainly of iron and silicates at a temperature of about 25,000 K, surrounded by a 35,000 km thick layer of mostly liquid metallic hydrogen, followed by a 25,000 km thick layer of liquid molecular hydrogen and a 1,000 km thick atmosphere. At the interface between the metallic and molecular hydrogen, the temperature was estimated to be about 11,000 K, at a pressure of about 3 million atmospheres. The liquid metallic hydrogen was thought to produce Jupiter's powerful magnetic field.

Whilst Pioneer 11 was *en route* to Saturn, Pioneer 10 produced yet one more interesting discovery, when in March 1976 its solar wind instrument temporarily registered zero over the period of a day or so. At the time Pioneer 10 was 690,000 million km from Jupiter, and beyond the orbit of Saturn. Interestingly, however, the spacecraft was approximately in line with Jupiter and the Sun, and so was in Jupiter's magnetic tail, which shields that part of interplanetary space from the solar wind. So Pioneer 10 had shown that the tail of Jupiter's magnetosphere extends even beyond the orbit of Saturn.

During the early discussions on the Grand Tour programme one issue of major consequence that had to be considered was how close to fly to Saturn. Should the spacecraft fly between Saturn and the inner edge of the inner ring, called ring C, or outside the outer edge of the outer ring, ring A? The main problem was that no-one knew whether there were any particles inside of ring C or outside of ring A.[12] Similar considerations now arose for scientists trying to decide on what trajectory Pioneer 11 should follow as it flew past Saturn in 1979. The problem had been compounded in the meantime, however, by Pierre Guerin who claimed in 1969 to have found a new ring (ring D) inside of ring C, and by Walter Feibelman who in 1966 had photographed and confirmed a large very faint ring (ring E) outside of ring A, which

[12] Rings C, B and A, in that order, were thought to extend almost continuously from 1.24 to 2.27 R_S from Saturn's centre, where R_S is Saturn's radius of about 60,000 km.

had been glimpsed in the early years of the century by Emile Schaer and others.

Initially, when Pioneer 11 was launched in 1973 it was intended to fly the spacecraft between Saturn and the inner edge of ring C, assuming that Pioneer 10 performed as expected in its Jupiter mission.[13] But in December 1977 NASA changed their minds, to lessen the risk of collisions with particles inside the main rings (rings A to C), and decided to send Pioneer 11 to cross the ring plane 29,000 km outside the outer edge of ring A, then fly under the rings to within 24,000 km of Saturn's cloud tops before crossing the ring plane once more outside the outer edge of ring A. It was also planed, if everything went well, that Voyager 2, which had been launched in the meantime (see later), would follow almost the same fly-by trajectory.

In August 1979 Pioneer 11 closed rapidly on Saturn crossing the orbit of Phoebe, Saturn's outermost satellite, at a distance of about 13 million km from the planet on 27th August. Jupiter had been known to possess a very active magnetosphere and radiation belts before the Pioneer intercepts, because it had been found to emit radio waves. No radio emissions had been detected from Saturn, however, so it was not known if it possessed a magnetosphere and radiation belts before Pioneer 11 arrived. Astronomers using intelligent guesswork, based on the Jupiter data, had estimated that Saturn would have a relatively strong magnetic field, but not as strong as Jupiter's, with a bow shock about 4 million km from the planet.[14] In the event Pioneer 11 did not cross the bow shock until 13.00 UT on 31st August when the spacecraft was only 1.44 million km or 24 R_S from the centre of Saturn, where R_S is Saturn's radius.

At 14.36 UT on 1st September Pioneer 11 crossed Saturn's ring plane about 30,000 km from the outer edge of ring A. The spacecraft then passed under the rings, passing within 21,400 km of the planet's cloud tops at 16.31 UT, before recrossing the ring plane at 18.35 UT. Not only did the spacecraft survive this close encounter with Saturn and its rings virtually unscathed, but modifications to NASA's DSN had allowed the spacecraft to transmit at up to four times the data rate anticipated when it was launched. On the day

[13] Otherwise Pioneer 11 may have to repeat Pioneer 10's intercept trajectory with Jupiter, and forgo the Saturn fly-by.

[14] This value of 4 million km for Saturn, compared with 7 million km for Jupiter, may seem to indicate that Saturn's magnetic field was not thought to be very much less than that of Jupiter. But Saturn is almost twice as far from the Sun as Jupiter, so the pressure of the solar wind at Saturn compressing its magnetosphere is only about 30% of its value at Jupiter.

after its closest approach to Saturn, Pioneer 11 flew past Titan, Saturn's largest satellite, at a distance of 354,000 km. Unfortunately, communications problems during the Titan closest approach resulted in some of the data being permanently lost.

The images of Saturn received during the Pioneer 11 encounter were disappointing, showing only the same subtle banded structure seen from Earth, but the images of Titan were even more frustrating as they were completely featureless. Saturn's atmosphere was found to be warmer than expected, and Iapetus, Rhea and Titan, its three largest satellites, were found to be of low density, indicating that they were composed mostly of ice. One new ring was found, the 500 km wide F ring, just 3,000 km outside the A ring, and a new small satellite, now called Epimetheus, was found in orbit about 14,000 km outside the A ring. In fact, the spacecraft passed only about 2,500 km from this satellite; a close encounter in astronomical terms.

The strength of Saturn's magnetic field was calculated from Pioneer data to be about 0.22 gauss at the planet's equator, which is a similar value to that of the Earth at its much smaller equator.[15] But Saturn's magnetic moment is about 540 times larger than that of the Earth because Saturn is so much larger. The magnetic axis of Saturn was found to be indistinguishable from its spin axis, and the magnetic field was dipolar. Although Titan was found to be generally within Saturn's magnetosphere, varying pressure from the solar wind drives the magnetopause back and forth across Titan, sometimes leaving Titan exposed to the solar wind.

Prior to the Pioneer encounter it had been anticipated that Saturn's ring system would prevent radiation belts from forming in the inner magnetosphere, and this is precisely what was found; the charge particle flux showing an abrupt cut-off as the spacecraft passed beneath the outer edge of ring A. Not only were the new F ring and the new satellite Epimetheus imaged, but their existence was also confirmed by the gaps that they had created in the electron population of Saturn's magnetosphere.[16] Pioneer 11 discovered similar features elsewhere in the magnetosphere, one of which was later found by Voyager 1 to be due to another ring (later called ring G).

---

[15] Table 3 in the Appendix compares Saturn's magnetic field with that of Jupiter.

[16] A similar effect had been observed by Pioneer 11 in 1974 at Jupiter, which Voyager 1 showed in March 1979 was caused by a ring around the planet (see Section 9.2).

Chapter 9 | THE VOYAGER MISSIONS TO THE OUTER PLANETS

9.1 The Voyager Spacecraft

We left the Voyager programme in 1972 when it had been approved as a two spacecraft programme, with both Mariner-class spacecraft due to visit Jupiter and Saturn after launches in 1977. The more sophisticated Grand Tour TOPS spacecraft had had to be cancelled, because of financial constraints. Then in April 1975 NASA announced that it was intending to launch a third Mariner-class spacecraft to the outer planets in 1979. This spacecraft would be almost identical to the two Jupiter–Saturn Voyagers and fly past Jupiter in 1981 and Uranus in 1985. Unfortunately, NASA were unable to get financial approval for this third Mariner-class, outer planet mission, and its objectives were eventually incorporated into the Voyager 2 mission, which would then fly-by Jupiter, Saturn and possibly Uranus. By the time that Voyager 2 was launched, Neptune had also been added as a possible fly-by target after Uranus.

Voyager 1 was due to fly past just two planets, namely Jupiter in 1979 and Saturn in 1980, but Voyager 2's itinerary at the time of its launch in 1977 included fly-bys of Jupiter in 1979, Saturn in 1981, Uranus in 1986 and Neptune in 1989. Voyager 2's visits to Uranus and Neptune were dependent, however, on Voyager 1's successful fly-by of Titan, Saturn's largest satellite, which was thought to have an atmosphere similar to that of the early Earth. If the Voyager 1's fly-by of Titan failed, Voyager 2 would be re-targeted to fly past Titan, even though this would mean that it could no longer go on to visit Uranus and Neptune.

The two Voyager spacecraft, each weighing 815 kg, were over three times as heavy as the Pioneer 10 and 11 spacecraft and were much more sophisticated than their predecessors. The main operational difference was the use of a much more powerful computer system that allowed all spacecraft functions, except for trajectory changes, to be controlled automatically on-board. This was essential as the round-trip time for signals between Neptune and Earth would be over eight hours.

The central spacecraft structure consisted of a large titanium tank that contained 104 kg of hydrazine fuel for attitude and orbit control, surrounded by a ten-sided structure (see Figure 9.1) that had thermally controlled compartments for the electronic units. A number of experiment units and an optical calibration target for the boom-mounted camera were mounted directly onto this structure. A fixed 3.7 m diameter, X-band communications dish, which was mounted on the top of the main structure, was designed to transmit data to Earth at rates of up to 115 kbps via a 21.3 W amplifier. Many of the experiments were mounted on a scan platform, at the end of a 2.3 m deployable boom, which allowed the experiments to scan their targets.[1] Three RTGs, which provided 384 W of power at Saturn, were mounted on another deployable boom, and two long, thin booms were used as radio antennae to detect radio emissions from the planets. Finally two low-field magnetometers were mounted on a 13 m long boom to minimise the effects of spacecraft interference on their measurements.

Among the experiment packages on the scan platform, which was controllable to 0.2 arcsec, were television cameras with lenses of 200 and 1,500 mm focal length (and filter wheels with various interference and colour filters), an infrared radiometer and interferometer–spectrometer (IRIS), an ultraviolet spectrometer and photopolarimeter. The television system, which took 48 seconds to read-out each image of 800 × 800 pixels at Jupiter, could be used to view simultaneously the same target region as the IRIS instrument. The latter was a true spectrometer capable of measuring at nearly 2,000 separate wavelengths over the range from 4 to 55 microns.

IRIS, whose design had previously flown on Mariner 9, was included to analyse the atmospheric composition of the planets and their satellites, to measure the energy balance of the planets (all of which may have internal heat sources), and to examine Saturn's rings. The ultraviolet spectrometer was to measure, *inter alia*, the hydrogen and helium abundances in the planetary atmospheres, and the photopolarimeter was to measure the amount of

[1] The dish antenna was fixed to the spacecraft, so this determined the orientation of the spacecraft during its mission, as the antenna had to point directly at the Earth. As a result, the whole spacecraft could not be panned to point the imaging experiments at their targets during the fly-bys, so these experiments had to be mounted on their own moveable platform.

Figure 9.1 Voyager's large high-gain dish antenna is seen mounted on top of the main spacecraft structure. The scan platform and its experiments are on the right-hand boom, the Radioisotope thermoelectric generators (RTGs) are at the lower left, and the magnetometer boom is seen extending out of the photograph to the top left. (Courtesy National Space Science Data Center, World Data Center-A for Rockets and Satellites, NASA; P-24653.)

methane, molecular hydrogen and ammonia in their upper atmospheres. Other instruments on each of the Voyagers included a cosmic ray detector system, to measure the energy spectrum of electrons in the planetary radiation belts, a pair of plasma detectors to analyse the solar wind and its interaction with the planetary magnetospheres, and a low-energy, charged particle experiment to analyse charged particles in planetary magnetospheres and, in particular, their interaction with planetary satellites. Finally radio emissions were to be detected from the planets over the range from 1.2 kHz to 40.5 MHz by the radio astronomy experiment.

9.2 Voyager 1 at Jupiter

At 14.29 hours UT on 20th August 1977 a Titan IIIE/Centaur TC-7 rocket launched Voyager 2 into an Earth-based parking orbit. A second burn of the Centaur upper stage, followed by a 43 second burn of a solid-propellant rocket then put the spacecraft on course to fly past Jupiter in July 1979. Two weeks later, Voyager 1 was also launched on its way to Jupiter on a slightly different trajectory that would see it overtake Voyager 2 and fly past Jupiter in March 1979. Unfortunately, Voyager 1's Titan first stage shut down early, but the Centaur upper stage was able to compensate and put the spacecraft on its correct trajectory to Jupiter.[2]

Voyager 2 gave spacecraft controllers a scare when, during take-off, the spacecraft automatically switched off a navigation gyroscope and switched on its backup. Then there was a problem with spacecraft stability after the solid rocket burn, which was not resolved until four days after launch, and finally there appeared to be problems with the 2.3 m experiment boom that would not lock in place. Happily, all these problems were eventually solved, and Voyager 2 settled into its long voyage to Jupiter and Saturn, and hopefully to Uranus and Neptune also, although not all of its problems were behind it. The most serious of these occurred in April 1978 when its radio receiver failed, and its on-board back-up unit was found to be faulty as it had only a very limited frequency scan capability.[3]

An extra spring was added to the 2.3 m experiment boom on Voyager 1 to try to ensure its full deployment, following problems with the deployment of the experiment boom on Voyager 2. This resulted in a five day delay in the launch of Voyager 1 to 5th September 1977. Although the early part of the Voyager 1 mission was less eventful than that of its predecessor, in February 1978 its scan platform stopped in mid-scan. It later responded correctly to commands, however, and by December 1978 the images of Jupiter received on Earth from Voyager 1 were already better than any images taken from Earth. Systematic imaging of Jupiter by the telephoto television camera started on 4th January 1979, when the spacecraft was still two months and 60 million km from Jupiter, with a new image being produced every two hours. Then over a 100 hour period starting on 30th January, Voyager 1 sent back to Earth one image every 96 seconds through each of three colour filters in turn. This allowed a full-colour image to be produced on the ground every 4.8 minutes.

---

[2] Had Voyager 2's Titan similarly underperformed, however, the Centaur would not have been able to compensate fully and the Uranus and Neptune fly-bys would not have been possible. So, by luck, Voyager 2 had the right Titan launcher.

[3] The receiver was designed to be able to scan over a frequency range of 100 kHz to compensate for frequency changes in the received signal due to the Doppler effect, as the spacecraft's velocity changed along the Earth–spacecraft line with time, especially at the planetary fly-bys. The back-up unit was found to have a frequency scan capability of only about 100 Hz.

Before the Pioneer missions to Jupiter, astronomers had detected a pattern of alternating easterly and westerly winds in zones parallel to Jupiter's equator. The only exception to this so-called 'zonal flow' that had been observed was the anticyclonic flow around the Great Red Spot (GRS). Unfortunately, the relatively poor resolution of the Pioneer images meant that they provided little dynamic information on Jupiter's cloud system, but the Pioneers did show a more complex pattern of clouds than seen from Earth. The resolution of the Voyager images was far superior to those from the Pioneers,[4] however, and even when millions of kilometres away the cloud motions became very clear. In place of the simple zonal motion, astronomers could see a planet-wide pattern of small-scale vortices (see Figure 9.2) that were forever changing on timescales of only hours. The most spectacular pattern was that surrounding the GRS, where the clouds were ripped apart in regions of very high shear. In addition there were numerous red, brown, white and blue spots scattered over the planet, only some of which had been seen from Earth. The movie sequence produced by splicing together the 4.8 minute images taken between 30th January and 3rd February, which covered 10 Jupiter 'days', were some of the most spectacular astronomical images ever seen.

On 22nd January 1979 Voyager 1 detected low energy charged particles from Jupiter when it was still some 600 R_J (about 50 million kilometres) from the planet. The spacecraft then crossed the bow shock on 28th February at 86 R_J and, as in the case of the Pioneers, crossed and recrossed that and the magnetopause several times over the next few days as the intensity of the solar wind varied. Then two days before closest approach NASA lost contact with Voyager 1! The problem turned out to be with the DSN receiving station in Australia which was being battered by a severe thunderstorm at the time. Fortunately, this problem only caused a three hour interruption in data, although such a loss two days later could have bordered on the catastrophic.

One of the most dramatic discoveries of Voyager 1's fly past of Jupiter was made as the spacecraft flew through the planet's equatorial plane some 17 hours before closest approach. For centuries Saturn had been known to have rings and in 1977, five months before Voyager 1 took off for Jupiter, faint rings had also been discovered around Uranus. So did Jupiter have rings? Although the Pioneers had not seen a ring system around Jupiter, in December 1974 Pioneer 11 had detected a reduction of high energy particles

---

[4] Thirty days before closest approach, when Voyager 1 was still 30 million km away from Jupiter, the spatial resolution of its images already exceeded the best achieved by the Pioneers.

Figure 9.2 Voyager 1 took this image of Jupiter whilst still some 33 million km from the planet, clearly showing the Great Red Spot south of the equator. The zoning seen from Earth is clearly visible, but the detailed, complex structure of the clouds at the edge of and within these zones was a revelation. (Courtesy National Space Science Data Center, World Data Center-A for Rockets and Satellites, NASA; Team Leader, Dr Bradford A. Smith; P-20993.)

Table 9.1. *The Voyager 1 closest approaches to Jupiter's satellites (in time order)*

	Dia. (km)[a]	Dist. from centre of planet (km)	Closest approach Time[b] (hr:min)	Dist. (km)[c]	Best surface resolution (km)
Amalthea	270 × 150	181,300	*J* − 6 h	420,000	8
Jupiter	142,984		*J*	277,000	8
Io	3,638	421,600	*J* + 03:05	19,000	0.6
Europa	3,126	670,900	*J* + 6 h	732,000	33
Ganymede	5,276	1,070,000	*J* + 13:53	112,000	2
Callisto	4,848	1,883,000	*J* + 29:04	124,000	2.3

Notes:

[a] For comparison our Moon is 3,476 km in diameter and Mercury is 4,880 km in diameter.

[b] The time *J* of closest approach to Jupiter was 12:04 UT spacecraft time on 5/3/79. The signals were received on Earth 38 minutes later.

[c] This is the distance above the surface of the satellites or the cloud tops of Jupiter.

near its closest approach distance of 0.6 R_J above Jupiter's cloud-tops. This led Mario Acuña and Norman Ness to suggest in 1976 that there was either an undiscovered ring or satellite there. Such an intriguing prospect could not possibly go unchecked, so Tobias Owen and Candy Hansen asked that an attempt be made using Voyager 1 to find the possible ring or satellite. As a result, 17 hours before closest approach a single 11 minute exposure was made looking to the side of Jupiter, and much to most astronomers' surprise a faint, thin ring was seen.

Jupiter was known to possess thirteen satellites at the time of the Voyager 1 encounter, but because of the spacecraft's intercept trajectory only five would show significant disks in the spacecraft's images. These were tiny Amalthea, which had been discovered by Edward Barnard in 1892, and the four Galilean satellites Io, Europa, Ganymede and Callisto (see Table 9.1). Although these satellites were imaged from time to time as Voyager 1 approached Jupiter, the best images were generally received during the spacecraft's closest approach.

Voyager 1's first closest approach was to Amalthea (see Figure 9.3), just six hours before the fly past of Jupiter. Amalthea is about the size of a large asteroid, and is about ten times the size (in linear dimensions) of the two small satellites of Mars, but about ten times smaller than Jupiter's Galilean

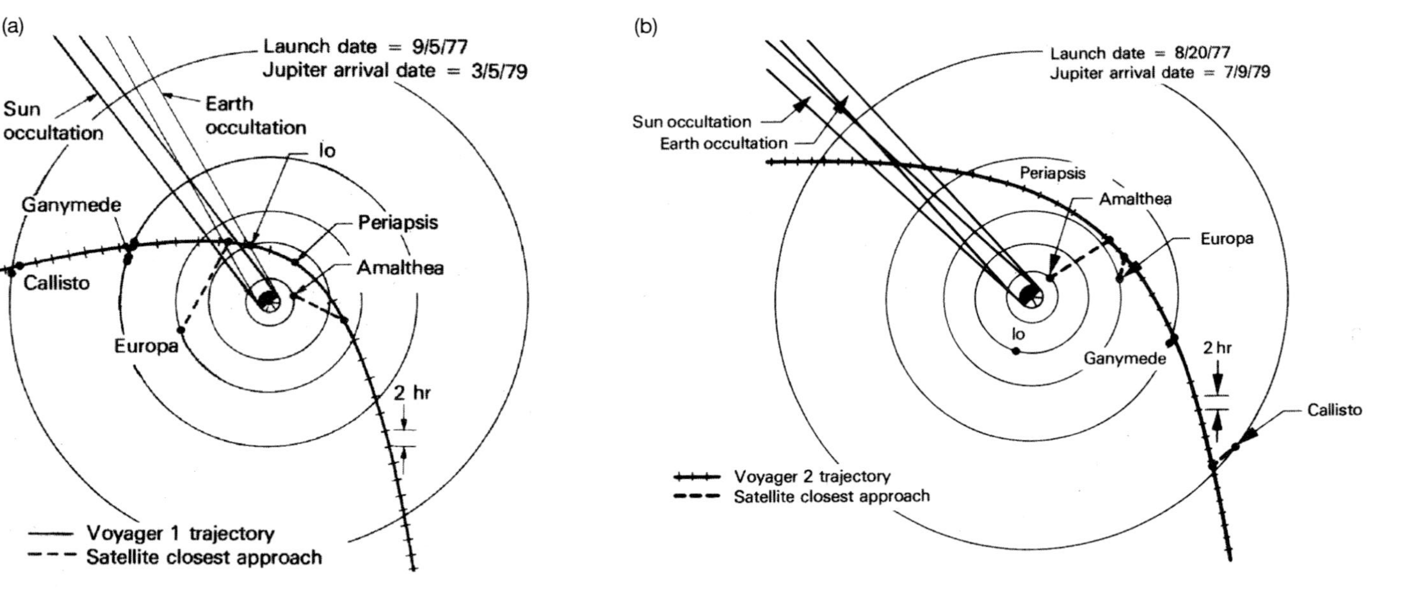

Figure 9.3 The trajectories of Voyagers 1 and 2 were chosen to optimise their imaging of the Galilean satellites. In the case of Voyager 1 (a) the spacecraft first flew past the tiny Amalthea before passing within 277,000 km of Jupiter, whereas in the case of Voyager 2 (b) Callisto was the first satellite to be observed at close quarters. The drawings are taken from a position perpendicular to the equatorial plane of Jupiter, but as neither the satellites nor Voyager travelled exactly in that plane, the closest approach distances indicated by the dotted lines are less than the true distances. (Note the dates are shown as month/day/year). (Courtesy NASA; the second image is a slightly modified version of NASA image 260-533A.)

satellites. It is so small and the distance of closest approach was so great, however, that all that could be seen was a very blurred, irregularly-shaped red image that showed evidence of cratering. Astronomers attributed the red colour to sulphur particles from Io.

Voyager 1 then flew past Jupiter at 12:42 UT[5] on 5th March 1979, followed three hours later by a very close encounter with Io, passing just 19,000 km above its surface. Three hours later was the closest approach to Europa, but the distance was too large to show much surface detail. Finally, Voyager 1 flew just 112,000 km above Ganymede, and then 124,000 km above Callisto 15 hours later.

Prior to the Voyager 1 encounter it was thought that Io would look like a reddish version of our Moon, but covered with sulphur-coated impact craters.[6] It was also thought that the other three Galilean satellites would have ice-covered surfaces with little surface relief and only a few impact craters, as the ice would have flowed and eliminated all but the most recent craters over time. Europa has the highest reflectivity and the highest density of these three icy satellites, so, although it was thought to have water ice[7] on its surface, it was thought that this is probably in a shell only about 100 km thick. Ganymede and Callisto, on the other hand, were thought to be richer in water ice than Europa because of their lower densities.

As mentioned above, it was generally thought, prior to Voyager 1, that Io's surface would be old and cratered like the Moon. But in the 2nd March issue of *Science*, published just three days before Voyager 1's closest approach, Stanton Peale of the University of California and Patrick Cassen and Ray Reynolds of NASA Ames came up with an alternative theory. They pointed out that since Io is subjected to resonant gravitational forces by the other Galilean satellites, its orbit would be eccentric, although over time it averages out as circular. As Io is so close to Jupiter, however, the planet's gravity would produce powerful tidal forces in Io's crust,[8] causing the satel-

---

[5] This is the time that the signals were received on Earth. The closest approach had actually taken place 38 minutes earlier (see Table 9.1).

[6] The presence of sulphur was deduced from Io's colour and from its ultraviolet reflectance spectrum.

[7] Water ice had been detected in Europa's near infrared spectrum.

[8] Io spins on its axis once at it orbits Jupiter once (that is, its spin is synchronous with its orbital period). If Io's orbit is circular, its Jupiter-facing bulge caused by Jupiter's gravity would always be at the same place on Io's surface, so there would be no surface flexing. As the orbit is not circular, however, this bulge moves on the surface of Io, causing it to flex and heat.

(a)

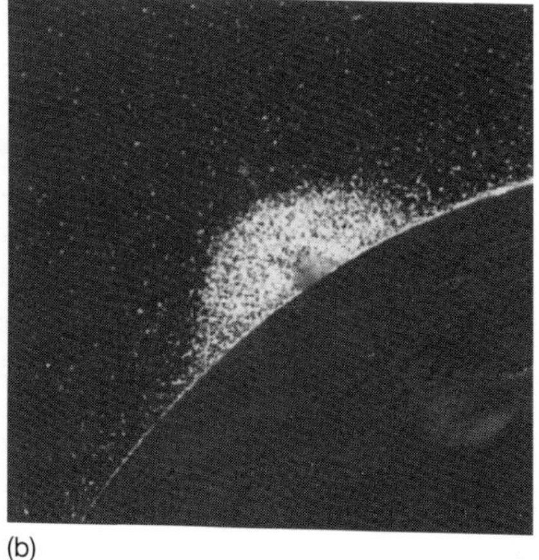
(b)

Figure 9.4 (a) This Voyager 1 image of Io shows its mottled, fresh-looking surface. The circular, doughnut-shaped structure at the centre is a volcano that was seen to be erupting in other images. Voyager 1 was still 860,000 km from Io when this image was taken. (Courtesy NASA and NSSDC; Voyager 1 FDS 16368.36.) (b) The second of the volcanic plumes on Io to be detected by Voyager 1 rises some 200 km above the surface of the satellite. (Courtesy National Space Science Data Center, World Data Center-A for Rockets and Satellites, NASA; Team Leader, Dr Bradford A. Smith; P-21334.)

lite to heat up. As a result, these three astronomers suggested that there would be widespread volcanism on Io.

Initially as Voyager 1 closed in on Io a number of small dark dots were seen surrounded by faint rings. These were thought at first to be the expected impact craters but later, high resolution images showed no such craters. Even at closest approach, with a resolution of 0.6 km, no craters could be seen, implying a surface age of less than about 1 million years. What could be seen was not the old, cratered surface generally expected, but a fresh-looking landscape (see Figure 9.4a) showing numerous volcanic calderas and rivers of lava. Peale, Cassen and Reynolds had been absolutely correct in their predictions published just three days before. With such a young surface (in astronomical terms) of less than 1 million years old, and with so many volcanic caldera, it was possible that some volcanoes were still active, although the chances of seeing one erupting was considered to be very slim. But the astronomers were in for another shock.

On 8th March, just three days after closest approach, Linda Morabito of JPL was looking at images of Io, taken from a distance of 4.5 million kilometres, that had been deliberately over-exposed to show stars as part of a navigational investigation. On one of these images she was surprised to see an umbrella-shaped plume on the limb of Io, reaching about 270 km above the surface. It was not an artefact, but was the plume of an erupting volcano. An immediate search of previous images was then undertaken by astronomers called in to help, and these showed even more plumes (see Figure 9.4b). Eventually it was found that there had been eight volcanoes active during Voyager 1's closest approach, producing plumes varying from 70 to 300 km in height with vent velocities as high as 1.0 km/s (2,200 mph). Rather than being as dead as the Moon, the surface of Io was found to be far more active than that of the Earth.

In contrast to Io, Europa was found by Voyager 1 to be very bland, with surface makings of low contrast. Very little surface detail could be resolved, because the fly-by distance was so large, but numerous dark stripes were seen, tens of kilometres wide and up to thousands of kilometres long, criss-crossing the surface. It was thought that they may be faults or fractures caused by tectonic activity. These tantalising glimpses of Europa whetted the appetite of astronomers for the much closer fly-by of Voyager 2, just four months later.

Ganymede, the largest of Jupiter's satellites, was seen to have a complex intersecting pattern of parallel grooves and ridges together with numerous rayed craters on a basically two-toned surface. Many of the rayed craters were seen to be very light in colour, probably because the ice there is fresher and more powdery than on the surface as a whole. Interestingly, some of the grooves showed lateral offsets, indicating fault lines where the surface had moved laterally.

The surface of Callisto appeared to be very old and relatively dark compared with Ganymede, and it is almost saturated with craters. The craters are quite shallow, however, and the limb of Callisto shows that there is virtually no surface relief. The largest feature is the large bright impact structure, now called Valhalla, which is surrounded by numerous concentric rings of ridges up to 3,000 km in diameter. Crater densities in the inner part of this ring system are less than elsewhere, indicating that the large impact was not the earliest feature still visible on the surface.

9.3 Voyager 2 at Jupiter

In July 1979, four months after the Voyager 1 encounter, Voyager 2 flew past Jupiter and its satellites along the trajectory shown in Figure 9.3b. In general the fly-by distances were not as close as for Voyager 1 (compare Tables 9.1

Table 9.2. *The Voyager 2 closest approaches to Jupiter's satellites (in time order)*

	Dia. (km)	Dist. from centre of planet (km)	Closest approach Time[a] (hr:min)	Closest approach Dist. (km)[b]	Best surface resolution (km)
Callisto	4,848	1,883,000	*J* − 33:16	213,000	4
Ganymede	5,276	1,070,000	*J* − 14:23	60,000	1
Europa	3,126	670,900	*J* − 03:46	205,000	4
Amalthea	270 × 150	181,300	*J* − 02:18	560,000	11
Jupiter	142,984		*J*	650,000	15
Io	3,638	421,600	*J* + 01:40	1,130,000	21

Notes:

[a] The time *J* of closest approach to Jupiter was 21:37 UT spacecraft time on 9/7/79. The signals were received on Earth 52 minutes later.

[b] This is the distance above the surface of the satellites or the cloud tops of Jupiter.

and 9.2),[9] but this did not prevent it returning stunning images of Jupiter (see Figure 9.5). In fact, a comparison between the Voyager 1 and 2 images enabled the relative velocities of its various belts and zones to be determined. The GRS, which appeared more uniform in colour than during the Voyager 1 intercept, was found to have drifted west by about 0.26° per day, relative to the rotation rate of Jupiter's core as deduced from variations in its radio emissions. The maximum wind velocity, relative to the core, was found to be about 150 m/s (340 mph) at low latitudes (see Figure 9.8, later).

Although Voyager 2 passed much closer to Europa than Voyager 1, the resolution was still not sufficient to show any structure in the numerous long dark stripes found all over the visible surface, and so help to determine their origin. No large craters were seen, but a few 20 metre diameter craters were observed near the terminator suggesting that, although the surface of Europa is younger than that of Ganymede or Callisto, it is still a few hundred million years old, on average. Voyager 2 also showed that the surface of Europa (see Figure 9.6) has virtually no surface relief above about 100 m high,

[9] Voyager 2 found that Jupiter's radiation environment during its encounter was three times stronger than during the Voyager 1 encounter. Had Voyager 2 flown as close to Jupiter as Voyager 1, it may well have been seriously damaged by this level of radiation. It was pure luck that the Voyager 2 fly-by distance was planned to be further away from Jupiter than that of its predecessor.

Figure 9.5 Part of Jupiter's Great Red Spot is seen top centre, together with a large white spot and associated turbulence in this Voyager 2 image. (Courtesy NASA/JPL/Caltech; PIA 00372.)

which is consistent with the pre-Voyager concept of a 100 km thick ice layer covering the surface. Tidal heating and/or radioactive heating could be sufficient to melt the lower levels of this ice crust, and so some astronomers wondered whether the surface of Europa was composed of pack ice floating on water, with the linear patterns showing where the ice had fractured.

During the Voyager 2 fly-by the spacecraft found that Pele, Io's most active volcano during the Voyager 1 intercept, had become dormant, although six of the seven other volcanoes[10] seen by Voyager 1 were still active, and two new volcanic vents were seen. Voyager images showed what appeared to be clouds or white surface deposits along scarps or faults, particularly in the

[10] The seventh volcano was out of sight of Voyager 2.

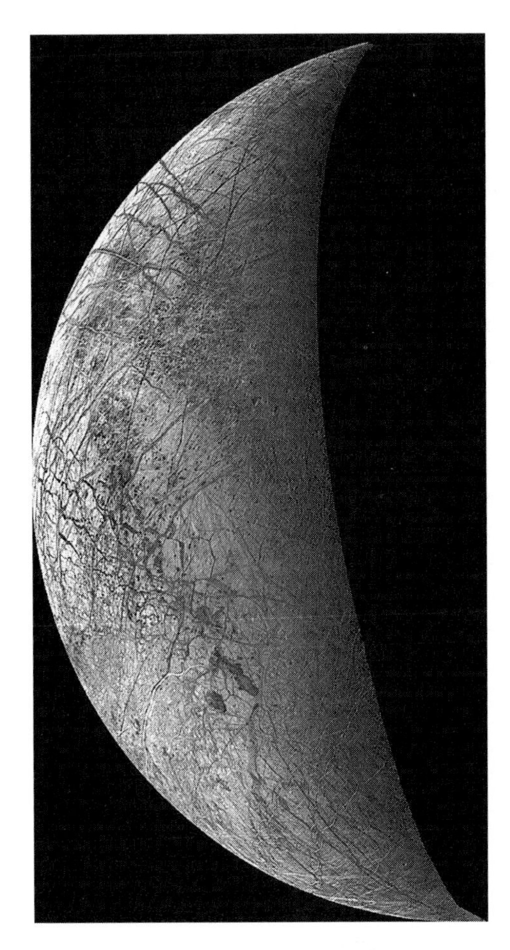

Figure 9.6 The highest resolution image of Europa taken by Voyager 2 shows enigmatic long linear features. The maximum surface relief observed was only about 100 metres. (Courtesy Paul M. Schenk, Lunar and Planetary Institute

south polar region, and spectrophotometric analysis indicated that these white particles were composed of sulphur dioxide and sulphur. Sulphur dioxide gas was also discovered near the volcano Loki. So both sulphur dioxide and sulphur appear to be present on Io's surface and the volcanoes could be powered by either or both of these materials.

Voyager 2 imaged Jupiter's ring as the spacecraft approached the planet and also repeated the observations as the spacecraft receded. Much to everyone's surprise the ring appeared more than twenty times brighter in forward-scattered light than in back-scattered light, implying that many of the ring particles are only about one to two microns in diameter. This was very interesting as such small particles can only exist in the ring for a short period of time, implying that there must be some resupply mechanism. It was thought that this could be volcanic dust from Io, or the result of interplanetary micrometeoroids colliding with larger ring particles. Four months after the Voyager 2 encounter Mark Showalter and colleagues found that they could distinguish three components in Jupiter's ring. A main ring was found to extend from 1.72 to 1.81 R_J from Jupiter's centre, with a broad faint 'halo' ring inside it, and a much broader faint 'gossamer' ring outside it. Two previously undetected small satellites (now called Metis and Adrastea) were discovered near to the outer edge of the main ring, possibly constraining its outward expansion and/or helping to supply it with material.

9.4 Voyager 1 at Saturn

In December 1979 radio contact was suddenly lost with Voyager 1, which was now well on its way to fly past Saturn in November 1980. As mentioned previously, one of the two redundant receivers on Voyager 2 had already failed in April 1978, and mission control were concerned that a similar or even worse problem had afflicted Voyager 1. Fortunately, the Voyager 1 problem turned out to be one of spacecraft orientation, and communications were slowly re-established via the low-gain antenna. Full radio contact was restored after three days of nail-biting anxiety, but it took three more days before spacecraft operations were back to normal.

Shortly after Voyager 1 was reconfigured, combined data from both Voyagers enabled the rotation rate of Saturn to be determined for the first time. Previous estimates of Saturn's rotation period had been made by timing the movement of infrequent, temporary spots on the planet as seen from Earth, but these produced results varying from 10 h 14 min at the equator to 10 h 40 min at mid- to high-latitudes. The new Voyager measurements, which yielded a period of 10 h 39.4 min, were based on variations in Saturn's radio emissions, which were thought to be linked to the rate of rotation of Saturn's core. This implied that the 10 h 14 min spot was moving around Saturn at the surprisingly high velocity of about 400 m/s (900 mph) relative to this rate of rotation.

Before the Voyager 1 fly-by of Saturn there was great a great deal of confusion as to how many satellites Saturn possesses. Phoebe, Saturn's ninth satellite, had been discovered in 1899, then in 1966, when Saturn's rings were edge-on as seen from Earth,[11] Andouin Dollfus photographed a tenth

[11] This is the best time to see faint satellites because light from the rings is then at a minimum.

satellite from the Pic du Midi Observatory in France. His discovery was confirmed by two independent observers and called Janus. Then in 1977 Stephen Larson and John Fountain of the University of Arizona announced that they had found another satellite at about the same distance from Saturn as Janus, when re-examining plates taken in 1966. The exact orbits of both Janus and the eleventh satellite, now called Epimetheus, were uncertain because of limited data, then in September 1979 Pioneer 11 accidentally flew within about 2,500 km of Epimetheus. But it was not until March 1980 that it became evident that Janus and Epimetheus were in essentially the same 16.67 hour orbit.

That was not all, as in March 1980 Jean Lecacheaux and P. Lacques at the Pic du Midi Observatory found yet another satellite, which was recorded just hours later by astronomers at the Catalina Observatory in the United States. This satellite, initially designated 1980 S6 but now called Helene, was subsequently found to be near the 60° Lagrangian point ahead of Dione and co-orbiting with it. So as Voyager 1 closed in on Saturn for its November 1980 fly past, it appeared as though Saturn had twelve satellites visible from Earth with known orbits,[12] of these it was planned to image eight of the nine largest during the Voyager 1 encounter and all nine during the Voyager 2 encounter (see Table 9.3 and Figure 9.7). The first of these to be visited, Titan, was by far the most important because of the similarity between its atmosphere and that hypothesised for the early Earth.

Systematic imaging of Saturn by Voyager 1 started on 25th August, some 80 days before the close encounter and, much to the relief of the waiting astronomers, these early images started to reveal cloud features on the planet after they had been computer-enhanced. So it looked as though the bland images returned by Pioneer 11 would give way to some that could be used to study Saturn's atmospheric dynamics. These and later images eventually showed that the wind velocity at cloud-top height reaches a maximum of almost 500 m/s (1,100 mph) practically on the equator (see Figure 9.8b). This peak velocity is about three times that on Jupiter, and the equatorial jet stream on Saturn, of which it is part, covers the region from about 30°N to 40°S (where the velocity falls to zero), compared with Jupiter's jet, which only covers the region from about 12°N to 16°S (see Figure 9.8a). Voyager 1 showed that Saturn emits about 1.8 times as much

[12] Two other satellites, 1980 S13 and S25, now called Telesto and Calypso, had been imaged by Bradford Smith and colleagues in March and April 1980, but their orbits had not been well defined by the time of the Voyager 1 fly-by of Saturn. They were found to be at the Lagrangian points 60° in front of and 60° behind Tethys and co-orbital with it.

Table 9.3. *The Voyager 1 and 2 closest approaches to Saturn's satellites (in time order)*

	Dia. (km)	Dist. from centre of planet (km)	Closest approach Time[a]	Closest approach Dist. (km)[b]	Best surface resolution (km)
Voyager 1					
Titan	5,150	1,221,850	$S-18{:}05$	4,400	0.5
Tethys	1,050	294,660	$S-01{:}30$	416,000	
Saturn	120,536		S	124,000	3
Mimas	390	185,520	$S+00{:}58$	104,000[c]	
Enceladus	500	238,020	$S+02{:}05$	201,000	
Dione	1,120	377,400	$S+03{:}42$	162,000[c]	
Rhea	1,530	527,040	$S+06{:}34$	73,000	1.5
Hyperion	410 × 220	1,481,000	$S+16{:}58$	880,000	
Iapetus	1,440	3,561,300	$S+2$ d	2,470,000	
Voyager 2					
Iapetus	1,440	3,561,300	$S-3{:}13{:}54$	909,000	
Hyperion	350 × 200	1,481,000	$S-1{:}13{:}57$	470,000	9
Titan	5,150	1,221,850	$S-17{:}46$	666,000	
Dione	1,120	377,400	$S-2$ h	502,000	
Mimas	390	185,520	$S-1$ h	310,000	
Saturn	120,536		S	100,000	2
Enceladus	500	238,020	S	87,000	2
Tethys	1,050	294,660	$S+3$ h	93,000	
Rhea	1,530	527,040	$S+4$ h	645,000	
Phoebe	220	12,952,000	$S+10$ d	1,473,000	

Notes:

[a] The time S of closest approach to Saturn was 23:43 UT spacecraft time on 12/11/80 for Voyager 1 and 03:23 UT on 26/8/81 for Voyager 2. The signals were received on Earth 1 h 25 min and 1 h 27 min later for Voyagers 1 and 2, respectively. Times shown in the table are hr:min or day:hr:min.

[b] This is the distance above the surface of the satellites or the cloud tops of Saturn.

[c] This is the closest approach when the satellite was still fully illuminated. The closest approach actually took place when the satellite was in Saturn's shadow.

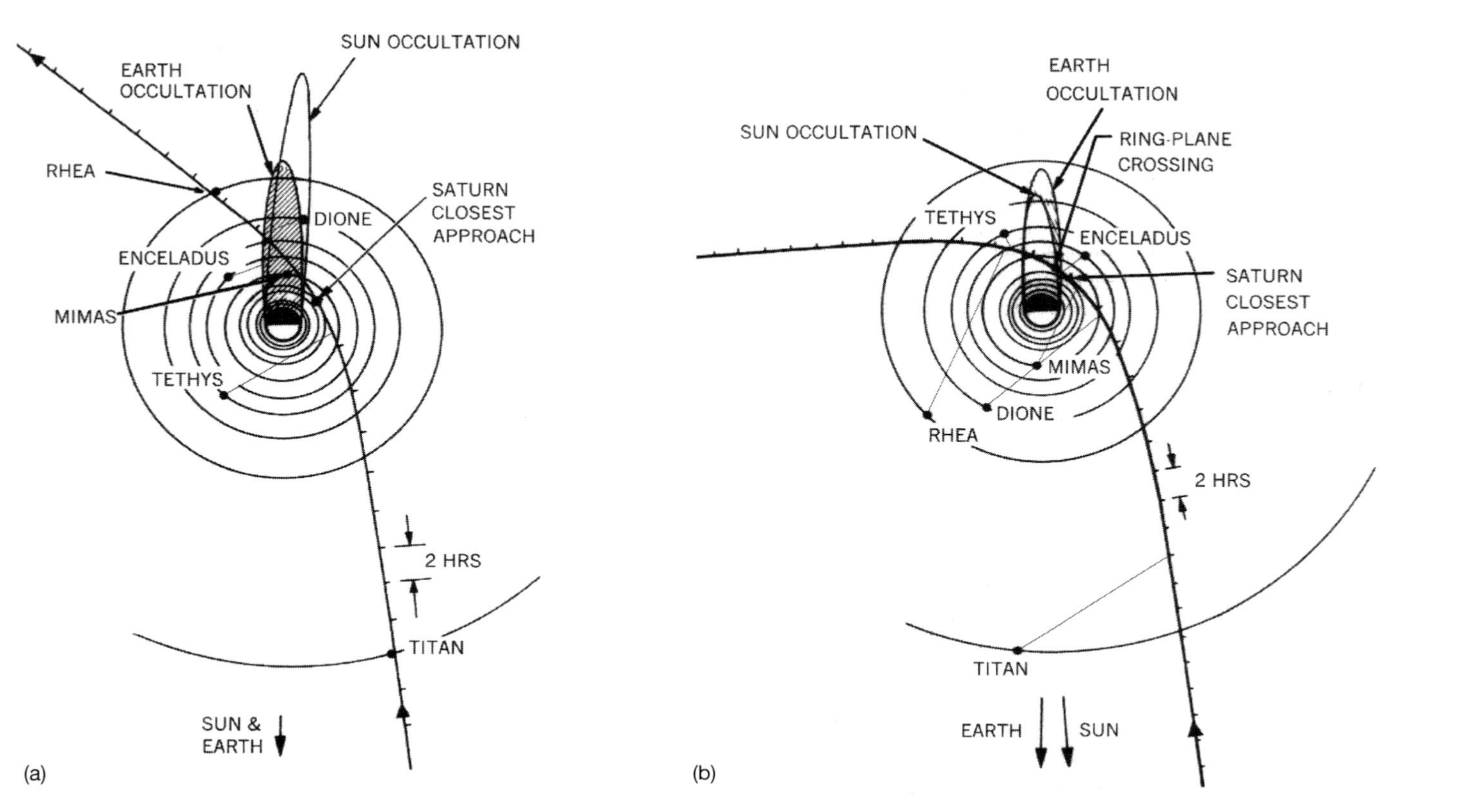

Figure 9.7 The trajectories of Voyagers 1 (a) and 2 (b) are shown during their fly past of the Saturnian system in November 1980 and August 1981, respectively. (Based on NASA drawings 260-845A and P-23349.)

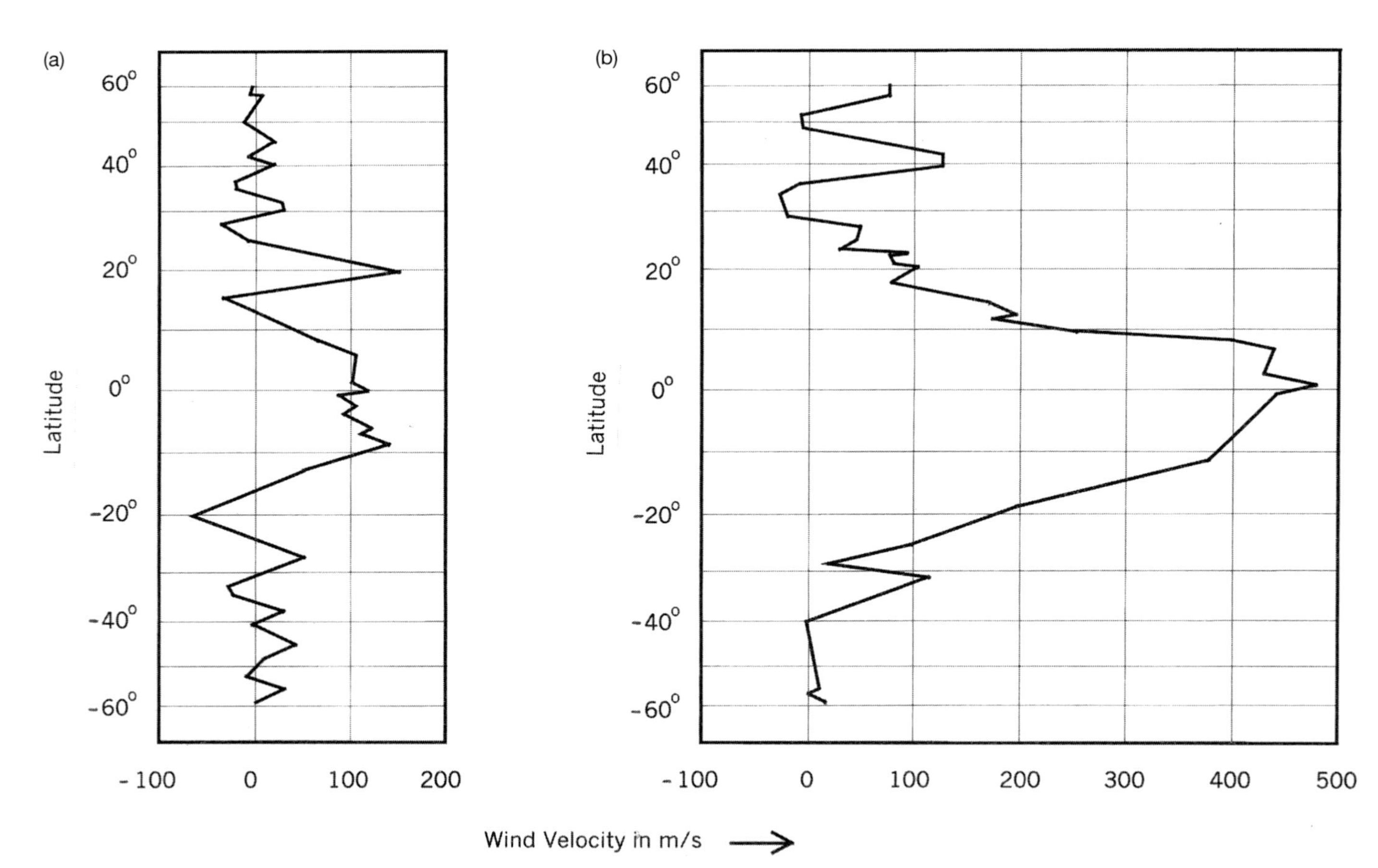

Figure 9.8 The cloud-top wind velocity versus latitude is shown for Jupiter (a) and Saturn (b). Saturn's equatorial jet is seen to be both broader and with a higher peak velocity than that of Jupiter.

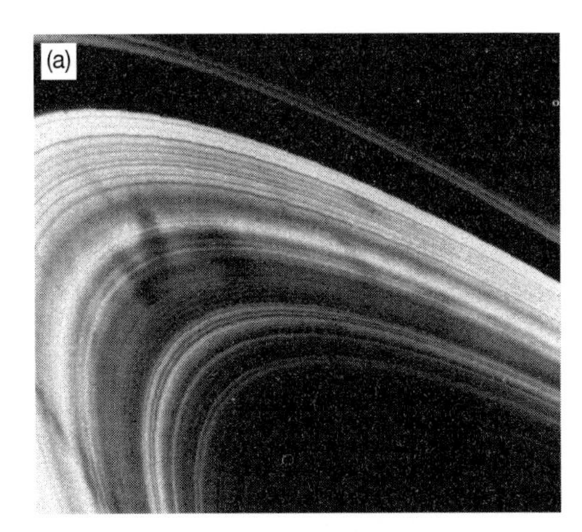

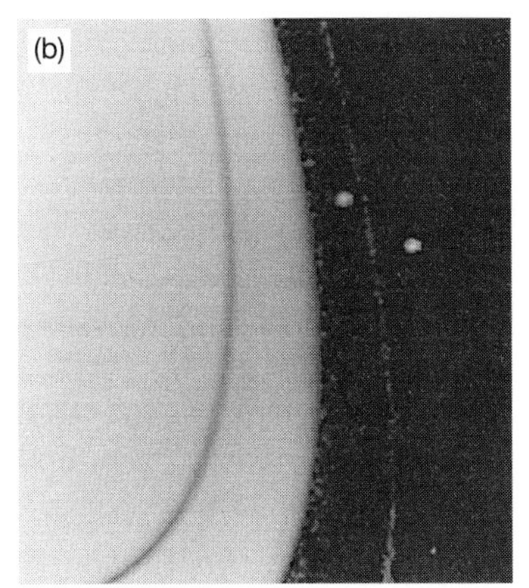

Figure 9.9 (a) The spokes in Saturn's B ring are shown in this image taken from a distance of 2.5 million kilometres. (Courtesy NASA/JPL/Caltech.) (b) The satellites Prometheus and Pandora shepherding the F ring. (Courtesy National Space Science Data Center, World Data Center-A for Rockets and Satellites, NASA; Principal Investigator, Dr Bradford A. Smith; P-23911.)

energy as it receives from the Sun, which is, within error, the same ratio as that deduced for Jupiter. So it is still not clear why Saturn, which receives much less solar energy than Jupiter and is somewhat smaller, should have much the more powerful equatorial jet.

The next surprise was the discovery of 'spokes' in the B ring of Saturn (see Figure 9.9a) by Richard Terrile whilst Voyager 1 was still over five weeks from closest approach. Radial shadings on ring A and occasionally on ring B had been observed from time to time in the last hundred years or so, but it had never been clear whether these shadings were real or not. About a week after the first images of the spokes, two new satellites designated 1980 S26 and S27, and now called Prometheus and Pandora, were found by Andy Collins and Richard Terrile, one just inside and one just outside the narrow F ring (see Figure 9.9b) which had been discovered by Pioneer 11. These two satellites appeared to be shepherding or stabilising the ring. Shepherding satellites had first been proposed by Peter Goldreich of Caltech and Scott Tremaine of the University of Toronto in a 1979 *Nature* paper to explain the narrow rings of Uranus, but these two shepherding satellites of Saturn were the first such satellites to be discovered. Then in early November Terrile found another small satellite, 1980 S28, now called Atlas, orbiting Saturn just 800 km outside the outer edge of ring A, and apparently restricting its outward expansion.

The excitement mounted as Voyager 1 approached closer and closer to Saturn and then, just four days before closest approach, a thunderstorm at the NASA DSN receiving station near Madrid caused a breakdown in spacecraft communications. Once more rain had stopped play[13] with the consequent loss of scientific information, but the break was strictly temporary and soon the data was being received again.

As Voyager 1 approached Saturn the smooth rings A, B and C seen from Earth began to break up into more and more individual ring components. Then 18 minutes after its closest approach to Titan, the spacecraft passed through the ring plane whilst still over 1 million kilometres outside the outer edge of ring A, before passing under the rings and recrossing their plane between the orbits of Dione and Rhea, about 240,000 km outside the outer edge of ring A. This enabled the structure of the rings to be analysed in detail in both reflected and transmitted light.

The ring system of Saturn was found to consist of about 500 to 1,000 narrow rings, and even the Cassini division between rings A and B, which is seen to be dark from Earth and was assumed to be empty, was found to contain over 100 narrow rings. The A ring was found to be bright in both reflected and transmitted light and to be relatively smooth and unstructured, whereas the B ring is bright in reflection but almost opaque to transmitted light. The B ring is also highly structured, consisting of hundreds of concentric rings with dusky radial, spoke-like markings. These spokes caused the theoreticians a real headache, as they appeared to contravene Kepler's Second Law. This requires the particles further away from Saturn to orbit the planet at a slower angular rate than those nearer in, thus destroying the spokes very quickly. But it was soon realised that the spokes rotate around Saturn at the same angular rate as the planet rotates on its axis and so they must, therefore, be associated with its magnetic field.

The relatively dark C ring was brighter in transmission than reflection, with a narrow gap between its outer edge and the inner edge of ring B, which showed up bright in transmission. A very faint D ring was found inside ring C, apparently confirming Guerin's observation of it in 1969, but it is unlikely that he saw the ring as it appears to be too dark to be seen from Earth. The narrow ring F, discovered by Pioneer 11 just outside ring A, was found to be three intertwined or braided rings, apparently contravening the laws of dynamics. Like the spokes in the B ring, however, it was suggested that this braided ring could also be responding to Saturn's magnetic field. A very faint, relatively narrow ring, ring G, was found between the F ring and the orbit of Mimas, at the position predicted by the absence of charged par-

---

[13] The same had happened with the Voyager 1 fly-by of Jupiter, as described earlier, but the problem in that case was caused by a thunderstorm in Australia.

ticles as detected by Pioneer 11 (see above). Finally, the very large and faint ring E was recorded stretching 300,000 km from about the orbit of Mimas, past the orbits of Enceladus, Tethys and Dione to almost that of Rhea.[14]

Titan, by far the largest of Saturn's satellites and the second largest satellite in the Solar System (behind Ganymede), was the first on Voyager 1's trajectory. The spacecraft flew by just 4,400 km above Titan's cloud tops some 18 hours before its closest approach to Saturn. Unfortunately, the images of Titan returned were featureless, with only the faintest shadings being visible after extensive computer processing, although images of the limb showed an extensive haze layer above the impenetrable cloud layer.

In 1944 Gerard Kuiper had detected methane lines in Titan's spectrum, but before Voyager 1's fly-by no other constituents of its atmosphere were known, although the methane lines were known to suffer from pressure-broadening. Some astronomers thought that Titan's atmosphere probably consisted of methane, with only traces of other gases, at a surface-level pressure of about 20 millibars. This would explain the pressure broadening, but Donald Hunten of the University of Arizona suggested that the same result could be obtained if the main atmospheric constituent was nitrogen, which was undetectable from Earth, with methane being present at a much lower percentage. Measurements with the partially completed VLA radio telescope implied that the surface temperature of Titan was about 87 K, and this led Hunten to conclude, using his model of Titan's atmosphere, that the surface pressure was about 2,000 millibars (or twice that of the Earth).

Voyager 1's radio occultation experiment showed that Titan's surface-level atmospheric pressure is actually about 1,500 millibars at a temperature of 94 K, thus vindicating Hunten's atmospheric model. The ultraviolet spectrometer on board the spacecraft also showed that Titan's atmosphere is approximately 90% nitrogen,[15] in line with Hunten's prediction. The minimum atmospheric temperature detected was about 68 K at a pressure of 380 millibars about 50 km above the surface. Nitrogen could condense into droplets under those conditions and form clouds.

The orange colour of Titan seen from Earth was found to be due to a layer of photochemical smog, which is produced when ultraviolet sunlight breaks up the atmospheric methane and nitrogen molecules into their constituent

---

[14] So the order of the rings is (from Saturn outwards) D, C, B, A, F, G and E.

[15] It was originally thought that Voyager 1's measurements indicated a nitrogen concentration of 99%, but after Robert Samuelson had reanalysed the data he suggested that Titan's atmosphere may contain up to about 12% argon, a gas that cannot be detected spectroscopically.

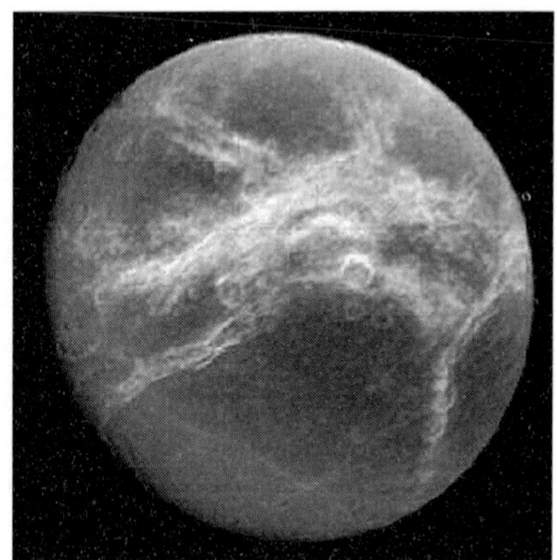

Figure 9.10 (left image) Mimas showing the prominent 135 km diameter crater, now called Herschel, which is up to 10 km deep. (Courtesy NASA/JPL/Caltech.)

Figure 9.11 (right image) Bright streaks are seen crossing the dark side of Dione in this contrast-enhanced image. (Courtesy NASA and NSSDC; Voyager 1 FDS 34933.38.)

atoms. These then recombine to form the various hydrocarbon and nitrogen compounds, including ethane, propane and hydrogen cyanide, found by Voyager 1 in small quantities in Titan's atmosphere. Finally, the surface temperature of 94 ± 2 K is tantalisingly close to methane's triple point of 90.7 K, at which methane can exist in solid, liquid and gaseous states. So the surface of Titan may possibly have oceans of methane, with cliffs of methane in its polar region.

After the very close fly-by of Titan, Voyager 1 flew past most of Saturn's other large satellites at distance ranging from 73,000 km for Rhea, up to 880,000 km for Hyperion and over 2 million kilometres for Iapetus (see Table 9.3). The first to be intercepted was Tethys, but this was only seen at a distance of over 400,000 km and little surface detail could be resolved, apart from its generally cratered appearance and a long north–south gorge about 4 km deep. Mimas, which was the next satellite on the itinerary, was found to have a 135 km diameter crater on its 390 km diameter surface (see Figure 9.10). Not only is the crater very large compared with the size of the satellite, but it is an incredible 10 km (33,000 ft) deep. Enceladus was, like Tethys, poorly seen by Voyager 1, but even at a distance of 201,000 km it was expected that some craters would be seen. In the event no surface topography could be observed on its very bright surface.

Dione was seen to have a dark- and a light-coloured hemisphere, with broad white streaks criss-crossing the dark side (see Figure 9.11) which appeared to be relatively smooth. The light side, on the other hand, was seen to have a number of sinuous valleys on its moderately cratered surface. Rhea was found to be similar in appearance to its smaller neighbour Dione, having one heavily cratered hemisphere and a darker hemisphere that was not nearly as heavily cratered. Also like Dione it has broad white streaks on its surface, although the contrast difference between the streaks and the adjacent surface is not as high. Hyperion and Iapetus were not well imaged by Voyager 1, but the images were good enough to show that Hyperion is very irregular in shape. A satellite of its size (410 × 220 km) should be spherical, so it appears to be the remnant of a body that has been broken up by an impact. Iapetus had been known for some time to have one hemisphere with about ten times the reflectivity of the other, and Voyager 1 showed that the bright hemisphere is heavily cratered with the dark hemisphere appearing to be featureless.

Voyager 1 then left the Saturnian system with no more planets on its itinerary, and on 19th December 1980 the two cameras were switched off, but some of the other instruments were left on to measure the interplanetary environment. Astronomers were also hoping that the spacecraft would still be operational when it passed through the heliopause into interstellar space, possibly about the end of the century.

9.5 **Voyager 2 at Saturn**

In the meantime, whilst Voyager 1 had been undertaking its Saturn mission, Voyager 2 had flown past Jupiter in 1979, as described above, and was now on its way to fly past Saturn in August 1981. Before it could do so, however, it was necessary to decide on whether to continue Voyager 2 on its present course, which would allow it to visit Uranus in 1986 and Neptune in 1989, after its intercept with Saturn, or whether to modify that trajectory to fly closer to Titan and forego the Uranus and Neptune fly-bys. In order to make a decision, NASA analysed the performance of Voyager 1 at Titan and the health of Voyager 2, and on 8th January 1981 announced that Voyager 2 would continue on its present course and visit Saturn, Uranus and Neptune.

Contrary to earlier proposals it was decided not to send Voyager 2 through the ring plane along the Pioneer 11 trajectory, but to undertake the whole approach phase to Saturn above the ring plane. The Sun would illuminate the top of the rings at an angle of about 8° to the ring plane during the Voyager 2 encounter, compared with an angle of 4° for the Voyager 1 mission, and the Voyager 2 intercept trajectory would make an angle of about 15° to the ring plane, which was also higher than for Voyager 1. As a result the Voyager 2 images of the rings should be better than those returned by its predecessor, and their fine structure should be determined to an even greater level of detail by analysing stellar occultations with the Voyager 2 photopolarimeter.[16] In addition, Voyager 2 would pass closer to Iapetus, Hyperion, Enceladus, Tethys and Phoebe than Voyager 1, and would also image some of the smaller satellites that had been only recently discovered. Unfortunately, the Voyager 2 trajectory, which was optimised for the Uranus intercept, would cross the ring plane only about 1,200 km outside of the newly-discovered G ring. Although this was not expected to cause a problem, it made everyone nervous.

It is difficult to believe it now, but before Voyager 1 arrived at Saturn it was thought that the ring system, which appeared to consist of three broad, featureless rings, was generally understood, with the gaps in the rings being caused by resonances with Saturn's satellites, Mimas in particular. For example, any particles in the Cassini division between the B and A rings would have a period of one half that of Mimas, and so would suffer a pull from Mimas on alternate orbits. As a result these particles would gradually vacate that orbit to produce the vacant Cassini division. There would be similar effects for particles at the inner edge of the B ring, which has a period of one third that of Mimas, or at the inner edge of the C ring, which has a period of one quarter that of Mimas. However, the fine structure of the main A, B and C rings found by Voyager 1, the non-empty Cassini division, the spokes on the B ring, the braided F ring, and the existence of other new rings, demanded a fundamental rethink of the theory of Saturn's rings.

By the time that Voyager 2 arrived at Saturn, nine months after Voyager 1, there had been tentative explanations proposed for many of these new phenomena, with the new rôle of shepherding satellites receiving particular attention. Some astronomers thought that, even though the television cameras on Voyager 2 had better detectors than those on Voyager 1 and the ring imaging conditions were better, the number of individual rings observed by Voyager 1 of 500 to 1,000 would not be exceeded, but this was not so. Voyager 2's images showed more fine structure to the main rings than Voyager 1, and the Voyager 2 photopolarimeter showed even finer structure, with some rings in the main rings being as narrow as a few hundred metres. Such fine structure implied that the rings were very thin, and the A ring, for example, which is about 15,000 km from its inner to outer edge, was estimated to have a thickness of only about 100 m at most.

---

[16] The Voyager 1 photopolarimeter had failed before the Saturn fly-by.

The Voyager 1 images had shown the small satellites Prometheus and Pandora shepherding the F ring, and Atlas apparently preventing the A ring from expanding. So one of the key investigations to be carried out with Voyager 2 was to look for other shepherding satellites to try to explain the fine structure of the rings. The fly-by period was going to be a hectic time, however, so the Voyager 2 investigators decided to concentrate, as far as the rings were concerned, on observing the fine detail of the Cassini division, where Voyager 1 had shown two clear gaps in the ring structure, and the highly-structured B ring, including the spokes.

Voyager 2 investigators found no shepherding satellites in the Cassini division. Had such satellites been found, of course, this would have given a great boost to the shepherding satellite theory of fine ring structure, and more shepherding satellites would have been looked for elsewhere in the system. A negative result does not disprove the theory, however, as the shepherding satellites may be too small to be detected by Voyager 2, but it did make astronomers look for other causes of the fine ring structure.

As mentioned above, the theory that the Cassini division is caused by a 2:1 resonance with Mimas had to be modified when Voyager 1 found that the Cassini division was not empty. But then Carolyn Porco showed, using Voyager 2 data, that the outer edge of the B ring, which is the inner edge of the Cassini division, is not circular (see Figure 9.12) but slightly elliptical, with the major axis precessing once every 22.6 hours. This is the orbital period of Mimas, so the outer edge of the B ring is clearly controlled by this 2:1 Mimas resonance.

In the 1970s Peter Goldreich and Scott Tremaine had shown that spiral density waves, like those thought to occur in galaxies (producing their spiral arms), would also exist in circumplanetary discs of material. In the latter case the density waves would be produced by a satellite orbiting near the disc, producing a pattern of condensations and rarefactions in the disc. Voyager 2 found about fifty such spiral density wave trains in Saturn's A ring, partially explaining its complex structure.

In 1986 Mark Showalter and three colleagues published a paper in *Icarus* analysing the 325 km wide Encke gap in the A ring, and deducing the possible mass, size and orbit of the satellite responsible for producing the gap. This enabled a search to be made for the satellite using the Voyager images. In 1990 it was successful, with the discovery by Showalter, then of NASA-Ames, of a 20 km diameter satellite, now called Pan, orbiting within the Encke gap.

As far as the large satellites were concerned, the main aim of Voyager 2 was to obtain more data on Titan, particularly with the photopolarimeter, to provide higher-resolution images of Iapetus, Enceladus and Tethys than

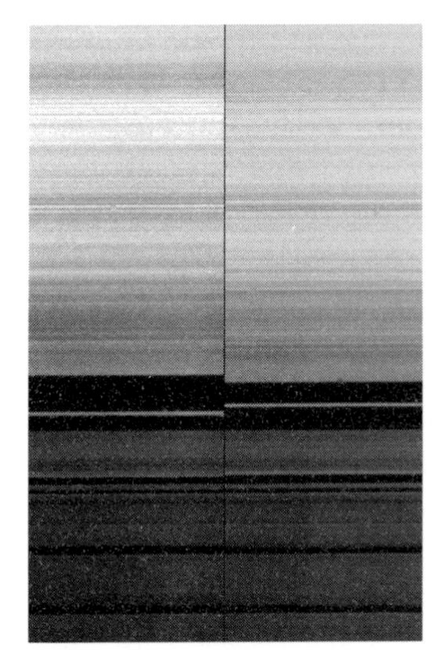

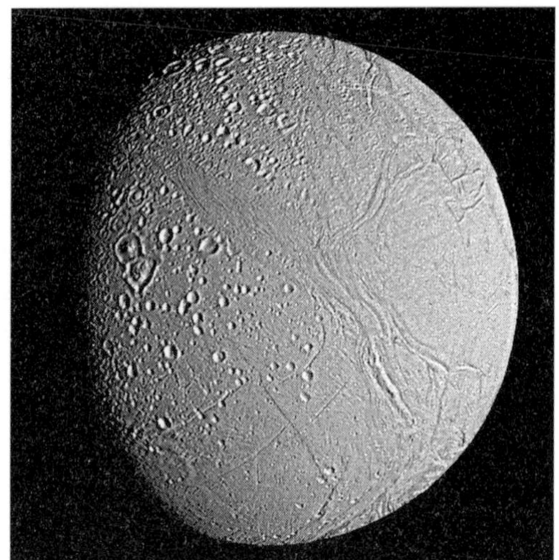

Figure 9.12 (left image) Two images of the B ring (top 60% of image) from opposite sides of Saturn are placed side by side to show its lack of circularity. Just below the edge of the B ring, a separate thin ring in the Cassini division is also seen to be non-circular. (Courtesy NASA/JPL/Caltech; PIA 01390.)

Figure 9.13 (right image) Enceladus is seen to have a young smooth surface (right part of image) and an older cratered surface (left), both of which are crossed by a broad set of ridges and valleys. (Courtesy NASA/JPL/Caltech; PIA 01367.)

possible with Voyager 1, and to obtain the first good images of Atlas (the shepherd for the outer-edge of the A ring), the two shepherds (Prometheus and Pandora) for the F ring, the co-orbital satellites Janus and Epimetheus, the two Tethys trojans[17] (Telesto and Calypso), and the Dione trojan (Helene).

Voyager 2's polarisation measurements in the near ultraviolet and infrared bands indicated that Titan's atmosphere at an altitude of about 200 km contained droplets of maximum size 0.05 microns, whereas some 30 km lower the droplets were larger, having a maximum size of 0.12 microns. It was hoped that a basic knowledge of these size distributions, linked to spectroscopic analysis of the atmosphere, would help scientists to disentangle the structure of Titan's clouds.

The Voyager 2 images of Iapetus showed that there were numerous craters on its light-coloured hemisphere, with some of these craters near to

[17] The term 'trojan' is a general term referring to small bodies at the $\pm 60°$ Lagrangian points co-orbiting with planets or satellites. The first trojans to be found were asteroids at the $\pm 60°$ Lagrangian points in Jupiter's orbit around the Sun.

the dark hemisphere having dark-coloured floors. Before Voyager 2 it was thought that, as the dark hemisphere is the leading side of Iapetus in its orbit around Saturn, the dark material could have been picked up as the satellite orbited the planet. The discovery of dark-floored craters in the light-covered hemisphere makes this unlikely, however. What is more probable is that the dark material has welled up from beneath the surface, but some astronomers feel uncomfortable with this explanation as it does not explain why the dark hemisphere happens to be the leading side in the satellite's orbit around Saturn, and why the reflectances of the two surfaces are so radically different. So a solution using both the welling up theory and the orbital intercept theory has been suggested as a possible answer.

Two hours after its closest approach to Saturn, and whilst it was still behind the planet as seen from Earth, Voyager 2 hit trouble. The first indication that anything was wrong was when the spacecraft reappeared from behind Saturn and none of the planned images were transmitted to Earth. It was quickly established that the scan platform was only scanning about one axis instead of two. Initial attempts to free the second axis failed, but a further attempt two days later solved the problem. As a result of the failure the highest-resolution images of Saturn's southern hemisphere and of the satellites Enceladus and Tethys were lost.

Although the best images of Enceladus were lost, those that were returned showed remarkable detail (down to 2 km resolution) on this the most reflective satellite in the Solar System. Large areas were found to be craterless (see Figure 9.13), suggesting a surface age of less than 100 million years, but even the cratered areas showed relatively few large craters, suggesting that that surface is not as old as that of Saturn's other satellites. In addition there are numerous valleys and ridges up to 1 km high cutting across the cratered regions. All this indicates that Enceladus has been geologically active until the relatively recent past (in astronomical terms) and it may even be geologically active today. Neither Mimas nor Tethys, the satellites with orbits on either side of Enceladus, showed anything like this level of activity. Enceladus is intermediate in size between these two satellites and all three have similar densities, so the cause does not appear to be due to flexure caused by its close proximity to Saturn. A more likely reason is thought to be the 2:1 resonance of its orbit to that of the larger satellite Dione, which will cause tidal flexing on alternate orbits.

The Voyager 2 images of Tethys showed an enormous 450 km diameter crater which, relative to the size of the satellite, is even larger than the 135 diameter crater on Mimas. In addition, Voyager 2 showed that the north–south gorge imaged by Voyager 1 extends 1,500 km from almost the north to the south poles, with an average width of about 100 km and depth of about 4 km. It is unique in its size relative to that of Tethys for bodies in the Solar System, as far as we know. It has been suggested that this gorge, now called Ithaca Chasma, could have been produced when the watery interior of Tethys froze, causing the satellite to expand (as water expands on freezing), but it is not clear why it appears to be the only large gorge on Tethys, and why similar gorges do not exist on the other icy satellites of Saturn.

With two very successful planetary intercepts behind it, Voyager 2 was now on its way to Uranus and Neptune to give us, if all went well, our only close-up glimpses of these planets and their satellites in the twentieth century.

9.6 Voyager 2 at Uranus

Although Voyager 2 had performed very well during its encounters with Jupiter and Saturn, it was not in the best of health. As mentioned before, the primary receiver had failed before the Jupiter encounter, and the redundant receiver had only limited frequency adjustment capability. One axis of the scan platform had jammed after Voyager's closest approach to Saturn and, although the fault had been corrected, it had been decided to restrict the scanning speed to avoid further problems at Uranus. This restriction, and the fast fly-by of the Uranian system, meant that an alternative form of motion compensation was required, and so it was decided to pan the cameras by slewing the whole spacecraft. Starting and stopping the tape recorder also caused a judder which affected image quality, but this was solved using very short compensatory bursts from the spacecraft's thrusters. Finally there was an irreparable problem with a failed chip in one of the two scientific computers but this, fortunately, only caused a small loss of memory. So although Voyager 2 was not working perfectly, spacecraft controllers knew its idiosyncrasies and, provided nothing else went wrong, the fly-by of the Uranian system should be completely successful.

During the Jupiter encounter the image data from Voyager 2 had been transmitted to Earth at a maximum speed of 115 kbps, taking just 48 seconds to transmit a complete image. During the Saturn encounter, however, Voyager 2 had been about 1.7 times as far away from Earth, so the power level of the signals received on Earth was only about 35%[18] of the level received during the Jupiter intercept. To compensate, the bit rate would normally have

[18] $\frac{100\%}{1.7^2} = 35\%$

Table 9.4. *The Voyager 2 closest approaches to Uranus' satellites (in time order)*

	Dia. (km)	Dist. from centre of planet (km)	Closest approach: Time[a] (hr:min)	Closest approach: Dist. (km)[b]	Best surface resolution (km)
Titania	1,578	435,840	$U-2$:49	364,400	7.2
Oberon	1,522	582,600	$U-1$:47	469,800	12.9
Ariel	1,158	191,240	$U-1$:38	126,400	2.5
Miranda	472	129,780	$U-0$:55	28,800	0.6
Ring plane crossing			$U-0$:43		
Uranus	51,100		U	81,500	
Umbriel	1,170	265,970	$U+2$:53	324,400	10.9

Notes:

[a] The time U of closest approach to Uranus was 17:59 UT spacecraft time on 24/1/86. The signals were received on Earth 2 h 45 min later.

[b] This is the distance above the surface of the satellites or the cloud tops of Uranus.

been reduced by about the same amount (to about 40 kbps), everything else being equal. But a maximum rate of 44.8 kbps had been used at Saturn, as the link margins had been shown to be too conservative during the Jupiter fly-by.

Uranus was 1.9 times further away from Earth than was Saturn at the time of their respective Voyager 2 encounters. So, if nothing else was done, the spacecraft bit rate would have to be reduced to about 12.4 kbps[19] for the Uranian intercept, which would have meant that it would have taken about ten minutes to transmit one image. This was too long, as the total encounter period with Uranus and its satellites was only a few hours (see Table 9.4). The spacecraft could not transmit more power; in fact the power budget was already critical because of the natural reduction in power available from the RTGs with time. So it was decided to improve both the ground receiving facilities and the efficiency with which the image data was encoded on board the spacecraft.

NASA planned to use three 64 metre diameter DSN antennae in succession, spaced around the world, to give uninterrupted coverage of the Uranus

[19] It should be noted, however, that not all of this bit rate is available for image transmission, as spacecraft housekeeping and other scientific data has also to be transmitted.

intercept. One of these antennae was near Madrid, one was at Goldstone, California, and the other was near Canberra, Australia. To help solve the communications problems associated with the weak signal, the 64 m diameter Madrid antenna was linked electronically to a 34 m diameter dish nearby, thus increasing the received power level by 28%. The Goldstone antenna was linked to two 34 m diameter dishes, giving an improvement of 56%, and the Canberra antenna was linked to one 34 m dish and the Parkes radio telescope to give a total improvement of 128%. The Australian system was particularly important as it was due to cover the key phase of the encounter around closest approach to Uranus.

In addition NASA reprogrammed the two on-board Flight Data Subsystem computers to perform image data compression (IDC). In this IDC system the absolute intensity of only the first pixel in a scan line was transmitted, with only the difference in intensity of the next pixel being transmitted, not its absolute intensity. Similarly, the difference in intensity of the third pixel compared with the second was transmitted, and so on. As the intensity of two adjacent pixels tended, *in general*, to be similar, this resulted in an increase in transmission efficiency of about 140%, allowing the simultaneous transmission of real-time imaging and tape recorder playback for the first time with Voyager 2.

Although IDC was efficient, it suffered from one major drawback. If noise corrupted the read-out of one pixel, this error would feed through to all the subsequent pixels in the scan line. To solve this problem a system called Reed–Solomon coding was used to encode the spacecraft data prior to transmission. This system allowed the detection on the ground of corrupted data and its subsequent correction.

Our knowledge of Uranus and its satellite system was very sketchy before the Voyager 2 encounter. The most striking feature of this system was that the orbits of the Uranian satellites, of which five were known, were almost perpendicular to the orbit of the planet around the Sun. The angle was actually greater than 90° (about 98° in fact), so the satellites were seen to orbit Uranus in the opposite sense to the orbital rotation of Uranus around the Sun.

It was very difficult to observe any markings on Uranus from Earth, but a few had been seen to show that the planet has a spin axis perpendicular to the orbit plane of its satellites. The markings were too sparse, however, to provide a reliable estimate of Uranus' axial rotation period. Until about 1975 a period of 10 h 50 min had been generally accepted, but starting in 1976 various periods had been deduced from measurements of the intensity fluctuations of the planet. These new estimates were generally in the range of from 15 to 17 hours, with a most likely value thought to be in the range 16.0 to 16.4 hours.

Methane and molecular hydrogen had been detected in the Uranian atmosphere before Voyager but, unlike Jupiter and Saturn, there was no evidence of any ammonia, presumably because it had been frozen out of the atmosphere by the intense cold. Far infrared measurements of Uranus showed, also unlike the case for Jupiter and Saturn, that there was no evidence for any internally generated heat.

In March 1977, just a few months before the launch of both Voyager spacecraft, Uranus had been found to have a ring system by James Elliot, Edward Dunham and Douglas Mink. They had been observing the occultation of a star by Uranus, using the Kuiper Airborne Observatory flying over the Indian Ocean, when they discovered five reductions in intensity of the star before it went behind Uranus. Similar reductions were also observed after the star had come from behind the planet. These reductions were found to be due to five narrow rings, which were initially designated rings α, β, γ, δ and ε. The number of rings was later increased to nine, by including stellar occultation data from observatories at Perth and Cape Town. The ε ring was found to be both eccentric and variable in width, and the narrowest three rings, numbered 4, 5 and 6, which are closest to the planet, were found to be only about one kilometre wide. This was a big surprise as at this time the only rings known were the broad rings of Saturn. The thin F ring had not been found at Saturn until 1979, the same year that the thin rings had been discovered around Jupiter. It was also not until 1980 that the complex structure of Saturn's main rings had first been recorded.

Because of Uranus' 98° spin axis orientation, the planet orbits the Sun spinning almost on its side, with the Sun getting to within 8° of the zenith at one pole during the summer, and then half a Uranian year (or 42 Earth years) later getting to within 8° of the zenith at the other pole (see Figure 9.14). During the Voyager 2 intercept the Sun was almost overhead at the south pole.

Voyager 2 approached the Uranian system in a trajectory almost perpendicular to the orbital plane of its satellites (see Figure 9.15), as this is almost perpendicular to the ecliptic. This meant that the whole close encounter sequence would last for little more than 5 hours, compared with 34 hours for the Jupiter system and the even longer time for Saturn. The trajectory of Voyager 2 would take it through the ring plane between Miranda, the nearest known satellite to Uranus, and the outermost ring. It would then continue to pass behind both the ring system and the planet, as seen from Earth, allowing more detailed measurements to be made of their constitution using radio occultation techniques.

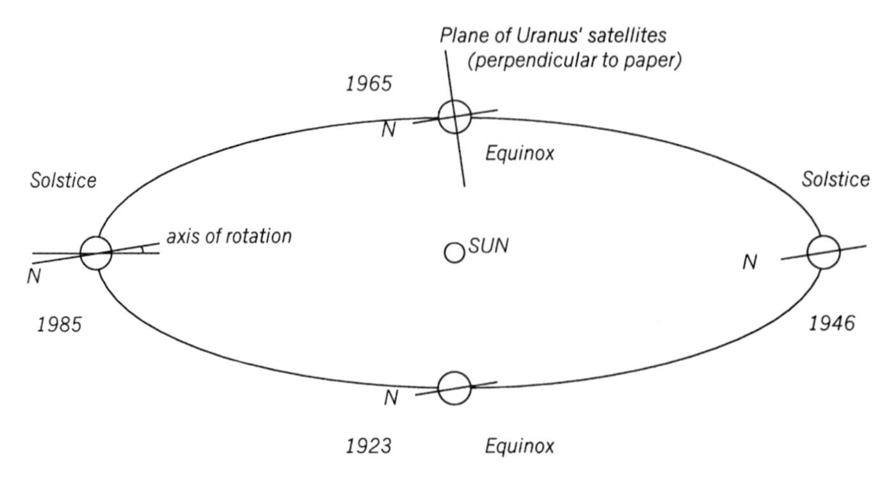

Figure 9.14 The position of Uranus' spin axis is shown as the planet orbits the Sun in 84 years. The Sun was overhead at Uranus' equator in 1923 and 1965, and almost overhead at its north pole in 1946 and at its south pole in 1985. The Uranian satellites orbit the planet in its equatorial plane which is perpendicular to the plane, of the paper.

The Voyager observatory sequence was planned to start on 4th November 1985, some two and a half months before closest approach, when the spacecraft would take images that could be used to produce time-lapse movies to determine the rotation rate of Uranus. Regular observing was due to start on 10th January with the start of the far encounter phase, but the crossing of the magnetopause, if one exists, was not expected until 23rd January, just one day before closest approach.

As Voyager closed in on Uranus in early December 1985, it was realised that the satellites were not exactly where they were expected to be because Uranus was about 0.25% heavier than previously thought. Consequently on 24th December the spacecraft's trajectory had to be slightly modified by a 14 minute rocket burn, to ensure that Voyager's closest approach to Uranus was some 300 km further out than previously planned, otherwise it would have missed Neptune, its next port of call, by about 4 million km!

At this stage in December 1985 no radio emissions had been detected from Uranus, indicating that it did not have a magnetic field, although measurements of excess ultraviolet light by the Earth-orbiting International Ultraviolet Explorer (IUE) spacecraft suggested the contrary.[20] It was not until five days before closest approach, however, much later than expected,

[20] This ultraviolet excess measured by IUE was attributed to auroral activity on Uranus, which was an indication that it has a magnetic field.

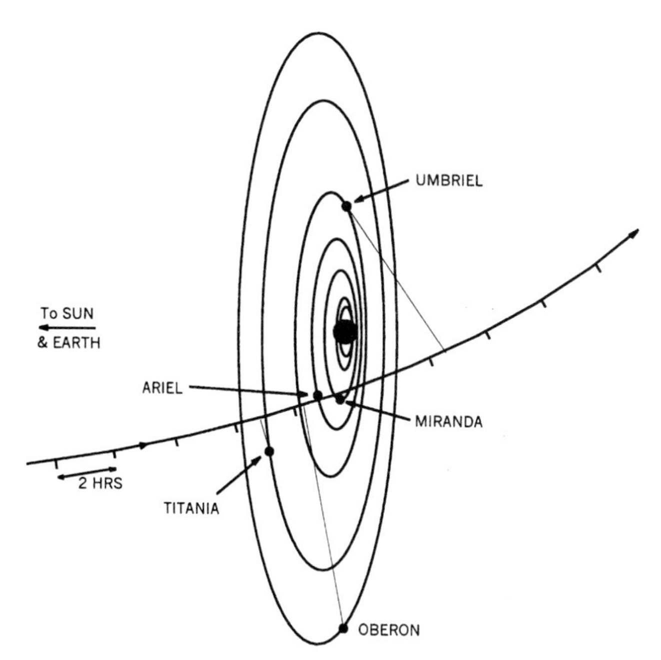

Figure 9.15 Voyager 2's trajectory is shown as it encountered the Uranus satellite system. The spacecraft intercepted the ring plane about 14,000 km inside the orbit of Miranda and 54,000 km outside the outermost ring, ring ε.

that Voyager detected a strong burst of polarised radio signals, confirming the presence of a magnetic field. Analysis of timing variations in the radio emissions over several weeks showed that the interior of Uranus rotates with a period of 17 h 14 min.

Ten and a half hours before closest approach the spacecraft crossed the bow shock some 600,000 km (or 23 R_U) from Uranus. A day later it became evident that Uranus' magnetic axis was tilted at an amazing 60° to the planet's rotational axis, and that the magnetic centre was displaced from the geometric centre by about 0.3 R_U (see Figure 9.16). The magnetic field at cloud-top height at the planet's equator was calculated to be about 0.25 gauss, which is similar to that at the Earth's equator, although because Uranus is larger, its magnetic moment is about 50 times that of the Earth (see Table 3 in the Appendix). Because of its intercept geometry Voyager did not cross the bow shock as it left Uranus until 77 hours after closest approach, at a distance of 4.1 million km from the planet. Over the next day or so the spacecraft crossed the bow shock three more times, as the bow shock contracted and expanded in response to the varying intensity of the solar wind.

The cloud motion movies of Uranus produced in November 1985 showed no clouds at all, even after computer enhancement, and it was not until late December that computer-enhanced images began to show a

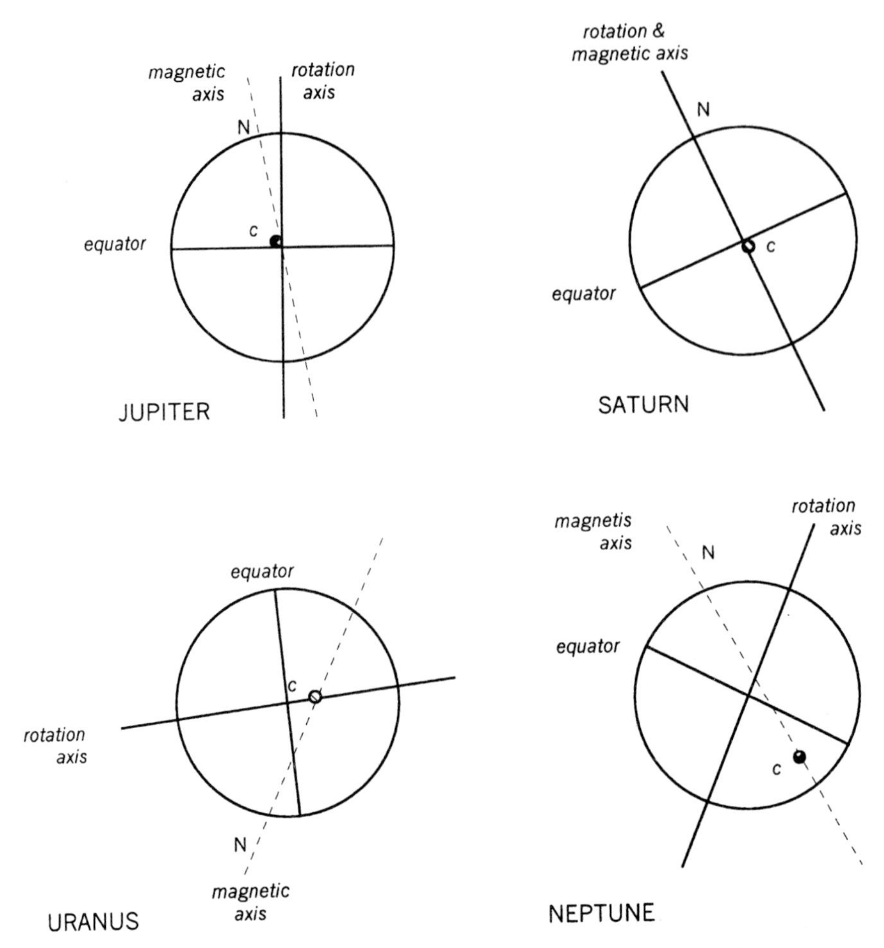

Figure 9.16 The magnetic axes and magnetic centres of the gas giants are shown relative to their rotational axes. (The magnetic axes are indicated by the dashed lines, and the magnetic centres by the letter 'c'. The position of the north magnetic pole is indicated by the letter 'N'. The plane of the orbits of the planets around the Sun is horizontal in this diagram and perpendicular to the plane of the paper.)

banded pattern. But it was not until ten days before closest approach that individual clouds could be seen, enabling the wind velocity to be deduced.

As the Sun was virtually overhead at the south pole during the Voyager 2 fly-by, clouds could only be seen in the southern hemisphere, and even here they were few and far between. Nevertheless, some wind velocity data was produced between 25° and −71° latitude,[21] with an additional single

[21] The minus sign indicates a southern latitude.

data point at $-6°$ produced by radio occultation measurements. The results showed a 100 m/s (220 mph) easterly[22] equatorial jet, with a westerly flow for most of the southern hemisphere. This is contrary to the situation for both Jupiter and Saturn where the equatorial jets (shown previously in Figure 9.8) are both westerlies.

Before the launch of Voyager 2 it was thought that the four large gaseous planets, namely Jupiter, Saturn, Uranus and Neptune, generally known as the Gas Giants, would have a constitution basically unaltered since the time that they were formed from the solar nebula. If that was the case, they should all have the same helium to hydrogen mass ratio, or helium abundance as it is called. It was known that the helium abundance of the Sun was 0.28 ± 0.01, but nuclear reactions in the Sun were continuously creating helium from hydrogen, so the helium abundances of the gaseous planets should be less than this.

Voyager measured the helium abundance of Jupiter to be 0.18 ± 0.04, which is less than that of the Sun, but the value measured for Saturn of 0.06 ± 0.05 was much less than the solar value. This value for Saturn caused theoreticians to reconsider their analysis, and it was then realised that, at the very high pressure in the interiors of Jupiter and Saturn, liquid helium, which is heavier than liquid hydrogen, would gradually sink towards the interior. The effect would be greater for Saturn than Jupiter because of Saturn's lower temperature. This would cause a depletion of helium in the atmospheres of Jupiter and Saturn, with a larger depletion for Saturn, as observed. Uranus and Neptune are much smaller than Jupiter and Saturn, however, (see Table 1B in the Appendix), with consequently much lower internal pressures, so the above mechanism should not operate there, and the helium abundance should be that of the original solar nebula. The key question was, therefore, whether the helium abundances of Uranus and Neptune were similar to, but less than that of the Sun, and higher than those of Jupiter and Saturn.

Just before the Voyager 2 encounter with Uranus, Glenn Orton of JPL deduced, using his analysis of its infrared spectrum observed from Earth, that more than half of Uranus' atmosphere was helium. This caused great consternation and excitement as, if he was correct, there must be an unknown mechanism either increasing the amount of helium or reducing that of hydrogen in Uranus' atmosphere. In the event the theoreticians need not have worried, as the helium abundance for Uranus was measured by Voyager 2 as 0.25 ± 0.05, or about the same as that of the Sun, within error.

[22] Westerly winds are defined as those that blow in the same direction as the planet rotates, as on the Earth.

The only gases detected by Voyager in the lower atmosphere of Uranus were methane, helium, and molecular and atomic hydrogen. The height of the cloud deck was determined by the radio occultation experiment as being at about the 1 bar level, and acetylene was detected in the upper atmosphere. The acetylene was thought to be produced by the action of sunlight on methane, producing a photochemical smog which raises the temperature of the upper atmosphere and reduces that of the lower atmosphere.

As mentioned previously, Voyager 2 arrived at Uranus when the Sun was almost directly overhead at the planet's south pole and, because of its long year of 84 Earth years, it was estimated that the illuminated south pole would be about 5 to 10 K warmer than the dark north pole, at cloud-top height. It was something of a surprise, therefore, when the temperatures over the two poles were found to be almost identical, indicating that there must be very good atmospheric mixing across the latitudes.

Before the Voyager intercept it had been expected that there would be pairs of shepherding satellites for each of the nine known Uranian rings. On 30th December 1985 Voyager discovered the first new satellite of Uranus, designated 1985 U1 (now called Puck), and on 3rd January 1986 the next new satellites, 1986 U1 and 1986 U2 (Portia and Juliet), were discovered. Cressida, or 1986 U3, was found on 9th January, and 1986 U4, U5 and U6 (Rosalind, Belinda and Desdemona) were discovered four days later. None of these satellites were shepherds for the rings, as all of these satellites orbited Uranus outside of its outermost ring, namely the ε ring. On 20th January, however, four days before Voyager's closest approach to Uranus, two small shepherding satellites 1986 U7 and U8 (Cordelia and Ophelia) were found on either side of the ε ring, and on the following day these two satellites were imaged together with the nine rings detected from Earth (see Figure 9.17). On the same day a tenth new satellite, 1986 U9 (Bianca),[23] was added to the list of Voyager discoveries. A later review of earlier Voyager images showed that all these satellites had been imaged previously, enabling their orbits to be calculated. Only Puck was large enough to have its diameter (of 155 km) measured directly; all the other satellite diameters (ranging from 25 to

[23] It was suggested that it would be a fitting memorial to the seven astronauts who lost their lives in the Challenger disaster, in same month as the Uranus encounter, to name seven of these newly discovered Uranian satellites after them. In the event, the International Astronomical Union (IAU) decided to follow the precedent established for previous space fatalities, and name some craters on the far side of the Moon after these seven astronauts.

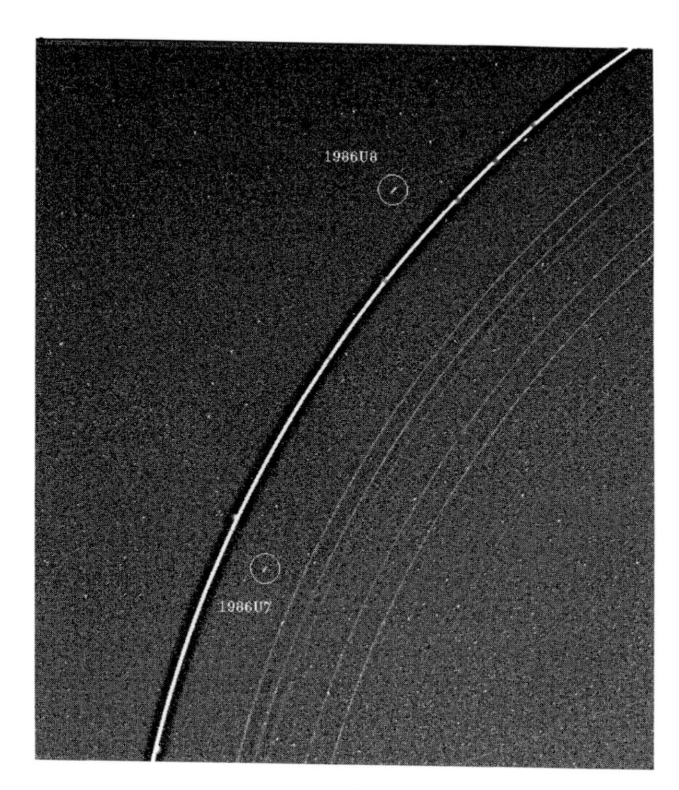

Figure 9.17 The shepherd satellites 1986 U7 and U8, now called Cordelia and Ophelia, are shown on either side of the ε ring. All nine of the rings detected from Earth in 1977 can be seen on the original image. (Courtesy NASA/JPL/Caltech; PIA 00031.)

110 km) being estimated from their brightness, assuming that their reflectivity was the same as Puck (a dark 7%). No further shepherding satellites were found, so only one of the nine previously known rings had been found to have a pair of shepherds.

The nine known rings of Uranus, which had first been imaged by Voyager 2 a few days before closest approach, looked narrow and dark, very much as expected. Then on the day before closest approach a very faint tenth ring, called the 1986 U1R or λ ring, was found between the ε and δ rings, and on the following day a faint ring (1986 U2R) about 2,500 km wide was found between the innermost ring (ring 6) and the planet. Over 200 images were taken of the rings after the spacecraft had passed through the ring plane, as it was expected that they would show up better in this forward-scattering arrangement, but all but one of these images was blank. Fortunately, the last exposure of 96 seconds, with Voyager 2 just 8° from the Sun line, showed a clear set of rings. Each of the ten known narrow rings were seen, with the newly-discovered λ ring being the brightest, together with numerous new dust rings (see Figure 9.18).

Figure 9.18 The difference in appearance of the Uranian rings before (top) and after closest approach is shown in these two images. The faint new ring (the 1986 U1R or λ ring) is the brightest ring in the back-lit configuration (bottom), and the eccentricity of the ε ring (at the far left) is evident by comparing the two images. The extra dust rings seen only in back-lighting are also clearly shown. (Courtesy NASA, composite of two images.)

Jupiter's rings, which had been found to be brightest in forward-scattering light, consist mostly of very fine dust of about 1 to 2 microns in diameter, whereas Saturn's rings had been found to consist of objects ranging from microns to tens of metres in diameter. The very low intensity of the Uranian rings in forward-scattering light indicated that they had relatively little dust in them compared with Jupiter's rings. Collisions of the larger particles in the Uranian rings and the impact of micrometeorites must produce dust. But Voyager found that there is probably enough atomic hydrogen at the altitude of the rings to limit the lifetime of these dust particles from less than a year for the inner rings to about 1,000 years for the outer rings, which is a very short time in astronomical terms.

The fine structure of the Uranian ring system was determined using the occultation of the star β Persei (Algol) as observed by Voyager's photopolarimeter. This provided a radial resolution of about 100 metres for all of the rings. A similar occultation of the star σ Sagittarii (Nunki) provided a radial resolution of about 10 metres for the ε, λ and δ rings. Occultation of the radio signals from Voyager, when the spacecraft passed behind the rings, as seen from Earth, provided a radial resolution of about 50 metres for all of the original nine rings. This data taken as a whole showed that the narrow rings have relatively well-defined edges, with the ε and γ rings, in particular, having sharp edges at both their inner and outer boundaries.

A density wave pattern, of the sort found in Saturn's A ring, was found in the highest resolution profiles of Uranus' δ ring. Unfortunately, the satellite causing the Uranian density wave has not been found, presumably because it is too small to be imaged by Voyager. Depending on where it is, it could assist in constraining either the inner edge of the δ ring and the outer edge of the γ ring, or the outer edge of the η ring and the inner edge of the γ ring.

The ε ring was estimated to have a reflectivity of only 1.4%, compared with the reflectivity of Saturn's A and B rings of about 60%, and of the C ring and Cassini Division of about 25%. The other Uranian rings were also very dark and, like the ε ring, colourless. No evidence of water was found in their spectra, and so it was concluded that they were probably composed of carbon, which may have been produced by the decomposition of methane ice by energetic protons.

Little was known about the five largest satellites of Uranus, that had been discovered using Earth-based telescopes, until the Voyager 2 intercepts. In fact it was not until the early 1980s that their diameters had been measured with any certainty, as ranging from 1,600 km for Titania and Oberon, to 1,300 km for Ariel, 1,100 km for Umbriel and 500 km for Miranda. This meant that they were smaller than Jupiter's Galilean satellites (see Table 9.2) and about the same size as the large satellites of Saturn, excluding Titan (Table 9.3). Because of their relatively small size and their great distance from the Sun, it was anticipated that these five Uranian satellites would show little in the way of geological activity, being covered instead with craters dating back to their early formation.

The satellites of Saturn are generally less dense than those of Jupiter, because they contain more water ice, and it was generally anticipated that those of Uranus would contain even more water ice, and be even less dense, as they are further from the Sun. They are darker than Saturn's satellites, however, and the ice spectral bands were not as clear. So it was assumed that the ice was more contaminated on the surface of the Uranian satellites. In the event Voyager 2 showed that the densities of the Uranian satellites were higher than those of the similar-sized satellites of Saturn, not lower, as they consisted of less ice, not more.

As previously explained, Voyager 2 intercepted the Uranian satellite system almost at right angles to the satellites' orbital planes (see Figure 9.15), passing between Uranus and Miranda, the nearest of the five large satellites to Uranus. Because of the geometry of this intercept, however, the only satellite to be seen close up was Miranda (see Table 9.4).

The outermost of the Uranian satellites, Oberon, was seen by Voyager 2 to have an ancient, heavily cratered surface, as anticipated, with some craters being more than 100 km in diameter. Some of the craters were seen to have dark floors, which is thought to be due to dirty water coming from beneath the surface. On the highest-resolution image, an Everest-sized mountain, which is probably the central peak of an impact crater, was seen on the edge of the disc.

Titania, with a diameter of about 1,580 km, the largest of the Uranian satellites, was seen to have fewer craters than Oberon, which was a surprise as the two satellites are of similar size and density and are in adjacent orbits around Uranus. Two very large, multi-ringed craters were imaged near the terminator, and a large rift valley system was seen that was at least 1,500 km long, with some of the valleys being up to 50 km wide and 5,000 m (17,000 ft) deep. It is thought that this rift valley system may have been produced when the ice in the interior froze, in the same way that the large rift valley was thought to have been produced on Tethys, one of Saturn's satellites. Clearly the ancient surface of Titania has been modified since formation by geological processes, although it is still relatively old.

Umbriel, with an albedo of 0.20, is clearly the darkest of the large Uranian satellites. It is somewhat surprising, therefore, that Ariel, whose orbit is just inside that of Umbriel, has an albedo of 0.40 and is clearly the brightest of the satellites. Umbriel and Ariel are of almost identical size and density, so why one is so much darker than the other is a mystery. It may be partially explained by the age of their surfaces, as Umbriel was seen to have an ancient, heavily cratered surface like that of Oberon, which is also quite dark. Umbriel was seen to have just one outstanding feature, a bright, doughnut-shaped ring near the edge of the disc.

Ariel was found to be similar to Titania in so far as both have large rift valleys, although those on Ariel cover more of the visible surface. Why Ariel and Titania should have extensive rift valleys, whilst Umbriel and Oberon do not, is not understood. Maybe there are rift valleys on the latter two satellites but they are in the hemisphere not imaged by Voyager 2. We will have to wait a long time to find out, however, as there are no spacecraft currently approved to re-visit Uranus.

Some of the valleys on Ariel, many of which are flat-bottomed, are over 20,000 m (60,000 ft) deep! Parts of its surface have fewer craters than Titania, because they appear to have been more recently resurfaced by volcanic activity. Detailed crater counts seem to indicate that there have been several periods of partial resurfacing extending up to quite recent times. The reasons for this are unclear as, unlike the case of Io, there are no orbital resonances between Ariel and any of the other large satellites, although there may have been such resonances in the past.

By a stroke of good fortune the highest-resolution images of all the Uranian satellites were those of Miranda, with a diameter of only 480 km,

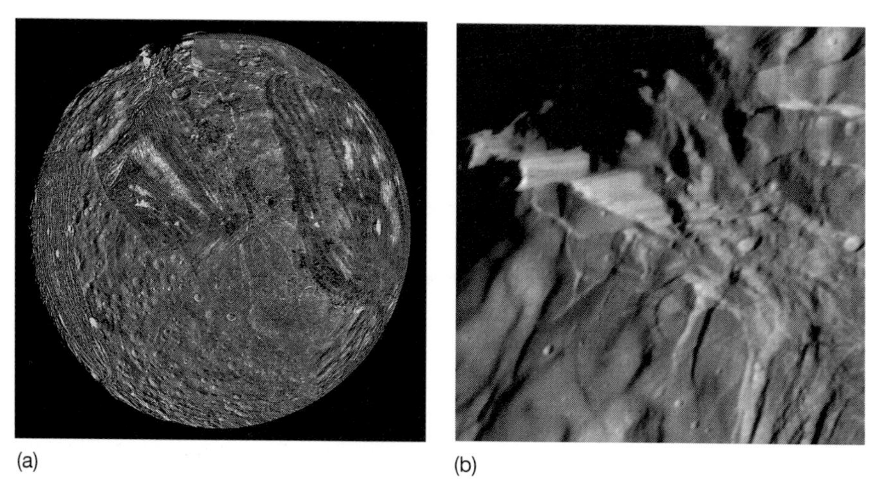

(a) (b)

Figure 9.19 Miranda's strange surface features are shown in this composite high-resolution image (a). Arden corona is the large feature on the right-hand side, Inverness corona and its chevron albedo feature is at the upper left, and Elsinore corona is on the left-hand side. The enormous cliffs of Verona Rupes at the top left are shown in more detail in the separate image (b). (Courtesy NASA/JPL/Caltech; PIA 01490 and PIA 00044.)

the smallest of the five large satellites and the closest of them to Uranus. Miranda proved to be the star of the show, with its chevron-shaped feature, two large ovoids and enormous cliffs (see Figure 9.19). The light-coloured chevron shape, which was detected when Voyager was over 1 million km from Miranda, was seen to be surrounded by a large trapezoid-shaped feature, now called the Inverness Corona, that has several parallel linear features that intersect at right angles. The two large ovoids, now called Elsinore Corona and Arden Corona, has concentric sets of ridges and grooves that make them look like large dirt racetracks. The Inverness and Arden coronae appear to be bordered by trenches often associated with parallel scarps. In the case of Inverness corona, these scarps continue to the terminator where the enormous cliff face called Verona Rupes is seen, ranging in height from 10 to 20 km. So Miranda was seen to be far removed from the ancient, cratered world that had been expected.

9.7 Voyager 2 at Neptune

Voyager 2 was now on course for its close encounter with Neptune, following this highly successful fly-by of the Uranian system. The first question that had to be settled was where exactly to target the spacecraft.

In August 1980, some nine years before the encounter, it had been decided to fly Voyager 2 over the north polar regions of Neptune to enable its trajectory to be changed sufficiently for a close fly-by of Neptune's large satellite, Triton. Initially, just in case Neptune had rings like the other three gas giants, it had been decided to aim for a trajectory that crossed Neptune's equatorial and hypothesised ring plane about 30,000 km above Neptune's cloud tops (or about 2.3 R_N from the centre of Neptune) to fly outside the possible rings.[24] This led to a closest approach to Neptune at about 23 hrs UT on 24th August 1989, followed by a fly-by of Triton five hours later at a distance of about 44,000 km. Then, during the next few years, evidence began to accumulate that Neptune may possess one or more partial rings, one of which may be beyond the planned intercept between the spacecraft and Neptune's equatorial plane. For spacecraft safety reasons, if nothing else, it was essential that the position of these rings be established as soon as possible.

Given the rings around Saturn, and the discovery of rings around Uranus in 1977 and around Jupiter in 1979, astronomers had then set out to see if Neptune, the fourth gas giant, also possessed a ring system. Neptune's equator was thought to be inclined at an angle of about 30° to its orbit around the Sun, so any ring system would be much less open and more difficult to detect than that of Uranus, which is almost perpendicular to its orbit. Neptune is also almost twice as far away from Earth compared with Uranus, so any rings would probably appear much smaller from Earth than those of Uranus. In addition Neptune moves more slowly than Uranus against the stellar background, making it more difficult to find a suitable stellar occultation. Nevertheless, in spite of all these difficulties a search was made.

Neptune was due to almost occult the star 52 Ophiuchi on 24th May 1981, and the encounter was observed by Harold Reitsema, William Hubbard, Larry Lebofsky and David Tholan of the University of Arizona. As Neptune approached the star, they found that the star was eclipsed for eight seconds, but there was no symmetrical eclipse on the other side of the planet. The eclipse could not be due to a ring, as that should have produced a second eclipse as Neptune receded from the star (as happened with the occultation by Uranus in 1977), so they concluded that they had probably discovered a new satellite of Neptune, designated 1981 N1, at least 180 km in diameter, about 50,000 km from the centre of the planet.[25] They weren't completely sure, however, as the chance that a small satellite of Neptune

---

[24] The outer ring of Uranus, which is a similar sized planet to Neptune, is 1.95 R_U from the centre of the planet.

[25] The distance was later updated to 70,000 km.

had been just in the right place at the right time to occult the star was less than one in a thousand.

Another one-sided event had previously been observed in 1968 by Edward Guinan and J. Scott Shaw of Villanova University (in the USA) from Mount John Observatory in New Zealand, but that event had lasted for 2½ *minutes*. It could not have been due to a satellite, however, because the long period of the occultation implied that the satellite would have been large enough to be easily visible, but if it was due to a ring or partial ring it was not the same as the one seen in 1981 as it was closer to the planet. So what had been detected in 1968 and 1981?

Another occultation opportunity occurred on 5th June 1983, but unfortunately no evidence of rings or of a satellite was found by the many teams of astronomers who observed the event. Then on 22nd July 1984 the first unambiguous discovery of a partial ring was found during the occultation of the star SAO 186001 by astronomers at both the European Southern Observatory (ESO) and 95 kilometres to the south at the Cerro Tololo Inter-American Observatory (CTIO). The team at the ESO, who were observing the occultation at the request of André Brahic of the University of Paris, found a 35% reduction in the intensity of the star that lasted for about one second. As before, the occultation was not observed on the other side of Neptune. It so happened, however, that William Hubbard and his team at the CTIO had actually recorded the same event, but they did not realise it because their quick-look data recording system did not have a fine-enough time resolution. Hubbard's team heard in October of the ESO discovery, and this caused them to re-examine other data that they had recorded at the time which had a much finer time resolution. Their results turned out to be identical to those of the ESO.

These observations of an identical partial occultation at observatories spaced 95 kilometres apart proved that the effect could not have been due to a satellite. It appeared to be due, instead, to a partial ring, or ring arc, about 20 km wide and 67,000 km (2.71 R_N) from the centre of Neptune. The ring arcs detected in 1968 and 1981 would be about 30,000 km and 70,000 km (1.21 and 2.83 R_N), respectively, from the centre of Neptune, if ring arcs they were.[26] A further occultation in 1985 also showed that there was probably another ring arc 56,000 km from the centre of Neptune.

As the evidence for ring arcs was gradually being accumulated, the Voyager mission analysts were also trying to optimise the spacecraft trajectory to take it closer to Triton, whilst avoiding the ring arcs. This resulted

[26] Reitsema still thought that the 1981 occultation was due to a satellite, however.

in a decision in September 1985 to fly Voyager 2 just 3,400 km above Neptune's atmosphere, followed by a closest approach just 10,000 km above Triton's surface.

Shortly after the new trajectory had been agreed, however, scientists working on the dynamics of the Neptunian system concluded that the planet was nearly 1,000 km larger than previously estimated and its mass was 1.5 % smaller. In addition, the estimate of Neptune's inclination was increased and Triton's orbit modified. This meant that a Voyager trajectory to fly-by Triton at a distance of 10,000 km would now take the spacecraft only 1,250 km above Neptune's atmosphere near its north pole. This trajectory was considered to be far too risky, and so an extensive evaluation of alternative strategies was undertaken to produce a new, definitive fly-by trajectory.

The problem in designing a safe trajectory so close to Neptune was that even in 1985 the structure of Neptune's upper atmosphere was only approximately known. Mission analysts were concerned, in particular, about both drag and high-voltage arcing produced by Neptune's upper atmosphere, and various possible trajectories were analysed and discussed. These ranged from a closest fly-by 27,300 km from the centre of Neptune, about 3,100 km above its cloud layer and 2,500 km above the 1 microbar level, which allowed a 23,000 km fly-by of Triton, to much more conservative trajectories. The Voyager Imaging Science Team requested a Triton fly-by distance of 25,000 km, but the Voyager Radio Science Team were naturally more interested in obtaining the best radio occultation data for Neptune and requested a trajectory resulting in a 55,000 km Triton fly-by. In the event Edward Stone, leader of the Voyager Science Steering Group, decided on a compromise Triton fly-by distance of about 40,000 km. This new trajectory (see Figure 9.20, and Table 9.5) would pass 4,800 km above Neptune's cloud layer near its north pole at 4:00 UT on 25th August 1989 and, just over five hours later, it would fly-by Triton at a distance of about 40,000 km. Unfortunately, Voyager would fly no closer than 4.7 million kilometres to Nereid, Neptune's only other known satellite, and at this distance Nereid would only be a few pixels across on Voyager's detector. This would only allow its approximate size and reflectivity to be determined.

Neptune was about 50% further away from Earth than Uranus during its intercept by Voyager 2. As a result, the signal power level received on Earth from Voyager at Neptune was only 44% of that received when it was at Uranus. Similarly the solar intensity at Neptune was only 41% of the level at Uranus, so longer image exposures would be required.

To solve the problem of low signal power received on Earth, NASA's 64 m diameter DSN antennae were increased to 70 m in diameter, and their

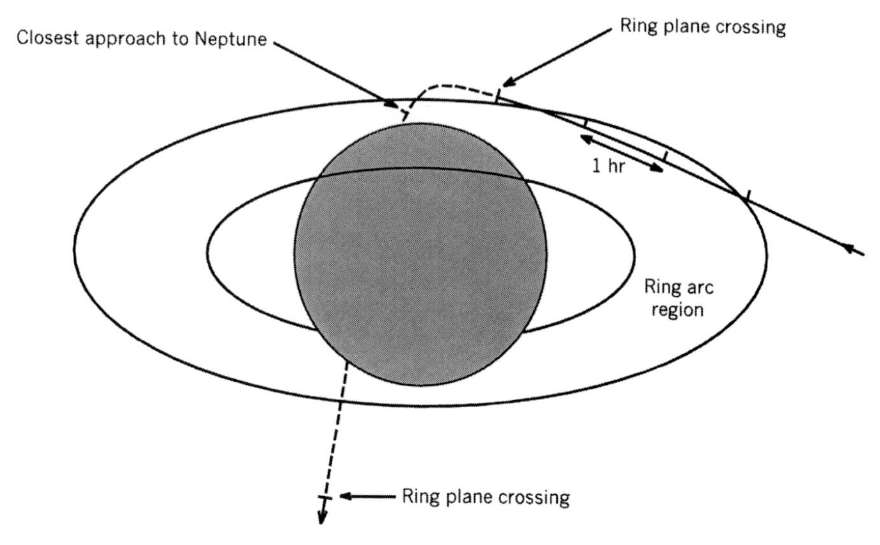

Figure 9.20 This diagram shows Voyager 2's trajectory as seen from Earth. It was optimised to enable a close fly-by of Neptune and Triton, whilst avoiding a possible collision with particles in the suspected ring arcs. This took the spacecraft over Neptune's north polar region, followed shortly afterwards by a period of 49 minutes when communications were not possible, as the spacecraft was behind Neptune as seen from the Earth.

shape and rigidity were improved to produce a 58% increase in their sensitivity. In addition the VLA, with its twenty-seven 25 m dishes, was linked to the Goldstone 70 m DSN antenna to assist in spacecraft reception. As a result the maximum bit rate transmitted by Voyager at Neptune was 21.6 kbps, which was identical to that at Uranus.

To reduce image smear during exposure, the thruster firing routine used at Uranus to compensate for judder, caused by starting and stopping the tape recorder, was refined. Scan platform scanning was used for wide-angle and infrared observations, where high resolution was not possible, and a new 'nodding' routine was introduced to avoid having to use the tape recorder to store images, thus saving valuable tape recorder capacity. In this system the whole spacecraft was slewed to follow the object being observed, as at Uranus, but the angle of slew was limited at Neptune, to avoid losing communications with Earth, and the spacecraft was then slewed back once the image had been obtained. This nodding system was used to search for rings and new satellites, as well as for some Triton observations, whereas the Uranus-type, large angle spacecraft slewing (which required the use of the tape recorder as contact with Earth was temporarily broken during each

Table 9.5. *The Voyager 2 closest approaches to Neptune's system (in time order)*

	Dia. of satellite or planet (km)	Av. dist. of sat. from centre of planet (km)	Closest approach	
			Time (hr:min)[a]	Dist. (km)[b]
Test Nereid imaging	340	5,510,000	$N-11{:}20$	4,690,000
Closest to Nereid			$N-4{:}00$	4,670,000
Inbound ring-plane crossing			$N-0{:}57$	
Closest to Neptune	49,500[c]		N	4,800
Outbound ring-plane crossing			$N+1{:}30$	
Best imaging of Triton	2,705	355,000	$N+4{:}40$	45,300
Closest to Triton			$N+5{:}14$	37,300

Notes:

[a] The time N of closest approach to Neptune was 4:00 UT spacecraft time on 25/8/89. The signals were received on Earth 4 h 6 min later.

[b] This is the distance above the surface of the satellites or the cloud tops of Neptune.

[c] Equatorial diameter. The polar diameter is less.

image-taking sequence) was used for imaging Neptune[27] and some Triton imaging.

Voyager 2 had experienced a number of problems prior to its fly-by of Uranus, but there were no other problems as it closed in on Neptune, apart from some loss of sensitivity of the infrared and photopolarimeter instruments.

Like the case of Uranus, our knowledge of Neptune and its satellites was very sketchy before the Voyager 2 encounter. Because of Neptune's and Uranus' similar size and mass (see Table 1B in the Appendix) and their adjacent orbits around the Sun, they were often considered to be twins. But Neptune, like Jupiter and Saturn, but unlike Uranus, was known to have an internal heat source (see Table 2 in the Appendix), and Neptune was also known to be significantly denser than Uranus. The atmospheres of both planets were known to contain both methane and molecular hydrogen, but as long ago as 1948 Bernard Lyot had detected cloud features on Neptune, whereas Uranus showed no clear features from Earth. But the most striking

[27] This method of scanning was essential for close-up images of Neptune because of the rapid change in viewing geometry with time.

difference between the two planets is the angle between their spin axis and their orbit around the Sun, which is an unusual 98° for Uranus but a relatively normal 29° for Neptune. So, although Uranus and Neptune were seen to be similar in some ways, they were clearly different in others.

For many years it was thought that Neptune's rotation rate was about 15 h 50 min, but in 1980 Belton, Wallace and Howard deduced a period of 18 h 10 min using spectroscopic and photometric measurements, and in July 1986 Heidi Hammel detected a large bright cloud, centred on −40° latitude, which rotated around Neptune with a period of 17 h 50 min. She found a similar cloud in 1988, centred on −30° latitude, that had a rotation period of 17 h 40 min, from images taken in the methane absorption band at 890 nm. These and earlier[28] ground-based images were most encouraging to Voyager scientists as they showed that Neptune appeared to have more cloud features than the bland Uranus.

Most theories of the origin of the Solar System assume that the Sun, the planets and their satellites condensed about 4.6 billion years ago from a large, spinning cloud of gas. This explains why the all planets orbit the Sun in the same direction, which is the same as that of the Sun spinning on its axis. It also explains why all the planets, with the exception of the outermost planet, tiny Pluto, orbit the Sun in about the same plane, in almost circular orbits. The satellites are also thought to have condensed from the proto-planets. In which case the satellites should all orbit their planets in the same direction as their planet's spin (in a so-called prograde orbit), in almost circular orbits, in their planet's equatorial plane. This is so for all but the Moon, the small outermost satellites of Jupiter and Saturn, and Triton and Nereid, the two satellites of Neptune that were known before the Voyager mission. The Moon is considered to be a special case, as it was probably formed by a collision with the proto-Earth, and the small outer satellites of Jupiter and Saturn are thought to have been captured after the planets and their satellite systems had condensed.

As far as Neptune's two satellites are concerned, Triton, which is the closer to Neptune, orbits the planet in a retrograde sense (i.e. in the opposite direction to the planet's spin), suggesting that it may have been captured by Neptune. On the other hand its orbit is almost circular, whereas that of Nereid, which is prograde, is highly elliptical. The satellites' orbits are inclined at 23° and 29°, respectively, to Neptune's equatorial plane, which

[28] In 1979 Bradford Smith, S.M. Larson and Harold Reitsema took near infrared images of Neptune with an early CCD camera, and found a prominent white cloud in each of Neptune's north and south hemispheres.

are unusually large inclinations, so maybe both satellites have been captured, Triton because of its retrograde orbit and Nereid because of its large orbital eccentricity. The discovery of new satellites of Neptune should help to clarify this.

Surprisingly, there was considerable uncertainty about the size of Triton before the Voyager 2 mission, with diameter estimates ranging from 2,200 to 5,000 km, with a most-likely figure of 3,600 km. The situation was even worse for Nereid, with estimates ranging from 200 to 1,000 km. In 1978 Dale Cruickshank (of the University of Hawaii) and Peter Silvaggio (NASA–Ames) discovered an absorption feature in Triton's spectrum from which they concluded that Triton had a methane atmosphere. If confirmed, Triton would be the third satellite after Titan and Io known to have an atmosphere. Then in 1982 Cruickshank, Carl Pitcher and Jay Abt completed a new set of observations from which they concluded that Triton had both methane ice on its surface and methane gas in its atmosphere. In the following year, Cruickshank (who had moved to NASA–Ames), Roger Clark (University of Hawaii) and Robert Brown (JPL) interpreted a shallow absorption feature at 2.15 microns as being due to liquid nitrogen on Triton's surface. If there is liquid nitrogen on the surface, there must also be nitrogen in Triton's atmosphere, as nitrogen is highly volatile. The exact amounts and condition of the methane and nitrogen on the surface and in the atmosphere would depend critically on Triton's surface temperature, which was thought to be in the range 50 to 65 K. The temperature would, in fact, depend on latitude and on the seasons on Triton, evaporating in the summer and condensing in the winter. As Neptune's equator is inclined at 29° to its orbit around the Sun, and Triton's orbit is inclined at 23° to Neptune's equatorial plane, the Sun can reach 52° altitude at Triton's poles at mid-summer in the extreme case, during a complex cycle of seasons that covers a period of about 680 years. In fact the Voyager fly-by would take place in early summer in the southern hemisphere.

The methane feature in Triton's spectrum was confirmed by further ground-based observations in 1986 by Cruickshank and colleagues, but these observations indicated that the features were weakening compared with earlier years. Similarly, whilst observations in 1977 and 1981 had shown that the intensity of Triton varied by 6% as it rotated, those in 1987 showed that the intensity variation was less than 2%. So it was thought that the methane atmosphere was becoming more hazy as the Sun was gradually heating up the southern hemisphere, and there were fears that the atmosphere may be too hazy during the Voyager fly-by for the spacecraft to see the surface.

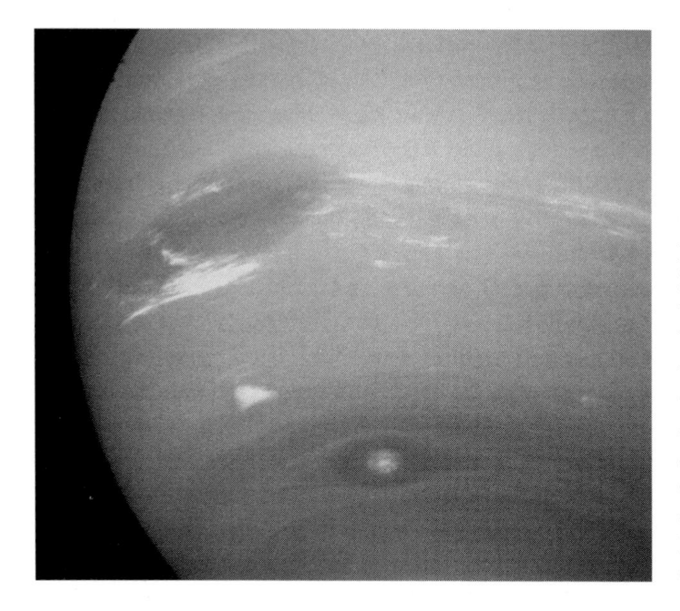

Figure 9.21 Neptune's Great Dark Spot (GDS) is seen at the left centre of this image, which also shows, near the bottom of the image, the white triangular cloud called the Scooter, and the dark D2 spot. White cirrus clouds are also clearly seen to the south of the GDS and in the centre of D2. (Courtesy NASA/JPL/Caltech; PIA 01142.)

Although the encounter with Neptune was not due to start until 5th June 1989, with a series of observations designed to produce a cloud movie, images returned as early as January 1989 gave astronomers a taste of what to expect. These images, taken when Voyager 2 was still over 300 million kilometres from Neptune, showed clear evidence of cloud structure with both a light and a dark spot visible. By April the light spot had faded but the dark spot was still evident, and as the spacecraft closed in on the planet this and other features became clear.

Voyager observations during the encounter phase showed that the dark spot, which was later called the Great Dark Spot or GDS (see Figure 9.21), is centred at about $-22°$ latitude and rotates around the planet once every 18.3 hours. Although at $12{,}000 \times 8{,}000$ km it is physically smaller than Jupiter's Great Red Spot (GRS), it is about the same size as the GRS relative to the size of their respective planets. Also like the GRS, the GDS rotates counter-clockwise about its centre south of the equator, making it a high pressure feature. On the other hand, whilst the GDS shares in the general planetary circulation at its latitude, the GRS is in a shear region on Jupiter where winds blow in one direction to its north and the other direction to its south. So there are similarities and differences between these two prominent features on Neptune and Jupiter.

Voyager 2 imaged numerous other features on Neptune during the encounter, including a second, smaller dark spot called D2 at about $-53°$

latitude, which like the GDS also appears to be a high-pressure system. When D2 was discovered at −55° latitude it had a rotation period around Neptune of 16.0 hours. Its period then slowed to 16.3 hours as it moved north to −51°, before returning to −55° with a period of 15.8 hours. It seemed to be less constrained in latitude than similar features on Jupiter and Saturn.

Light-coloured clouds were seen to be associated with both the GDS and D2, but those associated with the GDS were seen just to the south of the spot,[29] whereas D2's light-coloured clouds were at its centre. Detailed Voyager imaging of these clouds showed that they were produced when methane was forced into the higher atmosphere by these two high-pressure systems, where it condensed to form cirrus clouds. A small, fast-moving, light-coloured feature called the Scooter was seen at −41° latitude, which changed shape with time from a circular to a triangular feature. It was not as high in the atmosphere as the high-altitude cirrus clouds associated with the GDS and D2 spots, however. More high-altitude cirrus clouds were seen near the terminator of the planet, casting shadows on the blue atmosphere about 75 km below, and many other small features were seen which appeared and disappeared in one planetary revolution in Neptune's dynamic atmosphere.

The GRS rotates around Jupiter at about the same rate as the planet's interior, as deduced from the variation in its radio signals, and so it was assumed that Neptune's interior would be found to rotate every 18.3 hours, which was the rotation period of the GDS. It was something of a surprise, therefore, when eight days before closest approach Voyager 2 detected radio signals from Neptune[30] that were varying with a period of 16.11 hours,[31] which was assumed to be the spin rate of Neptune's interior. Relative to this, Neptune's clouds showed that there is an easterly equatorial jet, like that on Uranus, only that on Neptune is five times as fast at an incredible 500 m/s (1,100 mph). These Neptune winds are the fastest atmospheric winds in the Solar System, which is surprising considering that Neptune receives such a small amount of radiation from the Sun (see Table 2 in the Appendix), so the winds are thought to be driven by Neptune's internal heat source, which is significant. The maximum westerly winds were found to be at about −70° to −80° latitude where they reach about 200 m/s (440 mph).

---

[29] Simultaneous observations from Voyager and the Earth showed that a white cloud detected from Earth was that associated with the GDS.

[30] Later analysis showed that Neptune's varying radio signals had been detected by Voyager 2 some 30 days before its closest approach.

[31] It may or may not be a coincidence that this is about the period of the D2 spot, although the latter was not constant.

Voyager 2 had found that the magnetic axis of Uranus was inclined at about 60° to the planet's rotational axis (see Figure 9.16), with a magnetic centre about one third of the way from the centre of the planet. It was very difficult to explain how this strange arrangement could have been produced. One theory was that it was caused by the collision which was thought to have made Uranus spin on its side as it orbits the Sun. Alternatively, it was suggested that, by chance, we may be observing Uranus as its magnetic poles were reversing polarity, as it was known that the Earth's magnetic poles reverse their polarities about every 500,000 years. These two theories had to be abandoned when it was found by Voyager that Neptune's magnetic axis was inclined at about 50° to its spin axis, with the magnetic centre over half way from the centre of the planet (see Figure 9.16, and Table 3 in the Appendix). Not only was Neptune not spinning on its side like Uranus, because of a supposed impact, but it was not credible to suggest that Uranus and Neptune had been visited by Voyager just as both were suffering a reversal of their magnetic fields. Instead astronomers concluded that, unlike the cases of Jupiter and Saturn which have normal magnetic fields caused by a central dynamo, there is no such dynamo in the cores of Uranus and Neptune. Rather, Uranus and Neptune appear to have magnetic fields produced by convection in a highly compressed ionised water layer, which is thought to exist about 5,000 km below their cloud-tops.

Unlike Uranus, Neptune was known to have a significant heat source, and this was expected to produce more convection, which in turn would produce a larger magnetic field than on Uranus. It was surprise, therefore, when Neptune's magnetic field was found to be less intense than that of Uranus.

In mid June, about two months before closest approach, Neptune's first new satellite, initially designated 1989 N1 and now called Proteus, was discovered by Voyager 2 about 117,600 km from the planet's centre, in a circular, prograde orbit in Neptune's equatorial plane. This compares with Triton which is about 355,000 km from Neptune in a circular, retrograde orbit, inclined at 23° to the planet's equatorial plane, and Nereid which is an average of about 5.5 million km from Neptune in a highly elliptical, prograde orbit, inclined at 29°.

Over the next two months five other new satellites were discovered, all of which are in circular, prograde orbits between 73,600 km (for 1989 N2 or Larissa) and 48,000 km (1989 N6 or Naiad) from the centre of Neptune. Proteus has a diameter of about 400 km, which is a little larger than that of Nereid, with the diameters of the other new satellites ranging from about 200 km (for Larissa) to 50 km (for Naiad). Proteus and Larissa were the only new satellites to be imaged at high resolution, showing highly cratered surfaces with reflectivities of only 6%.

Contrary to expectations, the presence of these small satellites in circular, prograde orbits did not help to resolve the question as to whether Triton was an original or captured satellite of Neptune, as it was difficult to see how Triton could have been captured without disturbing the orbits of these much smaller satellites. When it was initially captured, Triton would have had a highly elliptical orbit that would have taken it close to these small satellites, causing severe disruption to their orbits. Equally, it was difficult to accept that Triton, in its inclined retrograde orbit, was an original satellite of Neptune. Maybe Triton was captured before the small satellites had condensed from Neptune's proto-satellite nebula, but it was not clear how a largish body like Triton could have condensed from the solar nebula before the very much smaller satellites could condense from Neptune's nebula, as the smaller bodies should have condensed first.

In early August, about three weeks before closest approach, Voyager confirmed that Neptune apparently had ring arcs of the type deduced from previous Earth-based observations of stellar occultations. But, as the spacecraft drew closer to Neptune, it became clear that the ring arcs were just the brightest segments of a complete ring, designated 1989 N1R and later called Adams,[32] about 62,900 km from the centre of Neptune (see Figure 9.22). A second complete ring, 1989 N2R and called Le Verrier,[33] was also seen inside the first and about 53,200 km from the centre of Neptune. After its closest approach to Neptune the spacecraft imaged the rings in back-lighting (see Figure 9.22b), showing a third, diffuse ring, 1989 N3R and called Galle,[34] about 42,000 km from the centre of Neptune, and a broad band of relatively large particles and dust, 1989 N4R or the Plateau ring, between 1989 N1R and N2R. In addition, as Voyager passed through the ring plane outside the outer ring, a large number of dust impacts were recorded, starting about one hour before the crossing. So there was also a thick cloud of dust orbiting Neptune.[35]

[32] After John Couch Adams, who was one of the two people to predict the position of Neptune before its discovery in the nineteenth century.

[33] After Urbain Le Verrier, the other astronomer to predict the position of Neptune before its discovery.

[34] After Johann Galle, the discoverer of Neptune using Le Verrier's orbital predictions.

[35] The peak impact rate measured at Neptune was 300 impacts per second, compared with 400 per second when Voyager crossed Saturn's ring plane, and only 50 per second when it crossed that at Uranus. Voyager passed much closer to the outermost ring of Neptune than to that of Uranus, however, which could possibly explain the higher figure at Neptune.

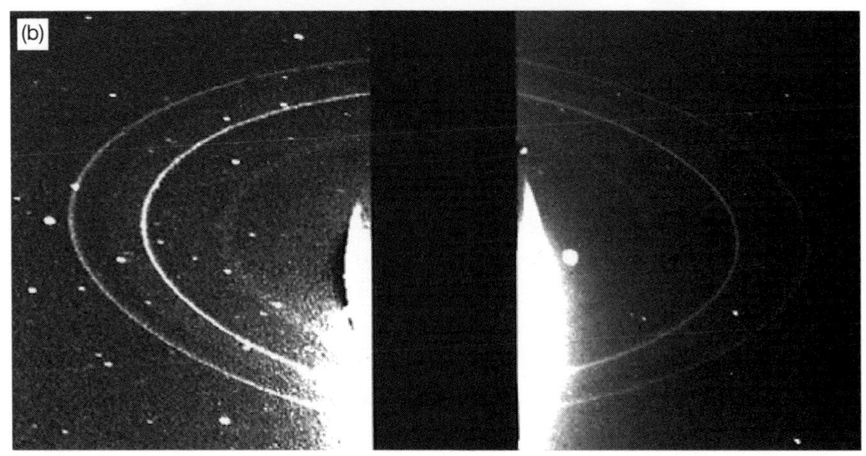

Figure 9.22 Neptune's rings are shown in front-lighting (a) and in back-lighting (b). The Adams and Le Verrier rings are the only rings seen in (a), with the clumpy nature of the Adams ring being evident. The faint Galle ring is seen in (b), about half-way between the Le Verrier ring and the overexposed image of Neptune. The brightness of these very faint rings is greatly exaggerated in (b), in particular, which was the result of a spacecraft exposure of almost ten minutes. (Top image courtesy NASA/JPL/Caltech; PIA 01493; bottom image courtesy NASA.)

The satellite Galatea (or 1989 N4) orbits Neptune about 900 km inside the clumpy Adams ring (1989 N1R), and the satellite Despina (1989 N3) orbits Neptune about the same distance inside Le Verrier ring (1989 N2R), helping to shepherd the rings. Galatea could not in itself cause the clumping of its associated ring, and neither Galatea nor Despina could stop either ring from expanding outwards. To do this a further satellite was required just outside each of the two rings, but no such satellites were found.

Further analysis showed that the ring arcs discovered in 1984 and 1985, at 67,000 and 56,000 km from Neptune, were part of the Adams ring at 62,900 km. The reason for the discrepancy in distance was because of a slight error in the earlier estimation of the angle of Neptune's spin axis to its orbital plane. More interesting, however, was the realisation that the stellar occultation observed by Harold Reitsema, William Hubbard and colleagues in 1981, at about 70,000 km from Neptune, was not due to a ring arc, but to the 200 km diameter satellite designated 1989 N2. Reitsema had been right all along when he maintained that the occultation was due to a satellite, not a ring arc, and in recognition of his team's prior discovery of this satellite, the IAU invited them to name the satellite from names connected with Neptune mythology. They chose Larissa, as both Reitsema and Hubbard had daughters named Laurie, a near match.

Voyager 2 flew only about 37,000 km above Triton, some five hours after the spacecraft's closest approach to Neptune. The satellite was found to have a highly reflective surface with a diameter of about 2,700 km, near the low end of the expected range. It density was estimated as 2.03 g/cm^3, greater than that of any of the satellites of Saturn or Uranus, and somewhat higher than generally expected. This indicated that Triton has less ice and more rock in its interior than anticipated.

Because of its highly reflective surface, with an average reflectivity of about 75%, Triton absorbs little solar energy, so its surface temperature was lower than anticipated at 38 K, which is far too cold for nitrogen to exist in liquid form, except in possible 'hot' spots. Because of the lower than expected temperature, the surface-level atmospheric pressure of 15 microbar was considerably less than expected, as most of the atmosphere was 'frozen out' on the surface. Fortunately, although there was a haze layer about five kilometres above the surface, the surface could be clearly seen. Voyager found that the atmosphere consists primarily of nitrogen, with a trace (at 0.01%) of methane.

Voyager arrived at Triton during early summer in its southern hemisphere and revealed a very bright, pinkish-white, southern polar cap of nitrogen ice extending three-quarters of the way from the pole to the equator (see Figure 9.23). The ice cap appeared to be slowly melting at its edges to reveal a darker, redder surface underneath. The edge of the ice cap was ragged, and at one place there was a small group of circular dark features with bright skirts.

The retreating and advancing polar ice caps have ensured that the surface of Triton is still subject to change, even though it is the coldest

Figure 9.23 Triton's prominent south polar cap of nitrogen ice is seen in the left half of this image, with the dark geyser-like plumes pointing to the lower right. The dimpled cantaloupe terrain is seen in the right half of the image, with fissures that extend right into the polar cap. (Courtesy NASA/JPL/Caltech.)

planet or satellite known in the Solar System. Dark streaks of up to 150 km long were seen all over the south polar cap and, as the ice retreats every summer, those streaks near the edge of the cap must have been produced recently. Their appearance, and the fact that all of these dark streaks point approximately north-east, suggested that nitrogen or methane had been ejected at one end of the streak in some sort of explosion, and carried by the wind in Triton's tenuous atmosphere.

Robert Hamilton Brown proposed a mechanism for these eruptions on Triton, likening them more to geysers than volcanoes. In particular, he suggested that sunlight penetrates the almost-transparent nitrogen ice of the polar ice cap, where about two metres down it is absorbed by frozen methane, which has been darkened by exposure to ultraviolet light. The heat is trapped by the nitrogen ice, as it is a poor conductor, and as the heat builds up the subsurface nitrogen ice is turned into gas. Eventually the gas pressure is too much to be resisted by the surface ice, and the nitrogen gas explodes through the surface, carrying with it the Sun-darkened methane, to produce a geyser-like eruption.

Crater counts across Triton's surface indicate that none of its original surface has survived, so there must have been considerable geological activity in its early lifetime, possibly caused by gravitational stresses as a result of its capture by Neptune.

In the equatorial regions there are large areas of relatively young, dimpled terrain quite unlike anything seen elsewhere in the Solar System. This so-called cantaloupe terrain was broken in places by shallow linear ridges up to 30 km wide and 1,000 km long, probably caused when water ice in Triton's mantle froze and expanded. Within this cantaloupe terrain and to its north there are areas of frozen lakes and calderas, showing evidence of liquid flows, possibly of water mixed with ammonia and methane,[36] as a result of volcanic activity. Some of the frozen liquid flows are ancient with many craters, but some look fresh and are almost crater-free.

A month after the encounter Lawrence Soderblom and Tammy Becker at the US Geological Survey were stereoscopically examining Voyager images of the dark streaks on the south polar cap, when they were amazed to discover an image of an eruption in progress. It produced a plume 8 km high, which extended horizontally for about 150 km downwind. They then found an image of a second eruption in progress. These two eruptions, and three other suspected eruptions, were all near to the sub-solar point, indicating that they are caused by solar heating. This gave some confidence that Robert Hamilton Brown's theory of geyser-like eruptions, or something like it, may well be correct.

---

[36] The addition of ammonia and methane substantially reduces the freezing point of water. The lakes and calderas cannot be made of nitrogen and/or methane, with no water, as the surface temperature is too near to the melting points of nitrogen and methane to be able to support the crater walls, which, in places, are over 1,000 m (3,000 ft) high.

So ended the remarkable mission of Voyager 2. It had been conceived in 1965 as part of a Grand Tour programme, approved in simplified form in 1972, and launched in 1977, with a Neptune fly-by in 1989, some twenty-four years after conception. Without doubt Voyager 2 has been by far the most successful spacecraft to investigate the Solar System.

Chapter 10 | THE SUN

10.1 Early Work

Early observations of the Sun and of its influence on the Earth's environment have already been touched on in Chapters 1 and 2. Before going on to examine the contribution of space research to this area, however, I will now summarise our state of knowledge as the space age opened in the late 1950s.[1]

The Sun was known to be a sphere of mainly hydrogen and helium gas, generating heat by nuclear fusion, and rotating about once every 25 days at its equator and once every 34 days near the poles.[2] Each of the sunspots on the Sun's surface (the photosphere) consists of a dark central area, called the umbra, surrounded by a lighter penumbral region. Sunspot numbers vary over a cycle of about eleven years, although it has been as long as seventeen years and as short as seven years. The cycle starts with spots appearing at about 30° to 40° latitude, north and south of the equator, immediately after sunspot minimum. The latitude gradually reduces as the cycle progresses, being about 15° at sunspot maximum, and about 6° at sunspot minimum, with very few sunspots being seen at the equator. After minimum, sunspots then start to appear at mid latitudes once more.

Sunspots, which are cooler than the general surface of the Sun, have intense magnetic fields of about 3,000 gauss,[3] whereas the general magnetic field of the Sun is only about 1 gauss. The spots tend to appear in pairs, with

---

[1] Excluding early work on the solar constant which will be considered in Section 10.8.

[2] These are its sidereal periods. Its apparent (synodic) periods as seen from the Earth are about two days more (because of the Earth's orbital motion around the Sun).

[3] This is about 10,000 times that of the Earth's magnetic field at its equator.

the preceding or 'p' spot of the pair, as it moves across the solar disc, almost always having a different polarity to the other spot. The p spots in one hemisphere of the Sun generally have the same polarity as each other, which is opposite to the p spots in the other hemisphere. When a new sunspot cycle starts, the polarities reverse between the hemispheres, so the sunspot cycle is really a 22 year rather than an 11 year cycle.

Large loops and filaments of rapidly moving gas were first seen as so-called prominences at the edge of the Sun during total solar eclipses.[4] The most spectacular (or eruptive) prominences, which last for a few hours and are associated with sunspots, sometimes produce bright solar flares which last for a few minutes and which are best seen in the monochromatic light of the hydrogen-α line. There are also cloud-like prominences associated with faculae or very bright markings on the surface of the Sun. The eruptive prominences contain hydrogen, helium and other elements, of which iron, titanium, calcium, and barium are the most evident, whereas the cloud-like prominences consist almost entirely of hydrogen, helium and calcium. Sunspots and eruptive prominences are generally limited to low to mid latitudes, whereas faculae and cloud-like prominences can be seen at any latitude.

Normally, the very bright photosphere overwhelms the light from the solar chromosphere and corona, both of which are visible during a total solar eclipse, the chromosphere as a thin pinkish band surrounding the Sun just above the photosphere, and the corona as an extensive white halo extending for several solar radii into space. The outer edge of the chromosphere is seen, on closer examination, to be made up of many fine, jet-like structures called spicules. The observed shape of the corona changes with the stage of the solar cycle from being roughly circular around solar maximum to roughly elliptical around solar minimum.

The chromosphere can be imaged across the total solar disc, without waiting for a total solar eclipse, by taking photographs in the monochromatic light of hydrogen alpha or that of the calcium H and K lines. These spectroheliograms, as they are called, show the filaments mentioned above as long dark structures, together with bright plages which are associated with sunspots.

It was known in the nineteenth century that there was a correlation between sunspots and disturbances in the Earth's magnetic field, which the Norwegians Kristian Birkeland and Carl Størmer attributed in 1896 to

---

[4] Prominences are not limited to the edge of the Sun in eclipses, of course, it is just easier to see them there at that time.

beams of cathode rays[5] emitted by the Sun. Then in 1904 the English astronomer E. Walter Maunder showed that the larger magnetic storms on Earth start about thirty hours after a large sunspot group crosses the centre of the solar disc, suggesting that the sunspots emitted some sort of particles that reached the Earth about a day or so later. Smaller storms did not seem to be generally associated with sunspots, however, but they had a tendency, not shown by the larger storms, to recur every 27 days, which is the synodic period of rotation of the Sun's equator. The German geophysicist Julius Bartels christened the invisible source on the Sun of these smaller magnetic storms 'M regions'.

In the 1930s Howard Dellinger of the American Bureau of Standards showed that problems with the reception of short-wave radio waves on Earth were often, but not always, associated with solar flares seen on the Sun a short time before. These radio waves are reflected from the F region of the Earth's ionosphere. So he suggested that some unknown solar phenomenon was producing solar flares and generating invisible radiation that penetrated the F region and modified the lower levels of the ionosphere, causing them to lose their transparency to radio waves. In 1939 T.H. Johnson of the Bartol Research Foundation and Serge Korff of New York University suggested that the cause of the disturbances was solar X-rays.

Many astronomical discoveries are made as the result of accidental rather than deliberate observations. An excellent example of this occurred in February 1942 when British army radars picked up radio noise which was initially attributed to enemy jamming. James S. Hey, a physicist at the British Army's Operational Research Group, was tasked with investigating the cause of this interference and found, much to his surprise, that the source was the Sun. In particular, he found that the appearance of the signals correlated with a large sunspot group crossing the solar meridian. Later in the 1940s it was also found that solar flares emitted radio waves but, unlike the case of the sunspot-generated radio signals, the flares could be anywhere on the visible solar disc and still produce radio signals detectable on Earth. In addition, the radio signals from flares lasted for only a few minutes, whereas those associated with sunspots lasted for hours. So from time to time the Sun not only interrupted short-wave communications on Earth by modifying the Earth's ionosphere, but it also produced radio noise of its own.

In the 1940s the American Scott Ellsworth Forbush found that the number of medium and high energy cosmic rays observed at the Earth's

---

[5] In the following year J.J. Thomson showed that cathode rays are particles, now called electrons.

surface varied with the 11 year solar cycle, reducing slightly as solar maximum approached. In addition, these cosmic ray counts often diminished abruptly during solar storms. Solar particles or X-rays could not in themselves have such an effect, but the Sun's magnetic field could, by deflecting these cosmic rays away from the inner Solar System. Hannes Alfvén suggested, therefore, that the Sun's magnetic field was increased during active periods by a magnetic field carried out from the Sun by charged particles.

In the early 1950s Horace and Harold Babcock of the Mount Wilson Observatory measured the magnetic field of the Sun across its disc with their newly developed solar magnetograph, and showed that there were both bipolar and unipolar magnetic regions on the Sun. The bipolar regions could persist for up to nine months, with sunspots forming and dispersing within them, whereas the unipolar regions, where the magnetic flux was either leaving or entering the Sun, did not appear to be connected with sunspots at all. In the bipolar regions the magnetic flux leaving the Sun was about equal to that entering it.

The Babcocks suggested that ions and electrons (called corpuscular radiation) leaving the Sun in bipolar regions would follow the field lines above those regions and collide over the Sun, generating radio noise and forming prominences and flares. Corpuscular radiation leaving the Sun in unipolar regions, on the other hand, would stream away from the Sun and some of it would reach Earth, modifying the ionosphere and causing problems with short-wave reception. The Babcocks suggested that these unipolar regions were Bartel's M regions. Corpuscular radiation would also leave the Sun from its polar regions and follow the general magnetic field lines away from the Sun.

The discussion above has generally concentrated on the disturbed Sun of sunspots, prominences, flares, and the like, and their effect on the Earth. Even when the Sun is relatively quiet, however, it has a major influence on the Earth that extends beyond its fundamental one of keeping the Earth in its orbit and providing it with heat and visible light.

Not only was the Sun found to emit radio waves from sunspots and solar flares, but in June 1942 George C. Southworth of Bell Labs. discovered that the Sun emitted radio waves continuously in the centimetre band,[6] and later

[6] For electromagnetic waves, whether they be visible light or radio waves, $\lambda\nu = c$, where λ is the wavelength and ν the frequency of the waves, and c is the velocity of light, i.e. 3×10^{10} cm/s. So if $\lambda = 3.2$ cm, where Southworth first detected these radio waves, then $\nu = (3\times10^{10})/(3.2) \approx 9\times10^{9}$ /s, or 9 GHz.

work in Australia and elsewhere also detected solar emissions in the metre band. These centimetre radio emissions were independently explained by the Australian David F. Martyn and the Russians Vitaly Ginzburg and Iosif Shklovskii in 1946 as coming from the solar chromosphere at a temperature of 10,000 K, whereas the metre-waveband emissions were thought to be produced by Bengt Edlén's very hot corona at a temperature of some 2 million K (see Section 1.2).

The Sun is not only a continuous source of visible light and radio waves, since in 1952 Friedman had shown, using sounding rocket experiments, that the solar corona also emits X-rays that are intense enough to sustain the E region of the Earth's ionosphere (see Section 1.2). Similarly the Sun's ultraviolet radiation, which does not penetrate to ground level, was found to sustain the D region of the ionosphere.

In the early seventeenth century Johannes Kepler had suggested that comets' tails point away from the Sun because material coming from the head of the comet is pushed away by particles of solar radiation. His theory fell out of favour in the nineteenth century, as light was then thought to be a wave motion and not a stream of particles. Kepler's theory gained a new lease of life in the early twentieth century, however, after it had been shown that light has both wave- and particle-like characteristics, exerting pressure on objects in its path.

As the twentieth century unfolded, however, it became evident that sunspots probably emitted charged particles (as explained above) and, if that was the case, some astronomers thought that there may also be a weak flux of charged particles evaporating continuously from the Sun. Some of these particles would be captured by the Earth's magnetic field, which would then channel them to the Earth's poles, causing the observed persistent faint aurorae and constant minor fluctuations in the Earth's magnetic field. So did the pressure of light or these solar-generated charged particles cause comets' tails to point away from the Sun?

In 1951, Ludwig Biermann of the Max Planck Institute in Germany showed that the radiation pressure suggested by Kepler could not exert enough force to create the observed comets' tails, but solar ions and electrons could if their velocities were between 500 and 1,000 km/s, and their density was between 100 and 1,000 /cm^3 at the distance of the Earth, resulting in an average ion (or plasma) flux of about 2×10^{10} ions/cm^2/s.

The English geophysicist Sydney Chapman was intrigued by Edlén's very hot solar corona of some 2 million K, and calculated that such a corona, consisting mainly of electrons and protons, would extend beyond the Earth's orbit, where its temperature would be about 200,000 K. Biermann's theory was of a dynamic solar environment, with the Sun emitting a continuous stream of charged particles, whereas Chapman's corona was in hydrostatic equilibrium, with the Sun's gravity balancing the corona's gas pressure. Chapman's corona was static, but Biermann's was dynamic, losing matter on a continuous basis which was replenished from below.

Even though Biermann's and Chapman's assumptions were so radically different, their predicted ion or plasma densities at the distance of the Earth's orbit from the Sun were the same. Chapman's theory had a problem, however, in that it predicted an ion pressure at the orbit of Pluto some 10^7 times that thought to exist in the interstellar medium. This implied that the solar corona extended far beyond the orbit of Pluto, which seemed unlikely, although it was not impossible.

In 1957 Eugene Parker of the Enrico Fermi Institute tried to reconcile these two apparently contradictory theories of Biermann and Chapman by adding dynamic terms to Chapman's static equations to represent the hypothesised continuous expansion and evaporation of the solar corona, and found that his results were now consistent with Biermann's predictions. In the following year Parker developed his theory of this so-called solar wind[7] and showed that the expanding coronal gas would draw the magnetic field lines in the corona far out into the Solar System, producing spiral magnetic field lines because of the Sun's rotation. But his theory was too efficient in transferring energy through the corona, even using Biermann's lowest estimate of solar flux, so a highly efficient mechanism was required to maintain the Sun's corona at the 2 million K near the Sun, as estimated by Edlén. Parker suggested as a possibility that hydromagnetic waves, propagating through the Sun's photosphere, could supply the required energy, but his theory was rejected by other physicists.

So as the space age opened there was a great deal of information on sunspot variabilities, prominences, solar flares, solar-generated radio noise, magnetic storms on Earth, short-wave radio fades, and the like, but the linkage between these was only imperfectly understood. The cause of sunspots, the reason for the sunspot cycle, and the cause of the ultra-high temperature of the corona were key problems still unsolved, and the nature and character of the suspected solar wind was also unknown. The space age would help to solve some of these problems, even if only partially.

---

[7] The words 'solar wind' were first used by Parker. In his theory he predicted that the velocity of particles in the corona would increase with distance from the Sun, such that at a distance of about 5.6 solar radii from the centre of the Sun their velocity would become supersonic. It was this supersonic, radial flow that Parker called the solar wind.

10.2 **The Solar Wind**

The very earliest spacecraft were Earth-orbiting with relatively low orbits, and so the *in situ* environment that they measured, although influenced by the Sun, was controlled by the Earth's magnetic field. These spacecraft and their results have already been discussed in Chapter 2. In this present chapter I will first outline the results of *in situ* measurements of the interplanetary environment, which is generally controlled by the Sun, and then discuss observations of the Sun carried out by Earth-orbiting and other spacecraft.

The first spacecraft to measure the plasma away from the influence of the Earth's magnetic field was Luna 2, *en route* to the Moon, in September 1959. The results indicated that the extended solar corona, at the distance of the Earth from the Sun, contained high-speed ions, thus confirming Biermann's theory. The flux measured was about 2×10^8 ions/cm^2/s, however, or about two orders of magnitude lower than Biermann had predicted. Unfortunately, it was not possible to measure the speed or direction of the ions in the solar wind directly using the simple ion traps on Luna 2. After analysing the results, however, Shklovskii suggested that the primary reason for the low flux was that the plasma density was lower than Biermann had predicted.

Meanwhile in America the Space Science Board of the National Academy of Sciences started considering in 1958 what areas of research should be undertaken in space. One suggestion coming from its Committee on Space Projects, chaired by Bruno Rossi, was a study of the interplanetary plasma or solar wind[8] but, although this suggestion was supported by the Space Science Board, it fell largely on deaf ears in the fledgling NASA. This was because the instrumentation would need developing and it was difficult at the time to send satellites sufficiently far away from the Earth to be outside the influence of the Earth's magnetic field. NASA's initial reluctance did not last long, however, helped by the political decision in the late 1950s to rapidly expand its space programme, following Russia's series of dramatic space successes. So Rossi's team at MIT, which was led by Herbert Bridge, was funded by NASA to design and build a plasma probe to fly on Explorer 10 (see Figure 10.1) into a highly eccentric orbit around the Earth with an apogee of 180,000 km.

Bridge's plasma probe was a considerable advance on that carried by Luna 2, as it had a higher signal-to-noise ratio and was able to measure the

---

[8] Another area of study suggested by Rossi's committee was X-ray astronomy (see Section 1.3).

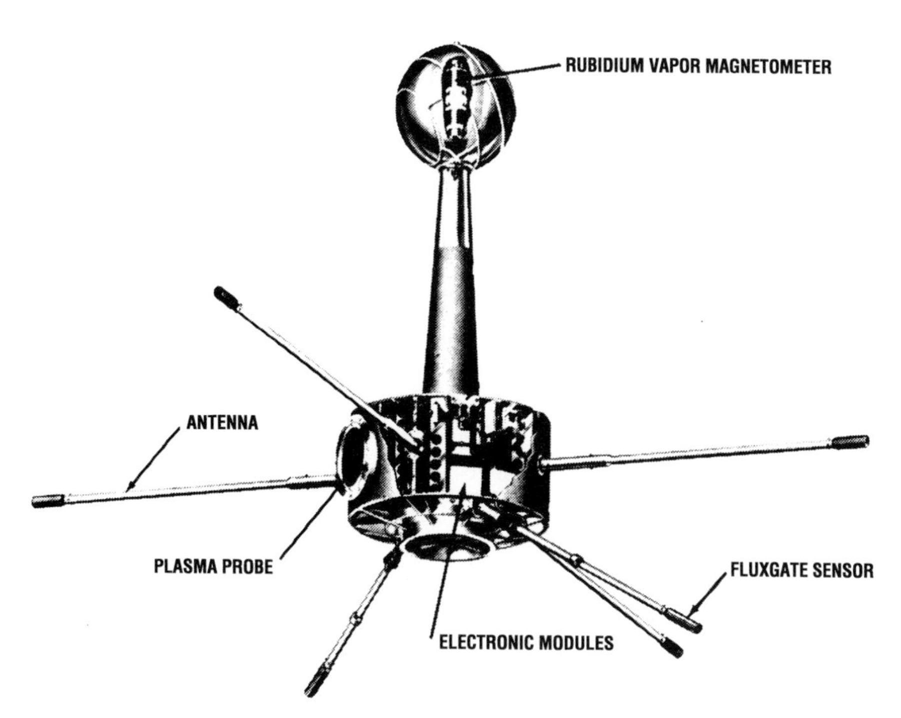

Figure 10.1 Explorer 10 is shown in this diagram in its orbital configuration. The rubidium-vapour magnetometer was capable of measuring fields as weak as 0.1 gamma, with the direction of the field being determined by the boom-mounted fluxgate magnetometers. Bridge's plasma probe (discussed in the text) was fitted in the main body of the spacecraft. There were no solar cells, as power was provided by silver–zinc batteries. (Courtesy NASA and *Sky & Telescope*.)

ion flux in six different velocity intervals up to 660 km/s. The 35 kg Explorer 10, which was launched on 25th March 1961, surveyed space on the night-side of the Earth and, although it was not clear at the time, it never completely escaped from the Earth's influence owing to the Earth's long magnetic tail (see Figure 2.2). Nevertheless, the plasma flux of 4×10^8 ions/cm^2/s measured around apogee[9] was similar to that measured two years earlier by Luna 2. The velocity of the ions, which were assumed to be protons,[10] ranged from 120 to 660 km/s, with an average of about 300 km/s, in the general direction away from the Sun, again supporting Biermann's

[9] Signals ceased one day after reaching apogee as the 16 kg of silver–zinc batteries were exhausted and there were no solar cells on the spacecraft to recharge them.

[10] The plasma trap of Explorer 10, like that on Luna 2, had a negative grid to repel all but the most energetic electrons.

theory. The plasma density varied from about 6 to 20 protons/cm^3, which was lower than predicted by Biermann. Unfortunately, the spacecraft measurements fluctuated a great deal, partly because the spacecraft explored space on the night-side of the Earth and did not escape completely from the Earth's influence, spending some time in the disturbed region between the magnetopause and the bow shock. So some sceptics still doubted Biermann's and Parker's ideas. What was really required was either data from space on the day side of the Earth or from interplanetary probes.

In parallel with Rossi's team at MIT, Marcia Neugebauer and Conway Snyder of JPL produced the outline design of a plasma spectrometer in 1959 to measure the energy spectrum of both electrons and protons in the solar wind, and pushed NASA for its inclusion in an early lunar or interplanetary probe. Their first launch opportunities were on board Rangers 1 and 2, which should have flown passed the Moon in 1961, but both spacecraft got stuck in their Earth parking orbits because of a problem with their launch vehicles. The next opportunities were on Mariners 1 and 2, which were launched towards Venus in 1962 but, unfortunately, the launch of Mariner 1 was again a failure. Fortunately, however, that of Mariner 2 on 27th August 1962 was a resounding success.

Neugebauer and Snyder's plasma spectrometer showed that the velocity of the solar wind, which was always blowing away from the Sun, was variable, being generally in the range from 400 to 700 km/s, but occasionally it reached the maximum measurable velocity of 1,250 km/s. The positive ions in the wind were generally assumed to be protons, at a measured density of about 1 to 5 protons/cm^3, but some helium nuclei were also apparently detected. A comparison with the geomagnetic index showed that velocity peaks in the plasma correlated with peaks in the index which tended to recur every 27 days. So the rôle of this corpuscular radiation or plasma in producing general magnetic storms on Earth was clear. Neugebauer and Snyder tried to go even further and find the M regions on the Sun responsible for these storms but were unable to do so.

Whilst the Mariner 2 solar wind results were still being analysed, NASA launched the first of a new series of spacecraft, the 62 kg Interplanetary Monitoring Platform, or IMP-1, on 27th November 1963. Two of the key scientists involved were Norman Ness of NASA's Goddard Space Flight Center, who was responsible for the satellite magnetometers, and John Wilcox of the University of California at Berkeley, who acted as the interface with Mount Wilson's solar observatory. Wilcox wanted to try to correlate Mount Wilson's measurements of magnetic fields on the surface of the Sun with interplanetary magnetic fields measured by the spacecraft.

IMP-1 (also known as Explorer 18) was the first of a series of ten IMP spacecraft that were planned to investigate the interplanetary environment over a complete solar cycle of about eleven years, the first three spacecraft of the series having identical designs. The experiment payload of IMP-1 consisted of three magnetometers, four cosmic ray experiments and four solar wind particle experiments. A rubidium-vapour-type magnetometer, an improved version of that carried on Explorer 10, was mounted at the end of a 1.8 metre axial boom to keep it as far away from spacecraft-generated disturbances as possible. Two fluxgate magnetometers were mounted at the end of two 2 metre radial booms for the same reason. The rubidium magnetometer, which had a measurement range of from about 0.1 to 1,000 gammas, was designed to measure the intensity of the interplanetary magnetic field, but it could not measure its direction. That was achieved by the two fluxgate magnetometers.

IMP-1 was to have been launched into a highly eccentric, six day orbit with an apogee of 278,000 km on the sunward side of the Earth, but unfortunately the launcher malfunctioned. This resulted in a four day orbit with an apogee of 198,000 km, although this was still on the sunward side of the Earth. Initially IMP-1 spent about 75% of each orbit in interplanetary space beyond the Earth's bow shock, but as time passed it spent less and less of each orbit in interplanetary space, because of the changing orientation of the Sun–Earth line as the Earth orbited the Sun. By mid February 1964 the time in interplanetary space had reduced to zero, but by then the Sun had rotated through almost three revolutions, providing a wealth of solar-related data.

As mentioned above, Parker had shown in 1958 that the Sun's magnetic field lines should spiral out from the Sun in a pattern similar to that produced by a rotating lawn sprinkler. He calculated that the magnetic field lines at the Earth's distance from the Sun should make an angle of about 45° to the Sun–Earth line. Norman Ness found, on analysing the IMP-1 magnetometer data, that the magnetic field lines did indeed make such an angle to the Sun–Earth line, but he was surprised to find that every few days the magnetic field suddenly changed direction by about 180°. Ness and Wilcox then began a detailed examination of the correlation between the spacecraft measurements and those of the Sun at Mount Wilson, and found that the interplanetary field's polarity was correlated with that of the Sun's photosphere at the Sun's equator with a lag of about 4.5 days. This implied an average solar wind velocity of about 380 km/s, which was consistent with that measured by IMP-1's plasma trap. So the interplanetary magnetic field is an extension of that on the Sun, changing as that on the Sun changed.

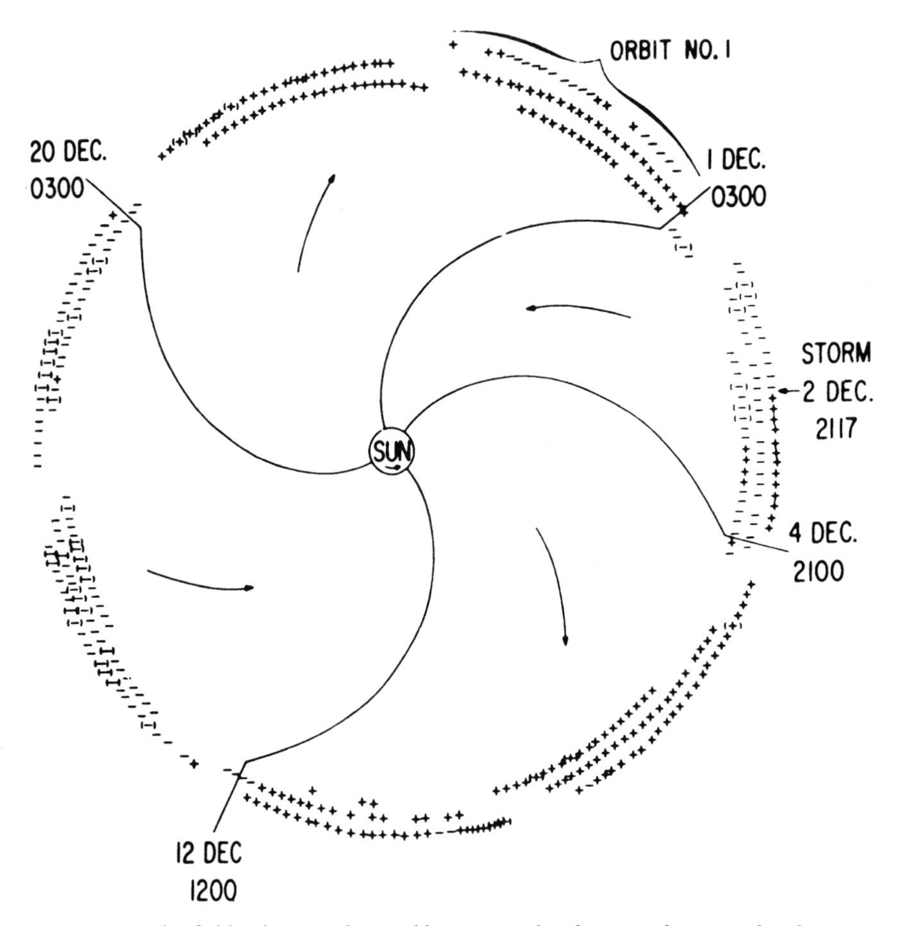

Figure 10.2 The field polarities observed by Ness and Wilcox are shown in this diagram that covers almost three solar revolutions. The sector pattern of the interplanetary field is clearly seen, together with the effect of the solar storm of 2nd December. (Reprinted with permission from *Science* 148 (1965), p 1592, by Ness, N.F., and Wilcox, J.M.. Copyright 1965 by the American Association for the Advancement of Science.)

When Ness and Wilcox analysed the results of almost three solar rotations they found that the abrupt, 180° changes in field direction occurred at about the same place relative to the Sun's surface on each revolution, except on 2nd December when a solar storm suddenly caused the field to change direction (see Figure 10.2). This led them to suggest that the Sun-generated interplanetary magnetic field is composed of organised sectors of alternately inward and outward pointing magnetic fields. They also found that the passage of IMP-1 across the sector boundaries was not only evident in the

interplanetary magnetic field but also in the plasma velocity measured by the spacecraft, and in the geomagnetic index measured on Earth.

Between February and May 1964 IMP-1, as it was no longer in the interplanetary field, investigated the Earth's magnetosphere bounded by the magnetopause (see Figure 2.2), and found that the magnetic tail on the night side of the Earth extends well beyond the satellite's apogee of 32 R_E. Field strengths of about 20 gammas were measured inside the magnetosphere, compared with the interplanetary field at the Earth's orbit of about 5 gammas. IMP-1 then discovered a narrow region in the magnetosphere at local midnight where the field fell to almost zero. This so-called neutral sheet was found to separate the region to its north, where the Earth's magnetic field lines were pointing sunward, from that to its south, where the field lines were pointing in the opposite direction.

The effect of the Moon on the solar wind was first investigated by IMP-6 (or Explorer 35), which was put into lunar orbit on 22nd July 1967. Unlike the Earth, Jupiter, Saturn, etc., the Moon has practically no magnetic field and, as a result, it has no bow shock. Instead IMP-6 found that the sunlit surface of the Moon appears to be directly bombarded by the solar wind, thus creating a shadow-shaped void in the solar wind on the antisolar side of the Moon. IMP-6 also investigated the interplanetary field near the Moon and the effect of the Earth's magnetospheric tail as it swept past the Moon every 29.5 days.

Cosmic rays in interplanetary space, the so-called primary cosmic rays, are modified when they impinge on the Earth's atmosphere to produce what are called secondary cosmic rays. As mentioned above, in the 1940s Forbush had found a slight decrease in secondary cosmic rays at the surface of the Earth as solar maximum approached. The average change between solar minimum and solar maximum was only 4%, in fact. Over the period 1967 to 1972 the spacecraft IMP-6 was able to measure the intensity of primary cosmic rays in interplanetary space, and found that the effect was much more marked, with a maximum figure 80% above the minimum. This is because at the Earth's surface we only detect secondaries from the higher energy cosmic rays, as the lower energy cosmic rays are completely absorbed by the Earth's atmosphere, and the higher energy cosmic rays are little affected by the solar magnetic field. In interplanetary space, however, there are cosmic rays of all energies, including the lower energy cosmic rays which are affected by changes in the solar magnetic field. Interestingly, since their launches in 1972 and 1973 Pioneers 10 and 11 have also found a gradual increase in the average cosmic ray intensity of about 1.5 to 2% per AU as these spacecraft leave the Solar System, confirming the solar influence on cosmic ray intensities.

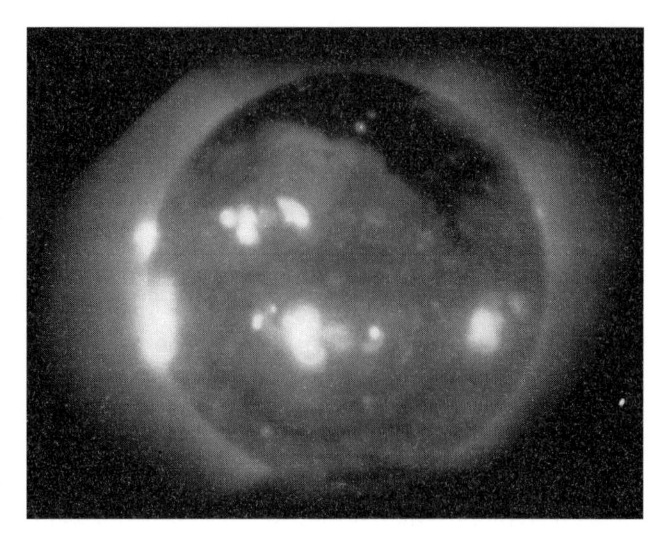

Figure 10.3 A large polar coronal hole is clearly seen as a dark region at the top of this X-ray Skylab image taken in 1973. Small X-ray bright points are also seen dotted over the solar disc, with one bright point in the coronal hole. (Courtesy Harvard-Smithsonian Center for Astrophysics.)

In the late 1960s and early 1970s there were two major problems still taxing solar physicists interested in the Sun's magnetic field and its extension into interplanetary space. Where on the Sun's surface are Bartel's M regions, the expected source of the high-speed particle streams, and what does the three-dimensional interplanetary field look like? Are the sectors discovered by Wilcox and Ness like the segments of an orange, each running in a North–South plane back to the Sun, or are they the intersections of a warped neutral sheet and the ecliptic? Unfortunately, at that time no spacecraft had ventured very far from the ecliptic, so the three-dimensional structure of the interplanetary field could only be deduced from indirect evidence.

In 1957 M. Waldmeier had discovered so-called coronal holes in visible light as gaps in the corona on the solar limb. These observations were then extended in the 1960s to the extreme ultraviolet with the OSO (Orbiting Solar Observatory)-4 spacecraft, which showed coronal holes in front of the solar disc. Coronal holes were most clearly seen at X-ray wavelengths, however, where they appear as dark (i.e. non-X-ray emitting) regions of the Sun (see Figure 10.3), sometimes extending from pole to pole. In 1973 Allen Krieger and Adrienne Timothy of AS&E and Edmond Roelof of the University of New Hampshire found a good correlation between low X-ray intensities at the solar equator and high solar wind velocities measured by the Vela and Pioneer 6 spacecraft, thus showing that coronal holes are Bartels' M regions emitting a fast solar wind. Subsequent observations of coronal holes by Krieger, Werner Neupert, and others with Skylab, OSO-7 and the solar telescope on Kitt Peak, and wind speed data with IMP-6 and -7 and the German spacecraft Helios 1 and 2 confirmed this relationship.

Coronal holes are more prevalent just before solar minimum and so it was fortunate that Skylab, with its two high-resolution X-ray telescopes, was operational just before the 1976 solar minimum.[11] The six low-latitude coronal holes,[12] most of which lasted a number of months, found in the Skylab images were observed to have a strong association with large unipolar magnetic field regions in the Sun's photosphere. In addition the coronal hole boundaries were marked by magnetic inversion lines where the magnetic field changes from one polarity to the other. So not only are coronal holes Bartels' M regions, but Babcocks' idea was also correct, that fast corpuscular radiation emanates from unipolar regions where the magnetic field lines run freely into space.

The second problem mentioned above, namely that of the three-dimensional structure of the interplanetary field in the region of the ecliptic, was solved in 1976 when Pioneer 11 found that the magnetic sectors disappeared when the spacecraft's heliographic latitude exceeded 15°. This clearly indicated that the warped sheet model was correct, although the location of the sheet in three dimensions and the structure of the interplanetary field as a whole was still unknown. The neutral sheet was then found to change a good deal around the solar maximum of 1979, when the Sun's general field changed polarity, and during this period secondary neutral sheets were also found to be present.

In addition to measuring the solar wind *in situ*, spacecraft also imaged the Sun in wavelengths not observable from the Earth's surface. The most important of these are in the ultraviolet, X-ray and γ-ray bands, with the results as outlined below.

10.3 **Flares**

Over the late 1940s and early 1950s Friedman's group at NRL had undertaken a number of sounding rocket experiments using ultraviolet and X-ray sensors (see Chapter 1), and in 1956 he undertook a campaign to try to observe X-rays from solar flares. Unfortunately, the liquid-fuelled Aerobee sounding rocket, which was his usual vehicle for high-altitude observations,

---

[11] Skylab was launched on 14th May 1973. Its last crew left on 8th February 1974. The design of Skylab and its other results are discussed later in this chapter.

[12] Skylab also showed that there were permanent coronal holes at both poles.

took too long to prepare for launch to enable its experiments to capture the effect of solar flares, which lasted for only a few minutes. A solid-fuelled sounding rocket was the obvious answer, as that could be fired at a moment's notice, and a suitable size rocket was available called a Rockoon.

The Rockoon[13] consisted of a 4 metre long, solid fuelled, Deacon rocket suspended from a Skyhook balloon that floated in the atmosphere at an altitude of about 20 to 25 km. Friedman planned to launch a Rockoon in the morning, track its drift in the atmosphere by radar, and fire its rocket motor the instant a solar flare was seen. Because the Rockoon's balloon would drift prior to firing, a large launch range was required to cover the likely impact sites of the spent rockets. The most practical solution was to use a ship to launch the Rockoon and then follow it as far as possible prior to firing the rocket. So in July 1956 Friedman and his team set out to sea on board the USS *Colonial* with a batch of ten Rockoons to be launched at daily intervals.

On the first three days there were no flares, on the next day, which was a Sunday, there were two flares but there was no balloon aloft, but on the next day the rocket was fired when a flare appeared. Strong X-rays were detected when the rocket exceeded 70 km altitude, showing that solar flares produce X-rays. In addition to solar X-rays, Friedman discovered evidence of possible non-solar X-rays but, as mentioned in Section 1.2, his attempt to find these during subsequent night-time launches failed.

Solar flares not only emit X-rays and radio waves (as mentioned earlier), but they also emit charged particles. This was shown conclusively on 23rd February 1956 when a flare was observed on the Sun that was so bright that it could be seen in white light against the solar photosphere. The flare not only disrupted radio communications on the daylight side of the Earth as usual, but an hour later it also disrupted communications on the night-side. This new phenomenon could not be due to X-rays, as they travel in straight lines, but must be due to flare-generated, charged particles travelling in the Earth's magnetic field. Observations made from balloons of subsequent events showed that most of the charged particles causing these night-time fades are low-energy protons in the range of about 30 to 100 MeV.

The first photograph of the Sun taken in the light of Lyman-α, at about 121.6 nm in the ultraviolet, was made in 1956 from a sounding rocket. Four years later the first solar image in X-rays was produced using a pin-hole camera on board an Aerobee sounding rocket. This X-ray image, which was

---

[13] Rockoons were the idea of James Van Allen who used the first series in 1952 to investigate cosmic rays.

produced by Friedman's group in the 2 to 6 nm wavelength range, was very crude and its resolution was further compromised as the rocket and its X-ray camera rotated during the exposure. Although these first ultraviolet and X-ray images were of limited scientific use,[14] they gave the first inkling of what may be possible with a stabilised satellite and more sophisticated ultraviolet and X-ray cameras.

The first of a series of solar observational spacecraft called SOLRAD-1 was launched in June 1960. Weighing just 19 kg, this NRL solar radiation spacecraft was launched with the larger Transit 2A Navy navigation spacecraft in the first successful attempt to orbit two spacecraft with one launch vehicle. SOLRAD-1 had on-board two detectors to monitor the Sun, one operating at Lyman-α region from 105 to 135 nm and the other at X-ray wavelengths below 0.8 nm. Like many of the early small spacecraft it had no tape recorder on board, relying instead on line-of-sight communications with a series of ground stations. It was somewhat fortunate, therefore, that on 6th August 1960 it happened to be in sight of the Blossom Point receiving station in Maryland when a solar flare was seen from Earth. Simultaneously there were problems experienced with Earth-based radio communications and the ground-level, cosmic ray intensity also decreased. Radio emissions from the Sun were detected on Earth four minutes after the start of the visible flare. SOLRAD-1 observed a rapid increase in X-ray intensity which lasted until after the end of the radio burst. In fact, in the short period of time that the flare lasted, the whole sequence of flare-related activity then known was observed.

In parallel with the US Navy SOLRAD series of ten very small spacecraft, Ball Brothers designed a much larger 200 kg Orbiting Solar Observatory spacecraft for NASA (see Figure 10.4), the first of which was launched by a Delta rocket on 7th March 1962 . OSO-1 consisted of a nine-sided equipment platform that was spun-up after launch to provide gyroscopic stability, and a de-spun, fan-shaped platform partially covered in solar cells. Eight experiments were mounted on the spinning equipment platform and a further five on the de-spun platform. OSO-1 measured short-wavelength visible radiation between 380 and 480 nm, ultraviolet radiation between 110 and 125 nm (including Lyman-α), X-rays in three wavelength bands between 40 and 0.01 nm and γ-rays of various energies between 0.05 and 500 MeV (i.e. with wavelengths below about 0.02 nm). Other experiments measured neutrons, protons, electrons and micrometeorites. In all a

[14] Although the X-ray image did confirm that the corona was the major source of solar X-rays.

Figure 10.4 This 94 cm high OSO-1 spacecraft consisted of a nine-sided spinning platform and a fan-shaped, de-spun platform mounted with solar cells. The three boom-mounted spheres (two of which are seen in this photograph) contained gas for spin control. (Courtesy NASA.)

total of 6 OSOs were successfully launched in this first series, ending with the launch of the 290 kg OSO-6 in 1969.[15]

Unfortunately, OSO-1, which was expected to last about six months, operated satisfactorily for only 76 days before a spin control system failure resulted in only intermittent operation. During its period of full operation OSO-1 observed more than 75 flares and subflares, mapped the sky in γ-radiation, and monitored the inner Van Allen belt at an altitude of about 550 km. Five years later one of its successors, OSO-3, measured the spectrum of a flare that implied that it had heated the local plasma to an astonishing temperature of 30 million K.

The Sun had been known for some years to emit intense but relatively short-lived bursts of radio energy at nearly all frequencies, particularly around sunspot maximum, which have been given the type numbers I to IV. Type I radio bursts appear to be associated with sunspots, rather than flares, and so will not be considered here, but the other three types are all associated with flares. The earliest to arrive at the Earth after a flare are type III (or 'fast drift') bursts, which are of medium to low frequency (i.e. $<$500 MHz, see Figure 10.5), with a frequency that reduces rapidly with time, at a rate of $\sim$20 MHz/s. Type II (or 'slow drift') bursts, which arrive later, have a low frequency ($<$100 MHz) which gradually reduces with time, at a rate of $\sim$1 MHz/s. Type IV bursts, which are sometimes associated with type II bursts, exhibit a broad-band, high frequency continuum, possibly caused by synchrotron radiation.

Ground-based radio telescopes helped to characterise these radio bursts in the 1950s and 1960s, but, as the ionosphere is not transparent below about

[15] Two larger OSOs, i.e. OSO-7 and -8, were launched in 1971 and 1975.

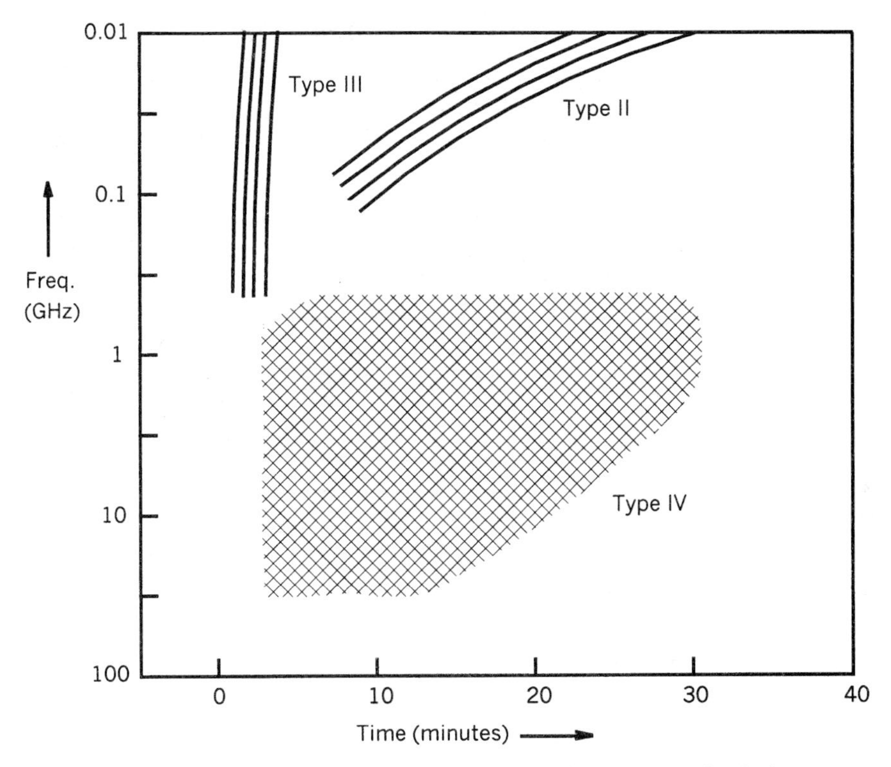

Figure 10.5 A schematic showing the different types of flare-associated radio bursts, according to their time of arrival at Earth after observation of the optical event and their radio frequency. In addition, the slopes of the lines for type II and III bursts show their change in frequency with time.

10 MHz, spacecraft have had to be used to cover these very important frequencies.

In the mid 1960s Canadian scientists detected type III radio bursts at frequencies of between 10 and 0.7 MHz using the Alouette 1 and 2 spacecraft. Similar observations were made a year or two later by scientists using the OGO-3 and the Russian Zond 3 spacecraft. It was known that the radio frequency of the burst is dependent on the electron density at the point of emission. Hence, knowing how the ambient electron density varies with distance from the Sun enabled an estimate of the source velocity to be determined, as it moves from the Sun, from the rate of change of its radio frequency with time. Using this technique, the spacecraft observations indicated that the velocity of particles producing type III radio bursts is about 30 to 40% of the velocity of light. This was confirmed in August 1968 when NASA was able to estimate directly the velocity of particles that produce a type III burst by tracking the source using the recently launched RAE-1 spacecraft.[16] The velocity seemed to be constant at about 0.35*c* (i.e. 35% of the velocity of light) up to a distance of 40 solar radii from the Sun. Further work with IMP-6 enabled the source of one type III burst to be tracked as far as the Earth's orbit (i.e. 215 solar radii from the Sun), by which time its velocity had been reduced by about 50%. Simultaneously with the arrival of this source, IMP-6 detected a burst of solar electrons, showing that type III radio bursts are produced by solar electrons travelling at a significant percentage of the velocity of light (see Figure 10.6).

The electrons producing type III bursts sometimes return to the Sun and produce what is known as a 'U burst'. On 29th November 1956 such a burst was observed at Fort Davis, Texas, when over a period of 8 seconds its radio frequency reduced from about 175 MHz to 125 MHz and increased again. A larger event of the same type was observed in the following decade by the RAE-1 spacecraft in which the radio frequency reduced from about 5 to 1 MHz and back to 3 MHz over a period of about 7 minutes. This was interpreted as being caused by a stream of electrons, travelling at about 0.3*c*, that reached a height of about 35 solar radii above the Sun before following the magnetic field lines back towards the surface.

A series of intense solar flares were seen on the Sun in August 1972, all associated with the same active region. Three flares were observed on the 2nd August, one on the 4th and one on the 7th. Although that of the 7th was the most intense, observations of the earlier ones produced interesting results, by far the most significant of which was the discovery of γ-rays from the 4th August flare by OSO-7. The main γ-ray emission was found at 0.5 MeV which was due to the mutual annihilation of electrons and positrons. As this mutual annihilation takes place very quickly, the positrons must have been produced in the solar flare, possibly by the decay of short-lived isotopes of carbon, nitrogen and oxygen, which had also been produced in the flare. The other strong γ-ray emission line was at 2.2 MeV, which was produced following the production of deuterium by the collision of a neutron and proton in the flare.

So flares seem to accelerate not only electrons, as shown by type III radio bursts, but protons, which produce various forms of electromagnetic

[16] The Radio Astronomy Explorer 1 (or Explorer 38) spacecraft was launched in July 1968 to measure the frequency and intensity of radio emissions from space. It could measure signals in the frequency range from 5.4 to 0.2 MHz, equivalent to distances of from 3 to 50 solar radii from the Sun.

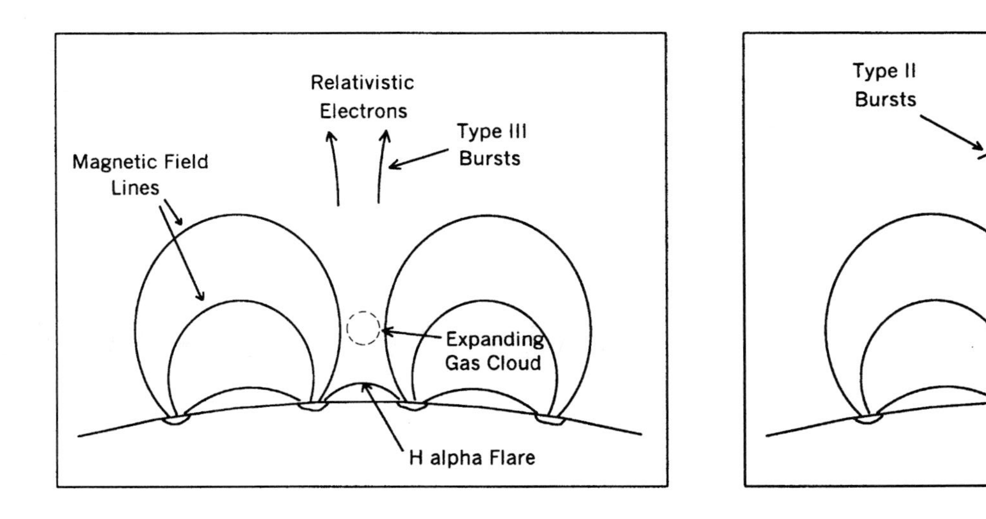

Figure 10.6 The situation above the surface of the Sun a few seconds after the flash phase of a solar flare (a), and about ten minutes later (b). 'Relativistic' electrons in (a) that travel at a velocity of about 0.35c produce type III radio bursts, whilst the shock wave in (b), in front of the slower-moving plasma, produces a type II radio burst.

radiation and participate in various nuclear reactions. Calculations showed that the observed effects of solar flares, particularly the shape of the X-ray continuum spectrum, can only be explained if the temperatures reached are over 100 million K in the flare region, assuming that the processes are purely thermal. As these temperatures seemed improbable, astronomers favoured the view that the processes are non-thermal, with the reactions being produced in a similar way to those in particle accelerators.

At 15:15 UT on 7th August 1972 the most intense flare of the series was observed on the Sun, which produced X-rays so intense that the sensors on SOLRAD-9 and -10 became saturated. Radio communications on Earth were simultaneously affected, γ-rays were again detected, and high-energy solar cosmic rays were observed at 15:40 UT. About 40 minutes later Pioneer 9 detected lower energy particles. A type II radio burst was detected by a radio telescope at Fort Davis, and as the radio frequency decreased the source of the burst was tracked by the IMP-6 spacecraft. Eventually over a day later it had been tracked all the way out to the Earth's orbit, by which time the radio frequency had reduced from about 180 MHz to about 30 kHz. This showed that the source of the radio burst had travelled outwards from the Sun at an almost constant velocity of 1,270 km/s. When the source arrived at the Earth at 23:54 UT on 8th August it produced a geomagnetic storm, thus confirming the hypothesis that terrestrial magnetic storms are caused by a shock wave produced by a solar flare, with the source of the type II radio burst being no other than this shock wave.

10.4 The Corona

The electrons and shock waves described above are generated by solar flares in the Sun's chromosphere, but they then have to travel through the Sun's corona to reach the Earth.

The structure of the corona had long been known to vary depending on the phase of the solar cycle, with the corona being more symmetrical around solar maximum. But it was not until the 1960s that it was possible to correlate the large coronal streamers, which are seen in white light[17] to stretch

---

[17] The white-light corona is best seen during total solar eclipses, but the lower and middle corona can also be seen without waiting for eclipses using a coronograph in which the bright image of the Sun is covered by a masking disc. Well before the spacecraft era it had been known that the white light of the corona is simply photospheric light scattered by electrons and dust.

for several solar radii above the chromosphere, with X-ray coronal activity. It was then found that the large white-light streamers nearly all have their bases in X-ray bright patches.

On 14th December 1971 the white-light coronograph on OSO-7 was observing a new, very bright coronal streamer when a solar telescope at Carnarvon, Australia, operating in Hα, recorded a faint spray of hydrogen gas in the streamer's vicinity. At about the same time a radio burst was detected at Culgoora, Australia, and other radio observatories in the Philippines and the USSR. Images from OSO-7 then showed several clouds of ionised hydrogen (i.e. protons) leaving the Sun in what are now called "coronal mass ejections" (see Figure 10.7). These clouds, which were about 300,000 km or so in diameter, and had temperatures of about 1 million K or so, were found to be leaving the Sun at about 1,000 km/s. It appeared as though the ionised gas cloud had been ejected from a solar flare just out of sight over the Sun's limb. Three and a half days later the flux of low-energy protons suddenly increased in the Earth's magnetosphere, and shortly afterwards Pioneer 6 in orbit around the Sun detected an increase in high-energy protons. The most energetic clouds of ionised hydrogen had missed the Earth. These OSO-7 images were the first white-light images to show coronal mass ejections.

The above sections summarise our knowledge of the Sun in about 1972,[18] that is just before the launch of Skylab with its tremendous range of instruments devoted to solar research. This major manned observatory will now be considered.

10.5 Skylab

On 10th September 1965 NASA had set up the Apollo Applications Program to use Apollo-designed hardware for extended manned Earth-orbiting and lunar surface missions, prior to the start of radical new post-Apollo missions. This followed earlier studies, under the Apollo Extension System, that looked at how to use hardware developed for the Apollo programme to construct a manned Earth-orbiting laboratory. By 1966 tentative plans included three orbital workshops fitted out in space, plus three fitted out on the ground, and four Apollo Telescope Mounts to attach to the workshops

[18] For convenience, the Skylab data on the source of the solar wind has already been considered in the solar wind section (Section 10.2). On the other hand, also for convenience, early work on the solar constant is considered later in this chapter in Section 10.8. So the 1972 date, which is generally valid, is not valid for all topics.

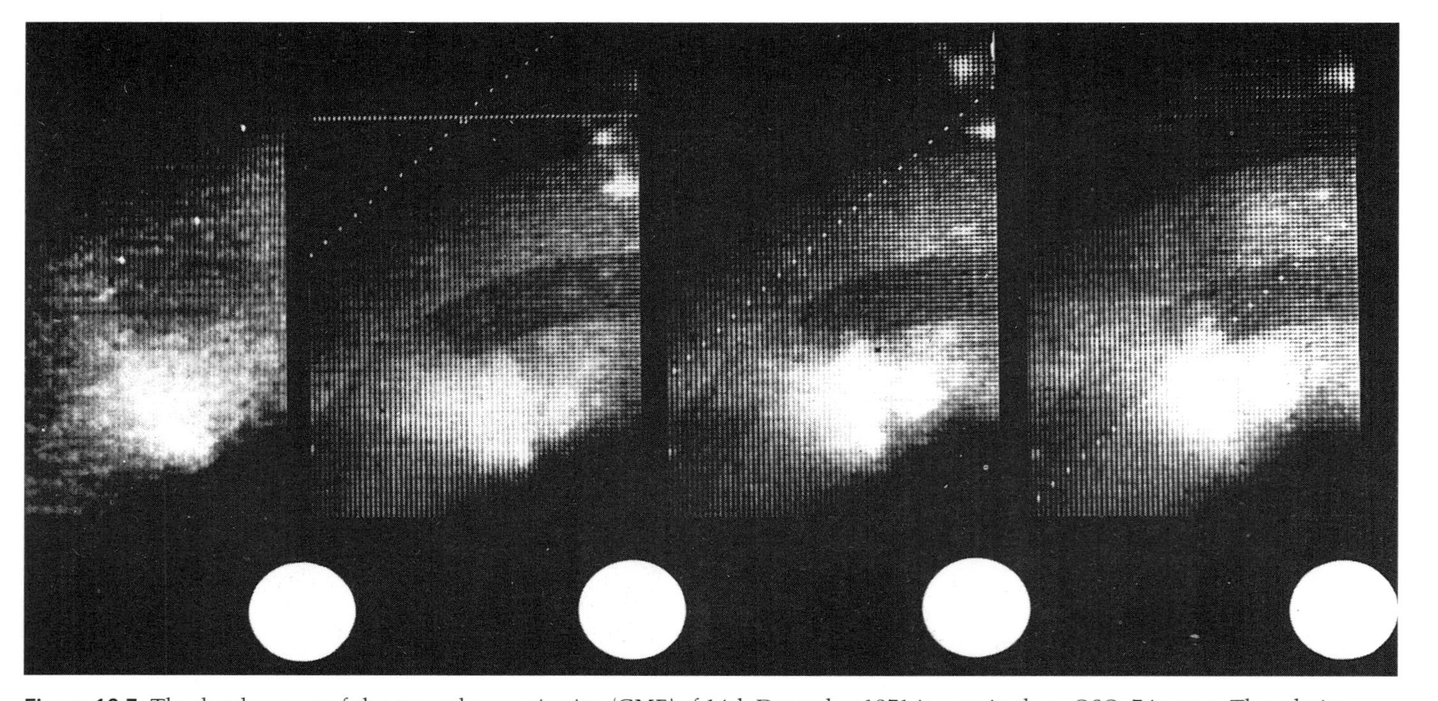

Figure 10.7 The development of the coronal mass ejection (CME) of 14th December 1971 is seen in these OSO–7 images. The relative position and size of the Sun is shown by the white circles at the bottom. The first image on the left shows a large coronal streamer and no CME. The next three images show the development of the CME, which OSO–7 had just captured on the top right-hand edge of its field-of-view. The total time elapsed between the second and fourth image is only 23 minutes. (Courtesy Naval Research Laboratory.)

when in orbit. Twenty six Saturn IB and 19 Saturn 5 launches were planned in this manned programme with a first launch in April 1968. In addition a new manned lunar programme was planned to follow the approved Apollo lunar missions.

Budget cuts saw a progressive reduction in the Apollo Applications Program over the next few years, with the lunar element being cancelled in 1968, and the six Earth-orbiting workshops being progressively reduced to one by 1969. In the event the one orbital workshop, now called Skylab, was launched by a Saturn 5 rocket on 14th May 1973, with three crews of three men occupying it for a total of 171 days between 25th May 1973 and 8th February 1974.

Skylab was a remarkable example of how spacecraft hardware designed for one mission could be adapted for another. It consisted of a modified S–IVB third stage of a Saturn 5, with a multiple docking adapter module at one end which gave access to the Apollo Telescope Mount (ATM) and the Apollo Command and Service Module (CSM) (see Figure 10.8). The modified S–IVB, the docking adapter and the ATM were all launched by a Saturn 5, and the CSM was launched by a Saturn IB with a three man crew a few days later. The S–IVB stage had been extensively modified to provide living quarters for the crew, together with laboratory facilities in the so-called orbital workshop (OWS), plus an airlock module. Power for the living quarters and workshop was to be provided by two large solar array panels attached one on either side of the S–IVB stage, and that for the ATM by four solar arrays attached in a cruciform, windmill-like assembly. Skylab's overall length (including the attached CSM) was 36 m and its overall mass was about 90 tons.

The Apollo Telescope Mount (ATM) contained most of the solar instruments (see Table 10.1). One of the big advantages of Skylab, compared with the previous un-manned, smaller solar observatory spacecraft, was the fact that most of the images were stored on film, rather than transmitted to ground electronically. The use of film allowed the recording and transmission to Earth of much more detail than would have been possible by electronic means. In addition, because of Skylab's large size, much more power and mass were available to the experiments than in any previous Earth-orbiting solar observatory spacecraft.[19]

---

[19] The largest such earlier spacecraft was OSO-7, which weighed only 635 kg in total and provided about 20 W of power. In contrast, Skylab as a whole weighed about 90,000 kg, and its solar experiments alone weighed more than 900 kg and used 2,000 W of power.

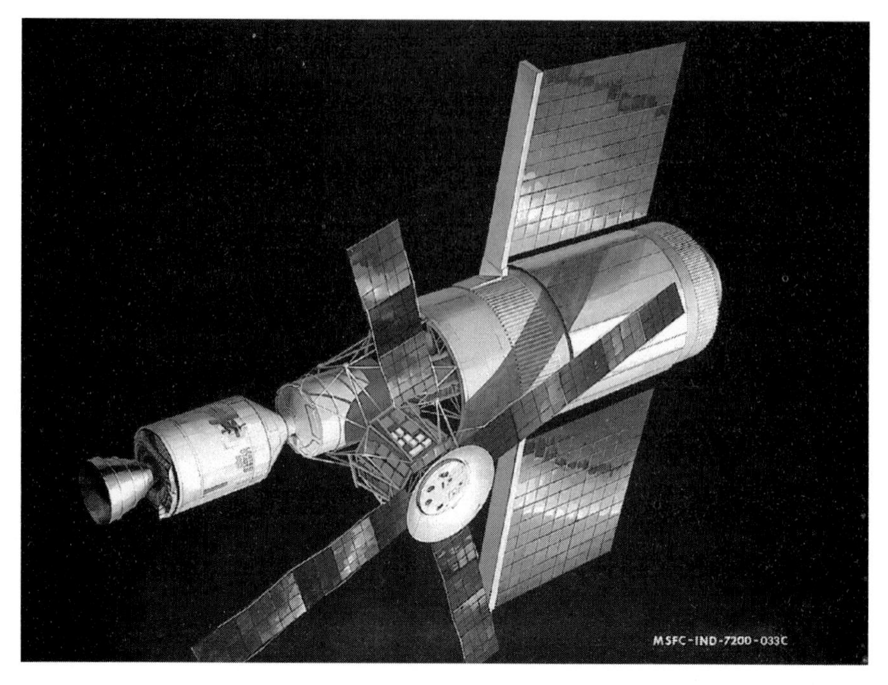

Figure 10.8 This is how Skylab should have looked in orbit after the astronauts had been ferried up in the Command and Service Module attached to the left hand side. In the event one of the two solar panel wings attached to the large diameter S–IVB stage on the right was ripped off during the launch sequence. The Apollo Telescope Mount is the module in the centre with the windmill-like solar arrays attached. (Courtesy NASA, MSFC 9801815.)

Skylab was launched after an almost faultless countdown on 14th May 1973. Sixty-three seconds later there was a slight vibration as the Saturn 5 vehicle passed through the point of maximum aerodynamic pressure. The launcher's protective nose-cone was then ejected, the rocket motors shut-down, and the ATM was rotated through 90° to make it perpendicular to the S-IVB stage. The ATM's four windmill-like solar arrays were then deployed, but the two solar-cell panels attached to opposite sides of the S–IVB stage failed to deploy correctly. Apparently, a heat shield and one of the solar panel wings had been ripped off during the launch sequence, and the heat shield had damaged the other solar panel wing in the process, which had only partially deployed. As a result the spacecraft was seriously short of power and the temperature increased rapidly to reach 190 °F (88 °C) inside the workshop, before attitude control manoeuvres reduced the temperature to 110 °F (43 °C). The launch of the astronauts, which had been due to take place on

Table 10.1. *Skylab's solar instruments*

Instrument	Wavelength range (nm)	Maximum[a] resolution	Solar region observed	Uses film?	Automatic capability?
(a) Fitted in Apollo Telescope Mount					
X-ray telescope	0.2–6	2″	Corona	✓	✓
Ditto	0.6–3.3	2″	Lower corona	✓	×
EUV spectroheliograph	15–61.5	2″	Chromosphere, transition region & lower corona	✓	×
UV spectroheliometer	30–140	5″	Ditto	×	✓
UV spectrograph	97–394	2″×60″	Ditto	✓	×
White-light coronograph	350–700	8″	Outer corona	✓	✓
Hα telescope No. 1[b]	656.3		Chromosphere	×	
Hα telescope No. 2[b]	656.3		Ditto	✓	
(b) Fitted in Orbital Workshop					
X-ray & EUV spectrograph	1.0–20		Chromosphere, transition region & lower corona		
UV and visual coronograph	245–600		Corona	✓	

Notes:

[a] The pointing accuracy of the Apollo Telescope Mount itself was 2″.

[b] Used for target identification, pointing and reference. One of these Hα telescopes was fitted with a camera which took a total of 64,000 images, thus providing an almost continuous record of the Sun over 9 months.

the day following the Saturn 5 launch, was postponed whilst NASA hastily designed a deployable sunshield for them to fix to the damaged area of Skylab. Finally, the astronauts Charles Conrad, Joseph Kerwin and Paul Weitz lifted off on 25th May to fix the spacecraft and operate it (as planned) for 28 days.

This is not the place, unfortunately, to describe the events of the next few days. Suffice it to say that the heat shield repair worked, bringing Skylab's internal temperature down to a reasonable level, and by 30th May the astronauts could start routine operations. All of the ATM's experiments were operational, but some of the workshop experiments were compromised as part of its sunward side was covered by the makeshift sunshield. On 7th June the astronauts succeeded in releasing the damaged wing, which provided about 3,000 W of power, but this compared with about 12,400 W that had been expected from both wings. Although the ATM solar arrays provided the 10,500 W anticipated, there was a severe power shortage on Skylab as a whole, and this placed restrictions on the number of experiments that could be operated simultaneously.

The operation on Skylab of the sophisticated set of instruments listed in Table 10.1 over a period of months added substantially to our understanding of the Sun. The key observations on coronal holes, where the magnetic field lines are open and stream into space, have already been mentioned. In addition large, bright, active regions in the chromosphere (imaged in ultraviolet light) were found to correlate with large, bright, arch-like features in the very high temperature corona (imaged in X-rays, see Figure 10.9), showing a strong linkage between the two. These arch-like features, called coronal loops, which account for nearly all of the electromagnetic energy emitted by the corona, were also found to link areas of opposite polarity in the photosphere.

It had been known for a long time that photographs of the Sun taken in visible light during total solar eclipses showed prominent plumes streaming away from the poles. These plumes were most obvious around solar maximum, but their source was not clear. Skylab resolved the problem, however, when its ultraviolet images of the polar regions showed that there were small bright points, associated with high magnetic fields, in the otherwise dark coronal holes; the bright points being the source of the plumes.

X-ray bright points, which are about an order of magnitude smaller than the above arch-like features, were discovered by G.S. Vaiana using a sounding rocket experiment in 1969, but their detailed study was first undertaken by Golub, Vaiana and others at AS&E using Skylab. They analysed images taken by one of the X-ray telescopes, and typically found about 100 bright points on each image spread evenly across the solar disc, unlike sunspots which are confined to just middle and low solar latitudes. The astronomers were surprised to find X-ray bright points even in coronal holes (see Figure 10.3 earlier), which are regions otherwise devoid of X-ray emission. The lifetimes of the bright points were typically about 8 hours, but a few of them showed exceptional brightening over a few minutes and disappeared just as quickly. Bright points were found to be associated with localised magnetic fields of opposite polarity in bipolar regions, and the total magnetic flux contained in bright

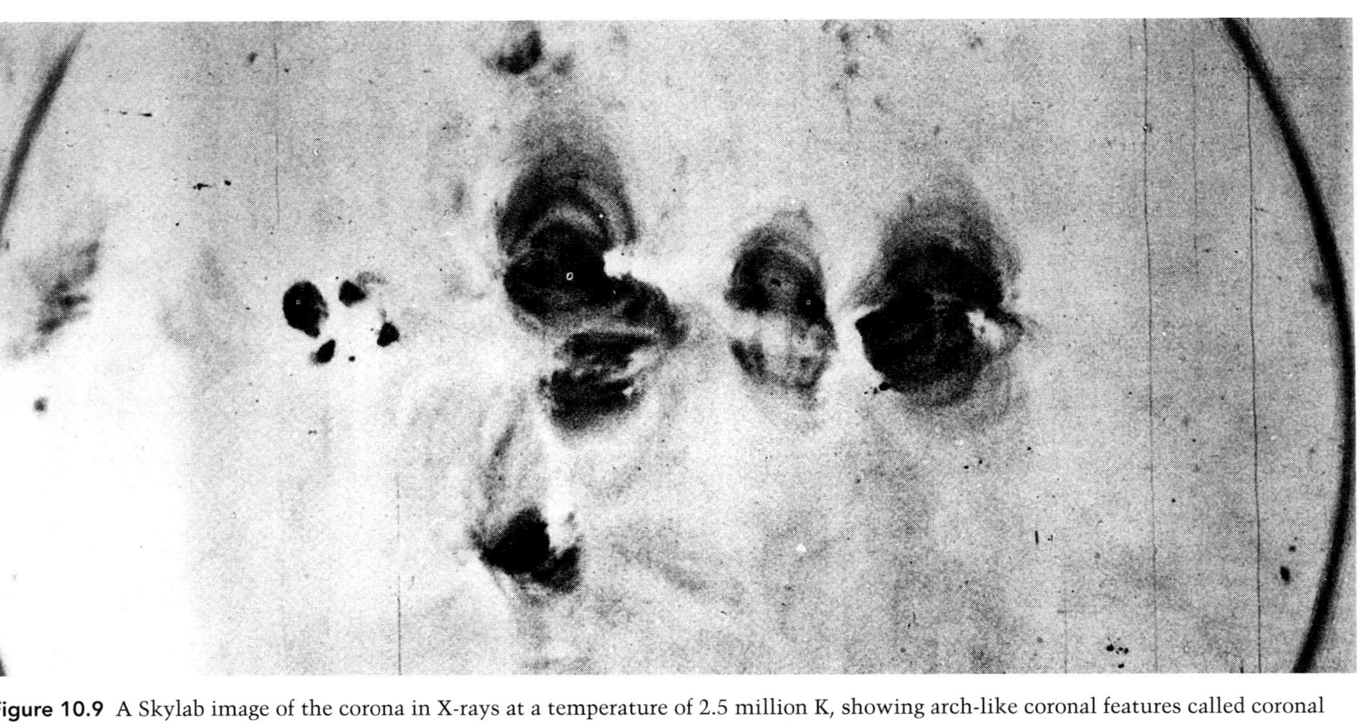

Figure 10.9 A Skylab image of the corona in X-rays at a temperature of 2.5 million K, showing arch-like coronal features called coronal loops. They correlate with active regions in the chromosphere. (Courtesy Naval Research Laboratory.)

points across the solar disc was found to be similar to that contained in sunspots and active regions. They were later found, using a radio telescope at the Clark Lake Radio Observatory, to be associated with type III radio bursts that are themselves connected with solar flares.

Skylab also allowed a detailed examination to be made of over 100 coronal transients, which are major, rapid events in the corona, of which 77 were thought to be coronal mass ejections (CMEs) of the sort first observed by OSO-7 (see above). Most of these CMEs, which eject material at velocities of up to 1,200 km/s, were found to be associated with eruptive prominences, and some were associated with flares, although whether the CMEs triggered the prominences and flares, or whether these triggered the CMEs was unclear. CMEs are much larger than prominences, often becoming larger than the Sun itself. Even though their density is very low, as their velocity is high, a large CME can be as energetic as a large flare (i.e. with an energy of about 10^{25} J).

10.6 **Summary**

So by the mid 1970s spacecraft had shown the existence of a solar wind of ionising particles, with an average velocity of 400 km/s, which was found to be emitted by the newly discovered coronal holes, where the magnetic field lines run freely into space. The Sun's equatorial magnetic field was found, at the distance of the Earth's orbit, to be composed of sectors of alternating polarity which could be traced back to the varying polarity of the photosphere at the Sun's equator. Solar flares were shown to emit both X-rays and charged particles, as well as radio waves. Type III (fast drift) radio bursts, which are associated with flares, were found to be caused by solar electrons travelling away from the Sun at about $0.35c$, whereas type II (slow drift) bursts, also associated with flares, were produced by a shock wave travelling at about 1,000 km/s from the Sun. When the shock wave reaches the Earth it causes a geomagnetic storm. Large coronal streamers which are seen in white light were observed to have their bases in X-ray bright areas of the corona, and polar plumes were found to be emitted by ultraviolet bright points. Large coronal mass ejections, which are generally associated with eruptive prominences, were seen leaving the Sun at velocities of up to 1,200 km/s. Large, arch-like coronal features observed in X-rays were found to be associated with magnetic arches linked back to the photosphere. Finally, X-ray bright points had been discovered which are associated with bipolar regions in the photosphere.

10.7 Solar Max

The next major solar observation spacecraft, Solar Max,[20] was launched in February 1980 to observe the Sun, and solar flares in particular, around the time of solar maximum. In all there were seven experiments in the 2,400 kg spacecraft, including the first telescope designed to image the Sun in high-energy X-rays (called HXIS for 'Hard X-ray Imaging Spectrometer'). Other instruments included a gamma ray spectrometer (GRS), an ultraviolet spectrometer and polarimeter (UVSP), a white light coronagraph/polarimeter (C/P), and a total solar irradiance monitor (ACRIM for 'Active Cavity Radiometer Irradiance Monitor'). Some of the instruments (i.e. GRS, ACRIM and HXRBS, for 'Hard X-ray Burst Spectrometer') had no imaging capability, but simply measured the total energy emitted by the Sun in their waveband, whereas others (i.e. HXIS, UVSP and XRP for 'X-Ray Polychromator'[21]) were designed specifically to image active areas and solar flares. Finally the coronagraph/polarimeter was to obtain images of the corona in visible light.

Although Solar Max was launched by a conventional Delta rocket, it was designed to be retrieved at the end of its useful lifetime in December 1982 by the Space Shuttle and returned to Earth for refurbishment and repair. But in November 1980 the spacecraft's third (of four) gyro failed, causing a loss of the spacecraft's fine-pointing capability and reducing Solar Max to a non-imaging rôle. In a much-publicised space shuttle mission, however, Solar Max was repaired and refurbished[22] in orbit in April 1984, which enabled it to operate until shortly before its burn-up in the atmosphere in December 1989.[23]

The repair and refurbishment of Solar Max was portrayed by NASA as a vindication of their policy of having a manned Space Shuttle that can undertake in-orbit repairs and maintenance of spacecraft, which was the reason for NASA's publicity of the event. The costs of a Space Shuttle mission (at about $300m a time) are such, however, that it may have been less expen-

---

[20] Otherwise known as 'SMM' for 'Solar Maximum Mission'.

[21] Consisting of two spectrometers.

[22] The coronagraph/polarimeter was repaired, as well as the attitude control system, and a baffle was fitted on the XRP. Unfortunately, the HXIS instrument which had malfunctioned in June 1981 could not be repaired.

[23] Although the repair was successful, problems early on in the repair mission meant that the spacecraft could not be boosted into a planned higher orbit. As a result, its mission was curtailed by an earlier than planned burn-up in the Earth's atmosphere.

sive to launch a replacement spacecraft with another Delta, although the Shuttle did launch another spacecraft during the Solar Max repair and refurbishment mission.[24]

Before the launch of Solar Max there were two main theories of the solar flare phenomenon. In one, flares are caused by the reconnection of magnetic field lines in the coronal part of an active region. The energy released accelerates electrons that then spiral down the field lines to the solar transition region and chromosphere. There, the high-energy electrons are suddenly decelerated, giving off short bursts of hard X-rays, radio and γ-ray radiation in the so-called impulsive stage of the flare process. Some energy released by the electrons also heats the chromospheric gas, causing it to rise into the almost empty magnetic loop[25] where it produces soft X-rays. In the alternative theory the flare process starts with the compression and heating of gas in the chromosphere by magnetic fields, with the gas then expanding along a magnetic loop into the corona.

Early images with the HXIS on Solar Max showed conclusively that, during the impulsive stage of a flare, hard X-rays[26] are generated at the foot of the magnetic loop, brightening simultaneously at both ends of the loop within about five seconds, which is the maximum time resolution of the HXIS instrument. The HXRBS and HXIS instruments showed that the emission of hard X-rays peaked before that of soft X-rays (see Figure 10.10), with the peak in hard X-rays being much sharper than the peak in soft X-rays. In addition, the XRP showed that the active regions in soft X-rays extend well into the corona. Interestingly, the UVSP and HXRBS instruments showed amazing correlations in the intensity variations with time between the ionised oxygen line O V[27] at 137.1 nm in the ultraviolet, the nearby ultraviolet continuum at 138.8 nm, and hard X-rays. As the O V line is emitted in the transition region, this correlation to within a fraction of a second shows

[24] The economics of using the Space Shuttle, rather than expendable launchers, to launch a scientific spacecraft are quite complicated, but the high cost of launching the shuttle, together with the extra cost of partially man-rating the spacecraft (for safety reasons), makes the shuttle an expensive vehicle for launching such a spacecraft. Unfortunately, as mentioned in Section 7.6, the Americans decided to phase out expendable launch vehicles as part of the political deal to get the shuttle programme approved in the early 1970s.

[25] These loops were first imaged by Skylab (in non-flare conditions); see Figure 10.9.

[26] The HXIS instrument could produce X-ray images at energies of between 3.5 keV (soft X-rays) and 30 keV (hard X-rays).

[27] O V is oxygen ionised four times, i.e. with four electrons missing.

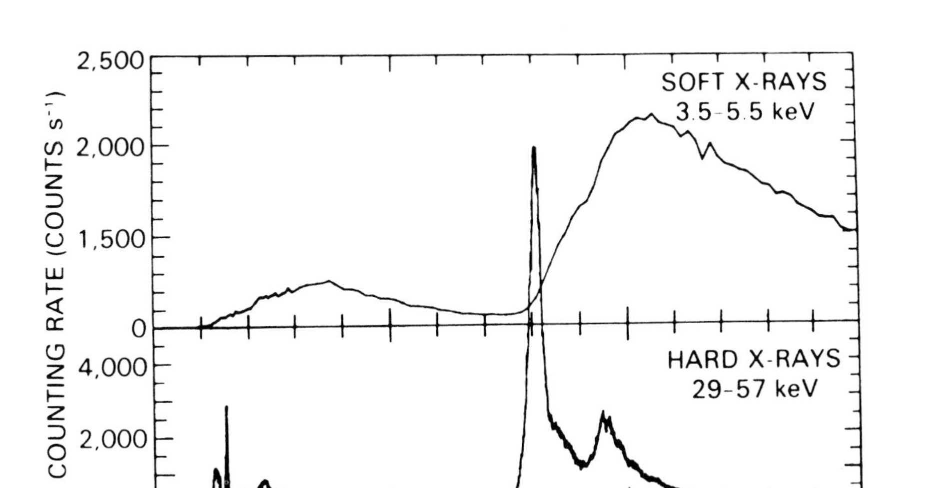

Figure 10.10 Hard and soft X-rays received from a flare on 5th November 1980 as measured by the HXRBS (lower) and HXIS (upper) instruments on Solar Max. The time on the horizontal axis is given in hours and minutes. The peaks of the hard X-ray emissions coincide with the start of a slower build-up to maximum of the soft X-rays. (Reprinted from *Solar Physics* 100 (1985), p 473, Figure 5, by Dennis, B.R., with kind permission from Kluwer Academic Publishers.)

that both the ultraviolet continuum and the hard X-rays are emitted there also. Ultraviolet emission in the chromospheric and coronal emission lines is not impulsive like that in the transition region lines, however, having a slower rise to maximum, like that of the soft X-rays.

The spacecraft OSO-7 had shown that some very energetic solar flares emitted γ-rays at energies of 0.5, 2.2, 4.4 and 6.1 MeV caused by electron/positron annihilation (0.5 MeV), the formation of deuterium (2.2 MeV), and de-excitation following the collision of protons with the nuclei of carbon (4.4 MeV) and oxygen (6.1 MeV). Prior to the launch of Solar Max it was thought that electrons were accelerated in a solar flare before protons, the latter being accelerated in a separate, second stage of the flare. If that was the case then hard X-rays should be produced before the proton-associated γ-ray lines, as hard X-rays are produced by electrons. OSO did not have a good-enough time resolution in its γ-ray detector to check on this, but the GRS instrument on Solar Max was quite capable of sorting out the problem. In the event Solar Max showed that electrons and protons are accelerated simultaneously. Furthermore, the γ-ray spectrum measured by Solar Max showed a series of peaks at 0.5, 1.3, 1.7, 2.2, 4.4 and 6.1 MeV (see Figure 10.11) superimposed on

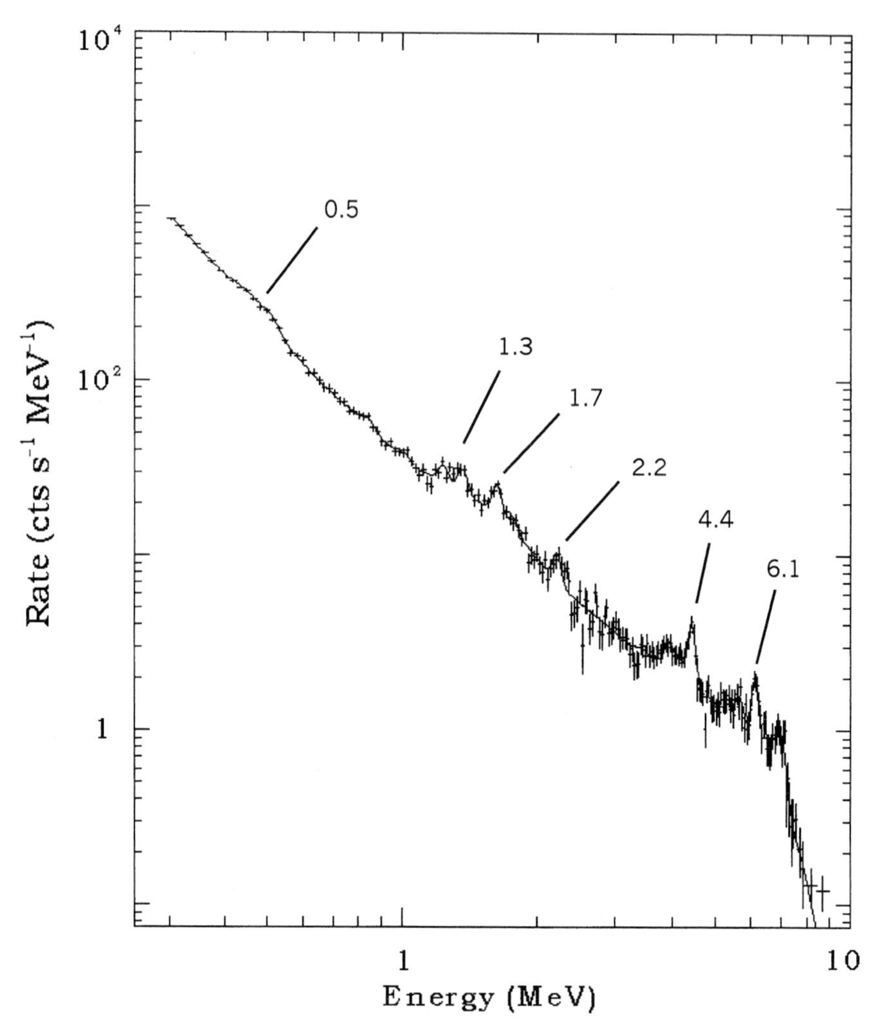

Figure 10.11 The γ-ray spectrum for a solar flare on 27th April 1981 as measured by the GRS instrument on Solar Max. The spectrum consists of a continuum sloping from left to right together with emission lines. The lines at 1.3, 1.7, 4.4 and 6.1 MeV are due to the de-excitation of nuclear excitation states in magnesium, neon, carbon and oxygen, respectively, the line at 0.5 MeV is due to the mutual annihilation of electrons and positrons, and that at 2.2 MeV is due to the formation of deuterium. (Courtesy NASA.)

a continuous background caused by the braking of electrons (to produce so-called bremsstrahlung radiation) in the solar atmosphere; the newly-observed 1.3 and 1.7 MeV lines being caused by the de-excitation of magnesium and neon nuclei following their collisions with protons.

During a large flare in 1980 and again in June 1981 the GRS instrument on Solar Max detected neutrons emitted by a solar flare about two minutes after the hard X-rays had been detected by the spacecraft. Assuming that the neutrons were emitted at the same time as the hard X-rays, they must have had a velocity of about $0.8c$. Neutrons of such high velocity could only have been produced by the collision of protons that had been accelerated to an extremely high energy by the flare, and photospheric protons and other atoms.

So after Solar Max it was generally thought that the solar flare process starts with the reconnection of magnetic field lines in the corona which then accelerates electrons and protons (see Figure 10.12). Most of them spiral down the field lines to the solar transition region and chromosphere and, in the case of the protons, the photosphere, but some very high energy electrons and protons escape from the Sun completely, and are sufficiently energetic to reach the vicinity of the Earth. The electrons that spiral down the field lines to the Sun are slowed rapidly in the transition region and chromosphere, producing short bursts of hard X-rays (see Figure 10.10), ultraviolet, type III radio and γ-ray radiation in the impulsive phase of the flare, heating the chromospheric gas. This gas then expands, rises and fills the magnetic loop, emitting soft X-rays in the process (see Figure 10.10), leaving behind gas in the chromosphere at the base of the loop that shines brightly at ultraviolet and optical wavelengths (particularly that of Hα). In the meantime, the high-energy protons have spiralled down the magnetic field lines to reach the photosphere, where they collide with photospheric protons to produce neutrons, which in turn are captured by protons to form deuterium, emitting 2.2 MeV γ-rays (see Figure 10.11) in the process. Similarly flare-generated protons impact magnesium, neon, carbon and oxygen nuclei which then emit γ-ray lines as they settle to a lower energy state. Short-lived isotopes of carbon, nitrogen and oxygen, also produced by flare-generated protons, decay, producing positrons which then interact with electrons in a mutual annihilation process to produce 0.5 MeV γ-rays. Some of the highest energy neutrons produced in the flare process also reach the Earth's orbit.

Although the processes just described sound plausible, solar flares do not all behave in the same way, with some processes appearing to occur in some flares and not in others. This caused doubt in the minds of some

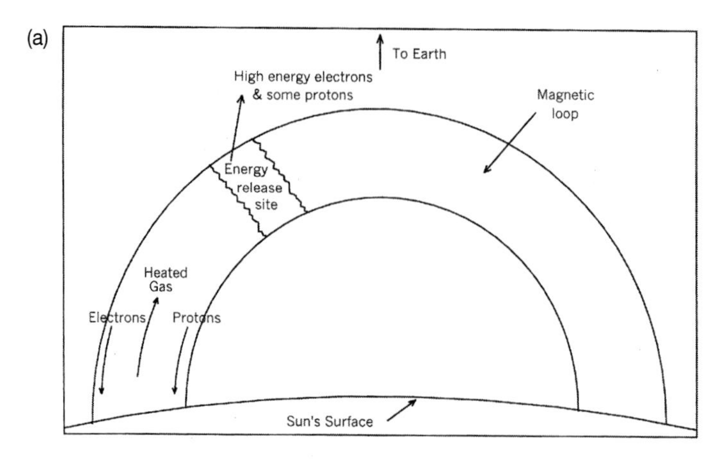

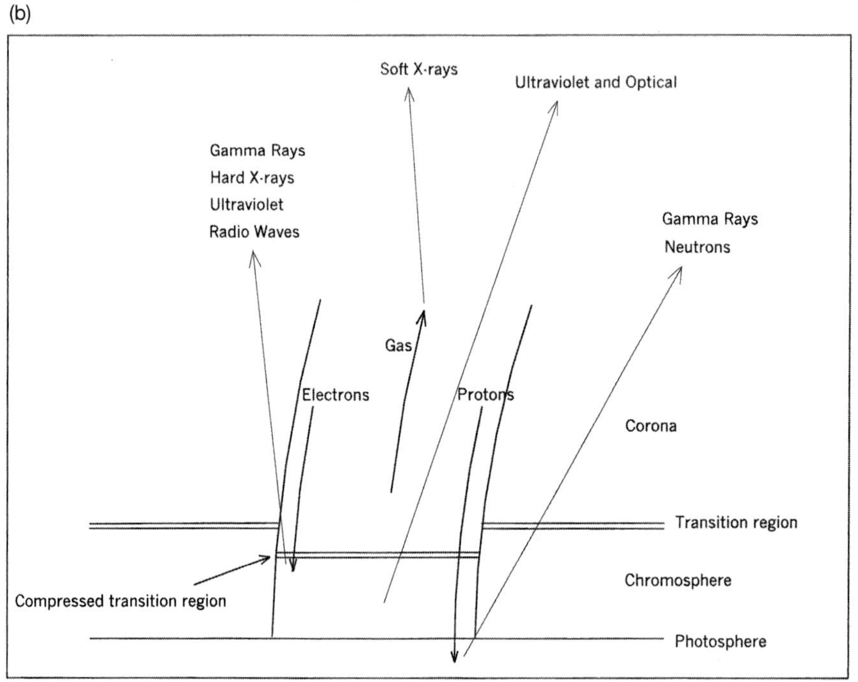

Figure 10.12 The energy release site shown in (a) is where the magnetic field lines are reconnected to start the solar flare process, accelerating electrons and protons, most of which follow the field lines down to the Sun. There they produce the effects shown in (b), which shows the base of the magnetic loop enlarged. See main text for more details.

astronomers, particularly on the question as to exactly how and where the reconnection of magnetic field lines occurs. It was expected that instruments of higher spatial and time resolution should help to clarify these and other matters.

Astronomers also used the white light coronagraph on Solar Max to observe the corona during both a period of high solar activity (in 1980 and 1989) and one of low activity (around 1986). One or two Coronal Mass Ejections were observed per day with the CMEs, like coronal streamers, occurring at most latitudes around solar maximum. Around solar minimum, however, CMEs were found to occur at only low latitudes, again like coronal streamers.

Solar Max allowed astronomers to study the relationship between CMEs and eruptive prominences. CMEs were often seen to consist of a bright loop of material moving away from the Sun with a dark cavity behind it, followed by a second loop that may sometimes be an eruptive prominence. At one stage it had been thought that eruptive prominences probably triggered CMEs, but the Solar Max images suggested that this may not be so, as many CMEs were seen without associated prominences, and in some cases the CMEs appeared before the prominences. Those CMEs associated with flares were also found to occur a few minutes before the flare itself, suggesting that CMEs also trigger flares.

10.8 The Solar Constant

Since the beginning of the nineteenth century attempts had been made to correlate variations in sunspot numbers with variations in the Earth's climate, on the basis that the heat output of the Sun would vary as the number of sunspots varied. Astronomers were divided, however, as to whether the Sun's heat output, measured by the so-called solar constant, would increase or decrease with increase in sunspot numbers. Such investigations into the effect of sunspots on climate were accelerated in the middle of the nineteenth century with the discovery of the eleven year sunspot cycle. Rudolf Wolf, in particular, researched historical records of sunspots back to 1700 and compared them with records of weather in Europe, but he could find no correlations. No one else managed to find an eleven year climate cycle either to match the sunspot cycle. The best correlation that could be found was a possible link between the Little Ice Age in Europe in the seventeenth century and the so-called Maunder Minimum between 1645 and 1715 when there were virtually no sunspots.

In the early years of the twentieth century the American A.E. Douglass started to amass climatic data for Arizona going back 1900 years by measuring the widths of tree rings. He thought that he had detected an eleven year cycle in the widths of the rings, but over the period from 1650 to 1740 this eleven year cycle was apparently missing. In 1922 Maunder pointed out to Douglass that this period was similar to that of the Maunder Minimum, when sunspots were virtually non-existent on the Sun, thus showing a linkage between sunspots and climate. This possible correlation caused another flurry of activity trying to link climate with sunspots but to no effect. In fact it was found that tree ring data around the world was inconsistent, ruining any possible world-wide linkage with sunspots.

Sunspot records were unreliable before about 1650, so to examine possible longer-term correlations between climate and solar activity another way of estimating solar activity was required that could provide data for earlier centuries.

As mentioned previously, in the 1940s Forbush had found a decrease in cosmic rays at the surface of the Earth around solar maximum. Cosmic rays were known to change atmospheric nitrogen into carbon-14, which is radioactive with a half-life of 5730 years. So a decrease in cosmic rays around solar maximum should be mirrored by a decrease in carbon-14 on the Earth. Ordinary carbon is carbon-12, which is taken up by trees in the form of carbon dioxide at the same rate as carbon-14, so the ratio of carbon-14 to carbon-12 in tree rings, for example, should give an indication of solar activity back over time. The effect would be delayed by about twenty years, however, as that is the time that it takes the radioactive carbon-14 to diffuse from the upper to lower levels of the Earth's atmosphere. When this radioactive carbon was measured in tree rings, the Maunder Minimum was clearly seen as a period of high radioactive carbon levels, as expected, showing that radioactive carbon levels in tree rings give a good indication of solar activity.[28] De Vries also found in 1958 that there was another peak in the radioactive carbon data from about 1460 to 1540, now called the Spörer Minimum,[29] which was later found to correlate approximately with another series of cold winters in Europe.

Attempts to measure variations in the heat output of the Sun directly, rather than rely on sunspot numbers or radioactive carbon data as a surrogate,

---

[28] The eleven year solar cycle cannot be seen in the radioactive carbon tree ring data, however, as the time taken for the radioactive carbon to diffuse from the upper to lower atmosphere is quite variable, thus smearing out any such correlation.

[29] Named after the German solar researcher of the nineteenth century who had independently discovered the Maunder Minimum in 1889.

were hampered by variations in the Earth's atmosphere, but the advent of sounding rockets and satellites offered a solution to that. At last there was a real possibility of measuring variations in the solar constant directly, which could then be compared with variations in the Earth's climate. It would also be of interest to solar researchers see how the solar constant varied with sunspot numbers. The first problem to be solved was how to build an accurate and reliable radiometer that was sufficiently small to be launched on a sounding rocket or spacecraft.

In the early years of the American space programme, the incentive to develop accurate radiometers to measure the solar constant was provided by spacecraft engineering problems, not by astronomical research. This was because NASA had found that spacecraft surface temperatures in orbit were inconsistent with pre-launch predictions based on ground simulation tests. The differences could be due to errors in the solar constant used, inaccurate radiometers in the space simulation chamber, poor thermal modelling, or in-orbit changes in the thermal properties of spacecraft surface materials. In 1964 JPL, which was carrying out investigations into the problem, gave a contract to Eppley Laboratory of Newport, Rhode Island, to measure the solar constant, to see if the correct value had been used in the ground simulation tests.

Andrew Drummond's group at Eppley built a twelve-channel radiometer, based on thermopile detectors, to remeasure the solar constant. It was flown in 1967 on the X-15 rocket aircraft at an altitude of about 80 km, and measured a solar constant of 1361 $W/m^2 \pm 1\%$. This was about 2.5% lower than that previously used in the American space programme, partly explaining the observed in-orbit temperature discrepancies.

In parallel with work at the Eppley Laboratory, Joseph Plamondon of JPL installed a cavity radiometer[30] on Mariners 6 and 7 in 1969 to monitor their thermal environment *en route* to Mars. Although these radiometers did not have the absolute accuracy of the Eppley radiometers, it was hoped that they were sufficiently stable to measure variations in the solar constant with time. Unfortunately, the instrumental variations were too large and masked any true solar variations.

Plamondon had realised before the Mariner 6 and 7 flights that more accurate radiometers were required, which is why Eppley had received their contract to develop such instruments. So in 1965 James Kendall, who worked for Plamondon at JPL, had produced a much more accurate cavity

---

[30] A cavity radiometer uses an insulated black cavity as a detector, instead of the thermopile used in the early Eppley radiometers.

radiometer, but this was only suitable for laboratory work. Richard Willson, also of JPL, used Kendall's design as the basis for a potential spacecraft instrument, and flew his first attempt on a balloon payload in 1968. It measured a solar constant of 1370 W/m^2, but the absolute accuracy was estimated at a disappointing ±2%, although the solar constant measured was only 0.7% higher than that obtained by the Eppley radiometer the previous year.

Four years of work had produced significant progress, but it was not until six years later that one of these new generation of Eppley or JPL radiometers flew on a spacecraft. This was the American low-orbiting meteorological spacecraft Nimbus 6 which flew an Eppley thermopile radiometer as part of an Earth Radiation Budget Experiment in June 1975. Much to everyone's surprise this radiometer measured a solar constant of 1392 W/m^2, with an estimated error of ±0.2%, which was 1.5% higher than the then-accepted value of 1370 W/m^2. It was possible that this new figure represented a genuine increase with time, and to check on this NASA launched an Aerobee sounding rocket on 29th June 1976 carrying a suit of five radiometers. They were an identical Eppley Nimbus 6 thermopile radiometer, which was calibrated against a Kendall cavity radiometer before launch, a prototype Eppley Nimbus 7 cavity radiometer (see below), two Willson cavity radiometers and one Kendall cavity radiometer. The results caused even more confusion as the two thermopile radiometers, one on the Nimbus 6 spacecraft and one on the rocket, gave identical solar constants of 1389 W/m^2, but the four cavity radiometers measured values of between 1364 and 1369 W/m^2. As the four cavity radiometers, of three different designs, had produced values that agreed to within ±0.2%, it appeared as if there was something wrong with the Nimbus 6 design of thermopile radiometer, but no one could be absolutely sure.

Eppley Laboratory had, in the meantime, started producing Kendall cavity radiometers under licence in the early 1970s, following its acceptance in 1970 as the instrument for measuring the International Pyrheliometric Standard. So when it came to designing a radiometer for Nimbus 7 John Hickey, who had taken over from Andrew Drummond at Eppley, proposed adding an Eppley–Kendall cavity radiometer to the payload.[31] Nimbus 7 with this new radiometer on board was launched in October 1978. The average solar constant over the first six months was measured as 1376 W/m^2, which Hickey mistakenly thought was a real increase from that measured by the cavity radiometers on the June 1976 rocket flight. The

[31] It was, in fact, a prototype of this Nimbus 7 cavity radiometer that flew on the sounding rocket flight of 29th June 1976 mentioned above.

maximum daily deviation from the mean over these first six months was 0.14%, but in August 1979 a 0.36% dip was recorded and a somewhat smaller reduction was recorded in October 1979. Although both reductions took place at a time of high solar activity, they could not, unfortunately, be attributed to a definite source on the solar surface.

Meanwhile in the early 1970s, the American John Eddy had started to review the historical evidence for the connection between solar activity and the weather on Earth. The eleven year solar sunspot cycle had already been well researched back to the early eighteenth century, but some astronomers thought that sunspot data earlier than that, which in particular showed the Maunder Minimum from 1645 to 1715, was highly suspect. This was even though such respected astronomers of the time as Johannes Hevelius, Giovanni Cassini, John Flamsteed and Jean Picard had regularly observed the Sun during the Maunder Minimum and had reported that sunspots were very few and far between, and in spite of the carbon-14 data from tree rings discussed above. Eddy's research unearthed more evidence for the Maunder Minimum and convinced most people that it did exist.

In 1976 Eddy produced an analysis of carbon-14 tree ring data going back to about 3000 BC, showing that the Sun went through a number of periods of high and low sunspot activity, which he was able to correlate with major climate changes, like those of the Maunder and Spörer Minima. This led him to conclude that the carbon-14 data not only showed changes in sunspot activity but also changes in the solar heat output, with low sunspot numbers producing a low solar constant and vice versa. Unfortunately, there was still no data to show that the Earth's weather followed the eleven year sunspot cycle. This did not concern Eddy, however, as he thought that the sunspot/climate correlation could only be long term.

In the 1970s whilst Eddy was looking at long-term solar variations, Peter Foukal and Jorge Vernazze were analysing pyrheliometric data collected by Charles Greeley Abbot and others at the Smithsonian Astrophysical Laboratory between 1923 and 1952, looking for a possible correlation between the solar constant and sunspots and faculae. Eventually, after a detailed statistical analysis of the 11,000 daily measurements, Foukal and Vernazze concluded in 1979 that sunspots reduce the solar flux whilst faculae increase it, producing a net short term variability of about 0.1% in the solar constant. Such small effects were difficult to disentangle from other influences, however, particularly from that of the Earth's atmosphere, whose transmission characteristics could possibly be influenced by solar activity. So their conclusions were generally treated with scepticism by other astronomers. Only measurements above the Earth's atmosphere would be conclusive. Unfortunately, the Mariner 6 and 7 radiometer measurements in 1969 were, as mentioned above, not stable enough to show any reliable changes in the solar constant with time. Likewise, the Nimbus 6 results of 1975 and 1976 were also thought to be too unreliable, but in 1979 Nimbus 7 had shown the first short-term correlations of the solar constant with solar activity, although the exact source of the energy reductions was unclear. Nevertheless, John Hickey's Nimbus 7 radiometer had shown the way.

Meanwhile, Richard Willson of JPL had not been idle, as he had his proposed ACRIM radiometer accepted for flight on Solar Max, which was launched in February 1980. Shortly after launch the instrument clearly showed a 0.15% reduction in the solar constant between 4th and 9th April and a 0.08% reduction between 24th and 28th May (see Figure 10.13), and *both reductions were clearly associated with large sunspot groups crossing the central solar meridian.* The average solar constant measured over the first six months of Solar Max was 1368.3 $W/m^2 \pm 0.1\%$, with a short-term instrument stability of about $\pm 0.002\%$. Willson's instrument was clearly the Rolls Royce of the time.

The question now arose as to what happened to the energy blocked by the sunspots. It was assumed that the Sun was still producing energy at a constant rate in its core, and that the energy blocked by sunspots was either stored in the Sun's convection zone below the photosphere and emitted later, or it was emitted immediately by nearby bright faculae. If it was stored it could either be emitted globally or by bright faculae, but over what timescales? If the emission of the stored energy was global, this could take hundreds of years to appear, whereas if the emission was local it could happen over virtually any timescale, although it was probably limited to typical faculae lifetimes of one or two months.

In 1986 Chapman, Herzog and Lawrence of California State University showed by analysing an active region over its whole lifetime that the facular emission from the region was between 70% and 120% of the energy blocked by the sunspots. If the figure had been 100% then this would have favoured the local re-emission theory, but with the large percentage error obtained this could still mean that the faculae were emitting energy stored years ago from other sunspots. If the solar constant could be shown to closely follow the sunspot cycle, however, with identical years for maxima and minima, it would show that any delayed emission would be delayed by the order of months at the most.

Solar Max had been repaired in April 1984 and there was considerable interest in comparing its high-accuracy solar irradiance data for 1984, which

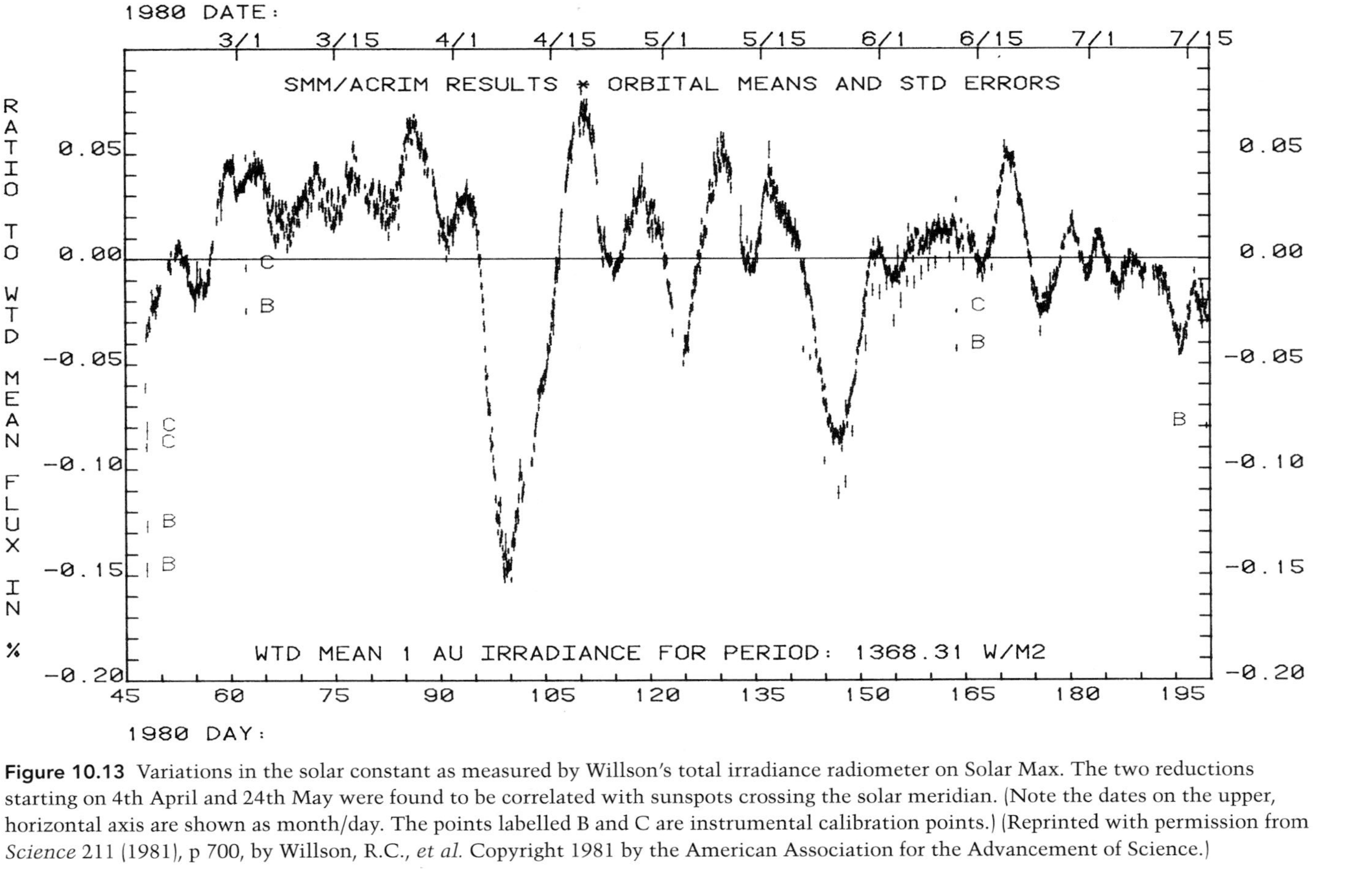

Figure 10.13 Variations in the solar constant as measured by Willson's total irradiance radiometer on Solar Max. The two reductions starting on 4th April and 24th May were found to be correlated with sunspots crossing the solar meridian. (Note the dates on the upper, horizontal axis are shown as month/day. The points labelled B and C are instrumental calibration points.) (Reprinted with permission from *Science* 211 (1981), p 700, by Willson, R.C., *et al.* Copyright 1981 by the American Association for the Advancement of Science.)

was expected to be just before sunspot minimum, with that of 1980, near sunspot maximum. Although Solar Max's fine pointing control system had been inoperative from November 1980 to April 1984, causing loss of its imaging capability, solar irradiance data had still been produced for this period, although it was of lower quality then before the failure. In 1985 Willson, Hudson, Fröhlich and Brusa reported, using Solar Max data, that there had been a reduction of 0.02% per year in the solar constant from 1980 to 1985. A similar reduction had also been observed using the Nimbus 7 radiometer, and so it seemed to be a real reduction. Three years later Willson and Hudson reported that the reduction in irradiance had levelled out, and it was increasing now that the sunspot cycle had passed through its minimum at the end of 1986 (see Figure 10.14). Shortly afterwards similar results were published by Hickey and R.A. Kerr for Nimbus 7 and the Earth Radiation Budget Satellite launched by the Space Shuttle Challenger in October 1984. *So although the measured solar constant is reduced by the passage of individual sunspot groups across the solar meridian, it is increased as the Sun approaches sunspot maximum.* Eddy had been correct in associating small numbers of sunspots with a low solar constant, although he had thought that the relationship only held over very long periods, not over the relatively short period of an eleven year solar cycle. But why, if individual sunspot groups caused a reduction in solar constant, did the many sunspots at solar maximum cause an increase in solar constant? Was it due to the over-riding influence of faculae which emitted more energy, on average, than that blocked by sunspots?

In 1988 Foukal and Lean corrected the solar irradiance measurements over the period 1981–84 from Solar Max and Nimbus 7 for the effect of sunspots, and then compared the resulting curve with that for a solar faculae index that they had devised. The match was excellent, showing that faculae over-compensate the effect of sunspots. So at last it was clear that the solar constant did reduce with reducing sunspot activity as sunspot-related faculae were less frequent at that time. On a finer timescale, however, individual sunspot groups caused a reduction in solar constant as seen from Earth when the sunspots crossed the solar meridian.

Five minute oscillations of the surface of the Sun had been discovered by Robert Leighton, Robert Noyes and George Simon at the Mount Wilson Observatory in 1960, and in 1973 Robert Dicke of Princeton University had found that this was not a local effect but represented vibrations of the Sun as a whole. It was apparent from a visual inspection of the Solar Max data over the period from March to July 1980 (see Figure 10.13) that the solar constant varied with a period of about 27 days (the solar synodic rotation rate), but

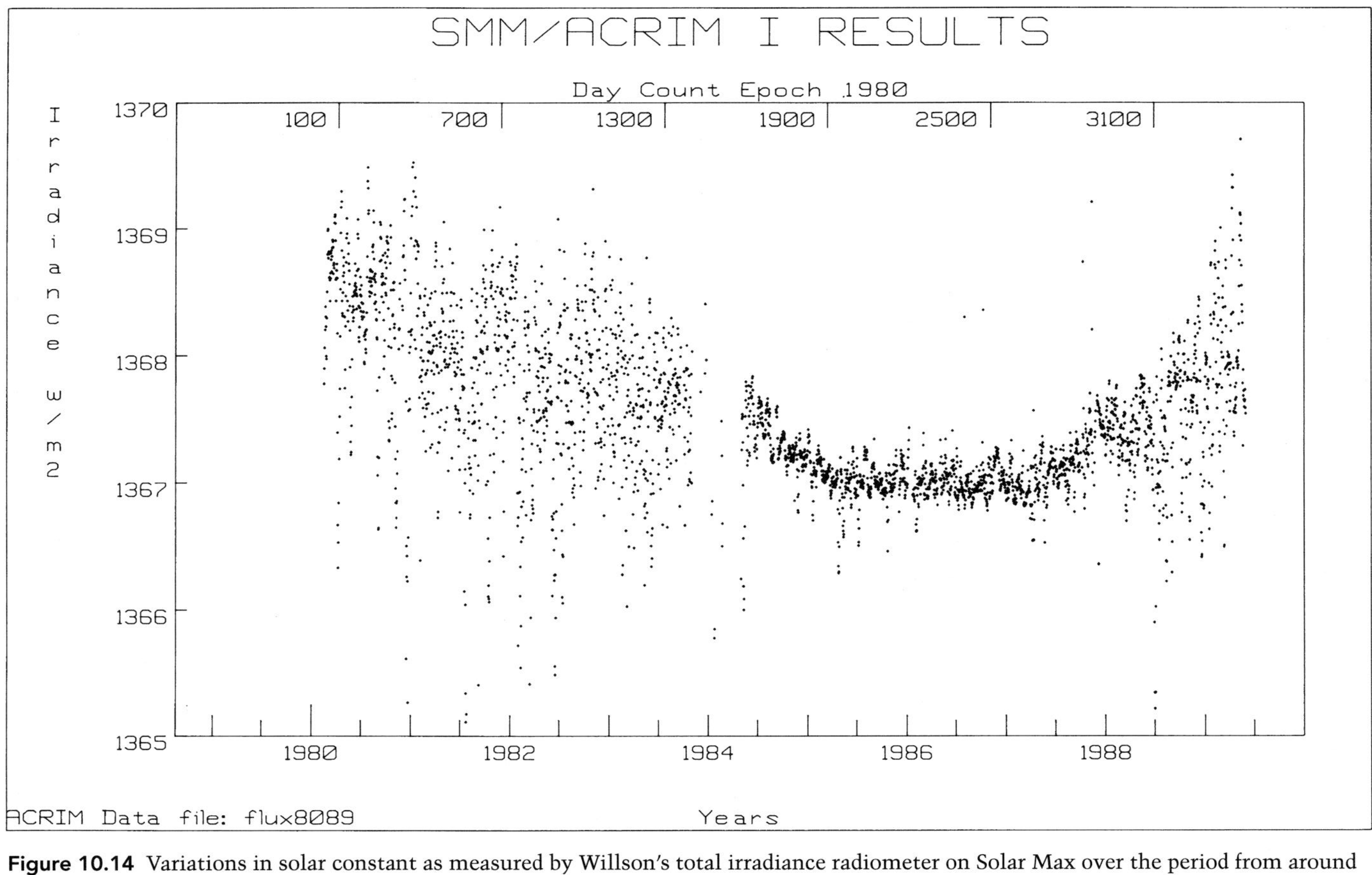

Figure 10.14 Variations in solar constant as measured by Willson's total irradiance radiometer on Solar Max over the period from around sunspot maximum in 1980 to beyond sunspot minimum in 1986. The general trend to a lower solar constant at sunspot minimum is clearly seen, as are the large, temporary decreases due to individual groups of large sunspots. The improvement in the instrumental performance is also evident as a result of the repair and refurbishment of Solar Max in 1984. (Courtesy R.C. Willson.)

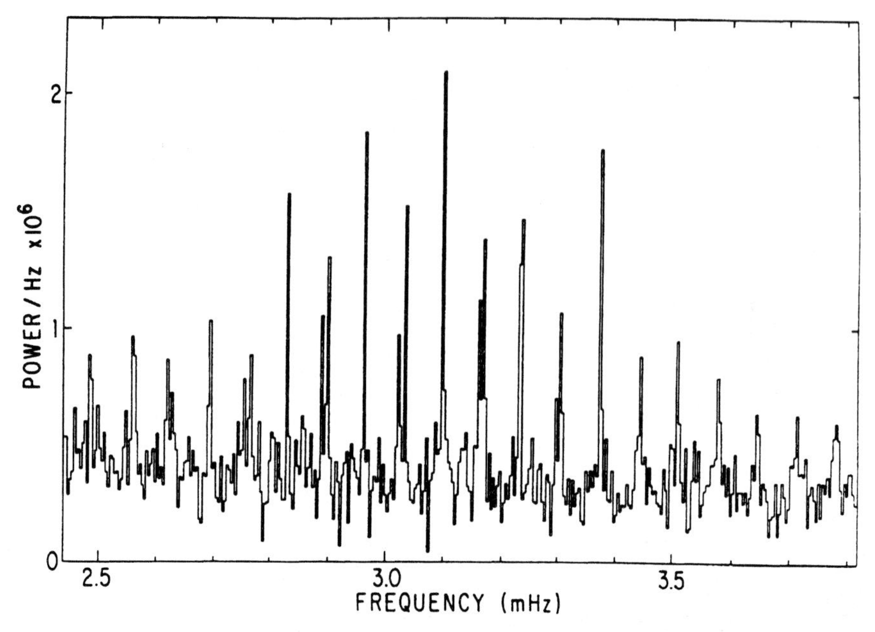

Figure 10.15 A Fourier (i.e. frequency) analysis of variations in solar constant as measured by the ACRIM radiometer on Solar Max showing that the solar constant varied over various discrete periods of around five minutes (3.3 mHz). (Courtesy R.C. Willson.)

Martin Woodard and Hugh Hudson of the University of California were interested in whether the solar constant would also show a 5 minute period. Their analysis was soon successful and in 1981 they announced their discovery of solar irradiance oscillations with various discrete periods of around 5 minutes (see Figure 10.15) and amplitudes of about 1 or 2 parts per million. It was hoped that analysis of this and other data would enable solar astronomers to understand the process occurring inside the Sun (see Section 10.11).

10.9 **Yohkoh**

The Japanese space programme had gradually expanded in the 1970s and 1980s and in August 1991 Japan launched a solar observatory spacecraft called Yohkoh (the Japanese word for 'sunbeam'). Weighing just 400 kg, Yohkoh was much smaller than Solar Max, but its payload, which was designed ten years after that of Solar Max, was more performant in many ways. It consisted of a Hard X-Ray Telescope (HXT) operating in the 15 to

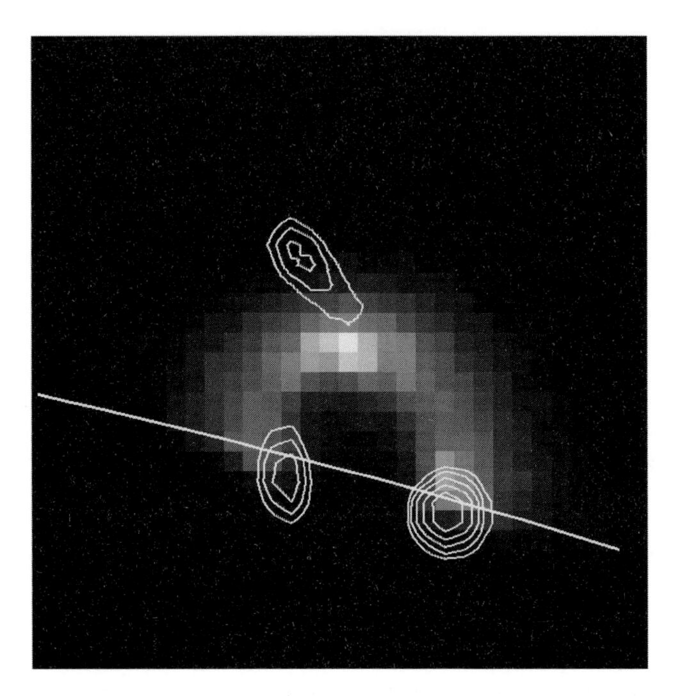

Figure 10.16 The soft X-ray image of a flaring loop with superimposed intensity contours for hard X-rays, as measured by the Yohkoh spacecraft. Hard X-ray emission is seen, as expected, at the foot of the loop, but an additional source of hard X-rays is evident above the apex of the soft X-ray loop. The position of the Sun's surface is indicated by the white line. (Courtesy S. Masuda.)

100 keV range, a Soft X-Ray Telescope (SXT) covering the X-ray range from 0.2 to 4 keV and visible light from 460 to 480 nm, a Wide Band Spectrometer (WBS) operating in the X-ray and γ-ray wavebands, and a Bragg Crystal Spectrometer (BCS) providing very high spectral resolution in four soft X-ray bands centred on 0.178 nm (Fe XXVI), 0.186 nm (Fe XXV), 0.318 nm (Ca XIX) and 0.506 nm (S XV). The main area of study was solar flares.

Yohkoh provided many valuable insights into the structure of solar flares showing for the first time, for example, that electrons beamed down magnetic loops into the chromosphere are the source of white-light flares. In addition, not only were 'bright knots' seen in soft X-rays at the top of the flaring loops, but Satoshi Masuda of the University of Tokyo and colleagues also found bright 'kernels' in Yohkoh's hard X-ray images *above* some of the soft X-ray loops (see Figure 10.16). This was a surprise as previously hard X-rays had only been detected at the foot of the loop. It was thought that the hard X-ray kernels coincided with magnetic field reconnection sites, but it was difficult to explain how a hard X-ray source could be confined to such a site near the apex of a loop for a significant period of time. So in solving one problem, another had been uncovered.

The soft X-ray telescope also imaged numerous small 'transient brightenings' or 'microflares', as they are sometimes called, that were clearly seen to take place in magnetic loops in some active areas. These microflares had energies ranging from 10^{29} erg (or about that of a small solar flare[32]) down to the instrument measurement threshold of 10^{25} erg. In some active regions more than one hundred were observed, brightening and fading on timescales of a few tens of minutes, whereas in other active regions no microflares were seen. These transient brightenings were found to contribute substantially to the soft X-ray intensity of active regions, although Shimizu showed that there were not enough of them to explain coronal heating, whose source was still unknown.

Before the launch of Yohkoh it had been thought that the Sun loses most of its mass along open coronal hole field lines, because it was thought that the much denser plasmas around active regions were contained by the closed magnetic loops. Although some mass was lost through CMEs, this was thought to be a relatively minor loss mechanism. Then Kazunari Shibata, Y. Uchida and colleagues found a new possible mass loss mechanism after analysing Yohkoh's soft X-ray images. They observed, above some active regions, long X-ray jets or columns moving outwards in the outer corona at velocities of up to 200 km/s. As with flares the initiating process was thought to be magnetic reconnection but, in this case, the material was apparently being ejected along open field lines.

10.10 **Ulysses**

The Ulysses spacecraft, which is still operational at the time of writing (1998), is unique in being the only spacecraft to fly over both poles of the Sun, but it was also important historically as its programme proved to be a watershed in ESA/NASA relations. As a result of the lessons learnt by both parties those relations would never be the same again, and because of this I will dwell more than usual on the early phases of this programme.

In 1965 Ludwig Biermann of the Max Planck Institute in Munich advocated building a spacecraft to measure the solar wind away from the ecliptic plane, and three years later Harry Elliot of Imperial College, London, advocated a mission to provide *in situ* measurements of the interplanetary environment out of the ecliptic. Elliot had been highly successful in having cosmic ray experiments flown on the British spacecraft Ariel 1 and the two ESRO spacecraft ESRO 2 and HEOS 1, but he had more trouble in persuading ESRO to agree to build an out-of-ecliptic spacecraft as it would be a much

[32] A large solar flare has an energy output of about 10^{33} erg.

more difficult and expensive proposition. Nevertheless, ESRO agreed to carry out a study of the proposed mission which was completed in 1972. It recommended that a detailed feasibility study should be undertaken into a spacecraft launched directly from Earth into an orbit inclined at about 40° heliographic latitude. ESRO also considered the possibility of using Jupiter's gravity to swing a spacecraft to a much higher heliographic latitude, but this concept was rejected for technical reasons.

Meanwhile NASA had been trying to develop a programme to visit the outer planets (see Chapter 8), and in 1969 it was suggested by their Outer Planets Working Group that a prototype outer planet spacecraft should be launched to Jupiter in 1974 where it could use the planet's gravity to fly out of the ecliptic. A study of the mission was undertaken by the NASA Ames Research Center in 1971, but budgetary problems put the proposed mission on the back burner.

In the early 1970s ESRO had the scientific spacecraft Cos-B and GEOS under development, and in 1973 the ESRO programmes ISEE-B and Exosat had also been approved, so this was not the time to press for approval of yet another ESRO astronomical spacecraft. But in 1971 the ESRO Council had agreed to participate in the NASA SAS-D project (which was to become IUE), so it was thought that an out-of-the-ecliptic spacecraft could probably be developed as a joint ESRO/NASA project, but this time with a more substantial ESRO rôle.[33] This was an ideal solution for the two agencies who were both strapped for money, and at a meeting at Estec in February 1974 ESRO and NASA agreed to a joint study of two new spacecraft missions, which were to become the Out-of-Ecliptic and Galileo programmes. During the course of these studies W. Ian Axford of the Max Planck Institute at Katlenburg-Lindau suggested that two Out-of-Ecliptic spacecraft could be built, one by ESRO and one by NASA. Both would use Jupiter fly-bys to take the spacecraft over the poles of the Sun, rather than use electric propulsion to send the spacecraft directly from Earth into a mid-latitude orbit. ESRO were not convinced about the Jupiter-assist mission and still favoured the electric propulsion approach, but NASA took the matter into their own hands and cancelled their electric propulsion programme that was to have supported the Out-of-Ecliptic mission. Although this unilateral NASA decision infuriated many European scientists at the time, since then engineers on both sides of the Atlantic have had difficulty in producing a reliable electric propulsion device, so the

---

[33] ESA's involvement in the IUE programme was strictly limited, with the solar arrays being the only ESA-provided spacecraft subsystem.

decision in retrospect was correct, but the way it was taken left a great deal to be desired.

Initially, it was intended to use a Titan launcher to launch the two Out-of-Ecliptic spacecraft to Jupiter, but in 1975 NASA unilaterally decided to use the Space Shuttle to put both spacecraft into Earth orbit, from which they would each be launched towards Jupiter by an Inertial Upper Stage (IUS). Jupiter's gravity would then be used to send one spacecraft over the Sun's north pole and the other over its south pole. A joint BAC/TRW feasibility study was completed on this basis in April 1976, and twelve months later ESA (who had by then taken over all ESRO programmes) and NASA issued an announcement of opportunity to the scientific community to propose experiments for the joint mission. In the meantime NASA had unilaterally changed the launch date from 1981 to 1983, which again upset ESA.

In fact 1977 was a difficult year for ESA/NASA relations for a number of reasons. In April 1977 an American Delta launch vehicle put the ESA GEOS spacecraft into the wrong orbit, and in September an even worse disaster overcame the ESA OTS spacecraft when its Delta launcher exploded shortly after lift-off, destroying the spacecraft. Naturally these problems, when added to those of the Out-of-Ecliptic programme, made ESA unhappy. On the NASA side, some NASA engineers were concerned at that time that the Europeans were developing their own launch vehicle (Ariane) that would make ESA largely independent of NASA. This would not only take revenue away from NASA, as ESA generally paid for NASA launchers at that time,[34] but Ariane may even compete with the Shuttle to launch commercial payloads, if the Shuttle failed to meet its launch cost target. So there were many reasons for tension between ESA and NASA engineers in mid 1977, but, in the case of the Out-of-Ecliptic programme, ESA would not have the capability for many years to launch such a spacecraft to Jupiter, and so they had to rely on a NASA launcher. Because of this ESA had to accept NASA's unilateral launcher decisions on the Out-of-Ecliptic programme whether they liked them or not.

In October 1977 NASA successfully launched the ESA ISEE-B spacecraft, and in November the launch of the ESA Meteosat spacecraft was also successful, so relations between ESA and NASA started to improve. The ESA participation in the Out-of-Ecliptic mission was also approved in November 1977, whilst NASA sought Congressional approval at the end of the year for the Out-of-Ecliptic mission as a 1978 new start, asking for just

[34] In 1977 ESA were forced by contract to pay for the launches of GEOS, OTS and Meteosat, even though the first two launches were failures.

$13 million for the initial budget. In the process NASA decided, unilaterally once more, to change the project name from the Out-of-Ecliptic to the Solar Polar Mission. Edward Boland, the chairman of the House subcommittee which was considering the programme, tried to limit the initial budget to $5 million, but Harold Glaser, the director of NASA's Solar–Terrestrial Division was concerned that this would be seen by NASA's European partners as only a partial commitment, causing them to pull out of the programme altogether. So he and a number of American scientists lobbied the various House subcommittee members hard, with the result that the project was finally approved, enabling President Carter to sign the appropriate authorisation in late 1978.

As mentioned above ESA and the European scientific community had been angered and irritated at what they saw as NASA's high-handed approach in unilaterally rejecting the electric propulsion approach, changing the launch vehicle, changing the launch date and finally changing the project name on what was supposed to be a jointly managed programme. So the Memorandum of Understanding (MOU) that was signed in March 1979 between NASA's Administrator, Robert Frosch, and ESA's Director General, Roy Gibson, specifically called for joint programme decisions in the future. The dual International Solar Polar Mission (ISPM) spacecraft, as they were then called, were planned to be launched on a Space Shuttle in February 1983, reach Jupiter in May 1984, and cross opposite poles of the Sun in September 1986. One spacecraft would be designed and built in Europe and one in the USA, but both spacecraft would carry European and American experiments.

This was not the end of major problems with the ISPM programme, however, as in 1980 NASA was forced by budgetary pressures to delay the launch yet again, from 1983 to 1985, and then in February 1981 NASA unilaterally cancelled their spacecraft due to lack of funds. This decision infuriated ESA and its Member States because, although NASA was still prepared to launch the ESA spacecraft,[35] the scientific value of the programme would be drastically reduced without the complementary NASA spacecraft. In addition the $15 million already spent by European experimenters on the payload of the NASA spacecraft would be wasted. Once more NASA had made the decision without any prior consultation with ESA, and ESA had only been told hours before the public announcement of the NASA decision on 18th February. This caused ESA to question what value to place

[35] NASA also agreed to continue to provide the radioisotope thermoelectric generator (RTG), and NASA's part of the scientific payload for the ESA spacecraft.

on any agreement with NASA, as ESA had thought that the ISPM MOU was legally binding on both parties, but it clearly wasn't as far as NASA were concerned.

The fundamental problem with the MOU was that there had been a get-out clause for both parties, in so far as the agreement was "subject to their respective funding procedures". In ESA once any programme had been approved it could not be cancelled unless it overspent by more than 20%, whereas in America a programme could be cancelled at any time, even if it was proceeding very well, because the funding was usually released by Congress on an annual basis. Hence the problem.

ESA and its Member States were furious about the ISPM decision and the way it was made and communicated to ESA. ESA protested to NASA at all levels, and an Aide Mémoire was delivered to the US State Department by three European ambassadors on 3rd March 1981, complaining about the decision and asking for it to be changed. Although this gained the support of the US Secretary of State, Alexander Haig, and made a number of Americans feel very uneasy, including many in Congress, the decision was not changed. The Europeans even offered to sell NASA a copy of the European spacecraft for $40m, compared with the $100m NASA had ear-marked for the American spacecraft, but this did not help either. The American ISPM spacecraft was dead.

The ISPM affair had a fundamental effect on ESA–NASA relations, so that when in 1983, for example, NASA tried to pressurise ESA to merge ESA's proposed Infrared Space Observatory (ISO) with the similar NASA SIRTF programme, the suggestion was roundly rebuffed.

At the time of the NASA ISPM spacecraft cancellation ESA had already spent about half of the budget for the European spacecraft, so ESA decided to continue with a one-spacecraft programme. In the event the ESA space-craft, re-named Ulysses, was completed in 1984 and put into storage awaiting a shuttle launch, which NASA further delayed from 1985 to 1986. Then came the tragic Challenger accident in January 1986 which eventually forced NASA to declare a further four year delay to the launch date, so the spacecraft was in storage awaiting launch for a total of six years.

Ulysses, which was finally launched by the space shuttle Discovery on 6th October 1990 (see Figure 10.17), carried five European and four American experiments. They were to investigate the solar wind, the heliospheric magnetic field, solar emissions, gas and dust in interplanetary space, and galactic cosmic rays. The 370 kg, spin-stabilised spacecraft carried 55 kg of experiments, and communicated with NASA's Deep Space Network on Earth via a 1.6 m diameter parabolic antenna using a tiny 5 watt transmitter.

Figure 10.17 An artist's impression of the Ulysses spacecraft after its deployment from the space shuttle in low Earth orbit. The spacecraft is mounted on top of the Inertial Upper Stage, which was to boost it on its way to Jupiter. (Courtesy NASA/ESA.)

Spacecraft power was provided by a NASA-supplied RTG generator that produced 285 W at start of life. In fact, the RTG generator provided yet another problem, as American environmentalists tried to get the launch of Ulysses cancelled on the grounds that the radioactive material in the RTG was potentially lethal in the event of a space shuttle explosion. Fortunately for the Ulysses programme the legal action failed.

Ulysses was planned to fly about 430,000 km from the centre of Jupiter on 8th February 1992, where it would use Jupiter's gravity to swing 330 million km above the Sun's south pole at a solar latitude of about 80° in August 1994 (see Figure 10.18). It would then cross the ecliptic at a distance of about 175 million km from the Sun in February 1995, and then pass 330 million km above the Sun's north pole in July 1995. Although Ulysses was launched at solar maximum, its first pass over the Sun's poles would take place nearer solar minimum.

Earth orbiting spacecraft and interplanetary spacecraft flying close to the ecliptic had found that the solar wind velocity was generally about 350 km/s, although every now and again a faster component had been detected with a velocity of about 750 km/s or so. This faster component was found more frequently just before solar minimum, when the polar coronal holes often had low-latitude extensions. Over the 13 month period from June 1992 to July 1993 Ulysses found that the solar wind oscillated between similar high and low values (of about 750 to 350 km/s), but after July the spacecraft only measured the high speed solar wind coming from the Sun's south polar coronal hole. The wind velocity stayed at this level as the spacecraft swung from over the south pole to over the north pole, except for a two month

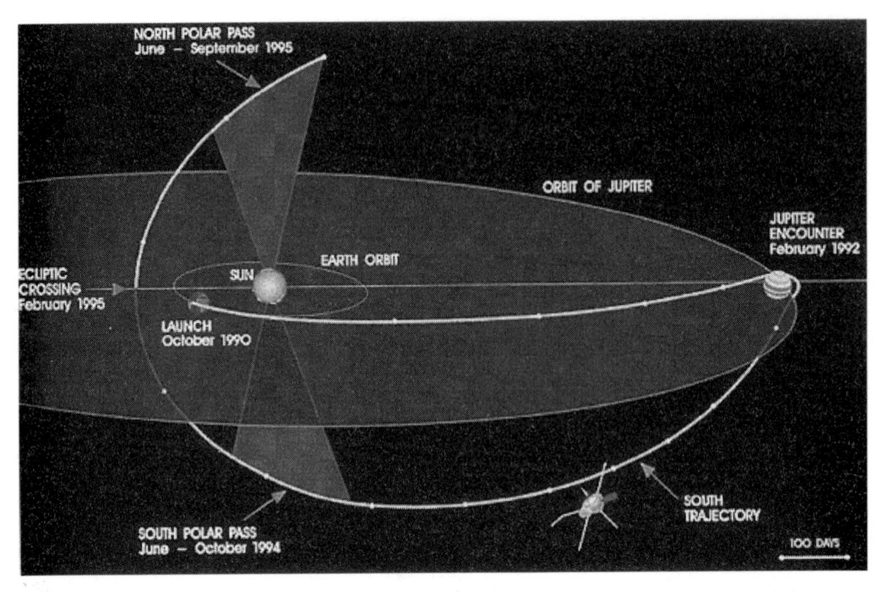

Figure 10.18 The trajectory of the Ulysses spacecraft took it over both poles of the Sun, after using Jupiter's gravity to take it out of the ecliptic plane. (Courtesy NASA/ESA.)

period in February and March 1995 when it crossed the ecliptic and solar equatorial plane going from 20° S to 20 °N solar latitudes (see Figure 10.19). At low latitudes, where the velocity of the solar wind was generally low, its density was found to be relatively high, whereas at high latitudes, where the velocity was high, its density was low.

Fast solar winds from the poles were found to have a lower ratio of seven to six times ionised oxygen atoms (O^{7+}/O^{6+}) than the slow solar winds at the equator. This reduction indicated that the fast solar wind from the poles originates in a region some 400,000 K or so cooler than the 1.8 million K region that produces the slow solar wind at the equator. The fast solar wind was also found to have quite a different composition from the slow wind, with a lower ratio of magnesium to oxygen atoms, for example.

Measurements were also made over nine solar rotations of the O^{7+}/O^{6+} and Mg/O ratios at middle latitudes, where both fast and slow solar winds were detected as the Sun rotated. These measurements showed that the changes in both of these ratios were more sudden than in the velocity of the solar wind but, interestingly, both ratios changed rapidly at virtually the same time. As the O^{7+}/O^{6+} ratio is indicative of the temperature of the corona where the solar wind originates, and the Mg/O ratio is indicative of the composition of the chromosphere, Ulysses appeared to show a correlation

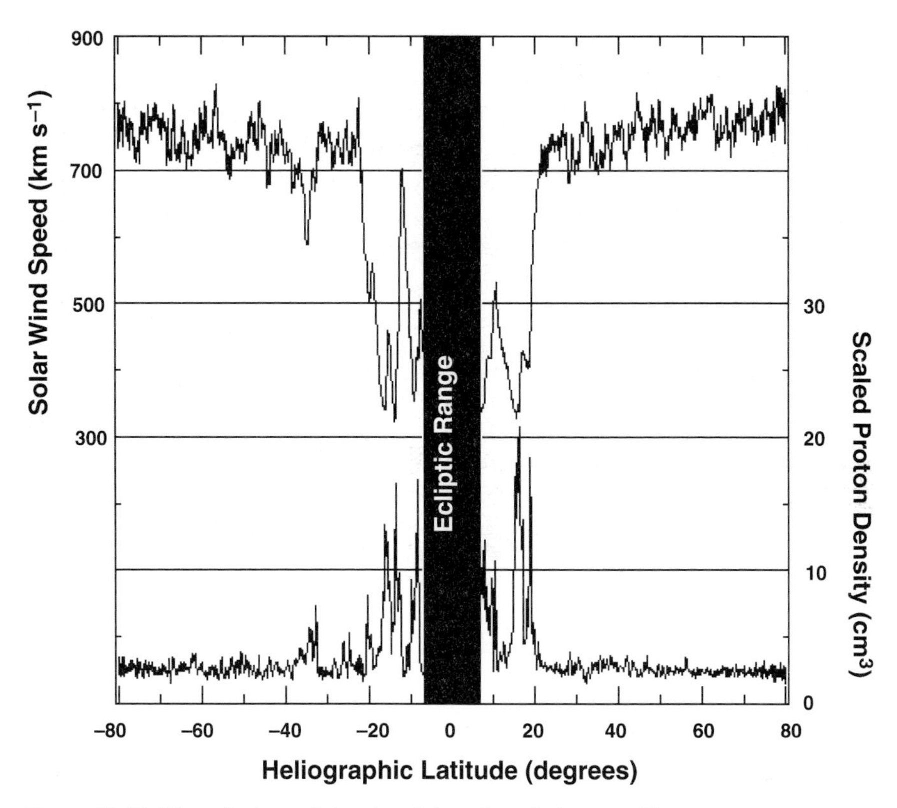

Figure 10.19 The velocity and density of the solar wind, top and bottom traces respectively, as measured by the Ulysses spacecraft as it orbited the Sun out of the plane of the ecliptic. The velocity was generally about 750 km/s except when the spacecraft was within about $\pm 20°$ heliographic latitude, where the solar wind velocity reduced and its density increased. The ecliptic region, which had been investigated by previous spacecraft, is shown by the black bar. (Courtesy J.L. Phillips.)

between chromospheric composition and the temperature of the overlaying corona. This was the first time that such a correlation had been detected.

It had been known since the 1940s that cosmic rays detected from Earth decrease around solar maximum (the so-called Forbush effect). This led astronomers, prior to the launch of Ulysses, to suggest that low-energy cosmic rays would be more impeded at low solar latitudes, where the solar magnetic field is complex because of sunspots and the solar rotation, compared with at high latitudes, where the cosmic rays would be 'funnelled' towards the Sun by its radial magnetic field lines. Ulysses found no such latitude effect for non-solar cosmic rays. They appeared to be equally rejected at all solar latitudes. It had also been thought, prior to the launch of Ulysses, that the Sun's general magnetic field would be stronger near the poles than at the equator, but Ulysses found no such latitude effect here either.

The original intention had been to terminate the Ulysses mission after the spacecraft had passed once over both solar poles, but Ulysses was in such good condition in late 1995 that its mission was extended for a second orbit of the Sun, with polar passes in 2000 and 2001. The first polar passes had taken place near solar minimum, but it is expected that the second pair will take place around solar maximum, so the results should be quite different.

10.11 SOHO

The SOHO (Solar and Heliospheric Observatory) spacecraft of the European Space Agency, which was launched by an American launch vehicle on 2nd December 1995, was designed to incorporate the objectives of two earlier missions that did not see the light of day, namely GRIST and DISCO. GRIST, which had been planned in the 1970s for multiple flights on Spacelab,[36] was to provide solar spectroscopy through grazing-incidence optics in the extreme ultraviolet. Unfortunately, it was cancelled in 1981 as part of the 'fall-out' from the ESA–NASA difficulties on the ISPM programme. DISCO, on the other hand, was to be a free-flying spacecraft undertaking solar seismology (by measuring the Sun's vibration modes), measurements of solar irradiance, and baseline in-ecliptic solar wind and other measurements in support of the ESA ISPM programme. It was important that DISCO should observe the Sun without interruption, that it was beyond the Earth's magnetosphere, and that it had a relatively low radial velocity with respect to the Sun to facilitate data reduction from the solar seismology package. As a result it was planned to position DISCO near to the L1 Lagrangian point between the Sun and Earth where the forces on the spacecraft balance about 1.5 million kilometres sunward from the Earth. In March 1983, however, ESA decided to develop the ISO spacecraft at the expense of DISCO, which was then cancelled.

Strangely, ESA had started consultation with scientists on ESA's next scientific spacecraft programme in June 1982 before they had decided whether to fly DISCO or ISO as an earlier mission. This resulted in an outline SOHO spacecraft mission being submitted to ESA in November

[36] The hardware produced in the Spacelab programme included a pressurised, manned laboratory module and attached, unpressurised instrument platforms called pallets that were flown in the shuttle cargo bay. It was Europe's contribution to the American post-Apollo programme.

1982 whilst the DISCO programme was still alive. At this stage SOHO was proposed as an Earth-orbiting solar observatory primarily dedicated to high-resolution ultraviolet spectroscopy.[37] In early 1983 DISCO was cancelled and ESA decided that helioseismological and solar wind experiments should be added to SOHO instead, and that its nominal orbit, like that of DISCO, should be around the L1 Lagrangian point. So SOHO inherited the main objectives of the cancelled GRIST and DISCO programmes.

Japan had launched a small (188 kg) solar observatory spacecraft called Hinotori ('Firebird') in 1981, and were planning in the early 1980s to launch one or two more solar spacecraft[38] over the next decade or so. NASA had a number of possible solar spacecraft in mind at that time, as did ESA. There was no point in each space agency going its own way and ignoring the other agencies' solar programmes, and so in September 1983 a meeting took place in the USA to try to rationalise these various programmes of NASA, ESA and ISAS (the Japanese Institute for Space and Astronautical Science). This meeting, which was a great success, agreed after a great deal of soul-searching on an International Solar–Terrestrial Physics (ISTP) programme, to which ESA agreed to contribute both the SOHO and Cluster spacecraft missions, assuming that they could both be financed.

Cluster was a proposed ESA programme to measure the solar wind simultaneously at a number of positions in near-Earth space by using four spacecraft. Its data would complement the solar wind information produced by SOHO. Normally in the ESA planning process one spacecraft mission is chosen from a list of possible missions, and that would have resulted in only SOHO or Cluster (at best) being selected. So Gerhard Haerendel of the Max Planck Institute suggested that the SOHO/Cluster mission be proposed to the ESA Member States as a single mission, but for that to be possible the costs to ESA had to be severely curtailed. This led to detailed discussions with NASA on their possible involvement in this proposed SOHO/Cluster programme. These discussions were still ongoing in February 1986 when the ESA Science Programme Committee accepted the joint SOHO/Cluster mission, now called the Solar–Terrestrial Science Programme (STSP), as the next ESA science programme, provided it could be accomplished within an ESA-funded ceiling of 400 MAU[39] at 1984 economic conditions.[40] This

[37] The 'H' in SOHO originally stood for 'high-resolution' not 'heliospheric'.

[38] The solar spacecraft mentioned in this paragraph include both solar observatories and spacecraft to investigate solar-terrestrial phenomena.

[39] MAU = Million Accounting Units. The AU/$US exchange rate varied over the years within a range of about 1 to 1.3 $US to the AU.

[40] i.e. prices and exchange rates.

compared with an estimated cost at that time of almost double that, including launchers.

ESA adopted two parallel approaches to meet their cost target, by persuading the experimenters to reduce their scientific requirements on the SOHO and Cluster programmes, whilst also trying to persuade NASA to provide a free launch for SOHO on either a space shuttle or expendable launch vehicle. Even these approaches would not solve the cost problem completely, however, as Cluster required a dedicated Ariane 4 launcher which would have cost about 100 MAU. After numerous discussions with the Ariane programme, it was agreed in 1988 that Cluster could fly as an experimental payload on one of the Ariane 5 qualification flights, provided ESA paid for the marginal costs of 13 MAU for accommodating Cluster on the launcher. In the event, on 4th June 1996 the first Ariane 5 qualification flight failed 36 seconds after lift-off, destroying all four Cluster spacecraft on-board.

SOHO, in the meantime, which was the other part of the STSP programme, had been successfully launched by an American Atlas II-AS from Cape Canaveral on 2nd December 1995, and had been inserted into its halo orbit around the L1 Lagrangian point on 14th February 1996, six weeks ahead of schedule. The launch and initial orbit operations were so successful that there was enough fuel on-board SOHO to maintain it in its halo orbit for at least ten years, or about twice as long as originally expected.

SOHO's scientific payload was provided by institutes in fifteen different countries, including the USA, Russia and Japan, and the launch vehicle was provided by the USA, so although the spacecraft bus was provided by ESA, this was a truly international mission. The payload consisted of three instruments (GOLF, VIRGO and MDI) devoted to helioseismology, five (SUMER, CDS, EIT, UVCS and LASCO) observing the solar atmosphere and corona, and four (SWAN, CELIAS, COSTEP and ERNE) measuring the solar wind. So far (1998) only interim results have been published, but it is already apparent that SOHO is yet another in a growing list of successful solar observatories.

There had been doubt expressed in the early 1980s by some scientists about whether it was a good idea to try helioseismology from space when ground-based measurements were producing high-quality results. The ACRIM results from Solar Max gave some indications of what could be done from space, however, and so the suite of instruments listed above was included on SOHO. Nevertheless, it was particularly pleasing to the advocates of helioseismology from space to see that the first set of results from the SOHO helioseismology experiments were excellent. The MDI (Michelson Doppler Imager) instrument, in particular, which measured the vertical

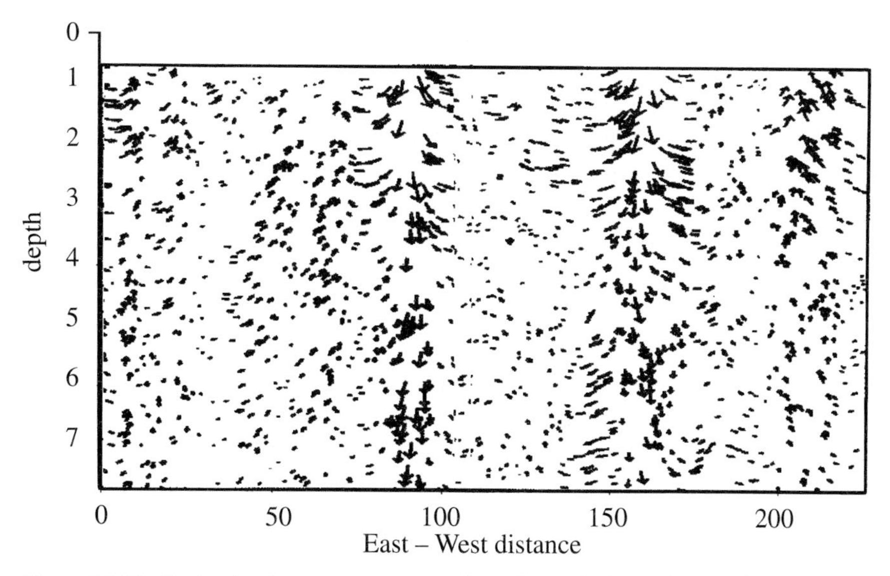

Figure 10.20 Motion in the upper convective layer beneath the surface of the Sun as deduced by helioseismic tomography using measurements from the MDI instrument on SOHO. The vertical and horizontal scales are in thousands of kilometres, so the total depth shown is about 8,000 km, which is only about 1% of the solar radius. The size of the arrows indicates velocity, with the largest velocities being approximately 1 km/s. (Courtesy ESA/MDI.)

motion of the Sun's surface at a million different points[41] once per minute, produced high-quality data of a type never seen before of the motions of gas in the convective zone just below the Sun's visible surface. This enabled solar physicists to produce the first maps of horizontal and vertical motions in the upper convective layer, which clearly showed vertical convection in approximately evenly spaced columns (see Figure 10.20).

High resolution, visible-light images of the solar photosphere had long been found to show granules covering the surface of the Sun. These granules, which are typically about 1,000 to 1,500 km across, are approximately 30% brighter than the intergranular lanes that separate them, indicating a temperature difference of about 400 K. Ground-based Doppler measurements indicated that the granules were the tops of ascending currents which then moved sideways to the intergranular lanes where the cooler gas descended. Individual granules last typically for about 20 minutes, with a velocity difference between the upwelling and descending currents of

[41] Using a 1024 × 1024 CCD array.

approximately 2 km/s. Opinions differed as to the likely depth of a granule from about 100 to 1,000 km. Unfortunately, the resolution of the MDI data was not sufficient to show the effect of such small surface disturbances.

In the 1950s A.B. Hart found a much larger surface structure called supergranulation, however, and that should be visible in the MDI results. Supergranulation is a much more subtle structure than granulation and cannot be seen in visible light, being only evident in spectroheliograms that show gaseous motion on the surface of the Sun. Later work by other observers showed horizontal flows of gas from the centre of these supergranules to their sides of about 0.4 km/s, with a weak upward flow in the centre, and a downward flow at their edges of about 0.1 km/s. The supergranules, which last for a day or so, were found to be about 30,000 km across. Initial results from SOHO's MDI instrument showed, however, that the downward convective motion seen in Figure 10.20 originated from the centre of the supergranules, rather than their edges. Whether later results will confirm this surprising result remains to be seen.

The Sun's surface was known to rotate once every 25 days at the equator and once every 34 days near the poles, but no-one could be sure how its interior rotated. It was initially thought that the Sun's rotation would be constant on cylinders parallel to the Sun's rotation axis. That is, if we drew a cylinder with its axis coincident with the Sun's rotation axis, and with a radius of half the solar radius, for example, all gas in the Sun at the surface of that cylinder would rotate in about 31 days, which is the period of rotation where the cylinder cuts the surface (at 60° latitude). Then in 1988 Kenneth Libbrecht of the Big Bear Solar Observatory found that this was incorrect using helioseismology, as his data implied that the angular velocity of the Sun is approximately constant along radial lines, for low and medium latitudes, down to the bottom of the convective zone.[42] The SOHO data confirmed this, and showed that below the convective zone the angular velocity is fairly uniform and similar to that of the surface at middle latitudes, so there is an intense shear layer at the base of the convective zone for other than mid-latitudes. It is thought that this shear layer would produce a large amount of turbulence and could be the site of the dynamo that generates the Sun's magnetic field.

It was hoped that a Fourier (i.e. frequency) analysis of the solar surface velocities measured by the GOLF instrument, and of the spectral and total irradiance measured by VIRGO, would show evidence of the so-called g-modes (gravity-modulated oscillations) caused by sloshing of material deep inside the Sun. So far, although the experiments have produced excellent data,

[42] The convective zone covers the outer 30% of the Sun's radius.

with a much higher signal-to-noise ratio than achieved with Earth-based instruments, no such g-modes have been identified.

The UVCS (Ultra-Violet Coronagraph Spectrometer) instrument measured the velocities of various ions in the outer solar corona, in an attempt to understand the mechanisms that accelerate them as they leave the Sun. It showed that in coronal holes, proton outflow velocities increased from 50 to 200 km/s between 1.5 and 3.5 solar radii, whilst the line widths of Lyman-α increased from 190 to 240 km/s, and those of O VI increased from 100 to 550 km/s. This new comparative data enabled theorists to conclude that a particular type of resonance in MHD[43] waves may well be accelerating the plasma.

Coronal mass ejections were studied in detail by LASCO (Large-Angle Spectroscopic Coronagraph), which consisted of three different coronagraphs with nested, overlapping fields of view, covering the corona from 1.1 to 30 solar radii (i.e. one seventh of the way from the Sun to the Earth). Some CMEs were found to have a constant velocity as they left the Sun, whilst others were continuously accelerated, and yet others had a sudden acceleration a few solar radii from the Sun. Amazingly, in a number of cases CMEs were seen on both sides of the Sun at the same time, showing their global nature (see Figure 10.21).

A particularly interesting CME was observed by the LASCO C2 coronagraph at 18:50 UT on 6th January 1997. The CME appeared to be heading towards the Earth, and was seen as a 'Halo' or 'Ring' event surrounding the Sun. Two hours later the CME had expanded enough to be visible in the LASCO C3 coronagraph that had a larger obscuring disc. The halo could be seen in the C3 for another 4½ hours before it became too faint to detect, but at 04:45 UT on 10th January both the SOHO and NASA's WIND spacecraft detected the cloud as it reached them moving at 450 km/s. NASA's Polar spacecraft then detected the effect of this cloud on the Earth's radiation belts, whose intensity was found to increase by a factor of more than 100 over the next few hours or so. Then on the following day the Telstar-401 communications spacecraft failed, possibly due to radiation-induced effects. Such solar storms have been known to have even more serious effects at high latitudes on the Earth, causing severe problems with electricity distribution networks. So an early warning system is now operating, based on spacecraft images, to warn electricity suppliers on the ground and operators of spacecraft in orbit of such likely events.

---

[43] MHD stands for magnetohydrodynamic. MHD theory is a way of simplifying the analysis of the corona by treating it as a magnetic fluid, but the complex details of this theory are outside the scope of this book.

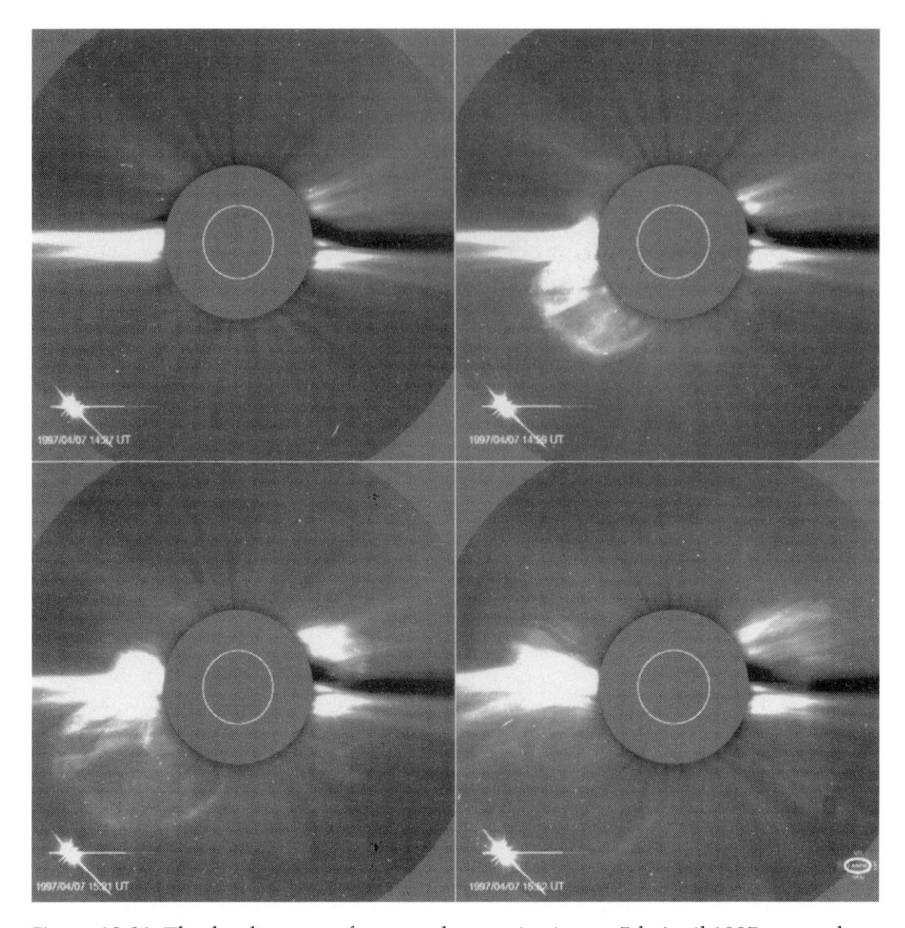

Figure 10.21 The development of a coronal mass ejection on 7th April 1997 as seen by the outer coronagraph in the LASCO instrument on SOHO. The size and position of the Sun is shown by the white circle superimposed on the black occulting disk. The first (top left hand) image shows the solar corona pre-CME, with the CMEs appearing in the second (top right-hand) image centred approximately at the 8 o'clock and 2 o'clock positions. Although the CME on the left hand side of the Sun is seen to be the brighter, the CMEs brighten on both sides of the Sun simultaneously, confirming that they are part of the same disturbance which extends at least 180° longitudinally around the Sun. (Courtesy ESA/NASA/LASCO, 97.07.010-003.)

The SUMER and CELIAS instruments on SOHO measured elemental abundances in the chromosphere and corona to help to determine the processes that feed and accelerate the solar wind. Elements were found by both instruments in these regions that had not been detected there before, showing an enrichment, compared with their photospheric abundances, of up to a factor of 20 for elements with a first ionisation potential (FIP) of less than 10 eV. Although the partial correlation observed between low values of FIP and high coronal enrichment had been known for some time, SOHO added significant new elements and isotopes to the database which should aid theoretical analysis. The high temporal resolution of CELIAS was also important in observing how the abundance enhancement changed with structural and velocity variations in the corona and solar wind.

Early solar wind results from the high-energy particle detectors in the COSTEP and ERNE instruments were disappointing because SOHO was launched at a time of solar minimum. On the other hand, this enabled the highest energy instrument, ERNE to detect and analyse galactic cosmic rays, finding significant amounts of hydrogen, helium, boron, carbon, nitrogen, oxygen, neon, magnesium, silicon and iron, the heavier nuclei probably coming from supernovae.

10.12 **Summary**

So over the last twenty years our knowledge of the Sun has increased in leaps and bounds due primarily, as far as space research is concerned, to Solar Max, Nimbus 7, Yohkoh, Ulysses and SOHO. Other spacecraft, both manned and unmanned, have also produced significant information over that period, but the above-mentioned spacecraft probably produced the most significant contributions. For example, Solar Max and Nimbus 7 enabled the linkage between the solar constant and sunspots and faculae to be finally resolved. In addition, Solar Max provided a great deal of new information on solar flares, which was further extended by Yohkoh. Ulysses is providing data for the first time at high solar latitudes on the solar wind, the Sun's magnetic field, and the effect of the Sun on galactic cosmic rays, in particular. And finally, SOHO is providing radically new information on the internal structure of the Sun, as well as on the solar atmosphere, corona and solar wind.

All space-based experiments are compromises between what is ideally required and what can be built within financial and technical constraints. Politics also intervenes from time to time. Something like ten years or so after the experimental concept is suggested, the spacecraft is finally launched, but when it is launched it is already out-of-date, in the sense that the experimenters had to freeze their requirements some years earlier. Astronomy has not stood still in the meantime, and if the experimenters could have been allowed to modify their experiment on the day of launch, they would have done so. Also the hardware was designed with the technology available some years before launch, and technology has not stood still either. So on the day of launch, if not before it, there is already pressure from scientists to build a better spacecraft with better experiments than those on the spacecraft just about to be launched. That pressure then builds up over the next year or so as the experimenters find out from the in-orbit results what they would really like from a new spacecraft. This gives the impression that astronomers are never satisfied and always want better spacecraft in orbit than the ones that they have.

There is a natural reluctance by the governmental authorities that provide the money for scientific spacecraft to approve another programme before they have got their money's worth out of the current one. But long programme lead times put pressure on the authorities, as outlined above, to make a commitment to a new one before the current programme has been completed. A number of the discussions outlined in this and other chapters of this book, between the holders of the purse strings and the scientists pushing for new programmes, reflect this permanent tension.

Ulysses and SOHO have already produced vast quantities of data, like many other current spacecraft, and they will no doubt provide much more before they reach the end of their useful lives. Because of this there is a case for saying that we will not launch another solar observatory spacecraft for ten years, until all the data have been used from these two spacecraft. "But neither spacecraft provides all the data we require on CMEs as they approach Earth, to enable us to give reliable warnings of geomagnetic storms on Earth", the scientists reply, for example, and so another spacecraft is designed and launched, and the story goes on.

Chapter 11 | EARLY SPACECRAFT OBSERVATIONS OF NON-SOLAR SYSTEM SOURCES

11.1 Introduction

We left the observation of non-Solar System sources from space in Chapter 1 after the discovery of the first X-ray sources by sounding rocket experiments in the early 1960s. As mentioned in that chapter, a number of people and organisations had recommended as far back as 1958 that spacecraft should be launched to observe the cosmos in all those wavelengths inaccessible from the surface of the Earth, but these suggestions did not meet with universal acclaim from the astronomical community. To understand why, it is necessary to go back a little further into history.

Observational astronomy before the Second World War had been limited to studying the sky from the Earth's surface using optical telescopes. As a result, observational astronomers were only familiar at that time with using photography, spectroscopy and photometry at optical wavelengths. After the war, radio engineers and physicists had started to observe the universe at radio wavelengths using army surplus equipment, but their equipment, observational techniques and method of presentation of their results were foreign to optical astronomers. As a result it took time for astronomers to realise that these people were actually producing some astronomically useful results. A similar situation occurred with astronomical observations using sounding rockets, as these were generally being undertaken by rocket engineers and physicists, and for many years after the war the amount of astronomically useful data produced by sounding rocket experiments was very limited. So the possibility in the late 1950s of entering yet another new area by using spacecraft as astronomical observatories was not greeted with universal enthusiasm by astronomers.

This lack of enthusiasm was firstly because there was the expectation that not much would be found using non-optical wavebands, although the early results from radio astronomy showed the more enlightened astronomers that this may not be the case. Secondly, astronomers at that time were used to working on their own or in small teams, and the idea of working in large, multidisciplinary teams with spacecraft engineers filled many with horror. Finally, astronomers expected that money would be taken from ground-based astronomical research and used in these 'misguided' space enterprises. Astronomers generally saw space programmes as nothing more than part of a macho, propaganda programme to catch up with, and then upstage, the Russians in space spectaculars. Notwithstanding all these concerns, as the Luddites found, you cannot stand in the way of progress, and so it was not a question of 'if' spacecraft would be used for astronomical research but 'how' and 'when'.

The Solar System was naturally easier to explore from space than distant stars and galaxies, and so it was the first to be observed, with spacecraft being launched to the Moon in 1958, Mars in 1960, and with the Sun being observed from Earth orbit in 1960. There was considerable doubt expressed in the late 1950s by many astronomers whether it would be possible to detect individual X-ray and γ-ray sources outside the Solar System, as they would be too dim to be observed with the detectors available at the time. It was clear that some stars would have strong ultraviolet emissions, however, and cosmic rays had been detected for many years, although their source was unknown. So these two areas of ultraviolet and cosmic ray research were thought to be the first areas of study outside the Solar System ripe for investigation by spacecraft.

The source and constituents of cosmic rays had been a mystery for many years. At first they had been thought to be very high energy γ-rays, but evidence of their particulate, non-photon nature gradually appeared in the 1930s. It was not until 1948, however, that the first unambiguous evidence was obtained, using high altitude balloons, that cosmic ray primaries consist mainly of hydrogen nuclei (or protons), with lesser amounts of helium nuclei (alpha particles) and other nuclei up to atomic number 40 (i.e. zirconium), but their origin was still unknown. Unfortunately, these nuclei are all charged, and so their direction as measured by spacecraft in Earth orbit would be influenced by the Earth's magnetic field, and their direction as measured by interplanetary spacecraft would be influenced by the Sun's magnetic field. However, it was thought that the collision of cosmic rays with interstellar hydrogen or dust would produce pi-mesons[1] in interstellar

[1] Pi-mesons are elementary particles. They can have either positive or negative charges or be neutral.

space, and Yukawa had pointed out some time before that neutral pi-mesons would decay to produce a pair of γ-rays with energies centred around 70 MeV. So observing the direction of these γ-rays should help to find the source of their parent cosmic ray particles. This was the objective of Explorer 11 when it was launched in 1961.

11.2 Explorer 11 and Gamma Rays

The payload of Explorer 11 consisted of a relatively simple γ-ray telescope, using a caesium iodide/sodium iodide scintillation detector system with an anti-coincidence arrangement, to discriminate against particles from the Van Allen belts. Power was provided by body-mounted solar cells when the spacecraft was in sunlight, and by a nickel–cadmium battery when the spacecraft was in eclipse. Data was recorded on a tape recorder for play-back when the spacecraft was in view of one of the NASA ground stations. The spacecraft spun around its longitudinal axis, but the fourth stage of the Juno 2 launcher was deliberately left attached to the spacecraft to cause it to tumble about a transverse axis also, and hence scan both the Earth (for attitude determination) and the sky.

Explorer 11, without the attached fourth stage, was about 1.1×0.3 m in size and weighed about 37 kg. It was launched from Cape Canaveral on 27th April 1961 into a 480×1780 km Earth orbit where it remained operational for five months. During that time it detected only 22 γ-rays, which was not a very encouraging start to γ-ray spacecraft astronomy.

11.3 Orbiting Astronomical Observatories and Ultraviolet Observations

NASA's series of Explorer-class spacecraft had been available since the start of the space age, but it was thought that these small spacecraft, which generally weighed between about 20 and 50 kg, were non-optimal for a spacecraft observatory where high pointing accuracy and stability were generally required. It was thought that it would be more efficient to stabilise a large spacecraft, and then attach as many experiments as possible, than go to the same cost and difficulty of stabilising a small spacecraft. In 1959 detectors at ultraviolet wavelengths were already of a sufficiently high resolution to require a high stability platform, and so NASA came up with the concept of a series of large Orbiting Astronomical Observatories which would each contain a number of modest-size telescopes operating at various wavelengths, generally in the ultraviolet.

The first such spacecraft, the 1,800 kg OAO (Orbiting Astronomical Observatory)-1, was launched into an 800 km altitude, circular Earth orbit on 8th April 1966. Its payload consisted of seven ultraviolet telescopes provided by the University of Wisconsin, two soft (2–150 keV) X-ray experiments, and a 50–100 MeV γ-ray experiment based on that flown on Explorer 11. The ultraviolet telescopes, which were designed to cover the range from 80 to 420 nm, consisted of one 40 cm diameter, f/2 instrument for making photometric measurements of nebulae, four 20 cm diameter, f/4 instruments for making photometric measurements of stars, and two spectrometers. Unfortunately, the spacecraft failed the day after launch when an electrical fault caused the battery to overheat. No astronomical data was received.

The failure of this two ton observatory well-illustrated the risk involved in launching one large observatory, rather than a number of smaller ones. In parallel with this catastrophe, NASA were also under increasing pressure to reduce their budgets, like other Federal agencies, to help to pay for both the escalating Vietnam war and the results of ethnic unrest in American cities.

NASA made a number of modifications before the launch of the next OAO spacecraft, OAO-2, which was a refurbished prototype. The payload consisted of an improved version of the University of Wisconsin ultraviolet experiment carried on OAO-1, together with a Smithsonian Astrophysical Observatory ultraviolet instrument called the Celescope (for CELEstial teleSCOPE). The two experiments observed the sky through opposite ends of the spacecraft.

As before, the University of Wisconsin experiment consisted of seven ultraviolet telescopes, with photometers but with no imaging capability, whereas the Celescope consisted of four 32 cm diameter telescopes, each with a field of view of 2.8°, designed to image the sky in the ultraviolet from 105 to 320 nm. The four Smithsonian telescopes each had a different filter to provide images centred on about 135 nm, 150 nm, 225 nm, and 265 nm. Development problems with the Celescope's ultraviolet-sensitive image tube, called the Uvicon, had caused the Smithsonian experiment to be deleted from the payload of OAO-1.

The 2,000 kg OAO-2, which had a design lifetime of one year, was launched on 7th December 1968 into a 770 km circular orbit, achieving a pointing accuracy of about half a minute of arc. One of the Uvicons was blinded by overexposure to sunlight, and the sensitivity of the other three decreased with time causing Celescope observations to be terminated in

April 1970, four months after the end of its nominal lifetime. The Wisconsin experiment did even better and remained operational until the spacecraft was switched off in February 1973.

Over its lifetime OAO-2 produced over 7,000 images and photometric data on over 5,000 stars, as well as detecting a huge cloud of hydrogen around the comet Tago–Sato–Kosaka and observing a nova in February 1970. The density of stars with an ultraviolet intensity greater than the Celescope's limiting magnitude of 9 (for a star of spectral type A0) was found to be very much lower near the north galactic pole than at the galactic equator, indicating that most of these stars are not nearby.[2] Although most of the stellar and galactic intensities in the ultraviolet were as expected from an extrapolation of their spectral curves in visible light, there were some objects that were appreciably brighter or dimmer than expected. Astronomers were loath to draw any conclusions from these anomalies, however, without further data. They thus eagerly awaited the launch of the next Orbital Astronomical Observatory in the series, OAO–B, which included a 90 cm diameter ultraviolet telescope with an attached spectrometer. Unfortunately the launch of OAO–B on 30th November 1970 was a failure, as the aerodynamic shroud that protected the spacecraft during the early launch phase failed to separate.

11.4 **X-rays**

Chapter 1 described the discovery of the first X-ray sources, Sco X-1 and that in the Crab (otherwise known as Tau X-1), by Giacconi and his group at AS & E using sounding rockets. That chapter also described Friedman's lunar occultation experiment of 1964 that showed that the Crab source was too large to be a neutron star.

In 1963, just after the discovery of Sco X-1, Giacconi sent a proposal to Nancy Roman, head of NASA's Astronomy Branch, suggesting that NASA should instigate a phased programme of X-ray research. He suggested that it should consist of:-

Phase 1 Continue rocket surveys of the sky to find new sources and develop new X-ray instruments

[2] The Milky Way is essentially a large spiral structure with a finite thickness. For stars near to the Sun, there are approximately the same number in each direction, as their distances are less than the thickness of the Milky Way.

Phase 2 Put an X-ray experiment on OSO-4
Phase 3 Build a small, dedicated X-ray spacecraft
Phase 4 Build a medium-size, X-ray spacecraft with a focusing X-ray telescope with Wolter-type optics (see Section 1.3) as its main instrument
Phase 5 Build a large, X-ray spacecraft to fly in the late 1960s.

NASA responded with interest, particularly to Phase 3, and so AS & E submitted a detailed proposal for an X-ray Explorer-class spacecraft to be launched just two years later in December 1965. As the proposed spacecraft did not have an imaging capability, its stability requirements were not too exacting. NASA were not prepared to, nor were they allowed to, give an instant approval to such a crash, X-ray programme, however. But it was eventually presented in March 1965 to the Congressional committee concerned with space matters for their approval, which it duly received. This X-ray spacecraft was to be variously called Explorer 42, SAS (Small Astronomy Satellite)-1, or Uhuru.

In the meantime more and more X-ray sources had been discovered using sounding rockets, and the more precise measurements of their position had enabled a number of them to be identified with optical sources. For example, the AS & E group, aided by MIT scientists, designed two large X-ray collimators that used the length of the Aerobee payload compartment to provide high angular resolution. Previously, the length of the collimator had been limited because the experiments had viewed out of the side of the Aerobee payload compartment, whereas with the new arrangement viewing was from the top of the rocket. The AS & E – MIT experiment was launched on 8th March 1966 to measure the size of the Sco X-1 source to see if it was extended, like the Crab X-ray source, or whether it was a star. The Crab nebula, in which the Crab X-ray source was located, was the visible remains of a star that had exploded as a supernova about 900 years ago and there was speculation that other X-ray sources may be supernova remnants also.

The AS & E – MIT instrument showed that Sco X-1 was too small to be a supernova remnant, and there was no nebula visible at that place in the sky. Extrapolating the X-ray energy curve into the visible, however, Hugh Johnson of Lockheed concluded that the object would be a 13th magnitude blue star. Its position had been determined to within about $\pm$ 1 arcmin by the Aerobee experiment, and this improved position estimate enabled Sco X-1 to be identified as a 12.6th magnitude blue star, remarkably similar to Johnson's prediction. Detailed observation of the blue star's visible spectrum then followed, showing that it had all the characteristics of an old nova. This was the breakthrough that the theoreticians had been waiting for, as it was known that most novae were binary systems in which a small star, such as a white dwarf, is being gravitationally bound to a large star. Two years earlier, Satio Hayakawa and Masaru Matsuoka had suggested that, in such an arrangement, gas ejected by the large star could produce a high-temperature shock wave when it hit the smaller star. The temperature could be even high-enough to produce X-rays! So there appeared to be a good fit between observation and theory, although Sco X-1 had not been shown observationally to be a binary, which left some doubt as to whether the X-ray production mechanism had been correctly identified.

If Sco X-1 was an old nova binary system, was this true of other X-ray emitters? At that time there were a number of X-ray sources known in the Cygnus constellation, and the AS & E team decided to examine them in more detail, and then try to find their optical counterparts, hoping to repeat their success with Sco X-1. They measured Cyg X-1, X-2, X-3, and X-4 with an Aerobee-launched, X-ray experiment of October 1966 and found that Cyg X-2 had all the characteristics of an old nova like Sco X-1. Unfortunately, however, although both Sco X-1 and Cyg X-2 flickered in their energy output, neither showed the periodic change expected of a binary system. The plane of their orbits could be such that there were no eclipses as seen from Earth, of course, but everyone would have felt much happier if they had both shown unmistakable binary behaviour.

In May 1967 astronomers at NRL discovered, using a sounding rocket experiment, that the quasar[3] 3C 273 is a very intense source of X-rays, and confirmed a previous suspicion from previous less accurate X-ray measurements that the bright radio galaxy M87 was also a source of X-rays. In the same year, the President's Science Advisory Committee endorsed the concept of a series of High Energy Astronomical Observatories, which was to become the very successful HEAO series. Then in 1968 the Vela supernova remnant was found to emit X-rays, and the Tycho supernova remnant and the Cygnus loop were also confirmed as X-ray sources.

[3] Quasars had first been discovered in the early 1960s to be very small, distant objects emitting enormous amounts of energy, assuming that their red-shifts were due to the expansion of the universe, and so were true indications of distance. The radio source 3C 273 was the second quasar to be discovered (in 1961), and the first to have its red-shift measured. The nature of quasars and their method of energy emission were unclear in the 1960s.

11.5 Pulsars

Light from ordinary stars twinkles at visible wavelengths because of instabilities in the Earth's atmosphere, but the light from planets does not twinkle because they are not point sources. In the mid 1960s it was found that a similar effect existed with the radio emission from quasars as detected on Earth, due to instabilities in the solar wind. The quasars that subtended the largest angle at the Earth did not scintillate at all, and the degree of twinkling, or scintillation of their radio signals was thought to be a strong indication of their angular size. Using this effect, Anthony Hewish of Cambridge University set out in the mid 1960s to estimate the angular diameter of quasars by measuring the degree of scintillation of their radio emission.

Shortly after observations had started in August 1967, Jocelyn Bell, one of Hewish's research students, detected a signal that had a regular pulse lasting only 0.016 seconds, with a frequency of 1.34 seconds. Moreover, the pulse frequency was incredibly regular, to about one part in ten million. Nothing like it had ever been seen before. The pulse width meant that the source could not be more than 0.016 light seconds or about 5,000 km across, but what could it be? The initial reaction was that it must be due to some sort of man-made interference, but this was quickly eliminated, and then three more sources were found at Cambridge with slightly different pulse periods. These objects are what we now call pulsars.

Various theories were advanced to explain these pulsars, including vibrating white dwarfs or vibrating neutron stars,[4] but Thomas Gold of Cornell University suggested in 1968 that they were rapidly rotating neutron stars about 10 km or so in diameter. Gold reasoned that if a large star collapsed at the end of its life to form a neutron star then, even if it lost an appreciable amount of mass in the process, it must still rotate very fast to conserve angular momentum. For example a star a little heavier than the Sun would end up spinning at about once per millisecond if it lost no mass, and maybe about once per 10 milliseconds if it lost an appreciable amount

[4] White dwarfs are the highly compressed stellar remnants of stars with a maximum mass of 1.4 solar masses, when they have stopped producing energy by thermonuclear fusion. They consist of atomic nuclei and electrons that have been completely stripped from atoms. Stars of greater than 1.4 and less than about 3 solar masses end their days as even smaller and more dense neutron stars, which consist almost entirely of neutrons. It is thought that stars of greater than 3 solar masses finally become black holes. White dwarfs are typically about 20,000 km in diameter and neutron stars about 10 km.

of mass. He also pointed out that not only would the star's rotation rate increase when it collapsed to form a neutron star, but its magnetic field would also increase substantially to about 10^{12} gauss. This would channel particles emitted from the neutron star's magnetic poles into highly directional beams. If the magnetic poles are not coincident with the poles of rotation, then the beams of particles would produce highly directional beams of radio emission as they are spun at very great velocity around the star. These radio beams would illuminate space like a light-house beam, and if the Earth were in such a direction as to be illuminated, we would see a pulse of radio energy at the rotation rate of the star, every time the beam swept past the Earth. Gold also pointed out that the spin rate would gradually slow down as the star lost energy through particle emission from it magnetic poles.

After the four Cambridge pulsars were found, more were soon discovered by radio astronomers in the USA, UK and Australia. Then in October 1968 a Sydney University group discovered a radio pulsar (now called PSR 0833-45) in the Vela X supernova remnant with a period of only 89 milliseconds, the shortest then known, and a pulse width of 10 milliseconds. This was a crucial discovery linking as it did a pulsar, or pulsating neutron star, with a supernova remnant, as many years before it had been suggested that the stellar remnant of a supernova explosion would be a neutron star. This was why in 1964 Friedman had measured the diameter of the X-ray source in the Crab Nebula, another supernova remnant, hoping that it would be a neutron star, which he found it was not. If the Vela X supernova remnant contained a radio pulsar, however, maybe the Crab Nebula did also, and this led David Staelin and Edward Reifenstein to search for such a pulsar with a radio telescope at the American National Radio Astronomy Observatory. In November 1968 they succeeded and found a pulsar in the Crab with a pulse rate of just 33 milliseconds, which took over from the Vela X pulsar as the shortest then known. Within a month of its discovery the Crab pulsar's pulse rate was found to have slowed down, as predicted by Gold. Then in January 1969 the Crab pulsar was also found to pulse in visible light.

The discovery of a pulsar (now called NP 0532) in the Crab Nebula that pulsed at the same frequency in radio and optical wavelengths, set off a race to find if it also pulsed in X-rays. It did not take long to answer the question as a rocket experiment launched by Friedman's NRL group on 13th March 1969 found the same 33 millisecond pulsations in X-rays. So the Crab Nebula has a large diffuse source of X-rays, first detected in 1963, and an X-ray pulsar. Calculations then showed that this X-ray pulsar provides the energy to sustain the diffuse X-ray source. The Crab pulsar's X-ray energy was 200 times its energy in optical wavelengths, and 20,000 times that at radio wavelengths where it was first detected. Finally, another series of X-ray occultation measurements in 1974 by satellite-, rocket- and balloon-borne experiments showed that the centre of the X-ray nebula was about 10 arcsec northwest of the pulsar, showing that the Crab X-ray source is not symmetrical.

11.6 Uhuru

The number of new X-ray sources discovered began to slow down in the late 1960s, with only twelve[5] being found in the period 1968–1970, because the observing periods possible with sounding rockets were too short, being limited to a few minutes per launch. It was, therefore, fortunate that the first X-ray dedicated spacecraft observatory was launched in 1970. This was the 140 kg SAS-1, or Uhuru[6] (see Figure 11.1), which was put into a 550 km circular orbit by an American Scout rocket from the Italian San Marco platform off the coast of Kenya. As the platform was only 3° south of the equator, this enabled the Scout to put the spacecraft into a near-equatorial orbit using the minimum of propellant.

The 65 kg payload of Uhuru consisted of two X-ray telescopes, mounted back-to-back, that scanned the sky as the spacecraft spun about its axis at about 30° per minute. The collimators of the X-ray telescopes produced $5^\circ \times 5^\circ$ and $\frac{1}{2}^\circ \times 5^\circ$ fields of view, and the detectors were argon-filled proportional counters sensitive to X-rays in the range from 2 to 20 keV. Once per day, during the early phase of operations, the spin axis of the spacecraft was adjusted to enable the X-ray telescopes to scan a different 10°-wide band of sky, enabling the whole sky to be observed about once every two months. All areas of the sky were not scanned with equal frequency, however, as the scan-patterns were arranged so that the Crab nebula and the region of the centre of the Milky Way near Sco X-1 could be scanned most frequently. Strong X-ray sources were located by Uhuru to within about 1 arcmin, and weak ones to about 15 arcmin. Data was transmitted in both real time and, following recording by a tape recorder, during a data dump once per orbit to

[5] One of these, Centaurus X-4, was discovered by the American Vela 5A and 5B military spacecraft. They first detected Centaurus X-4 as a variable source in May 1969, but it then increased dramatically in intensity, before reducing to below the detection threshold in September 1969. It was the first continuously observed X-ray nova.

[6] It was called Uhuru, the Swahili word for freedom, as it was launched on 12th December, which is Kenyan Independence Day.

Figure 11.1 Engineers are seen checking the Uhuru spacecraft that was to be launched in December 1970 by an American Scout rocket. The large black rectangle near the top of the spacecraft is one of the X-ray collimators, with a star and sun sensor mounted above it. (Courtesy NASA.)

the ground station at Quito, Ecuador. Unfortunately, six weeks after launch the tape recorder stopped working, and NASA were forced to bring other receiving stations on-line to pick up as much data as possible in real time, which resulted in about 50% of the data being received.

Uhuru was able to detect weaker X-ray sources than sounding rockets, and was able to measure their positions more accurately also. As a result, over its lifetime of just over two years Uhuru was to detect a total of 339 discrete sources (see Figure 11.2), 100 of which had accurate-enough locations determined to enable visible and radio counterparts to be identified.[7] Unlike stars at optical wavelengths, Uhuru found that X-ray sources tended to be highly variable in their X-ray intensity.

Uhuru helped to solve a number of problems. For example, Centaurus X-3 had been discovered by a team from Lawrence Livermore Laboratory in 1967, but later attempts by Ken Pounds and his team at Leicester University in the UK failed to detect it in X-rays. This led to doubt as to whether Cen X-3 really existed. Shortly after launch Uhuru showed that Cen X-3 certainly did exist, and the reason for the previous confusion was that its X-ray emissions were extremely variable. Firstly, Uhuru measured a regular periodicity of 4.84 seconds (see Figure 11.3), showing from its short period that Cen X-3 was a rotating neutron star. Then it was found that this period varied slightly with a period of 2.09 days. Finally, the X-rays disappeared completely for 11 hours every 2.09 days, indicating that Cen X-3 was part of an eclipsing binary system. This was confirmed when Krzeminski found a blue supergiant of 20 solar masses at Cen X-3's location that exhibited the same 2.09 day binary variations.

Astronomers now started to use Uhuru to look for other eclipsing binaries, and they were rewarded in late 1971 when Giacconi's AS & E group detected periods of 1.24 seconds and 1.7 days in the X-ray intensity of Hercules X-1. Again these pulses were found to be Doppler-shifted with a period of 1.7 days. The visible counterpart was found to be a blue star, HZ Herculis, of two solar masses, which also exhibited variations at the binary period. But there was more to it than that, as at optical maximum the spectrum was that of a B8 star, but at minimum its spectrum had changed to that of a cooler F0 star. This strange optical variability of HZ Herculis had been known for some time, and the discovery that its companion is a pulsating

[7] A Uhuru catalogue had been published at the end of 1972 that contained 150 sources and their positions. By 1974 optical counterparts had been found for 35 Uhuru sources, and then in 1978 the fourth Uhuru catalogue was published containing the 339 sources mentioned above.

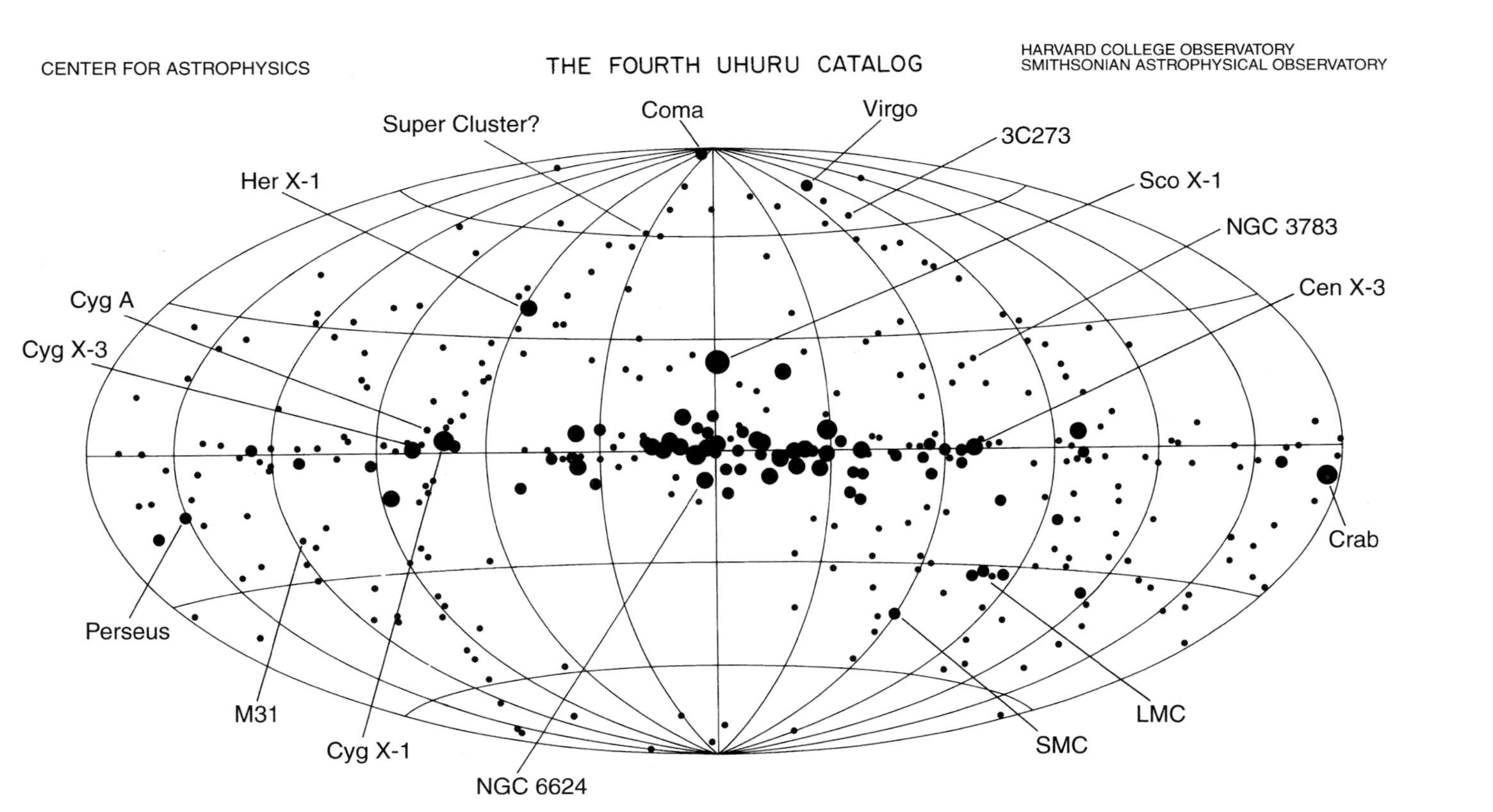

Figure 11.2 An X-ray map of the sky showing the sources detected by Uhuru. This map is in galactic coordinates, with the Milky Way crossing the plot at 0° latitude and with the Milky Way's centre at 0° lat. 0° long. (Courtesy Harvard-Smithsonian Center for Astrophysics.)

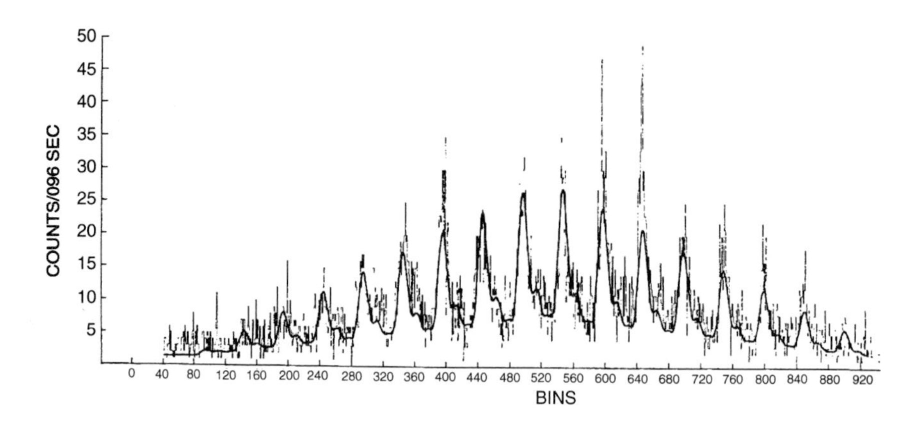

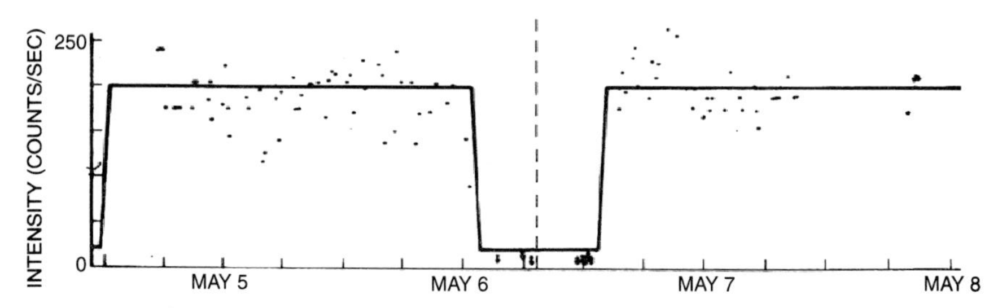

Figure 11.3 The pulsations in the X-ray output of Cen X-3, as measured by the Uhuru spacecraft. The top figure shows the 4.8 second period of the pulsar, whilst the bottom figure shows the 2.09 day period caused by the binary motion. The gradual increase to maximum intensity and its gradual decline in the top figure, superimposed on the 4.8 second period, is due to the effect of Uhuru scanning the source as the spacecraft slowly rotated. (Courtesy Harvard-Smithsonian Center for Astrophysics.)

neutron star explained its behaviour. X-rays from the neutron star Her X-1 are apparently heating up one side of HZ Herculis to 20,000 K, compared with the temperature of the un-radiated side of 7,000 K. So as the system revolves every 1.7 days we see alternately the hot and cool sides of HZ Herculis, hence the strange variability in its spectrum.

But that is not all, as Her X-l's X-ray emission could only be detected for about 12 days in every 35, in two periods of about 9 and 3 days in duration. During the 9 day period Uhuru detected one strong X-ray pulse every 1.24 seconds, but during the 3 day period it detected two X-ray pulses of approximately equal intensity about 0.62 seconds apart. The optical source continued to show its 1.7 day variability even when the X-ray source was not detectable, however. As Kenneth Brecher of MIT pointed out, the X-ray source still seemed to be heating up its optical companion, even when the

X-ray source was not detectable from Earth. Eventually it was concluded that the orbit of the binary is precessing with a period of 35 days, such that one of the two X-ray beams from the poles of the neutron star intercepts the Earth over a 9 day period, whilst both of the X-ray beams intercept the Earth over the 3 day period.

Uhuru also helped to explain the nature of the variable source Cygnus X-1, which had first been detected as an X-ray source by an NRL rocket experiment in 1964. Measurements in the following year showed that its intensity had decreased by 75% from its 1964 level. Further measurements over the next five years confirmed that Cyg X-1's X-ray intensity varied over timescales of months. Because of this it was one of the first objects to be measured by Uhuru, being observed on 21st and 27th December 1970 and again on 4th January 1971. The results mystified astronomers, as they seemed to show intermittent pulsations at variable frequencies, with the frequency of the pulses varying in no systematic way. In addition, the source flickered at the fastest rate measurable by Uhuru. It was obviously important to try to observe Cyg X-1 at optical wavelengths, to try to unravel what type of source it was, but Uhuru's position estimate was not sufficiently accurate to enable its optical counterpart to be identified unambiguously.

Shortly after the launch of Uhuru the MIT group observed Cyg X-1 using a rocket experiment that had a maximum time resolution of 1 millisecond, compared with the 96 millisecond resolution of Uhuru. The flickering of Cyg X-1 could just about be resolved by this rocket experiment, implying that the source could be no larger than 1 light millisecond or 300 km across. In addition the rocket experiment was able to pin-point the position of Cyg X-1 sufficiently accurately for Paul Murdin and Louis Webster of the Royal Greenwich Observatory to suggest that the blue supergiant HD 226868 was its optical counterpart. They then analysed the spectrum of this blue supergiant and found that it had a 5.6 day periodicity, thus suggesting that Cyg X-1 and HD 226868 were a binary pair. The mass of the blue supergiant was estimated as about 30 solar masses, and the dynamics of the binary showed that Cyg X-1 had a mass of at least 9 solar masses, which was too heavy for a neutron star. It must be a black hole! Or was it? HD 226868 was clearly a binary, but was the other star of the system really Cyg X-1?

Whilst the optical observations of HD 226868 had been under way, astronomers had had a stroke of luck that increased their confidence that the blue supergiant was the optical counterpart of Cyg X-1. Firstly, Braes and Miley found a radio source within 1 arcsec of HD 226868 using the Westerbork radio interferometer in the Netherlands. Then over the period from March to May 1971, as this radio source increased in intensity by at least a factor of 4, the X-ray intensity of Cyg X-1 in the energy range 2 to 6 keV was found by Uhuru to decrease by a factor of 3. Clearly Cyg X-1, HD 226868 and the radio source are all part of the same system. So Cyg X-1 really did appear to be a black hole. Then in 1976 S.S. Holt and colleagues found, using the Ariel 5 spacecraft, a 5.6 day modulation in the X-ray intensity of Cyg X-1, confirming that it was in a binary system with the blue supergiant HD 226868.

So in the course of its first year after launch, Uhuru had helped to show that Cen X-3 and Her X-1 were both rapidly rotating neutron stars, each in a binary system with a blue star, and Cyg X-1 was a black hole also in a binary system with a blue star. The theoreticians now had data from three binary systems to enable them to develop a theory of what caused the high-intensity X-ray emission in these systems.[8]

At first it looked as though astronomers were simply seeing, in the cases of Cen X-3 and Her X-1, a pulsar in a binary system with a blue star. But it soon became evident that the spin rate of these two neutron stars was not slowing down like the Crab pulsar, but speeding up. So rather than losing mass, these two neutron stars appeared to be gaining it. This led theoreticians to hypothesise in 1972 that mass is lost from the binary companion to the neutron star via a spinning accretion disc around the neutron star; the accretion disc being created as the infalling matter has angular momentum. The material or plasma in the accretion disc would be channelled by the neutron star's intense magnetic field onto its surface at its magnetic poles. This would create X-ray hot-spots there, powered by the gravitational potential energy of the infalling plasma. The situation with Cyg X-1 is similar, except that Cyg X-1 is a black hole, not a neutron star. An accretion disc would still form around the black hole, but X-rays would be produced as the inner edge of the accretion disc loses mass to the black hole. We would not see regularly pulsed X-rays, however, as, unlike the case of a neutron star, the black hole would not have a magnetic axis misaligned with the spin axis.

Uhuru measured X-ray emission from numerous other types of object (see Figure 11.2) such as globular clusters, the Large and Small Magellanic Clouds (LMC and SMC in Figure 11.2), some spiral galaxies (including M31, the great galaxy in Andromeda), the radio galaxy Centaurus A,[9] some

[8] By 1974 three other X-ray binaries had been found using Uhuru, all being neutron stars with O- or B-type supergiant companions.

[9] Centaurus A had first been identified as an X-ray source by Stuart Bowyer using a sounding rocket experiment in 1969.

Seyfert galaxies (including NGC 4151), and some clusters of galaxies (including the Virgo cluster). The SMC was found to be dominated in X-rays by the source SMC X-1 or 3U 0115-73, which was found to be the X-ray binary companion of the blue supergiant Sanduleak 160, but the four sources in the LMC could not be clearly correlated with any objects in other wavebands. Interestingly, the X-ray emission from a number of clusters of galaxies seemed to be greater than would be expected from the sum of its individual galaxies. The angular resolution of Uhuru was not sufficient to show the precise location of X-ray sources in galaxies or clusters of galaxies, however, and for that astronomers would need a more sophisticated spacecraft.

Uhuru, as the first dedicated, non-solar, X-ray observatory, heralded the end of the first era in X-ray astronomy, and the start of a new one. No longer did experimenters have to rely on just a few minutes of observational data from a sounding rocket, as they could observe objects with Uhuru over long durations, in the same way that they had used ground-based telescopes for centuries. This enabled them to measure the variations of X-ray intensities in timescales ranging from 0.1 seconds to many months and, in the process, begin to understand the causes of some of these variations.

In 1962, when the first non-solar X-ray source had been discovered, the universe had looked a relatively predictable place with only the occasional nova or supernova causing any significant changes. Ten years later, however, following the early detection and identification of X-ray sources, the universe was seen to have a number of highly variable, highly energetic sources, such as neutron stars and pulsars. Furthermore, at least one X-ray source appeared to be a black hole.

The investigation into the nature of Cyg X-1 requiring, as it did, data from spacecraft, sounding rockets, and ground-based optical and radio observatories, showed that astronomy had become an integrated subject by the early 1970s, where observers felt that they could use any waveband and any ground or space type of observatory to solve a problem. An analogous situation had occurred with radio astronomy in 1963, where radio and optical observations had both been required to help explain the nature of quasars. It is interesting, in fact, to compare the gradual integration of radio and X-ray astronomy into mainstream astronomy. Such a comparison is summarised in Table 11.1, where the development of X-ray astronomy is seen to closely mirror that of radio astronomy with a delay of about 10 to 15 years.

After the discovery of the first non-solar, X-ray source in 1962, there was a veritable avalanche of new results, leading to the number of X-ray papers

Table 11.1. *The gradual integration of radio and X-ray astronomy into mainstream astronomy*

	Radio Astronomy	X-Ray Astronomy
First tentative observations	1932	1948
Sun discovered as a source	1942	1949
First non-solar source discovered	1946 (Cyg A)	1962 (Sco X-1)
First optical identification of a source (other than the Sun)	1949 (NGC 5128 M87 & The Crab)	1963 (The Crab)
First major discovery of a new type of source by working with mainstream astronomy	1963 (Quasars)	1971 (black hole candidate)
Change from the discovery phase, where the emphasis is on finding new sources, to the understanding phase, where research is integrated with mainstream astronomy	Early 1960s	~1970 (when Uhuru launched)

recorded by the *Astronomischer Jahresbericht* and *Astronomy and Astrophysics Abstracts* increasing from one in 1962 to 311 ten years later. In parallel, the percentage of American astronomers working in the X-ray field increased from 0.8% in 1962 to 11.2% in 1972, placing it in third place below optical and radio astronomy in the list of key fields of astronomical research. X-ray astronomy was clearly coming of age.

Chapter 12 | A PERIOD OF RAPID GROWTH

12.1 Copernicus

The post-Uhuru period from about 1972 to 1979 saw a rapid growth in the launch and operation of astronomical spacecraft observatories, including the Copernicus, Einstein, and European Cos-B spacecraft. Thirteen such spacecraft were launched over this eight year period, all of which were successful, compared with just five over the previous six years, with two of these being complete failures (see Table 12.1). In the post-Uhuru period of the 1970s, virtually all wavebands were covered with significant instruments, except the Visible, which was to rely on the development of the Hubble Space Telescope in the 1980s (see Chapter 17).

The fourth and last satellite in the Orbiting Astronomical Observatory series, since called OAO-3[1] (see Figure 12.1), was launched on 21st August 1972 into a 740 km Earth orbit by an Atlas-Centaur launcher. More usually called Copernicus, after the famous Polish astronomer whose 500th anniversary took place the following year, this 2,220 kg spacecraft was the heaviest unmanned spacecraft yet orbited by NASA. It was to outlive its design lifetime of one year by seven years.

The main instrument on Copernicus, provided by Lyman Spitzer's group at Princeton University, was designed to measure absorption lines in the interstellar medium. It consisted of a 80 cm (32″) diameter, 450 kg ultraviolet telescope mounted in the 3 m long central tube of the spacecraft. Its f/20 Cassegrain optical system was used to feed a scanning spectrometer which covered the spectral range from 71 to 330 nm. The telescope had its own fine error sensor which ensured a star pointing accuracy of 0.1 arc seconds.

---

[1] It was called OAO-C before launch.

Table 12.1. Astronomical observatory spacecraft launched 1966–79[a]

Spacecraft[b]	Launched	Stopped using	Mass (kg)	Launcher	Wavebands observed
OAO-1	1966	Failure	1,770	Atlas-Agena	
OAO-2	1968	1973	2,000	Atlas-Centaur	UV
Explorer 38, RAE-1	1968		280	Thor-Delta	Radio
SAS-1, Uhuru, Explorer 42	1970	1973	140	Scout	X-ray
OAO-B	1970	Failure	2,000	Atlas-Centaur	
OAO-3, Copernicus	1972	1980	2,220	Atlas-Centaur	UV, X-ray
SAS-2, Explorer 48	1972	1973	190	Scout	γ-ray
TD-1A (ESA s/c)	1972	1974	470	Thor-Delta	UV, X-ray, γ-ray
Explorer 49, RAE-2	1973		330	Thor-Delta	Radio
ANS (NL/USA)	1974	1977	120	Scout	UV, X-ray
Ariel 5 (UK/USA)	1974	1980	140	Scout	X-ray
SAS-3, Explorer 53	1975	1979	200	Scout	X-ray
Cos-B (ESA)	1975	1982	280	Thor-Delta	γ-ray
HEAO-1	1977	1979	2,700	Atlas-Centaur	X-ray, γ-ray
HEAO-2, Einstein	1978	1981	2,900	Atlas-Centaur	X-ray
IUE (USA/UK/ESA)	1978	1996	670	Thor-Delta	UV
HEAO-3	1979	1981	2,700	Atlas-Centaur	γ-ray
Ariel 6 (UK)	1979	1982	150	Scout	X-ray

Notes:

[a] Excluding solar and Solar System spacecraft. All are NASA spacecraft unless otherwise stated.

[b] Where more than one name is given these are alternative names.

In addition to the main UV payload, Copernicus also included an X-ray experiment provided by University College, London, to observe already known X-ray sources at longer wavelengths than had previously been possible.[2] It consisted of three small X-ray telescopes operating over the wavelength range from 0.3 to 7.0 nm, and a collimated proportional counter operating between 0.1 and 0.3 nm (10 to 3 keV). Originally, the X-ray experiment was allocated 10% of the viewing time but, because of its success, this was later increased to 20%.

Prior to the launch of Copernicus, deuterium, or heavy hydrogen as it is sometimes called,[3] had only been observed on Earth, but the Big Bang theory

[2] Uhuru, for example, operated in the energy range from 2 to 20 keV, which is equivalent to wavelengths from 0.5 to 0.05 nm.

[3] The deuterium nucleus consists of a proton (i.e. a hydrogen nucleus) plus a neutron.

Figure 12.1 The Copernicus or OAO-3 spacecraft undergoing pre-launch tests at the Kennedy Space Center in 1972. Its UV telescope was the largest civilian telescope launched up to that time. The spacecraft had a pointing accuracy of 0.1 arcsec, with a stability of 0.02 arcsec. (Courtesy NASA.)

predicted that deuterium should have been formed about three minutes after the initial expansion of the universe. It was thought that the amount of deuterium would have changed little since then, so it should still be present all over the universe at the same level as three minutes after the Big Bang. What is more, the theory predicted exactly the amount of deuterium that was produced, and, unlike the amount of helium, this was strongly dependent on the density of the universe.

The Big Bang theory predicted that the universe will continue to expand for ever if the density of the universe is less than the so-called critical density. Such a universe is said to be open. If, on the other hand, the density of the universe is greater than this critical density, gravity will gradually slow down and then reverse the expansion, causing the universe to collapse in on itself. This is a closed universe. The borderline between these two possibilities is called a flat universe. The deuterium abundance for this flat universe had been calculated from the Big Bang theory to be about 1 part per million (by mass). For a higher-density, closed universe, however, the deuterium would almost all have been converted to helium, producing an abundance of less than one part per million, whereas if the universe is open there should be more deuterium.

Copernicus first detected deuterium in the interstellar medium by finding its spectral signature superimposed on the ultraviolet spectrum of the star β Centauri. The amount detected was about 15 parts per million (by mass). So, assuming that the deuterium abundance measured by Copernicus is that for the primeval universe, and that the universe is made of ordinary matter, the amount of deuterium observed is too large for a closed or flat universe, and so the universe must be open. But are these two assumptions reasonable?

Considering the first assumption: although deuterium is produced in stars, it is converted to helium almost immediately so ordinary stars will not add to the deuterium abundance in the universe. But deuterium could be produced in exotic reactions in supernovae, quasars or similar objects. So the concentration of deuterium in the primeval universe could possibly be lower than that measured by Copernicus. Similarly it is possible that not all of the matter in the universe is ordinary matter, but some may be in the form of invisible 'dark matter'. The amount of this dark matter may also be sufficient to close the universe, or at least stop it expanding for ever. So although the deuterium abundance measured by Copernicus is of undoubted importance, it does not give the definitive answer on whether the universe is open, flat or closed.

Copernicus also found that the abundance of some elements in the interstellar gas was lower than that observed in newly formed stars and in the Sun. For example, in some directions the ratio of calcium to hydrogen was a factor of 1,000 lower than in the Sun. The missing proportions of calcium and other elements were thought to be present in interstellar dust, which Copernicus could not detect. The spacecraft also measured thin, highly-ionised gas at a temperature of about 200,000 K in the regions between the previously observed interstellar hydrogen clouds. In the Orion region fast-moving clouds (with a velocity of 100 km/s) were detected having a temperature of about 10,000 K, which were thought to have been produced either by old supernovae, or by shock fronts created by high-velocity (up to 3,000 km/s) stellar winds from hot O- and B-type stars.

The University College X-ray experiment also provided valuable data on a number of known X-ray sources. The enigmatic source Cygnus X-3 was one example.

Cygnus X-3 had been discovered as an X-ray source from a rocket flight in 1966, and it was later found to be the source of very hard γ-rays. Cyg X-3 had also been detected at infrared wavelengths, but it could not be seen in the visible waveband, owing to absorption by interstellar dust. Then in 1971 the Uhuru spacecraft detected a variation in its X-ray intensity with a period of 4.8 hours, indicating that Cygnus X-3 is a binary star, although no such regular periodicity was found at radio wavelengths, where the source was highly variable. Then on 2nd September 1972 Philip Gregory at the Algonquin Radio Observatory in Ontario noticed that Cyg X-3 had brightened over 1,000 times, to become one of the brightest radio sources in the sky. Gregory immediately telephoned Hjellming and Balick at the American National Radio Astronomy Observatory at Green Bank, who confirmed the increase which had occurred in just a few days. Shortly after launch in 1972 Copernicus confirmed[4] the 4.8 hour period of Cyg X-3's X-ray emission, and in the following year simultaneous observations were made between Copernicus and those in the infrared made by Gerry Neugebauer and his colleagues of Caltech. Both the X-ray and infrared signals showed the same 4.8 hour period, proving that they were from the same binary source. Later the British Ariel 5 spacecraft detected an emission feature in the spectrum of Cyg X-3 at about 7 keV, which is due to Fe XXV (i.e. iron that has lost all but two of its electrons), indicating a source temperature of about 10 million K. Then the solar observatory OSO-8 showed in 1975 and 1976 that the emission feature varied slightly in energy with the 4.8 hour period. The exact nature of Cyg X-3 was still unclear, however, in the absence of a visible image, but it appeared to be a double star system consisting of a dwarf star and a neutron star.

---

[4] Uhuru's previous detection of a 4.8 hour period was not as clear as some astronomers would have liked, because the data was affected by the slow spin rate of the spacecraft. The Copernicus detection was clear and unambiguous, however, as the spacecraft operated in a pointing mode.

12.2 The European Dimension

Two weeks after the formation of NASA in October 1958, a staff report of the Congressional Select Committee on Astronautics and Space Exploration recommended that the United States should explore the possibility of international collaboration in space research. The report noted that the UK were the most advanced Western European country in this area at the time, with their development of the IRBM Blue Streak and the research rocket Black Knight. It so happened that at this time the UK were considering the possibility of building and launching their own scientific satellites, using a combination of these two rockets as the basis of a launch vehicle. In fact, at a meeting in late October 1958 a group of Fellows of the Royal Society backed the idea of an independent British Scientific satellite programme. The group was unanimous in their support for such a programme, except for the dissenting voice of the Astronomer Royal who wanted the money spent on more conventional astronomical facilities. But there was a problem, as not only would the programme be expensive, it would require the use of a military rocket as the launcher's first stage with consequent security problems. So collaboration with the United States or British Commonwealth countries was suggested as an alternative to a go-it-alone approach.

Early in 1958 it had been clear that space research, using sounding rockets and spacecraft, would continue after the International Geophysical Year finished at the end of that year. To handle the international aspects of this extended work, the International Council of Scientific Unions, the umbrella organisation of the IGY, agreed to the setting up of COSPAR, the Committee on Space Research, with members from many countries including the USA and USSR. The first meeting of COSPAR, which took place in London in November 1958 was very much a preliminary affair, but on 14th March 1959, at its second meeting at the Hague, the USA offered to launch scientific experiments or complete spacecraft free of charge for other countries, provided the experimental results were shared with the Americans. The USA proposed to use a Scout launch vehicle that they were developing for launching their own small scientific satellites. This offer came at an ideal time for the British, and led to an acceptance in principle by the British Prime Minister, Harold Macmillan, which he announced to the House of Commons just two months later. A cooperative programme was then agreed between the USA and UK in the following year, and the first spacecraft in the series, called Ariel 1,[5] was

[5] Ariel 1 was an American-built spacecraft with British-built experiments. The first all-British spacecraft was the 90 kg Ariel 3 launched in 1967.

successfully launched in 1962. This and the next three spacecraft were devoted to geophysical research, but the next spacecraft, Ariel 5, launched in 1974, was an X-ray spacecraft. This will be discussed later.

In the meantime, other European countries were developing their own scientific satellite programmes. For example, in 1962 Italy and the USA agreed on the so-called San Marco programme, which resulted in the launch of the first all-European spacecraft San Marco-1 two years later by an American Scout rocket. This spacecraft, which measured the density and temperature of the upper atmosphere, was launched from the USA, but the next three spacecraft in the series were launched by Scout rockets from the San Marco platform off the Kenyan coast. It was this platform that was to be the launch site of the American Uhuru X-ray spacecraft in 1970 (see Section 11.6).

The French adopted a different approach to the British and Italians, as the French developed their own launcher called Diamant, whilst accepting the American offer to launch a French spacecraft. Diamant launched a 42 kg test satellite called Astérix in 1965, making France the third 'space power',[6] and the Americans launched another French spacecraft, the 76 kg FR-1, ten days later. In parallel with these British, Italian and French initiatives, things had also been progressing at the European level.[7]

The 1950s had been a time when the countries of Western Europe had started to develop collaborative political arrangements. This resulted in six countries signing the Treaty of Rome in 1957, which set up the European Economic Community, or the Common Market as it was more popularly known. Five years earlier a different group of countries had set up CERN, the European high-energy physics laboratory. So it was natural, after the launch of the first Russian and American spacecraft, that European scientists and politicians should consider setting up a European space programme, in parallel to or instead of national programmes or bilateral collaborative programmes with the USA.

Numerous meetings took place over the next few years, culminating in 1962 in the conventions setting up the European Launcher Development Organisation (ELDO) and the European Space Research Organisation (ESRO) being opened for signature. ELDO's rôle was to develop a European satellite launcher (called Europa) with a British Blue Streak first stage, a French Coralis second stage, and a German Astris as the third stage, but after a number of launch failures ELDO was wound up in 1973. ESRO was quite a

---

[6] i.e. the third country to launch its own spacecraft.

[7] European in this chapter really means Western European.

different matter, however, managing a number of successful scientific programmes, before its functions were taken over by the newly founded European Space Agency (ESA) in 1975.

The initial ESRO programme envisaged in 1961 consisted of:

- A sounding rocket programme of about 20 launches per year, investigating the Earth's upper atmosphere and ionosphere.
- Small spacecraft to study the Earth's environment and Sun–Earth interactions. Two or three such spacecraft would be launched per year, starting four years after the start of ESRO.
- Large spacecraft and space probes, with launches beginning six years after the start of ESRO at the rate of one or two per year.

As time progressed it became clear that such a programme was over-ambitious, although the three basic elements were retained.

The first ESRO spacecraft were envisaged as being small, spin-stabilised spacecraft, similar in mass to the UK Ariel series. Because of this it was suggested in 1962 that the USA may be willing to launch the first two ESRO spacecraft free of charge, using their Scout launcher, under the same conditions as for the bilateral programmes with the UK, Italy and France mentioned above. This the Americans agreed to do, and in 1968 ESRO-I and ESRO-II[8] were successfully launched from the Vandenberg Air Force Base in California, followed by ESRO-IB[9] the following year.

12.3 TD-1

The first European spacecraft dedicated to extra-Solar System research was Thor-Delta-1 or TD-1, so-called as it was the first European spacecraft to be launched by the American Thor-Delta launch vehicle, which could launch larger payloads than the Scout. Initially, it was hoped by ESRO to launch a series of these three-axis stabilised TD spacecraft at the rate of one or two per year, but by 1965 this had been reduced to six TD spacecraft over an eight year programme. The number of spacecraft was further whittled down in

[8] The first ESRO-II spacecraft was launched in 1967, but it failed to reach orbit owing to a fault in the third stage of the Scout rocket. The ESRO-II spacecraft that was successfully launched in 1968, which was also launched before ESRO-I, was the back-up spacecraft.

[9] ESRO paid for the launch of this extra spacecraft, which was the flight spare of the first ESRO-I.

1967 to a total of two, namely TD-1 concentrating on stellar research and TD-2 devoted to the Sun and Sun–Earth interactions. By early 1968 even these two spacecraft could no longer be afforded in the total ESRO budget.[10] A trade-off was then undertaken between these two spacecraft, with a view to choosing one to continue to completion. Fortunately for the stellar community, TD-1 only needed to be stabilised when in sunlight, whereas some of the experiments on TD-2 required stabilising during solar eclipses, adding extra complications to the design. So TD-1 was chosen and TD-2 cancelled. As a compromise, however, it was agreed that some of the TD-2 experiments would be flown on a new satellite, called ESRO-IV, which was launched in 1972, just eight months after TD-1.

The 470 kg TD-1 or TD-1A (see Figure 12.2), it was called both, was launched from Vandenberg on 12th March 1972 into a near-polar, Sun-synchronous orbit at an altitude of about 540 km. It scanned a great circle of the sky once per orbit, and produced a complete survey of the sky once every six months. Its primary scientific mission was to undertake a sky survey in the ultraviolet, X-ray and γ-ray wavebands.

TD-1's main payload consisted of two ultraviolet telescopes. One was a 28 cm diameter telescope, developed jointly by the UK and Belgium, with a spectrometer designed to cover the waveband from 135 to 255 nm at a resolution of 3.5 nm, together with a photometer with a response peak at 274 nm. In orbit it measured the ultraviolet fluxes of about 31,000 stars down to about ninth visual magnitude, and enabled the absorption by interstellar dust to be estimated. These stars, which generally have surface temperatures in the range 15,000 K to 50,000 K, were found to be clearly concentrated in the direction of Cygnus and Orion, marking the position of the local spiral arm of the Milky Way. The other ultraviolet instrument was a stellar spectrometer, provided by the Utrecht Astronomical Institute in the Netherlands, with a 26 cm diameter primary mirror, operating in the range from 210 to 280 nm with a resolution of 0.18 nm. It had a 50 arcmin field of view at right angles to the scan direction, and as TD-1's orbit precessed about 4 arcmin per orbit, this meant that each of the 200 bright stars that it analysed could be observed on a minimum of 12 consecutive orbits.[11]

TD-1 also carried a proportional counter designed to measure the spectra of X-ray sources in the range 3 to 30 keV, but the experiment caused

---

[10] This was because the cost estimate for building the TD-1/TD-2 pair had doubled to about 220 million French Francs (MFF).

[11] Near the pole of the orbit this scan number would clearly increase, such that a star at that pole would be viewed on every orbit.

Figure 12.2 The European TD-1 spacecraft, launched in 1972, which was designed to carry out research in the ultraviolet, X-ray and γ-ray wavebands. (Courtesy ESA.)

electrical interference with the spacecraft's telemetry system and had to be switched off. There was also a spark chamber to detect γ-rays above 50 MeV, but it suffered from interference caused by the spacecraft's particle environment and, although it did detect γ-rays, no γ-ray source could be identified.

It had been planned to store data on two on-board tape recorders, which would be read out when the spacecraft was within sight of one of the ESA ESTRACK ground stations But one of the tape recorders failed in April 1972, just six weeks after launch, and the other failed a month later on 23rd May. This limited TD-1 to downloading data to Earth in real time, which resulted, at first, in only 15% of the data being received. Initially, ESA requested help with extra receiving stations from NASA, CNES and some other organisations, which increased the coverage to about 60%. But eventually, with the aid of further receiving stations, including mobile ground stations in Hawaii, Tahiti, Easter Island, and Marambio (in Antarctica), the coverage was steadily increased to 95% by February 1973.

Just over seven months after launch, at the end of October 1972, TD-1 finished its nominal mission as it passed into its four months eclipse season. But the problem with the tape recorders had severely compromised the amount of experiment data received on Earth up to that time. So in February 1973 the spacecraft was reactivated and continued to provide data for a further six months or so, until it again entered its eclipse season near the end of October 1973.

12.4 Dwarf Novae

In 1973 X-rays were detected from a new type of X-ray source, which had first been recognised almost a hundred years before in visible light. The star U Geminorum had been observed in the second half of the nineteenth century to spend most of its time at minimum brightness, but then to increase and decrease in intensity at irregular intervals as if it was showing large-scale eruptions. Starting in 1896 a gradually increasing number of such stars were found that increased in intensity by 2 to 6 magnitudes for a few days before returning to normal. One of the earliest of these was SS Cygni, which was found to increase in intensity by about 4 magnitudes every one or two months. This type of star became known as a dwarf nova, which was found to be a binary system consisting of a main sequence star and a white dwarf, with mass being transferred to the white dwarf from its companion via an accretion disc. On 30th March 1973 Saul Rappaport and colleagues of

MIT discovered soft X-rays from SS Cygni, using a sounding rocket experiment, the first of a number of such detections from dwarf novae.

12.5 ANS

On 30th August 1974 an American Scout rocket launched the 120 kg Dutch ANS[12] spacecraft from the Vandenberg Air Force Base into a 280 × 1,150 km polar orbit, instead of the planned 500 km circular orbit, because of a fault with the launch vehicle's first stage. The three-axis stabilised spacecraft was designed to observe the universe in both the ultraviolet and X-ray wavebands, but both were compromised by the incorrect orbit. Nevertheless, it made important X-ray observations of flare stars and X-ray bursters, as outlined below.

Almost thirty years earlier, Carpenter had found that the star Luyten L726-8 suddenly increased in brightness by almost three magnitudes over a period of just three minutes, and then faded gradually to its previous brightness. Luyten and Hodge concluded, as a result, that a violent flare had occurred on the star, similar in type but much greater in magnitude than those seen on the Sun. It was eventually found that Luyten L726-8 is a binary, one element of which, now called UV Ceti, produced the flare. Over the years before the launch of ANS a number of these so-called flare stars[13] had been found, all of which are red dwarfs with surface temperatures of only about 3,000 K and with masses of about 10% that of the Sun.

It may be thought that it would be pointless trying to observe X-ray emission from such cool red dwarfs, but solar flares were known to produce X-rays, and the flares on these flare stars are much more violent than those on the Sun. So a survey was made of flare stars using the ANS and other spacecraft to see if they emitted X-rays. The ANS search was successful when John Heise and colleagues discovered X-ray emissions from the flare stars UV Ceti and YZ Canis Majoris.

Another new type of X-ray source was found in 1975 by John Grindlay of the Harvard Smithsonian Center for Astrophysics, John Heise of Utrecht and colleagues using ANS, when they discovered that the X-ray source 3U

---

[12] ANS stands for Astronomical Netherlands Satellite.

[13] Novae, recurrent novae, dwarf novae and flare stars are all examples of what are now called 'cataclysmic variables'. These different types can be differentiated by whether there is one or more eruptions, the rapidity and magnitude of the intensity increase, and how long the eruptions last.

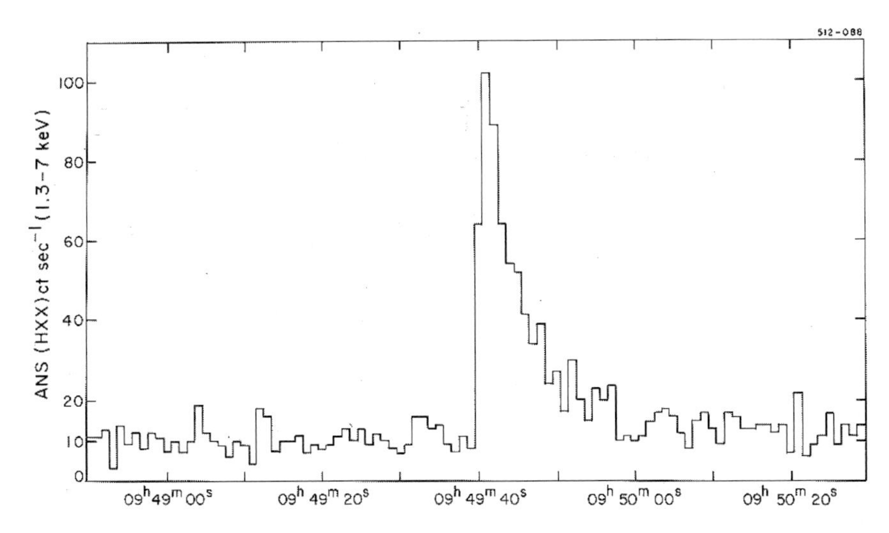

Figure 12.3 One of the two X-ray bursts from the source 3U 1820-30 is clearly seen in this plot taken by the ANS spacecraft on 28th September 1975. This was the first X-ray burster discovery to be announced. (J. Grindlay *et al.*, ApJ 205 (1976), p. L128. Reprinted with permission from *The Astrophysical Journal.*)

1820–30 emitted two very brief bursts of X-rays in the 1.3 to 7 keV range (see Figure 12.3). The source appeared to be in the globular cluster NGC 6624. This was the first X-ray burster to be found.[14] They typically emit 10^{39} ergs of X-ray energy in about 10 seconds, which is the amount of X-ray energy emitted by the Sun's corona in about 3,000 years. As soon as the discovery of this X-ray burster was announced, astronomers quickly tried to find other examples using the Ariel 5, SAS-3 and OSO-8 spacecraft, as well as ANS. Ariel 5 in particular found a number of these X-ray bursters, which showed a marked concentration towards low galactic latitudes, indicating that the sources were in, or associated with the Milky Way. Then in 1978 John Grindlay, Jeff McClintock, Walter Lewin and colleagues reported that they had detected simultaneous X-ray and optical bursts from the X-ray burster called MXB 1735-44, using the American SAS-3 spacecraft and the 1.5 m

[14] X-ray bursters had actually been found earlier by Belian and colleagues at Los Alamos using the military Vela spacecraft, but their discovery was not announced until later for security reasons. (The Americans did not wish the USSR to know about the capabilities of these sophisticated spacecraft, which were designed to look for evidence of clandestine Russian nuclear explosions after the Partial Nuclear Test Ban Treaty of 1963.)

telescope at the Cerro Tololo Inter-American Observatory. Further analysis showed that the optical burst had occurred about three seconds after the X-ray burst, possibly due to the time taken for X-rays to travel from their source to the absorbing material that produced the optical flash.

A few months after the discovery of 3U 1820-30, Water Lewin and his co-workers at MIT, using the SAS-3 spacecraft, discovered an even stranger object, the so-called 'Rapid Burster' (or MXB 1730-335). It had rapid X-ray bursts ranging from about once every 8 seconds to once every 5 minutes, with the energy in the strongest bursts being as much as 1000 times that in the weakest. The stronger pulses were generally followed by a longer gap than the weaker ones, in a similar way to that observed previously for recurrent novae and dwarf novae in visible light. The pulse effect in dwarf novae was known to be due to mass transfer from a main sequence star to a white dwarf, via an accretion disc, as described above. So the behaviour of the Rapid Burster was thought to be due to a similar process, except that the compact source was thought to be a neutron star, rather than a white dwarf.

12.6 **Ariel 5**

The first spacecraft in the UK series, Ariel 1, had been launched in 1962, the year that the first non-solar source of X-rays had been discovered by an American sounding rocket. Six years later British scientists felt sufficiently confident to propose to NASA that the UK should devote the fifth UK spacecraft, Ariel 5, to the study of X-ray sources. This was two years before the launch of the first such American spacecraft, Uhuru.

Ariel 5 was, in fact, the first UK spacecraft to be devoted to non-geophysical research. It was, in addition, the first to be controlled in orbit by UK engineers, using a control centre at the Appleton Laboratory of the Science Research Council linked to NASA receiving stations at Ascension Island and Quito, Ecuador.

The choice of launch site for Ariel 5 needed considerable thought. The trajectory of a spacecraft launch vehicle at lift-off is determined by range safety considerations, which in America has meant that the launch sites are situated on the coast, so that spent and faulty rockets can fall into the sea. Launch trajectories for all but polar orbiting spacecraft are usually in an easterly direction, to take advantage of the Earth's spin. In addition the spacecraft's orbital plane is usually inclined to the Earth's equator at an angle equal to the latitude of the launch site, because to change the orbital plane requires a significant amount of launch vehicle fuel, thus reducing the available spacecraft mass. So in the 1960s the usual launch sites for Scout rockets were at the Vandenberg Air Force Base on the West coast, which allowed the launch of polar-orbiting satellites, and at Wallops Island or Cape Canaveral on the East coast, which were used for 35° or 28° inclined orbits, respectively. The USA had (and still has) a problem in launching equatorial spacecraft as they have no launch site near the equator. This problem was solved for Uhuru, however, by launching it from the Italian San Marco platform off the coast of Kenya, and so the same solution was adopted for Ariel 5, as its experimenters also preferred an equatorial orbit.

Ariel 5, which was launched on 15th October 1974, contained six experiments, five of which were British, with the sixth being provided by NASA's Goddard Space Flight Center. The experiments were to measure the position, spectrum and polarisation of X-ray sources, generally in the energy range from about 1 to 30 keV, and sources of high energy X-rays up to about 1.2 MeV. The experiments continued to work well until the spacecraft re-entered the atmosphere on 14th March 1980, although it ran out of propane gas used for attitude control about 2½ years after launch.

One of the key areas investigated by Ariel 5 was that of the so-called X-ray transients that flare up suddenly and then disappear after a few weeks, the first of which had been discovered almost ten years previously in 1967.

On 4th April 1967, J. Harries and colleagues had discovered an X-ray source Cen X-2, where none had existed before, with a sounding rocket experiment launched from Australia. Six days later, Cen X-2 was independently observed by Brin Cooke and colleagues from Leicester University near to its maximum X-ray intensity, during an X-ray survey of the southern sky. Although the source was still observable on 18th May, it fell below threshold visibility in September and was never seen again. This first X-ray transient was followed by the discovery of Cen X-3, which also appeared to be a transient, as its X-ray intensity fell to below detectable levels when the Leicester group tried to observe it in the late 1960s. As related in Section 11.6, however, Cen X-3 later reappeared and was found to be an eclipsing binary, thus explaining why it was sometimes not detectable.

The second confirmed X-ray transient to be found was, in fact, Cen X-4, which reached peak intensity on 9th July 1969. The light curve was then measured by Doyle Evans and colleagues using the Vela 5A and 5B military spacecraft over a period of almost three months. A second outburst was observed almost ten years later on 12th May 1979 by Lou Kaluzienski and colleagues of the Goddard Space Flight Center using their all-sky monitor experiment on Ariel 5. This enabled its visual counterpart to be identified as a 13th magnitude red dwarf which had increased in intensity by at least six visible magnitudes. Further analysis showed that the source was a binary, with a period of just over 15 hours, but the mechanism that had led to the two sudden outbursts about ten years apart remained unclear.

Another transient was discovered by Martin Elvis and colleagues at Leicester University using the Ariel 5 spacecraft. They found that a strong source of X-rays in the constellation of Monoceros, first observed on 3rd August 1975, rapidly increased its X-ray intensity, so that five days later it was stronger than Sco X-1, and was thus the strongest X-ray source in the sky. It reached maximum X-ray intensity on 15th August, at four times that of Sco X-1, and over 500 times its intensity twelve days earlier, before starting a gradual decline. This X-ray transient, called A0620-00, was found during its outburst by F. Boley and R. Wolfson in the USA to coincide with Nova Mon 1975, which was then a blue 11th magnitude star, whereas before outburst it had been an 18th magnitude reddish star. A further search of old photographic plates showed that the visible source, now called V616 Mon, had also exploded in 1917, when it had reached 12th magnitude.

Eight months after its 1975 outburst, V616 Mon was found to have become an 18th magnitude orange dwarf star of spectral type K5. Then in 1983 Jeff McClintock of the Harvard Smithsonian Center for Astrophysics announced that the orange dwarf's intensity was varying regularly with a period of 7.75 hours. McClintock and Ronald Remillard of MIT later measured the Doppler shift in its spectral lines, and in 1986 concluded that the mass of its compact, X-ray emitting companion must be at least 3.2 solar masses, making it a black hole candidate. Interestingly, the Ariel 5 spectrometer had found that the intensity of the 30 keV X-rays was still increasing when the low energy flux had decreased to 20% of its peak value. A similar effect had previously been observed for the other black hole candidate Cyg X-1 during its May 1975 outburst.

By the end of the 1960s a number of Supernova Remnants (SNRs) had been found to emit X-rays, and the Crab pulsar in the Crab SNR had been found to pulse in X-rays (see Section 11.5). In addition, the Crab pulsar had been found to be the 'engine' that supplies the energy to keep the Crab nebula shining in X-rays,[15] but neither the Tycho nor Cas A SNRs, both of which emit X-rays, seemed to have a pulsar embedded in them. So what was the source of their X-rays? Once more Ariel 5 came to the rescue.

---

[15] The pulsar provides the energy by injecting high energy electrons into the nebula. There, these electrons spiral in the strong magnetic field and emit synchrotron radiation in all wavelengths from γ-rays to radio waves.

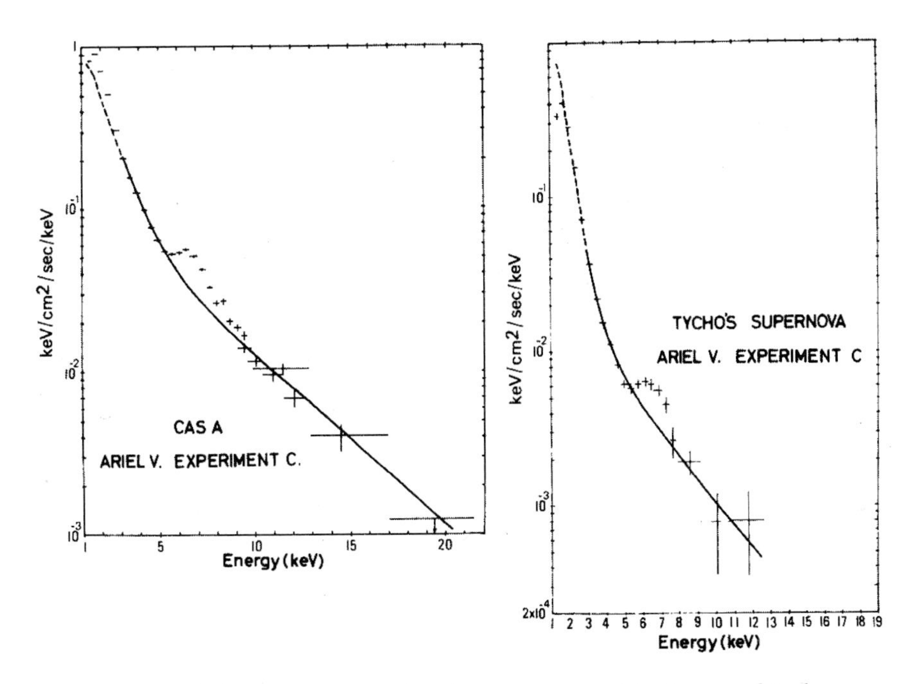

Figure 12.4 Ariel 5's measurements of the X-ray spectrum of the Cas A and Tycho supernova remnants. The 6.7 keV 'line' is shown by the energy peak above the gradual curve of energy decrease. (Courtesy J.L. Culhane.)

The Tycho and Cas A SNRs were thought to have ages of about 400 and 300 years respectively, compared with the 900 years of the Crab,[16] yet the absolute X-ray intensity of these younger sources was over an order of magnitude less than that of the Crab. This, together with their shell-like structure seen at radio wavelengths, and their lack of a central pulsar, suggested that the source of the X-rays in the Tycho and Cas A SNRs was shock-heated interstellar gas, rather than synchrotron emission. P. Davison, J.L. Culhane and colleagues confirmed this theory using the Ariel 5 X-ray spectrometer, which showed the existence of a line at an energy of 6.7 keV in the spectrum of both sources (see Figure 12.4). This is the resonance line of Fe XXV, which clearly indicates the existence of gas at very high temperatures.

[16] The Crab SNR is almost certainly the remains of a supernova recorded by Chinese and Japanese astronomers in the year 1054. The Tycho SNR is the remains of the supernova observed by Tycho Brahe in 1572, and Cas A is probably the remains of a star recorded by John Flamsteed in 1680 that can no longer be observed.

In addition to its observation of such sources as X-ray transients and supernova remnants in the Milky Way, Ariel 5 also made extensive observations of extragalactic sources, over half of which were found to be associated with either active galaxies or clusters of galaxies. Most of the active galaxies observed were Seyfert galaxies, some of which exhibited X-ray variations over the period of a day or so, indicting that the X-ray source was about 1 light day or so in diameter.[17] This is the same order of magnitude as the diameter of the Solar System, which is incredibly small when compared with the diameter of a galaxy.

Ariel 5 also measured the X-ray spectrum of the Perseus, Centaurus, Virgo and Coma clusters of galaxies. A 7 keV feature was found in the Perseus spectrum, together with a similar but weaker feature in that of the Centaurus cluster, due to the resonance lines of Fe XXV and XXVI, indicating a plasma temperature in these clusters of the order of 50 million K.

12.7 AM Herculis – The First Polar

A source recorded as 3U 1809+50 in the third Uhuru catalogue was largely ignored until Richard Berg and Joseph Duthie of the University of Rochester suggested that its optical counterpart may be the variable star AM Herculis. Although its position was just outside the error box for the Uhuru source, Berg and Duthie's photoelectric observations of AM Her had shown that it flickered continuously, in a similar way to other known X-ray sources. But in order for the optical and X-ray sources to be positively identified as one and the same, a better position fix was required for 3U 1809+50, and the variations of the two sources had to be shown to be consistent.

In 1976 Hearn, Richardson and Clark of MIT used the SAS-3 spacecraft to obtain a better position fix for 3U 1809+50 which showed, to the accuracy of measurement, that this X-ray source, which was the brightest soft X-ray point source in the sky, and AM Her were the same. Furthermore, SAS-3 showed that 3U 1809+50 exhibited similar erratic variations in its soft X-ray intensity to those of AM Her in visible light. In the same year Santiago Tapin of the University of Arizona found that the visible light of AM Her was linearly polarised, indicating the presence of a strong magnetic field, with its degree of linear polarisation varying with a period of 3.1 hours. Hearn and colleagues then re-examined their SAS-3 data and found that the

[17] Variations in Seyfert galaxies over this sort of timescale had first been observed in the 1960s at radio wavelengths.

soft X-ray source was totally eclipsed every 3.1 hours, proving that 3U 1809 +50 and AM Her were part of the same binary system. But there was more to it than that, as Tapin then found that AM Her appeared to show a 0.5 magnitude eclipse every 3.1 hours in yellow and red light, but no eclipse at all in the ultraviolet! It appeared as though the X-ray source was modifying the normal star in a similar way to the X-ray source Her X-1 which heated up its main sequence companion HZ Herculis (see Section 11.6).

Tapin originally detected linearly polarised light in the AM Her system, but he subsequently also found variable circularly polarised light as well. Circularly polarised light was known to be emitted only by white dwarf stars at that time, thus indicating that the compact, X-ray emitting star in the AM binary was most likely a white dwarf. In August 1976 William Priedhorsky of Caltech measured the optical spectrum of AM Her and found radial velocities varying from 200 km/s in recession to 400 km/s in approach, and back again, over the 3.1 hour period of the binary. But the maximum approach velocity occurred when the soft X-ray source was in eclipse. This could not be so if the velocities measured were those of the visible star, as the maximum recessional and approach velocities should be when the stars are spaced at their maximum extent across our line of sight, not when the X-ray source is being eclipsed by the visible star. So the velocity measured cannot be that of the visible star, but must be that of the gas falling onto the white dwarf. There is no accretion disc in this case, but the gas falls directly on to the white dwarf's surface at its magnetic poles, heating it up and causing it to emit soft X-rays. When the gas is coming towards us, the white dwarf's soft X-ray emitting pole is facing its companion and is no longer visible to us, thus explaining why the maximum speed of approach of the gas correlates with a minimum in the soft X-ray intensity. AM Her was the first so-called 'polar' source to be identified.

It may be wondered why the white dwarf's spin rate does not increase as it gains angular momentum from the material falling on to its surface. Apparently, the white dwarf's magnetic field is so strong that it acts as a brake, and keeps the spin of the white dwarf synchronised to the orbital period of the binary.

12.8 Gamma-rays

One of the problems with γ-ray astronomy is that because γ-rays are extremely energetic they are extremely scarce, so γ-ray photons are counted individually like high energy charged particles. It is also very difficult to collimate γ-rays and determine their direction of entry into a γ-ray telescope, and, as a result, the angular resolution of the early γ-ray telescopes was very poor. This enabled astronomers to identify only those sources that showed up predominantly in other wavebands.

Although Explorer 11 had detected a few γ-rays in 1961, the detection was too inaccurate and their numbers were too few to establish where they came from. The first significant γ-ray detections were made, in fact, by OSO-3, which was launched in 1967 and which enabled George Clark, Gordon Garmire and William Kraushaar of MIT to deduce that the Milky Way is a strong source of γ-rays. Throughout its lifetime, OSO-3 only detected a total of just over 600 γ-ray photons, however, which made it impossible to distinguish any point sources, but SAS-2, launched in 1972, fared somewhat better collecting 8,000 photons over its seven month lifetime. Like SAS-1 (or Uhuru), SAS-2 was launched from the San Marco platform and undertook a sky survey, although this time it was in γ-rays with energies in excess of about 25 MeV. Unfortunately, a fault cut short the mission when only about half of the sky had been surveyed. Nevertheless, even with a resolution of only about 1½° it was able to show that the galactic centre, the Vela and the Crab pulsars[18] are all strong γ-ray sources, together with an object called Geminga[19] in the constellation of Gemini which correlated with no known object. The X-ray source Cygnus X-3 was also detected.

The Vela military spacecraft have already been mentioned a number of times in this book, latterly as the first spacecraft to detect X-ray bursts (see footnote 14 above), but the Vela spacecraft are most well-known for their discovery of γ-ray bursters. Four Vela spacecraft detected about twenty short, intense γ-ray bursts over a three year period starting in July 1969, but it was not until 1973 that the data was declassified and the discovery announced by Ray Klebesadel, Ian Strong and Roy Olsen of Los Alamos. Typically a γ-ray burst lasts only for about one second,[20] and with the poor angular resolution of the detection system, it was impossible to determine the location of the sources to any degree of accuracy. Once the existence of γ-ray bursters had been announced, however, other spacecraft were pressed into service to try to detect and locate them. If more than two

[18] The Crab pulsar had first been detected as a γ-ray pulsar in 1971 by a high-flying balloon experiment.

[19] Geminga means 'it is not there' in the Milanese dialect of its discoverers. The name is also taken from *Gemini gamma* ray source.

[20] The range was from 0.1 to 30 seconds.

spacecraft could detect the same event, then its source location could be determined by simple triangulation, but even this failed to identify γ-ray bursters at other wavelengths.

The next civilian γ-ray observatory of note, after SAS-2, was the European Cos-B spacecraft, which had started off life in 1969 as one of eight competing spacecraft to be undertaken by ESRO in a second-generation scientific spacecraft programme. These competing eight projects were rapidly whittled down to three, and finally Cos-B and GEOS (a magnetospheric spacecraft) were chosen, and the third spacecraft called UVAS, which was to have measured the high-resolution ultraviolet spectra of stars, was rejected (see Section 14.1).

Because of the size and mass of γ-ray telescopes, γ-ray spacecraft tend to carry just this one instrument. This was essentially true of Cos-B, although it did have a small X-ray telescope to allow simultaneous monitoring of pulsars in both X-rays and γ-rays. Cos-B, which was launched in August 1975 by an American Thor-Delta, weighed about 280 kg, 120 kg of which was taken up by the γ-ray experiment. The spacecraft continued to operate until 25th April 1982, well beyond its two year design life, and finally ceased operation only when its γ-ray detector ran out of neon gas, by which time it had detected about 100,000 γ-ray photons. Although this sounds a great deal, this is still only two photons per hour, which was about the same rate as that detected by SAS-2. Cos-B's advantage over SAS-2 was its long lifetime, however, that allowed repeated observation of key areas of sky, so producing more γ-ray photons per square degree. As a result, Cos-B was able to locate the position of Geminga, for example, to within about ½°, although it still could not be clearly identified as any previously known source.

Cos-B, which covered the energy range from 30 to 5,000 MeV, mapped γ-ray emissions from the galactic plane and located about 25 point sources, most of which were unidentified at other wavelengths. But the quasar 3C273 was one clearly identifiable source, along with the Vela and Crab pulsars, the Cygnus spiral arm, the giant molecular cloud in Orion, and the ρ Ophiuchi dark cloud. The Milky Way and the Orion and ρ Ophiuchi dark clouds were not thought to be primary sources of γ-rays, however, as they are far too cold. It was thought, instead, that the γ-rays are produced when cosmic rays collide with their interstellar gas molecules. This was consistent with the observed correlation between the density of this interstellar gas and the intensity of the γ-rays. Alternatively, the γ-rays may come from numerous individual sources that could not be resolved by the Cos-B telescope.

Before the launch of SAS-2 and Cos-B, it was thought by many astronomers that they would find no point sources[21] of γ-rays, because of the tremendous amount of energy required to produce them, so the observation of some intense point sources was something of a surprise. But maybe just as surprising was the fact that the two most intense γ-ray sources, namely the Vela pulsar and Geminga, were relatively weak emitters in other wavebands.

12.9 Summary

So the period of the mid-1970s had seen the beginnings of ultraviolet and γ-ray astronomy and dramatic advances in the understanding of the nature of X-ray sources. European spacecraft had, for the first time, begun to make significant contributions to astronomical research.

Interstellar gas had been found at temperatures of up to about 200,000 K, and hot O- and B-type stars had been found to emit stellar winds of up to 3,000 km/s. X-ray and γ-ray bursters had been discovered that emit bursts of energy only a few seconds long, and flare stars and dwarf novae had both been found to emit X-rays. X-ray transients had been discovered that showed large intensity increases over a period of days, and then declined again over a period of months, in a similar way to novae in visible light. A number of these X-ray transients had been found to be binary stellar systems, one of which (A0620-00/V616 Mon) was thought to harbour a black hole.

The Crab pulsar had been shown in the 1960s to be the engine that keeps the Crab nebula shining in X-rays, and the Vela pulsar had been shown in the mid 1970s to be remarkably bright in γ-rays. The Tycho and Cas A SNRs are clearly different, however, having no energetic pulsar at their core, and these had been shown in the mid 1970s to be emitting X-rays because of the continued shock-heating of interstellar gas by the SNRs' expanding gaseous shells.

Some intense X-ray emitters had been found to be binaries in the 1960s, with energy being transferred from a normal stellar companion to a compact source, via an accretion disc around the latter. Then in the mid 1970s AM Herculis, the first polar, was found. Here, energy is transferred directly to the surface of a highly magnetised white dwarf at its magnetic poles, without having to go via an intermediate accretion disc. The magnetic field

[21] 'Point sources' is a relative term for an observatory with only ½° resolution, of course. They could include galaxies.

of the white dwarf is so large, however, that its spin rate is synchronised to its orbital rotation.

Finally, early results from γ-ray spacecraft, particularly Cos-B, indicated that γ-ray astronomy may well mimic X-ray astronomy in finding new types of sources, but significant improvements were clearly required in the angular resolution of γ-ray telescopes before more significant progress could be made.

Chapter 13 | THE HIGH ENERGY ASTRONOMY OBSERVATORY PROGRAMME

13.1 Programmatics

It may be remembered that in 1963 Riccardo Giacconi had proposed an extensive X-ray astronomy[1] research programme to NASA that included an X-ray telescope with focusing, Wolter-type optics (see Section 1.3). He suggested that this should follow a small, dedicated X-ray spacecraft that had conventional, non-focusing sensors. NASA eventually agreed to the latter in 1965; this was to become the Uhuru spacecraft launched in 1970 (see Chapter 11). Never satisfied, however, Giacconi continued to press for a spacecraft to follow Uhuru with a focusing X-ray astronomical telescope but, unfortunately, he got little support from either NASA or the astronomical community as a whole. On the other hand, NASA were prepared to fund him to build such telescopes to observe the Sun, one of which was to fly eventually on Skylab in 1973, so he could develop the design and manufacturing techniques on these solar programmes. In the meantime Giacconi continued to push the case for such focusing X-ray optics on a non-solar programme.

In 1967 NASA created the Astronomy Missions Board of distinguished scientists to advise it on possible future astronomy missions. The board consulted widely and eventually recommended a greatly expanded programme of high energy astronomy, with top priority being given to a series of large High Energy Astronomy Observatories (HEAOs), one of which would be devoted to X-ray research. A feasibility study was then undertaken into the design of these HEAO missions, which resulted in an enormous 11 ton spacecraft being proposed. The usual lobbying process then took place to get support for the

[1] 'Astronomy' in this chapter means the cosmos excluding the Solar System.

programme from NASA, the OMB and Congress, and in May 1970 competitive preliminary spacecraft design studies were let to Grumman and TRW.

In parallel with the industrial studies, NASA issued an Announcement of Opportunity calling for experiment proposals, and after examining the responses NASA decided on a programme using four HEAO spacecraft, two of which were to be slowly rotating survey spacecraft (like a larger version of Uhuru), and two were to be pointable. The first two were to be designed to discover faint new X-ray sources, to measure their positions accurately, and to measure their spectral properties. These two spacecraft were also expected to carry γ-ray and cosmic ray instruments. The second two HEAOs, which were pointable, were to be designed to enable short-period variations to be measured in the X-ray intensities of known sources. These second two spacecraft would also have focusing X-ray optics to provide X-ray imaging, and to enable position measurements to be made to a very high accuracy.

In 1972 a committee of eminent astronomers published their analysis, for the Academy of Sciences, of the ground- and space-based facilities needed by astronomers for the next ten years. Chaired by Jesse Greenstein of Caltech, it strongly supported the four spacecraft HEAO programme, but hardly was the ink dry on their report than NASA was obliged to cancel the fourth spacecraft to save money. Then in January 1973 NASA announced a reduction of $200m in the budget for the HEAO programme, along with measures on other programmes designed to meet an anticipated major reduction in the NASA 1974 budget. All significant work on the HEAO programme was suspended whilst ways were examined of substantially reducing costs. Most astronomers were concerned that the programme was about to be cancelled.

The NASA announcement was swiftly condemned by the Council of the American Astronomical Society, and an equally rapid response came from NASA suggesting that the mass of each spacecraft should be reduced from 11 to 3 tons. This would enable the original Titan III launcher to be replaced by the much less expensive Atlas-Centaur, and would result in a cost reduction for the total programme of 70%. After much discussion this was the programme that was finally approved.

In retrospect, it seems incredible that NASA had originally proposed launching four 11 ton HEAO spacecraft in as many years, when one considers that the mass of the Hubble Space Telescope is only 9 tons. But the HEAO programme had been conceived in the days of the late 1960s when NASA still had grandiose and far-reaching plans, including a reusable space shuttle *and* a twelve-man space station by 1975, a base in lunar orbit in

Figure 13.1 An artist's impression of the HEAO-1 spacecraft in orbit. The six proportional counters, designed to measure the X-ray background, are grouped on the left. One of the seven large-area proportional counters is in the upper centre of this equipment face, with the other six being on the non-visible, opposite side of the spacecraft. (Courtesy TRW.)

1976, and a 100-man space station by 1985. The space shuttle was approved in 1972, but the financial crisis of 1973–74 showed that these other plans had no chance of being adopted, at least in the short term. As far as the HEAO programme was concerned, however, once it had been de-scoped it proved to be very successful.

The de-scoped programme was to consist of three 3-ton spacecraft. HEAO-1 was to be a scanning mission with four X-ray instruments, HEAO-2 was to be an X-ray imaging mission based around Giacconi's focusing X-ray telescope, and HEAO-3 was to be a γ-ray and cosmic ray mission. In the event the three spacecraft were launched in August 1977, November 1978, and September 1979, respectively.

13.2 HEAO-1

The HEAO-1 spacecraft (see Figure 13.1) with its non-focusing X-ray detectors was the last of its type. Even though it was 5.8 metres long and weighed about 2,600 kg,[2] this spacecraft reached the limit of what was practically

[2] This was the heaviest un-manned spacecraft launched up to that time by NASA.

possible with large-area proportional counters, as their sensitivity increases only slowly with detector area.

HEAO's payload consisted of four experiments, the largest of which was NRL's Large Area X-ray Survey Experiment that consisted of seven proportional counters, six on one face and one on the opposite side of the spacecraft, which were sensitive over the range 0.25 to 20 keV. They were designed to detect weak sources and to measure rapid fluctuations of black hole candidates like Cyg X-1. A second experiment, consisting of six proportional counters, covering the range 0.2 to 60 keV, was designed to measure the X-ray background radiation to try to understand its origin. The position of 1 to 15 keV sources was measured by the Scanning Modulation Collimator Experiment to an accuracy of about 5 to 30 arcsec (depending on brightness), or ten times better than measured by Uhuru. Finally, the Hard X-ray and Low-Energy Gamma-ray Experiment was designed to detect sources and measure their spectra in the energy range from 10 keV to 10 MeV.

HEAO-1 rotated once every 33 minutes, allowing the experiments to slowly scan the sky, and once every 12 hours the spin axis was moved by 0.5°, to enable the spacecraft to complete a survey of the whole sky in six months. After this first survey was completed a second survey was started, interspersed with a number of pointed observations of discrete sources. Although the spacecraft was designed for a one year lifetime, it continued to operate for 17 months before its attitude control gas was exhausted.

Whilst HEAO-1 did not produce the spectacular images of its illustrious successor, HEAO-2, it nevertheless provided much interesting new information and detected new types of X-ray sources, because not only were HEAO-1's detectors more sensitive than Uhuru's, they also reached into the very soft X-ray region below about 1 keV (or above 1 nm), which Uhuru did not. So within a few months a number of new strong, soft X-ray sources had been discovered which had not been evident at the higher energies measured by Uhuru. Unfortunately, the position of these soft X-ray sources could only be determined to an accuracy of about 0.1°, but as a number of similar sources were found, this enabled their optical counterparts to be identified. In particular, the HEAO-1 source H0324+28 was found to be near to the variable star UX Ari, which is a well-known binary of a type called RS CVn, after the prototype in the constellation Canes Venatici (or CVn for short). Similar associations were then found between eight other high intensity, soft X-ray sources and RS CVn-type binary stars which were known to emit radio bursts. Such binaries usually consist of one K- and one G-type star, with the K star being a subgiant that has started to evolve off the main sequence of

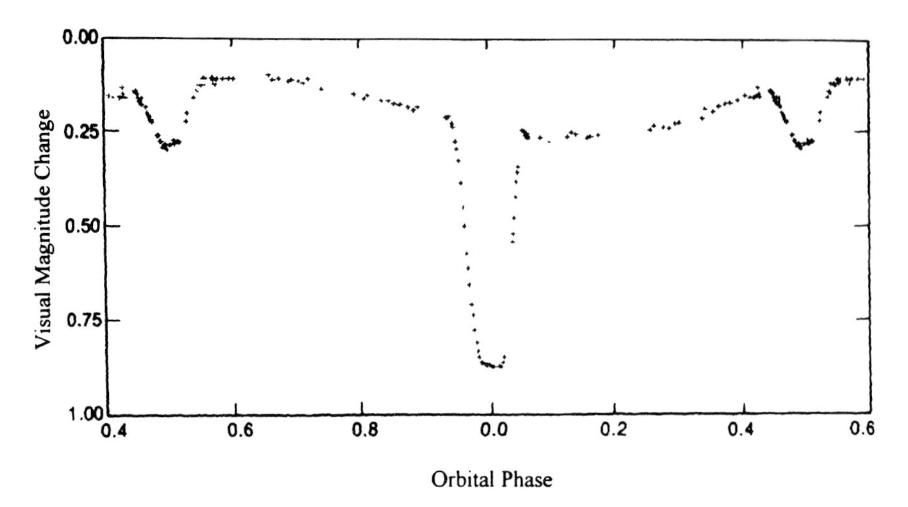

Figure 13.2 The yellow light curve of the binary star RS CVn, which shows both the primary and secondary eclipses and the gentle undulation out of eclipse that has been attributed to star spots.

the Hertzsprung–Russell diagram.[3] The light curve of such RS CVn systems shows both the primary and secondary minima when the stars eclipse each other, as well as showing gentle undulations out of eclipse (see Figure 13.2). These undulations and the radio bursts have been attributed to star spots on the K subgiant, and it is the coronal disturbances associated with these star spots that are thought to be the origin of the soft X-ray emission.

In 1976 the SAS-3 spacecraft had helped astronomers to understand the true character of AM Herculis, the first polar to be discovered (see Section 12.7), in which matter is lost directly from a red star to the magnetic poles of its white dwarf companion, and in which the spin rate of the white dwarf is synchronised with the orbital period of the binary by the very strong magnetic field of the white dwarf. Shortly afterwards the HEAO-1 spacecraft found a binary system H2252-035 in which the compact object also seemed to be a white dwarf, as the ratio of the X-ray to optical luminosity is more like that of a white dwarf than a neutron star. The optical and ultraviolet spectra also supported this conclusion. J. Patterson and C. Price found that the light of H2252-035's optical counterpart, called AO Psc, pulsed every

[3] The main sequence consists of a broad band of stars on the Hertzsprung–Russell diagram, in which absolute magnitude is plotted against spectral type (or colour). Most stars lie on this main sequence, but they start to move off it as they get towards the end of their life.

858 seconds and showed a longer period modulation of 3.6 hours. So it appeared as though the white dwarf's spin rate was 858 seconds and the binary period 3.6 hours. In which case H2252-035 is unlike AM Her as the spin rate of the white dwarf is not the same as the orbital period of the binary. Things were not as simple as that, however, as Patterson, and N. White and F. Marshall then independently measured the X-ray periodicity as 805 s. Clearly, the different optical and X-ray periodicities cannot both be the spin rates of the white dwarf, so which is its spin rate and why do the two rates differ? The situation became clear when it was realised that the X-ray period is 1/16th of the binary period, whereas the optical period is 1/15th of the binary period, so every one binary period the X-ray and optical pulses are temporarily back in step. The X-rays must be coming from the white dwarf or its accretion disc, so the X-ray period is the true spin period of the white dwarf, whereas the optical period is the result of the X-rays emitted by the white dwarf being absorbed and re-emitted by its companion as optical radiation. As the white dwarf rotates in the same sense as it orbits its companion, there are only 15 sets of optical pulses received via its companion, for 16 sets of X-ray pulses received directly from the white dwarf.

The model that best fits H2252-035/AO Psc is of a normal star losing mass to the magnetic poles of its white dwarf companion via an accretion disc around the white dwarf. In this case the magnetic field of the white dwarf is not strong enough to synchronise its spin rate to the orbital rate of the binary, so the system appears to be intermediate in type between the polar source AM Her and dwarf novae (see Section 12.4). For this reason it is called an 'intermediate polar'.

In 1979 Webster Cash of the University of Colorado and Philip Charles of the University of California were analysing HEAO-1 sky survey data taken with the sensors designed to detect the X-ray background radiation, when they noticed what appeared to be an unknown supernova remnant in the constellation of Cygnus. As they began to trace this supposed X-ray remnant they found that it was an incredible $18° \times 13°$ in size, and appeared to have a temperature of about 2 million K. The centre of the source was not detectable, however, although this was attributed to the X-rays being absorbed by the intervening Great Rift dust cloud, which is seen as a dark band cutting across the Milky Way in visible light. Further analysis showed that thin filaments of hydrogen detected in the light of $H\alpha$ coincided with the newly discovered ring. There are a number of massive O-type stars still in this region, so it is thought that this ring, now called the Cygnus Superbubble, was probably produced by a number of supernova explosions that occurred in a group of similar O-type stars about 3 million or so years ago.

HEAO-1 also discovered a new black hole candidate when it measured rapidly fluctuating X-rays from the source GX 339-4, reminiscent of Cyg X-1, but possibly this spacecraft's most interesting results concerned the X-ray background radiation. The measurements were difficult to interpret, as both the known X-ray sources and the galactic X-ray component had first to be eliminated from the data. When this was done there were naturally large areas of the sky where the X-ray background could not be detected because of these masking sources, but there was still sufficient data remaining to enable an analysis to be made. This showed that the intensity variations from point to point in the sky of the X-ray background was greater than expected from normal statistical fluctuations of a smooth background, implying that the background is not truly isotropic, but is made up of faint unresolved sources that will produce some or possibly all of the background radiation.

In the process of undertaking the above analysis, X-rays from the Milky Way had to be eliminated from the data. This Milky Way data showed that our galaxy probably had a number of unresolved, discrete sources near to its central plane, which was also seen to be surrounded by a diffuse X-ray halo. It was unclear, however, whether the X-rays in this halo were caused by discrete sources, thermal emission or other processes.

13.3 HEAO-2 or the Einstein Observatory

The second spacecraft in this series, HEAO-2 (see Figure 13.3), which was the first X-ray imaging astronomical spacecraft, was launched from Cape Canaveral by an Atlas-Centaur on 13th November 1978. Its main payload consisted of a Wolter-type telescope of nested cylinders (see Chapter 1) that was able to focus X-rays with energies of between about 0.25 and 4 keV. This telescope had four alternative detectors that could be placed at its focus by a rotating platform. One called the High Resolution Imager (HRI) had a field of view of about ½° and was able to produce X-ray images with about 2 to 3 arcsec resolution, but it had no spectral capability. The second instrument, the Imaging Proportional Counter (IPC), was able to produce both spectral and positional information and, although it was five times more sensitive to X-rays than the HRI, it had a poorer spatial resolution of about 1 arcmin. Spectral information of the whole field of view could be produced by using an Objective Grating Spectrometer (OGS) with either the IPC or HRI, but

Figure 13.3 An artist's impression of the highly-successful Einstein Observatory, or HEAO-2 spacecraft, in orbit. The length of the spacecraft was dictated by the 3.4 m focal length of its X-ray telescope, whose entrance aperture is the largest circular structure under the protruding solar array at the top left. (Courtesy TRW.)

the best spectral data were produced by the Solid State Spectrometer (SSS), which had an energy resolution of 0.15 keV, or the Focal Plane Crystal Spectrometer (FPCS), which, although less sensitive than the SSS, had an energy resolution of about 0.001 keV. Finally, the Monitor Proportional Counter (MPC), which was mounted separately from the main X-ray telescope, was used to examine the variability and spectra of the brightest sources.

After launch there was one heart-stopping period when it was thought that the star sensors had failed that were to be used to control the spacecraft, but it turned out that they were picking up the reflection of the Moon off the Pacific Ocean. Once this had been realised, the in-orbit check-out was completed and HEAO-2, or the Einstein Observatory as it is now called, was operational. It continued to work well for 2½ years until April 1981 when its attitude control system ran out of gas.

Prior to the launch of the Einstein Observatory it had been thought that red dwarf stars are too cool to have an extensive corona and so would not emit X-rays, except if they happened to be flare stars during outburst (see Section 12.5). Einstein surprised astronomers, therefore, by showing that normal red dwarf stars, and flare stars in their quiescent state, are strong X-ray emitters. They were found to emit of the order of 10% of their total energy in the X-ray band, implying that they had hot coronas like the Sun. Likewise, prior to Einstein it had been thought that young, hot[4] O-, B- and A-type stars would not have hot coronas. That was because these stars do not apparently have a turbulent convective zone just below the surface, which was thought to be the essential source of coronal heating. Again, the spacecraft proved the theory wrong, as some of these stars were found to be significant X-ray emitters. In fact Einstein found that all types of stars are X-ray emitters, with the sole exceptions of very cool giants and supergiants. Moreover, within any one star category the X-ray intensities covered a range of three orders of magnitude (i.e. a factor of 1,000 to 1), which caused another surprise as stars of a similar type should have a similar corona. Clearly the theories of stellar structure needed a radical overhaul.

So what causes red dwarfs and young, hot O-, B- and A-type stars to emit X-rays? It is thought that the X-ray flux from red dwarfs may be produced when magnetic field loops reconnect in their highly convective atmospheres. In the case of the young, hot stars, however, it is thought that the X-rays are probably produced when stellar winds produce shock waves in the interstellar medium.

There was also the question to be answered as to why do similar ordinary stars in the middle of the Main Sequence not have similar X-ray outputs? It was suggested by some astronomers that this could be a function of stellar rotation rates and age. Theoretically, it was expected that fast-rotating stars would, via the dynamo effect, produce stronger magnetic fields than more slowly rotating stars, and these stronger magnetic fields would retain more plasma and thus generate more X-rays. It was known that younger stars rotate faster than older stars, and so Zoleinski, Stern, Antiochos and Underwood used Einstein observations of stars in the relatively young Hyades cluster to see how their X-ray emission compared with that of other Main Sequence stars. The Hyades cluster was known to be about 600 million years old, compared with the Sun's 4.5 billion years, and Zoleinski and colleagues found that, on average, solar-type stars in the cluster were about 50 times more luminous in X-rays than the Sun. Jean-Pierre Cauillat also found that the average X-ray luminosity of stars in the even-younger Pleiades cluster was about ten times greater than those in the Hyades, thus adding credibility to the theory of an age/rotation/X-ray linkage.

---

[4] Although the surface of these stars is hot compared with other stars, it is not hot enough to emit X-rays. Even these stars, which have the hottest surfaces, need a hot corona, significant flare or similar activity to produce X-rays.

The Einstein Observatory provided much new data on X-ray transients. In particular, it helped astronomers to understand the structure of the transient A0538-66 which had been found in 1977 by Nick White of MSSL and Geoff Carpenter of Birmingham University using Ariel 5. Unfortunately, Ariel 5's lack of spatial resolution meant that the X-ray source could not be located accurately enough for an optical identification, but it seemed to be in the direction of the Large Magellanic Cloud. The Ariel 5 data soon showed that not only was A0538-66 highly variable, but it had a regular 16 day period between peaks, indicating that it was a binary. Mark Johnson (MIT), Richard Griffiths (Harvard) and Martin Ward (Cambridge) then used HEAO-1 to get a more accurate location of the X-ray source, which they found coincided with a variable star. Its optical period of 16.6 days was deduced by Gerry Skinner of Birmingham University using old Harvard Observatory plates, but he also found that there had been periods in the preceding fifty years when the variability had disappeared. This strangely variable optical source was clearly the X-ray source's binary companion.

In 1980 Philip Charles of Oxford University and John Thorstensen of Dartmouth College measured the optical spectrum of A0538-66 and found that the normal spectrum of a B-type star, with hydrogen and helium absorption lines, was completely transformed during an outburst period by the addition of many strong, broad emission lines. These red-shifted emission lines and blue-shifted absorption lines[5] indicated that the normal star was surrounded by a shell of gas expanding at velocities in excess of 3,000 km/s. Fortunately, the spectrum of the star in quiescence enabled the source to be definitively placed in the Large Magellanic Cloud, as it showed that the star was receding from the Sun at the same velocity as the LMC. This enabled A0538-66's absolute X-ray luminosity to be determined as about 10^{39} erg/s at its peak, making it the most luminous X-ray stellar source known at that time. Gerry Skinner and Marty Weisskopf of the Marshall Space Flight Center and colleagues then detected X-ray pulsations with a period of 69 milliseconds using the Einstein Observatory, indicating that the X-ray source is a rapidly rotating neutron star, and the IUE spacecraft (see Section 14.1) showed that the optical source is redder when it is brighter. Normally, the binary companion of an X-ray star becomes hotter and bluer when irradiated by the X-ray source, so why does this binary behave differently? The answer appears to be that at outburst the optical source is larger, because of the gaseous emission, so although it is brighter, the energy emitted per unit area is lower, and the source appears to be cooler and redder.

[5] This is called a P Cygni profile.

Figure 13.4 The Tycho supernova remnant, imaged by the Einstein Observatory spacecraft, showing its clear shell-like structure. This was caused by the expanding gas of the supernova shock-heating the surrounding interstellar medium. (Courtesy NASA.)

The widths of the spectral lines indicate that the optical source is rotating rapidly, probably close to the break-up speed of the star. So the hypothetical model of the system has an optical source, with a large equatorial disc of material, in a binary system with the X-ray source A0538-66. The orbit of the latter is highly eccentric and once every orbit it passes through the disc surrounding the optical source. This creates a burst of X-ray energy that heats and partially depletes the disc, as material from this disc is accreted onto the surface of the neutron star at its poles. After a number of orbits the material in this disc is not dense enough to overcome the X-ray star's magnetosphere, which expands as the disc pressure reduces, and the active period of the binary ends. The optical star continues to emit material, however, and eventually the disc accumulates enough for the process to start again.

In addition to observing transients, the Einstein Observatory was able to image supernova remnants for the first time. This showed, for example, that the Crab nebula in X-rays, which is due to synchrotron emission, is much smaller than the nebula in visible light. The Crab pulsar, which was also imaged, was seen to be clearly off-centre compared with the X-ray nebula. Whether this is due to an asymmetric supernova explosion or differences in the density of the surrounding medium is not clear. Einstein also imaged the Tycho and Cas A supernova remnants, and confirmed that their shell structure in X-rays (see, for example, Figure 13.4) is due to their exceptionally high temperature, as deduced from Ariel 5 measurements (see Section 12.6). Both Tycho and Cas A show outer shells at temperatures of about 40 to 50 million K, due to their expanding shells shock-heating the

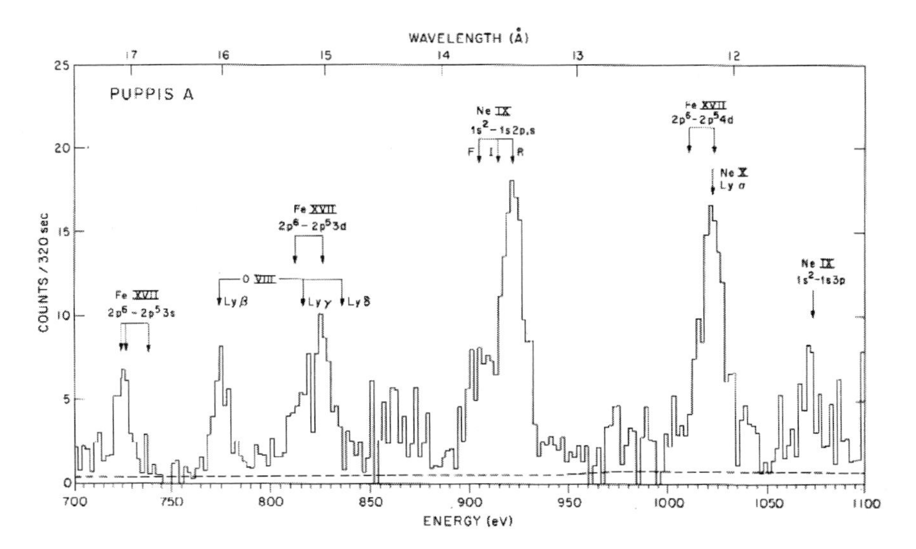

Figure 13.5 The X-ray spectrum of that part of the supernova remnant Puppis A which is brightest in X-rays, taken by the Einstein FPCS spectrometer, showing highly ionised lines of oxygen, neon and iron. The dashed line is the X-ray background level. (P.F. Winkler *et al.*, *ApJ* 246 (1981), p L29. Reprinted with permission from *The Astrophysical Journal.*)

interstellar gas, but the Cas A remnant also shows an inner shell of slightly cooler, denser gas. The outer shell is being slowed down by the interstellar gas, and the inner shell is catching up and starting to interact with it, causing the inner shell to emit X-rays also.

The Puppis SNR was analysed by Frank Winkler, Claude Canizares and colleagues using the FPCS on Einstein with the previously unheard-of resolution in X-rays of 1 eV. Highly ionised emission lines of oxygen, neon and iron were found (see Figure 13.5) which enabled the amounts of these gases to be determined. It appears that Puppis A has a much larger amount of oxygen and neon than normal, indicating that the progenitor star was a 25 solar mass star that ended its days in a type II supernova explosion. The current temperature of the gas was estimated to be in the region of 2 to 5×10^6 K.

When the Einstein Observatory was launched, only the Crab and Vela remnants were known to contain pulsars that emit X-rays, and only the Crab source was known to pulse in X-rays.[6] Using Einstein, astronomers

[6] Later the ROSAT spacecraft found a 4% modulation in the X-ray output of the Vela pulsar, which had previously been found to pulse strongly in the radio (1968), γ-ray (1975) and visible (1977) wavebands.

discovered three more X-ray pulsars in SNRs,[7] helping to establish the link between SNRs and X-ray pulsars. One of these SNRs, numbered 0540-693, which was discovered by Rick Harnden, David Helfand and Frederick Seward in the LMC, is remarkably similar to the Crab Nebula, having blue nebula emission due to synchrotron radiation. Its pulsar, which is bright, has a period of 50 milliseconds, compared with 33 milliseconds for the Crab, and it pulses in both X-rays and visible light. So the Crab Nebula is not unique.

The radio galaxy Centaurus A (NGC 5128) is normally seen in visible light to be a disc of stars of about 6 arcmin in diameter crossed by a prominent dust lane. At Cen A's distance this 6 arcmin equates to about 15,000 light years. Sophisticated image processing techniques have shown that the galaxy extends to about 30 arcmin in visible light, but in radio waves it was found to stretch an astonishing 9° across the sky in an orientation perpendicular to the dust lane.

In 1973 the OSO-7 spacecraft found that Cen A is an X-ray source varying in intensity over a few days,[8] whilst the subsequent Ariel 5 and OSO-8 spacecraft detected X-ray variations over less than one day. This indicated that the X-ray source was less than one light day in diameter, compared with the 75,000 light year optical source, and the 1¼ million light year radio object. Later, the SAS-3 spacecraft confirmed that the X-ray source was very small, being less than 1 arcsec in diameter, and lying within a few arcsec of the centre of Cen A.

The Einstein Observatory was the first to image the X-ray source of Cen A, showing that most of the X-ray energy came from a very small region at the galactic centre. It also showed that there was an X-ray emitting jet just on one side of the nucleus, with an orientation similar to that of the radio image of Cen A but very much smaller. Although the X-ray source is very small, its luminosity was found to be about 10^{42} erg/s, which is about 100,000 times that of the centre of the great M31 spiral galaxy in Andromeda. Such a high intensity from such a small source was thought to indicate that there is a supermassive black hole at the centre of the galaxy Cen A.

---

[7] A fourth pulsar (in the SNR CTB 80) was discovered by Einstein, but the pulsations could only be detected at radio wavelengths, as the X-ray source was too weak for pulsations to be clearly detected.

[8] The Vela 5A and B military spacecraft had independently detected a variable X-ray source in the area of Cen A between 1969 and 1973, but their position measurements were not sufficiently accurate to link Cen A unambiguously with the X-ray source.

M87, another radio galaxy, also known as Virgo A or 3C 274, had been found to be an X-ray source in the 1960s (see Section 11.4). A small optical jet had been found extending from the nucleus of M87 as long ago as 1917, but the jet had been largely forgotten until it was rediscovered by Walter Baade and Rudolph Minkowski in 1954. The polarisation of the light from the jet indicates that it is produced by synchrotron radiation from high energy electrons spiralling in the galaxy's magnetic field. A small optical counter-jet was found by Halton Arp in 1966. Sophisticated image processing of optical images then showed that M87 is much larger than originally thought, having a diameter of about 1 million light years, or about ten times that of the Milky Way.

The Einstein Observatory clearly showed that both the nucleus and jet of M87 are strong emitters of X-rays. At about the same time optical astronomers concluded, from the density and velocity of stars around the nucleus of M87, that there must be about 5 billion solar masses of material in a volume just ten light years in diameter at the core of this galaxy. This indicated the presence of a supermassive black hole there. If that were so, however, the intensity of X-rays from this nucleus should be higher than that observed by Einstein. On the other hand, the presence of the jet, which is seen in optical, radio and X-ray wavelengths, indicates the presence of a large accretion disc around the galactic nucleus, with the accretion disc focusing and emitting the jet along the accretion disc's axis. So at this time it was unclear as to whether or not there was a supermassive black hole at the centre of M87.

In 1983 D. Fabricant and P. Gorenstein analysed the X-ray emissions from M87, as measured by the Einstein Observatory, which extended beyond the edge of the visible light image of the galaxy. From this they concluded that there was ten times more mass in the central region of M87 than seen in visible light, and possibly 200 times as much when integrated over the whole of the X-ray source. So the relative amount of matter that cannot be seen, the so-called 'dark matter', apparently increases with distance from the centre, the total mass of the galaxy (including the dark matter) being at least 200 times that of the Milky Way. It is no coincidence, therefore, that M87 is seen to be stationary at the centre of the Virgo cluster of galaxies, with the other galaxies in the cluster rotating around it at velocities of up to 1,500 km/s.[9]

In 1967 the quasar 3C 273 had been found to be an intense X-ray source, and ten years later the Cos-B spacecraft had found that it was also an intense

[9] As long ago as 1933 Fritz Zwicky had also concluded that the mass of the *Coma* cluster of galaxies must be about ten times that of its visible stars to explain the velocity of the galaxies in the cluster.

source of γ-rays. In 1979 the Einstein Observatory found that the X-ray intensity, which is about a million times that of the Milky Way, is varying over periods as short as half a day, indicating that the energy is being emitted from a source no larger than the diameter of the Solar System. This again indicated the presence of a supermassive black hole. This idea is backed up by the presence of a jet apparently coming from the centre of 3C 273 as seen in the radio, visible and X-ray wavebands (as imaged by the Einstein Observatory). So the galaxies Centaurus A, Virgo A (M87 or 3C 274), and 3C 273 all seemed to contain supermassive black holes at their centre.

Two extensive deep sky surveys were undertaken in the 0.3 to 3.5 keV range using the Einstein Observatory, to see if the X-ray background radiation in this range can be explained by discrete sources. In the deepest survey (DS) the IPC and HRI instruments were used to image eight fields at high galactic latitudes that had no known X-ray sources. Exposure times extended up to 24 hours. As a result a total of 100 new sources were detected, of which 25 were selected for detailed analysis. Nine were found to be quasars, seven were individual stars, one was a cluster of galaxies and one a normal galaxy, with the remaining seven being unidentified extragalactic sources. In the second Einstein survey, called the medium sensitivity survey (MSS), previously unknown objects that appeared in normal IPC frames were analysed by Gioia, Maccacaro, Stocke and colleagues. Of the 800 or so objects discovered, 53% were found to be quasars and active galactic nuclei (AGN), 27% were stars, 13% were clusters of galaxies, 5% were BL Lac objects, and 2% were normal galaxies. Interestingly in the less deep survey carried out using HEAO-1 (see above), 51% of the sources were clusters of galaxies, so with successively deeper surveys the cluster figures reduced from 51% (HEAO-1), to 13% (MSS) to 6% (DS). As a result, clusters of galaxies are not expected to make a major contribution to the X-ray background radiation. In the case of quasars, however, the percentage increased substantially between the HEAO-1 survey and the deeper Einstein surveys. Further detailed analysis of some hundreds of quasars measured by the Einstein Observatory led Tananbaum and colleagues to conclude that almost all of the diffuse X-ray background radiation in the range 1 to 3 keV is due to unresolved quasars.

13.4 SS 433

One of the most remarkable supernova remnants that was investigated using the Einstein observatory was W50, which had first been discovered as a large (2° × ½°) radio source in the 1950s. In 1972 the English astronomers David Clark and James Caswell had started a radio survey of all radio-emitting supernova remnants visible from Australia, and they found that W50 had an elongated shape with a point radio source nearby. Shortly after they published their results, the Indian radio astronomer Thangasamy Velusamy found that Clark and Caswell had only imaged part of the remnant. In reality it was oval in shape with the point radio source at its centre, so it was beginning to look as if the point radio source was the remains of the star that had exploded to produce the remnant W50.

Unknown to these radio astronomers, both optical and X-ray astronomers were also becoming interested in a point source in the area of W50. Firstly, in the 1960s the Americans Bruce Stephenson and Nicholas Sanduleak had started looking for stars with emission line spectra, and in 1977 they published their Stephenson–Sanduleak (SS) catalogue of such sources. But this catalogue was not known to Clark, Caswell or Velusamy, and so the link between SS 433 in the catalogue and W50's point radio source, which were in fact one and the same, was missed. Secondly, amongst the detailed Uhuru data there was a weak source of X-rays in the region of this point radio source. It was not picked up as anything unusual, however, until in 1976 the American Frederick Seward noticed that the weak Ariel 5 source A1909 + 04, which was the same as the Uhuru source, was an X-ray variable. The position of A1909 + 04 was not known with sufficient accuracy for it to be correlated with any source in other wavebands, but it was thought that it may be the same as the radio source at the centre of W50. Then John Shakeshaft at Cambridge University found that the point radio source flared erratically at radio wavelengths, giving more circumstantial evidence that the A1909 + 04 X-ray source and this point radio source were one and the same.

In preparation for the attempt to find the optical counterpart of this point radio source, David Clark and David Crawford had produced a much more accurate position fix of the radio source, and this enabled Clark and Paul Murdin to locate a likely optical candidate on a UK Schmidt image. They then used the Anglo Australian Telescope to produce a spectrum of this optical candidate in June 1978, and were delighted to see an emission spectrum reminiscent of that of another ex-supernova source called Cir X-1. Shakeshaft then produced an even more accurate position fix of the point radio source in W50, but unfortunately it appeared to be 1 arcsec from that of the optical source, leaving some doubt as to whether they were really the same. Whilst trying to resolve this discrepancy in position, and in order to try to understand this source more fully, another optical spectrum was produced in July 1978.

Some of the emission lines of the June spectrum were still present, but others had disappeared and new ones had appeared.

In the meantime, Seaquist, Gregory and Crane had, unknown to Clark and Murdin, been trying to detect radio emissions from SS stars, and in June 1978 they found highly variable radio emission from SS 433. Clark instantly recognised the coordinates given in the IAU circular by Seaquist and colleagues as those of the point radio source in W50. So the optical source SS 433, the X-ray source A1909+04 and this point radio source[10] were all the same.

But what was the object? It was not unusual to find X-ray and optical sources that varied in intensity, but there was one strange aspect of the data already available on SS 433, and that was the large-scale changes in its spectrum. Investigations into this would show that SS 433 was a unique source at that time.

Clark and Murdin's first spectrum of SS 433 had been taken on 29th June 1978, but when the next spectrum had been taken just two weeks later, as outlined above, many of the features had completely changed. Bruce Margon of the University of California also found unusual lines in his spectrum of SS 433 taken in September 1978, and Remington Stone, also of the University of California, found over four nights in late October that prominent lines on either side of the Hα line were drifting in frequency. In parallel, Augusto Mammano and colleagues in Italy found the same effect. Normally, such frequency changes would be put down to the Doppler effect, but in this case, if the lines were displaced hydrogen lines, it would require hydrogen gas to move at velocities of up to 40,000 km/s, which seemed ridiculously high, and it would also imply that the gas velocity changed by up to 1,000 km/s per day, which seemed equally unlikely. In addition the gas appeared to be moving both towards and away from Earth at the same time, as the spectral lines appeared to be both blue- and red-shifted.

In January 1979 Andy Fabian and Martin Rees in the UK suggested that SS 433 was emitting two beams of gas, one towards and one away from us, at a varying speed. At about the same time the Israeli astronomer Mordehai Milgrom independently suggested that SS 433 had either a rotating pair of jets, or the spectral lines came from a ring of gas orbiting a central object. Interestingly, Milgrom observed that the blue- and red-shifted lines were moving about a fixed mean frequency that was slightly lower than the

[10] Later it transpired that Shakeshaft's Cambridge position of the radio source (numbered 4C 04.66 in the Cambridge catalogue) had been in error.

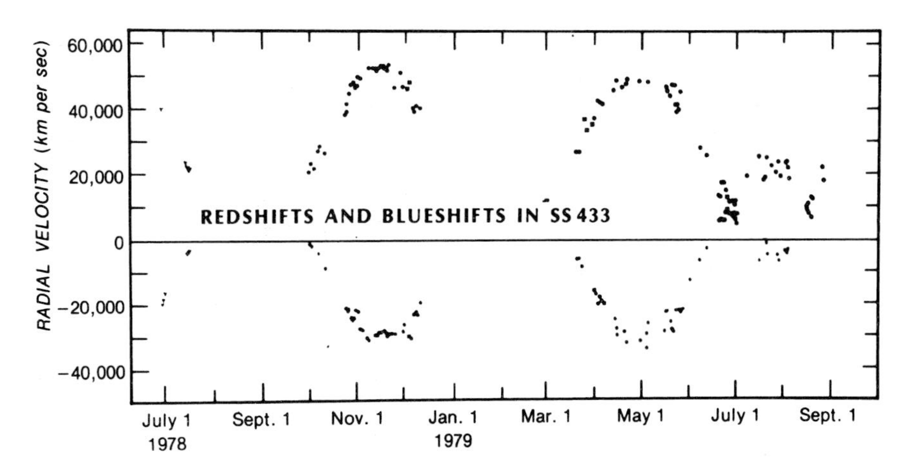

Figure 13.6 The red- and blue-shifts in the spectrum of SS 433 converted to radial velocities as measured by Bruce Margon *et al.* Positive velocities indicate recession. (Courtesy B. Margon and *Sky & Telescope.*)

frequency of the un-shifted line. This he attributed to the relativistic, transverse Doppler effect.[11]

In early 1979 Bruce Margon and James Liebert independently found that all the emission lines had identical blue-shifted, red-shifted and stationary components, proving that the lines were shifted due to the Doppler effect. In addition Margon found that the lines showed a periodicity (see Figure 13.6) of 160 days, which he later modified to 164 days. The deviation of the median position of the blue- and red-shifted lines from their rest position enabled the transverse velocity of the emitting gas to be determined by George Abell and Bruce Margon as 0.27 times the velocity of light, or about 80,000 km/s. So the jets of SS 433 could not be pointing directly at or away from the Earth at any stage, as their maximum velocity of approach or recession was only 40,000 km/s. Analysis of the geometry showed, in fact, that the lines were being emitted by a source which was moving in a conical trajectory with a cone of half-angle 17° (see Figure 13.7), and with an angle between the axis of the cone and our line of sight of 78°. This precessional motion had a period of 164 days.

So far the analysis of SS 433 and its radio and X-ray counterparts had been undertaken using data from numerous optical and radio telescopes and just

[11] When a source of energy is moving perpendicularly across our line of sight at normal velocities no Doppler shift is observed. If the source is moving at relativistic velocities, however, its emissions appear to be red-shifted due to the time dilation effect.

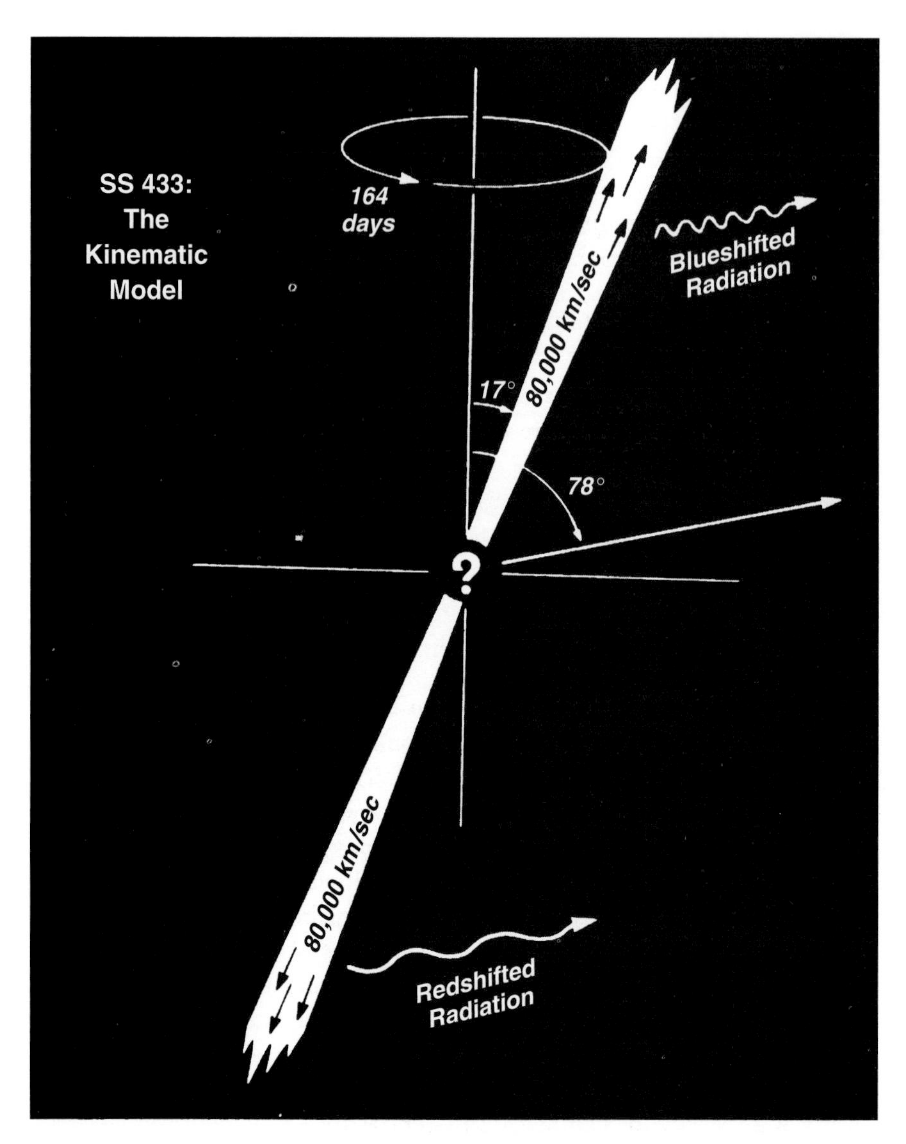

Figure 13.7 A diagram showing the configuration of the SS 433 jets, which precess over a period of 164 days. (Courtesy B. Margon and *Sky & Telescope*.)

one X-ray spacecraft, Ariel 5. All the data seemed to fit reasonably well with the precessing jet hypothesis, but astronomers would have felt much happier if they had had an image of that jet. There was also the faint possibility that the X-ray source A1909 + 04 was not the same as SS 433 as the position of the X-ray source was not known very accurately. These two issues were resolved by the Einstein Observatory spacecraft in 1979. Firstly, in April the HRI instrument on Einstein measured the position of A1909+04 much more accurately than Ariel 5, and confirmed that it was indeed the X-ray counterpart of SS 433. Then in October Seaquist, Seward and colleagues used the IPC instrument on Einstein and produced a spectacular image of the jets stretching out about 35 arcmin or about 100 light years in both directions. Interestingly, the X-ray emitting jets seemed to be pointing straight at the lobes of the W50 shell, suggesting that W50 may not be a supernova remnant after all, but is, instead, a hole blown in the interstellar medium by material from the jet.

In the meantime radio astronomers had also been attempting to resolve SS 433 with radio interferometers. In 1979, Ernst Seaquist and William Gilmore found, using the partially completed VLA, that SS 433 appeared to be elliptical in shape, with a semi-major axis of about 3 arcsec at a wavelength of 6 cm. At 20 cm wavelength, however, there appeared to be a narrow filament, aligned with the major-axis at 6 cm, extending about 20 arcsec on either side of the point source. These filaments were also approximately aligned with the much larger X-ray jets.

In 1979 and 1980 Hjellming and Johnson observed SS 433 at a number of frequencies, as more and more dishes were added to the VLA, and found that SS 433's lobe structure varied from week to week. In addition the radio intensity was found to increase rapidly on 7th December 1979 and 20th June 1980, to produce blobs or knots of radio-emitting material on both sides of the nucleus. These knots were observed to move from the central source at about 9 milliarcsec per day which, given the jet velocities calculated above, implied that SS 433 was about 16,000 light years from Earth. Over the next year or so further radio observations using the VLA, MERLIN and other radio interferometers showed that knots of material were emitted at different angles from the central source due to its precession. Once set in motion the knots of material continued to move from the central source in a straight line, so the radio 'jets' appeared to oscillate from side to side, rather like water coming from an oscillating garden hose.

So SS 433 was emitting jets, and was the first stellar source to be observed to do so, but what was the source of these jets and what caused them to precess?

David Crampton and colleagues at the Dominion Astrophysical Observatory in Canada produced numerous optical spectra of SS 433 over a period of three months in mid 1979, and found that the so-called 'stationary' lines move slightly backwards and forwards in frequency over a period of 13.1 days. The velocity curve deduced from these moving spectral lines was typical of that for a binary containing a B-type and a compact star. The binary nature of the system was confirmed in 1980 by the Russian Anatol Cherepashchuk, working in Australia, who detected both the primary and secondary minima in the light curve with a 13.1 day period, and by Bruce Margon who found that the intensities of the stationary emission lines varied over the same period.

So SS 433 appeared to be a 13.1 day binary system in which a large B-type star is losing mass to a compact source via an accretion disc around the latter. The orbital plane of the binary, as seen from Earth, is almost side-on, so that we can detect both the eclipse of the accretion disc by the B-type star and of the B-type star by the accretion disc. X-rays are emitted by material as it falls from the accretion disc to the compact source. But not all of the material ends up on the compact source, as some of it is ejected from the system completely in two oppositely directed jets perpendicular to the accretion disc. The plane of the accretion disc and of the binary system are not the same, so the gravitational pull of the B-type star on the accretion disc causes both the accretion disc and the jets to precess with a period of 164 days.

Some questions still remained, however. For example, is the compact source a neutron star or a black hole? SS 433 was the only stellar source known to have relativistic jets, and yet its X-ray intensity was relatively low for a binary containing a compact source. So where did the energy come from to accelerate the jets to such high velocities?

Mike Watson of Leicester University and colleagues used the GSPC instrument on Exosat (see Section 14.3) in the mid 1980s to measure the X-ray spectrum of SS 433's central source between about 4.5 to 11 keV, which covered the 6.7 keV line of highly ionised iron. Interestingly, although they detected the change of energy of this line over the 164 day precession period, they could only detect the blue-shifted beam, i.e. the one that is pointing towards us, and not the red-shifted one (see Figure 13.8). This was thought to be because the accretion disc itself, which is almost edge-on to us, is thick enough to prevent us detecting X-rays from the red-shifted beam near the core of SS 433.

The rate of mass loss of SS 433 along the observable beam could be calculated, assuming that the X-rays are being produced by hot gas. Then knowing the beam's velocity, the rate of energy loss of the source along the beam could be calculated. This turned out to be about 10^4 times higher than the energy being radiated in X-rays by SS 433, leading to the suspicion that the true X-ray intensity of the central region of SS 433 is much higher than that measured. If the true energy really is a number of orders of magnitude

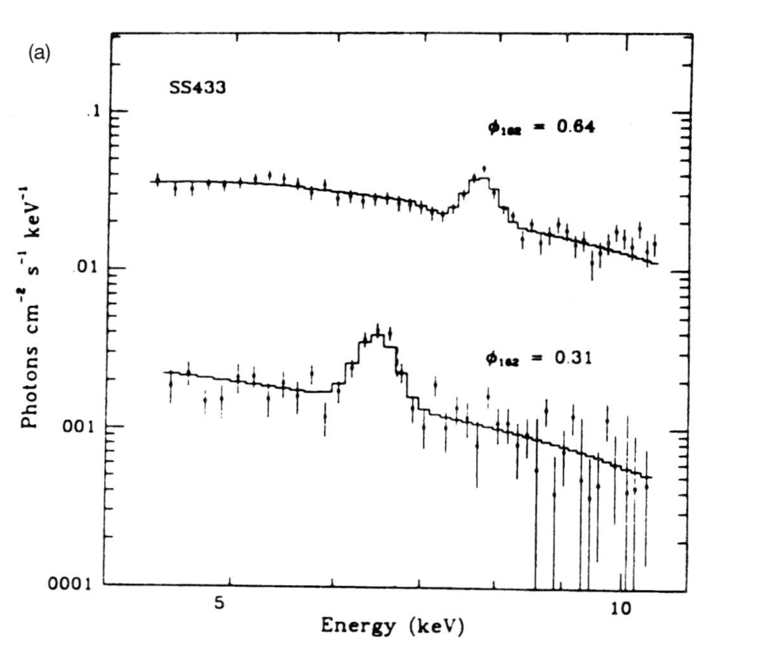

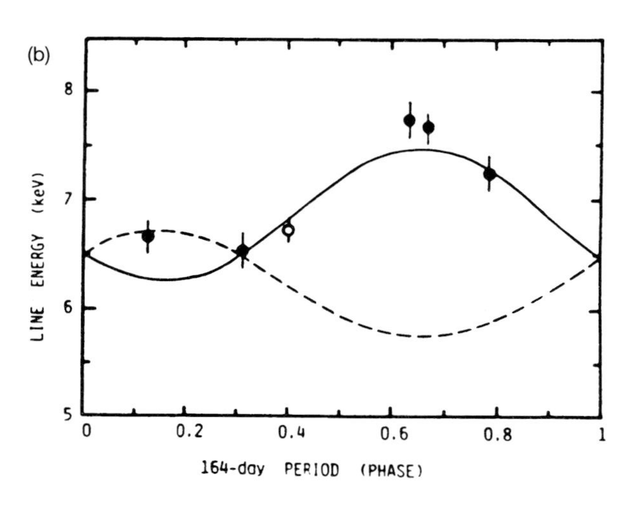

Figure 13.8 X-ray spectra (a) of SS 433 obtained by the Exosat spacecraft at two different phases of SS 433's 164 day period, showing the movement of the 6.7 keV ionised iron feature with phase ϕ. The energy of this feature is plotted with phase as circles (with error bars) in (b), showing that only the blue-shifted (high energy) beam is detected. (Courtesy ESA.)

higher than that measured, this could explain the very high speed of the jets and their high degree of collimation.

Finally, in 1991 Sandro D'Odorico and colleagues observed SS 433 using the New Technology Telescope of the European Southern Observatory, and measured the Doppler shifts of ionised helium within the spinning accretion disc. As a result he and his team estimated the mass of the compact source to be about 0.8 solar masses and of its companion to be about 3.2 solar masses. There is far too much energy in the system for the compact source to be a white dwarf, and yet it is not massive enough to be a black hole, so if D'Odorico and colleagues are correct the compact source must be a neutron star.

13.5 **HEAO-3**

The third and last High Energy Astronomy Observatory, which was launched on 20th September 1979, had a similar spacecraft 'bus' to HEAO-1 and -2 but its experiments were devoted to γ-ray and cosmic ray research. It operated until 30th May 1981 when its attitude control gas was depleted. HEAO-3's results were much more modest than those of its HEAO-2 predecessor, partially because it is more difficult to detect and locate γ-ray sources rather than X-ray sources.

The intense radio source called Sagittarius A* had been discovered at the dynamical centre of the Milky Way in 1974 by the Americans Bruce Balik and Robert L. Brown. Sag A* had been found to have a diameter of only about 20 astronomical units, and yet within this small volume it was emitting about 10,000 times the energy of the strongest known radio pulsar. This was taken to indicate that there may be a supermassive black hole of 1 million solar masses at the centre of the Milky Way. Meanwhile in 1970 a balloon experiment by Robert Haymes and colleagues of Rice University had detected 511 keV γ-rays that appeared to come from the Milky Way, and seven years later a group led by Marvin Leventhal of Bell Labs. confirmed the measurements, indicating that the 511 keV γ-rays were coming from the central region[12] of the Milky Way. It was thought that hot gas surrounding the supermassive black hole was emitting photons that collide with other particles to produce electron-positron pairs, which in turn annihilate each other to produce the observed 511 keV radiation.

---

[12] The instrument, although better than that of Haymes, still had a field of view of 15°, so the source location and size could only be detected to that accuracy.

The γ-ray spectrometer on HEAO-3 was expected to observe these 511 keV photons, but the instrument produced so many 'false positives' that it required two years of patient analysis by a team led by Allan Jacobson of JPL before the 511 keV line be unambiguously detected from the central region of the Milky Way. Interestingly, the intensity seemed to vary over a period of about six months, limiting the size of the γ-ray source to about half a light year. At about the same time, the Einstein Observatory also detected X-rays from a source within about 1 arcmin of the centre of the Milky Way. This varied over a period of about three years, giving an upper bound to the size of the X-ray source.

Jacobson and his team also detected 1.80 MeV γ-rays from the central region of the Milky Way using the HEAO-3 spectrometer. These γ-rays were most likely produced by the radioactive decay of aluminium 26 to magnesium 26 with a half-life of 740,000 years. This observation is interesting as it implies that, as the half-life of aluminium 26 is relatively short, it is probably still being produced in the central region of our galaxy, probably in supernovae and novae explosions.

One of the most interesting results was the detection of low energy γ-rays from the black hole candidate Cyg X-1. HEAO-3 observed Cyg X-1 for 170 days starting on 27th September 1979 and found that it was emitting flickering X-ray emission at about 100 keV. But it was not until the HEAO-3 data was reanalysed in the mid 1980s by James Ling and colleagues at JPL that low energy γ-rays were discovered. This strong γ-ray emission in the band from about 400 keV to 1.5 MeV was only present in the first two weeks of observations, however, as it then disappeared as the hard X-ray emissions increased.

13.6 **Summary**

So these three HEAO spacecraft had added significantly to our knowledge of the universe, finding both new types of object and helping us to understand the processes under way in previously known objects.

The first intermediate polar, H2252-035, had been found in which energy is transferred to the magnetic poles of a white dwarf via an accretion disc, and in which the magnetic field of the white dwarf is not intense enough to force its spin rate to be synchronised with the orbital rate of the binary. X-ray emission had been discovered from RS CVn-type binary stars that appeared to be because of very bright coronas and coronal loops associated with very large star spots on one of the stars in the binary.

Surprisingly, most types of star were found to be X-ray emitters, even cool red dwarfs, and young and apparently rapidly rotating stars were found to be emitting more X-rays than older stars of the same spectral type. Three more X-ray pulsars were found associated with supernova remnants, bringing the number then known to four,[13] including one that appeared to be in an SNR similar to the Crab Nebula. The Einstein Observatory was the first spacecraft to image X-ray sources and, as a result, provided valuable new information on the structure and processes going on in supernova remnants. This spacecraft also helped us to understand, *inter alia*, the most luminous X-ray transient (A0538-66), the source SS 433 in the supernova remnant W50, the active galaxies Cen A and Virgo A (M87), and the quasar 3C 273. These active galaxies and 3C 273 all seemed to contain a supermassive black hole.

Finally, the diffuse X-ray background radiation was found to be not completely isotropic, but appeared to be produced by numerous, faint, unresolved sources. Analysis using the HEAO-1 and Einstein Observatory spacecraft gave evidence that the X-ray background in the range 1 to 3 keV, in particular, was mainly due to unresolved quasars.

So ended the work carried out by these three large American, high-energy spacecraft. The next high-energy spacecraft were non-American, but before describing these we need to change frequency band and consider the International Ultraviolet Explorer which was launched in January 1978, some five months after HEAO-1.

---

[13] Excluding the pulsars that emitted X-rays, but did not appear to pulse in X-rays.

Chapter 14 | IUE, IRAS AND EXOSAT – SPACECRAFT FOR THE EARLY 1980s

14.1 The International Ultraviolet Explorer

The International Ultraviolet Explorer (IUE) spacecraft that was launched in 1978 is the longest-serving astronomical spacecraft to date. It was turned off in 1996 after 18 years of largely trouble-free service, but its conception and birth was far from trouble-free.

IUE started off life in the early 1960s as the Large Astronomical Satellite (LAS) of the newly-formed European Space Research Organisation (ESRO), but escalating cost estimates of this ultraviolet observatory forced its cancellation in 1967. The design was then substantially modified and simplified, and it reappeared in ESRO in the guise of the Ultraviolet Astronomical Satellite (UVAS), but this again lost out in 1969, this time in competition for funds with Cos-B and GEOS (see Section 12.8). Not to be put off, the UK astronomer Robert Wilson, one of the key figures in the LAS/UVAS project, wrote to Leo Goldberg, Chairman of NASA's Astronomy Missions Board, with the suggestion that UVAS could be adopted by NASA to fill the gap between the end of the OAO series of spacecraft, and the proposed Large Space Telescope (due to become the Hubble Space Telescope). His proposal was favourably received and, after a NASA review, it was suggested that, rather than being placed in a low Earth orbit, the spacecraft should be put into a geosynchronous orbit to allow continuous communications with one ground station. This would not only simplify spacecraft operations and control, but would allow the spacecraft to be operated more like a conventional observatory in 'real time', allowing last minute changes in the detailed observing schedule if required. A geosynchronous orbit was also better than a low Earth orbit as the Earth covers less of the observable sky, allowing more long exposures to be undertaken. Because of this a low-dispersion capability was added to the spectrograph design to allow spectra to be recorded from faint objects.

The idea at this stage was that IUE, or SAS-D as it was then called, should be an American spacecraft in the Small Astronomy Satellite series, with some British involvement in the scientific instrument, which was to be a 45 cm diameter ultraviolet telescope operating down to about 120 nm. There was to be no imaging capability, but spectra of bright objects would be produced with a spectral resolution of 0.01 nm, and those of dimmer objects would have a resolution of 0.6 nm. It was then agreed to invite ESRO to participate in the project, which eventually they agreed to do by providing the solar arrays and a European ground station.

The plan was for IUE to be placed in a geosynchronous orbit over the Atlantic and to be operated for 16 hours/day from an American ground station at the Goddard Space Flight Center, and 8 hours/day from an ESRO ground station at Villafranca in Spain. UK and ESRO[1] astronomers would initially share the 8 hours/day European observing time. In addition, both the Americans and the Europeans agreed to allow other astronomers to use the facility under a guest observer programme. Formal approvals were given to the project in the USA, UK and ESRO in 1971, and initial funding was included in the NASA 1973 budget.

The Vidicon tube that was used as a detector was only sensitive to visible light, and so an ultraviolet-to-visible conversion device had to be fitted to the front of the tube. The development of this conversion device was probably the most difficult task in spacecraft programme, and problems with it became so severe that a back-up contract had to be awarded to ITT of Indiana. They successfully produced the final device on a very short timescale. There was also a last-minute panic when it was thought that the coating of the telescope's main mirror was faulty, but happily this proved not to be so.

Mass was a continuing problem during spacecraft development and, in spite of a strict mass control régime, it proved impossible to launch IUE into its planned near-circular geosynchronous orbit. As a result the plan was changed and the spacecraft was launched into an elliptical geosynchronous orbit with an apogee of 45,000 km and a perigee of 26,000 km. Although IUE would still be permanently visible from the American ground station, it was now only visible from the Spanish ground station for about 12 hours per day around apogee. Fortunately for the Europeans, the spacecraft was in a more benign radiation environment around apogee than around perigee, so they

[1] The UK was also a member of ESRO.

had less trouble with particle-induced interference during their allocated 8 hour part of the orbit.

IUE (see Figure 14.1) was 4.2 m long and weighed about 670 kg, including the apogee motor that was to place the spacecraft into its final orbit. Its specified lifetime was three years, with a target of five years, so enough attitude control propellant (hydrazine) was loaded on-board to cover the full five year period. The solar array was designed to produce about 400 W at start of life, and data was transmitted to ground at 40 kbps. Star trackers, a sun sensor and six gyroscopes were included to provide attitude measurement and control to the high degree of accuracy required by the ultraviolet telescope. Although only three gyroscopes were required to control the spacecraft's attitude, because it was known that gyros were unreliable items[2] the spacecraft design included six, any three of which could provide full attitude control.

The IUE telescope had a 45 cm main mirror made of beryllium to save weight, and a quartz secondary because it needed to be less sensitive to temperature changes. There were two detection systems, one covering the range 115–195 nm and the other 190–320 nm, and both systems had redundant Vidicon tubes. As mentioned above, either a high-resolution spectrum could be produced for bright objects or a low-resolution one for dim objects. The Vidicon tube had a storage capability, but the read-out of the spectrum was a destructive process. So if there was any problem in transmitting the data to ground during read-out of the spectrum, the data was permanently lost and the exposure had to be repeated. It was also impossible, for the same reason, to see how the exposure was progressing and stop it when the lines were sufficiently strongly recorded.

IUE was successfully launched by a Thor-Delta on 26th January 1978. The initial in-orbit checkout went well and the first spectrum, that of η Ursae Majoris, was obtained on the third day. There was an early panic when the first image from the short wavelength prime (SWP) camera showed so much noise as to render it useless. But the problem was traced to an errant panoramic attitude sensor and, when this was corrected, the SWP noise disappeared. The onboard attitude control computer also accidentally slewed the spacecraft towards the Sun, but after a short break in communications, control was restored and everyone breathed easily once more.

As mentioned above, there were six gyros on-board to ensure that there were at least three still operational at the end of the spacecraft's target lifetime of five years. In the event three gyros did fail in this five year period.

[2] Even today the Hubble Space Telescope, for example, has had trouble with its gyros (see Section 17.5).

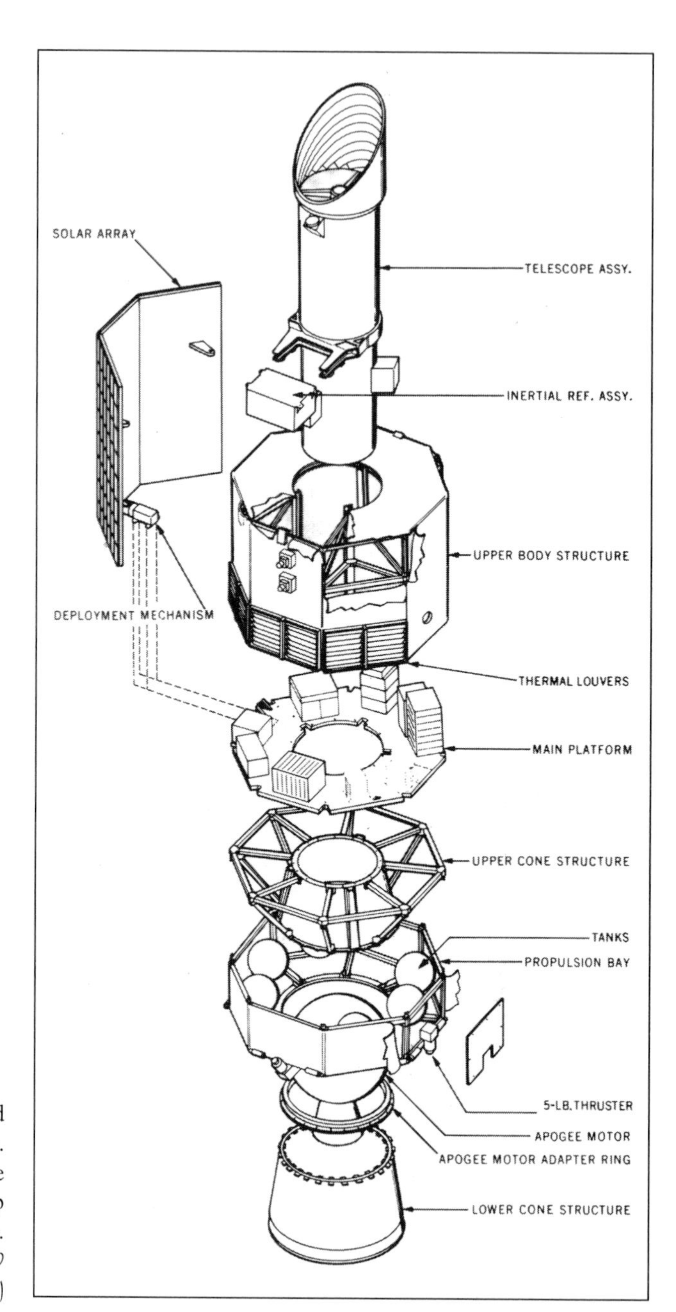

Figure 14.1 An exploded view of the IUE spacecraft. The sun shield is the asymmetric baffle at the top end of the UV telescope tube. (Courtesy NASA and *Sky & Telescope*.)

Then in August 1985 a fourth one failed, but the mission controllers devised an attitude measurement strategy involving the sun sensor and the two remaining gyros which proved perfectly satisfactory. One of the redundant pair of Vidicons in both the short and long wavelength paths failed, but the others continued working. In fact when IUE was finally turned off after 18 years service it was still working well, but NASA and ESA wanted to close it down for financial reasons, and to allow more time to be spent on current and future programmes. About 114,000 spectra were produced and archived over IUE's lifetime. It was a remarkable observatory.

IUE made numerous observations of comets and other Solar System objects, but these will not be discussed here as this chapter is devoted to non-Solar-System research.

The Copernicus spacecraft had shown that massive, hot stars (i.e. categories O, B and A) emit powerful stellar winds which are much more powerful than the solar wind. IUE was the first ultraviolet spacecraft to observe such stars in the Magellanic clouds, and enable a detailed analysis to be undertaken of two particular types of massive, hot stars, namely Be stars and Wolf–Rayet stars.

Be stars have been known as a group for almost fifty years as being stars with hydrogen emission lines superimposed on the normal spectrum of a B-type star (the 'e' in 'Be' stands for emission). Copernicus had shown that, based on measurements of the S IV line, a somewhat larger percentage of Be-type stars appeared to have stellar winds than ordinary B-type stars, but the results were not conclusive. Copernicus had also enabled mass-loss rates to be estimated as being generally between 10^{-11} and 10^{-9} solar masses/year, but there was no clear dependence of mass-loss rate on stellar parameters like luminosity. This worried astronomers as they were trying to understand this mass-loss phenomenon. IUE provided much better spectra than Copernicus, particularly of the N V and C IV lines, and on many more stars. Unfortunately, this data still did not clearly differentiate between the stellar winds in Be and B-type stars as groups, although it appeared that all Be stars hotter than B5 have stellar winds, whereas not all B stars hotter than B5 do.[3] There was also a suggestion that the mass-loss rate for Be stars is higher than for B-type stars.

The first Wolf–Rayet (WR) stars had been found in the nineteenth century as stars with broad emission lines on a continuous background.

---

[3] The star categories B, A, F, etc. (see 'main sequence stars' in the Glossary) are subdivided by numbers running from 0 to 9. So, for example, type B0 is hotter than B1, B1 than B2, etc. down to B9, with A0 coming next. The subdivision of O-type stars, which includes Wolf–Rayet stars, is different, however.

Then in 1930 Beals separated Wolf–Rayet stars, which are all very hot O-type stars, into those with intense nitrogen lines, and those with very strong carbon and oxygen lines. These are now called WN and WC stars. It is thought that WR stars are the helium-burning cores of stars whose initial mass was in the range of 30 to 40 solar masses, and which have lost their hydrogen-rich outer layers due to mass outflow in a very strong stellar wind.

IUE data was used to determine the terminal wind velocities from WR stars by analysing their high-resolution spectra. These velocities were found to vary from about 3,500 km/s for WN3 and WC5 stars[4] down to about 1,500 km/s for WN8 and WC9 stars. Radio or infrared data was used to determine the density of the outermost regions of the gaseous outflow, and this information, together with the velocities deduced from IUE spectra, gave average mass outflow rates of about 4×10^{-5} solar masses/year. This does not sound much, but at this rate a star would lose 1 solar mass of material in just 25,000 years.

The temperatures of WR stars deduced from their visible and IUE ultraviolet spectra appeared to be in the range 25,000 to 50,000 K, but these temperatures were thought to be too low to produce the observed mass loss rates. These temperatures may be in error, however, as they were deduced only after various corrections had been included to take account of interstellar reddening and other factors.

Huber and colleagues analysed the high resolution IUE spectra of the WR star HD 192163 (WN6), which is one of the WR stars that are surrounded by a ring nebula (NGC 6888 in this case). The spectra showed strong lines of Si IV, C IV and Al III with, in each case, a weak, blue-shifted absorption line, displaced by an amount equivalent to the expansion velocity of the ring as determined by optical data. This was the first detection of absorption lines in a nebula associated with a hot star. The temperature of the gas in the ring nebula was estimated to be about 60,000 K, and its chemical composition was like that of the Sun, implying that the nebula is shock-heated interstellar gas rather than ejecta from the WR star; the shock-heating being caused by the ejecta interacting with interstellar gas.

In 1943 Carl Seyfert discovered a number of galaxies, which we now call Seyfert galaxies, with small, intensely bright, very blue nuclei. In short photographic exposures only the nucleus could be detected, looking very much like a star, but in long exposures the spiral arms of the galaxies could also be seen. The spectra of these bright Seyfert galaxies were found to have

[4] These refer to WR stars in the WN or WC sub-categories, arranged according to spectral type.

strong, broad emission lines on a continuum background, the emission lines being attributed to rapidly moving gas clouds. Some of these galaxies were later found to be radio sources, with some of these having variable radio emission. This led astronomers to see if the light emissions also varied, and in 1967 Fitch, Pacholczyk and Weymann found a 0.25 magnitude variation in the intensity of the central region of Seyfert galaxy NGC 4151. Later observations of the nucleus of other Seyfert galaxies in visible light showed that such intensity variations were not unusual.

A plot of energy versus frequency over the visible waveband for the point-like nuclei of Seyfert galaxies showed a steeper slope (or power law) than for high-redshift quasars. This was something of a surprise as it was thought that Seyfert galaxies were similar in nature to quasars, as both have emissions lines on a continuum, although Seyfert galaxies are not as luminous as quasars. This apparent inconsistency in the slopes of the power law graphs was solved by IUE when it showed that the energy/frequency graph for Seyferts was not linear across ultraviolet and visible frequencies, but had a lower slope in the ultraviolet. If the Seyferts were put at the much greater distances of the high-redshift quasars, their ultraviolet spectra would be red-shifted to the visible and would show the lower slope seen in the quasars.

Astronomers used IUE to observe the brightest Seyfert galaxy NGC 4151 on a number of occasions, and found that its continuum emission doubled over periods varying from 5 to 30 days, implying a source size of the order of 1 light week. In X-rays such an increase sometimes takes only ½ day, so clearly the X-ray source is much smaller than the UV source. IUE also found that when the continuum of NGC 4151 is faint, all the ultraviolet emission lines are relatively narrow, but when the continuum brightens, the C IV line brightens and broadens about one to three weeks later. The breadth of this line implies gas velocities of up to about 16,000 km/s. The same happens with the Mg II line about one to three weeks after the brightening and broadening of the C IV line, implying a Mg velocity of about 11,000 km/s, whereas the C III line hardly changes at all. So it was concluded that the Active Galactic Nucleus (AGN) of the Seyfert galaxy NGC 4151 has a small ~½ light day X-ray emitting core surrounded by a ~1 light week source that emits the ultraviolet continuum. Surrounding this is a C IV gas cloud some ~1 light week away, followed by a Mg II cloud a further ~1 light week away and a much more distant C III cloud possibly ~1 light year away. Absorption lines of H I, C II, III & IV, N V, O I, and Si II, III and IV were also measured by IUE from the region surrounding the above nucleus structure, with line widths implying gas velocities of up to about 1,000 km/s. Similar but less complete observations were made by IUE of a number of other Seyfert galaxies which were found to have a broadly similar structure to that of NGC 4151.

14.2 The Infrared Astronomical Satellite

Ground-based infrared astronomy suffers from two basic limitations. Not only is the Earth's atmosphere generally opaque to infrared radiation, but the peak of the black body curve for both the sky itself and for ground-based objects like telescopes at ambient temperature is in the infrared band. This means that ground-based infrared observatories need to be at as high an altitude as possible[5] and the telescopes, and the detectors in particular, need to be cooled to the lowest possible temperature. The obvious solution to the atmospheric problem, which is much more severe than for optical wavelengths, is to place an infrared telescope in orbit around the Earth, but the development of infrared detectors lagged behind X-ray and gamma-ray detectors, so it was not until 1983 that the first infrared (IR) astronomical spacecraft called IRAS was launched.

The American astronomers Gerry Neugebauer and Bob Leighton had carried out the first infrared sky survey in the late 1960s using a simply constructed, Earth-based 1.5 m telescope. It was fitted with a lead sulphide detector cell cooled by liquid nitrogen. They worked in the 2.2 μm (micron), near-infrared window[6] and produced their Two Micron Survey catalogue in 1979 which listed 5,612 infrared sources, most of which were red giant or supergiant stars. This survey was followed in the early 1970s by a series of nine rocket flights under the project name HISTAR, using a 16.5 cm diameter telescope with liquid helium cooled detectors, undertaken by the US Air Force Cambridge Research Laboratory. The survey concentrated on observations at 10 and 20 μm, although they were also made at 4 and 27 μm. The total observing time was only about 30 minutes, but in this short time 2,000 stars and nebulae were detected, as well as interstellar clouds, ranging from very cool stars at the shortest wavelength with temperatures of about

---

[5] Much of the infrared atmospheric absorption is due to water vapour in the atmosphere, and so ground-based infrared observatories also need to be in exceptionally dry areas.

[6] The atmosphere is generally opaque to infrared wavelengths above about 1.1 μm, but there are so-called windows at 2.2 and 3.6 μm where it is almost transparent. Less transparent windows occur at about 1.6, 5, 10 and 20 μm.

1,500 K, to interstellar gas and dust clouds at the longest wavelength at temperatures of about 100 K.

In 1973, with the launch of ANS imminent (see Section 12.5), the Dutch were considering what would be their next scientific spacecraft, and hit on the idea of building a small observatory spacecraft to cover the infrared band from 10 to 100 μm. But as the outline design progressed, it became evident that it was too expensive to be undertaken by the Dutch on their own. In parallel Gerry Neugebauer and Frank Low in the USA had also suggested that NASA build an Explorer-class, infrared spacecraft and, although NASA were interested, they were also short of money. So the Dutch were favourably received when they approached NASA with the suggestion that an infrared observatory spacecraft be produced as a joint NASA/Dutch programme. The Dutch also suggested that the UK be involved to increase the European contribution. This was eventually agreed, and the work share was divided between NASA, who would contribute the infrared telescope, cooling system and launcher, the Netherlands who would build the spacecraft bus, and the UK who would build the ground station and 'quick-look' data analysis facility.

The development of the IRAS spacecraft seemed to have more than its fair share of problems. First an accident at Perkin-Elmer resulted in a large scratch in the primary mirror, then various problems with the MOSFET[7] amplifiers resulted in their having to be replaced by JFETs[8] in a crash programme. The 100 μm detectors were faulty and had to be replaced, and there was a problem with a microprocessor. There was also a fault with the 25 μm detectors, and there was evidence that the Dewar containing the liquid helium coolant was leaking, resulting in a reduced lifetime estimate from the planned 460 days to about 220 days. Then a problem was found with a batch of capacitors that had been found to have a completely unpredictable short circuit failure mode.[9] Things were so bad that at one stage, just two months before launch, the experimenters were split 50/50 on whether to advise the spacecraft authorities that IRAS be stripped down and completely repaired, with a consequent major launch delay. A number of people were concerned, however, that the very act of stripping down the

[7] MOSFET stands for Metal Oxide Semiconductor Field Effect Transistor.

[8] Junction Field Effect Transistor.

[9] The first capacitors to exhibit this short circuit failure mode were on the ESA Meteosat 1 spacecraft which caused a major in-orbit malfunction in November 1979, one day after the end of the spacecraft's design lifetime. Prior to then capacitors were only known to fail open circuit, not short circuit.

spacecraft may cause further problems. NASA were also under considerable pressure to launch 'as is', as not only had their costs increased from an initial $37m to $92m, but their Dutch and British partners had finished their work two years earlier, and it was difficult to keep their teams together whilst waiting for in-orbit data. So the decision was taken to launch.

The IRAS payload consisted of a 57 cm f/9.6 Ritchey–Chrétien reflecting telescope, with a field of view of about 30 arcmin, which was cooled to a temperature of less than 10 K by 72 kg of superfluid helium contained in a large vacuum flask or Dewar. An array of 62 rectangular semiconductor detectors in the telescope's focal plane was cooled to just 2.5 K by the same system. The detectors, which were designed to cover broad wavebands centred on 12, 25, 60 and 100 μm, were arranged in such a way that each source crossing the field of view would generally be seen by at least two detectors in each waveband. This was to help to discriminate real sources from spurious signals.

The main mission of IRAS was to undertake an all-sky survey in the four wavebands from 12 to 100 μm with a resolution of 4 arcmin, and to make detailed observations of selected sources using the Low Resolution Spectrometer (LRS) or the Chopped Photometric Channel (CPC). The LRS was designed to measure the spectrum of bright sources over the waveband from 8 to 23 μm, and the CPC was to map extended objects in the 41–63 μm and 84–114 μm bands with a resolution of about 1 arcmin. Source positions were to be determined by an array of optical detectors to within about 20 arcsec.

The one ton IRAS spacecraft was launched on 26th January 1983[10] from the Western Test Range in California into a 900 km altitude, Sun-synchronous, polar orbit. The telescope's field of view was such that any part of the sky was observed on two consecutive orbits, enabling spurious detections to be eliminated. Other repeat passes were also undertaken during the spacecraft's lifetime to further eliminate such spurious sources, and a complete scan of the sky was completed in 6 months.

Because IRAS was in a low polar orbit, communications with the ground station at the Rutherford Appleton Laboratory (RAL) in the UK was limited to brief intervals 12 hours apart. As a result, an on-board computer was updated from the ground every 12 hours with the required observing programme, and observations were stored on one of two on-board tape recorders and transmitted to ground at a speed of 1 Mbps every 12 hours.

---

[10] 26th January Universal Time, or 18.17 hrs on 25th January local (Pacific) time.

Quick-look analysis was undertaken at RAL, but final processing was carried out at JPL in the USA.

As anyone who has used a reflecting telescope knows, it is very difficult to avoid dust contamination of optical surfaces. It is even worse for space-based telescopes, however, as although they can be protected in a clean-room environment when on the ground, any cold surface in orbit can easily be contaminated by outgassing products from the spacecraft, and the IRAS mirror was very cold. To avoid this contamination, the IRAS telescope was fitted with an aperture cover which was ejected after a few days in orbit, and the spacecraft's attitude control was provided by momentum wheels rather than gas jets.

Shortly after launch it was found that, although IRAS responded correctly to commands when over the RAL ground station in the UK, when it returned one orbit later it had put itself into a 'fail-safe' mode. This pattern was repeated on the next orbit, so to try to understand what was happening NASA temporarily used its ground stations to pick-up the spacecraft's housekeeping data when IRAS was out of view of the UK ground station. This data, which gave the status of the spacecraft's subsystems, showed that spurious signals from the sun sensor were causing the problem. It was subsequently solved by modifications to the on-board computer software.

Eventually on 22nd November, after 300 days, the helium coolant was exhausted and the mission was terminated. Ninety-five per cent of the sky had been observed at least four times, and thousands of point observations had been made. The main IRAS catalogue, which was published in 1984, contained about 245,000 sources, and other specialist catalogues, which were published over the next few years, provided details of asteroids, comets and extended infrared sources.

In the early 1970s the American astronomer Allan Sandage had discovered filaments of dust at high galactic latitudes from his Earth-based observations, but this discovery was not well-known. It was, therefore, something of a surprise when, using IRAS, Frank Low and his colleagues of the University of Arizona detected streaky patches of dust, radiating at about 35 K, all over the sky. Observed mainly at 100 μm (see Figure 14.2), these clouds of 'infrared cirrus', as they are now called, were thought to consist mainly of small carbon particles or large hydrocarbon molecules. Many of the clouds were found to correlate with regions of neutral hydrogen in the Milky Way seen at radio wavelengths, thus suggesting that the cirrus is in our galaxy.

The star Vega is a well-studied, young, stable star about fifty times as luminous as the Sun and about 2.5 times as heavy. It had been extensively

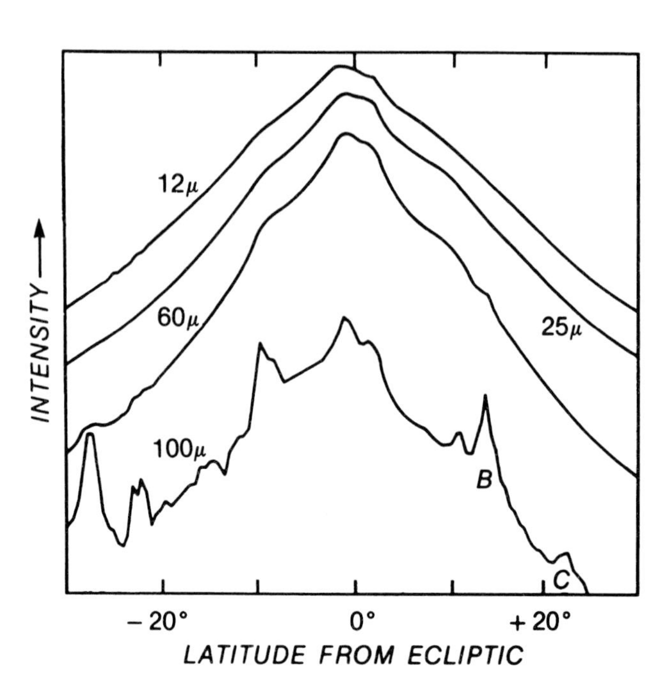

Figure 14.2 Sharp peaks in the 100 μm scan shows evidence of infrared cirrus, which is not much evident in the scans taken in the other infrared wavebands. The peaks labelled B and C coincide with known regions of neutral hydrogen. The broad peak in all wavebands is due to dust along the ecliptic which is the cause of the zodiacal light seen from Earth. (Courtesy NASA and *Sky & Telescope*.)

observed at ultraviolet, optical and infrared wavelengths up to 20 μm, and for these reasons Vega was chosen as one of the calibration sources for IRAS. It was expected that it would be able to extrapolate the measured intensities of Vega and other calibration stars from the UV, visible and near infrared into the far infrared up to 100 μm. All the other calibration stars behaved as expected, but H.H. (George) Aumann and Fred Gillett found that Vega seemed to be too bright in the infrared. The excess emission appeared to be coming from a thin shell of dust at a temperature of about 80 K. The surface temperature of Vega was known to be about 10,000 K, so the IRAS measurements implied that the dust was about 80 a.u. (astronomical units) away from its surface, assuming that the dust was only being heated by the star. IRAS measurements also indicated that the dust grains were rather large, being at least 1 mm in diameter, which is at least 1,000 times larger than the size of interstellar dust grains. This was the first firm evidence of a dust cloud around another star and, as Vega is relatively young (a few hundred million years old), it was thought that we could be observing a cloud of dust which may be in the process of forming planets.

The discovery of Vega's dust cloud created a great stir, and other similar stars were examined with IRAS to see if they too had an infrared excess. A number were found, of which the most interesting was β-Pictoris, which was found to have a shell about 800 a.u. in diameter that could be imaged by ground-based telescopes. Further ground-based observations showed that there appeared to be a region within about 30 a.u. of β-Pictoris which is dust free, possibly because the material that had been there had already formed into planets.

New stars form in dense clouds of interstellar dust and gas that are opaque at visible wavelengths. Fortunately, however, infrared and radio waves can penetrate these clouds, and as cool protostars have their peak energy in the short infrared region, that is the key waveband to find and study such objects. Because of this IRAS was able to detect such protostars for the first time. For example, a small, dark cloud in Perseus, called Barnard 5, was found to contain a very bright object, now called IRS 1, which appeared to be a newly-formed star associated with dust at temperatures ranging from 30 to 800 K. There was also another, much cooler and less luminous object, IRS 2, which appeared to be a star at an even earlier stage of formation.

A number of ordinary stars in dark interstellar clouds were also found by IRAS to be surrounded by dust concentrations. For example, in the dark cloud called Chamaeleon 1 there were large amounts of dust around a number of stars, the most massive being the A-type stars HD 97048 and HD 97300. The dust cloud around the first of these was found to be at a temperature of 1,500 K near the star, but at about 40 K some 20,000 a.u. away from it. The cloud around HD 97300 was seen to be similar but about twice as large.

IUE had provided much valuable information on stellar winds emitted by hot stars, whilst IRAS examined matter ejected from cool red stars near the end of their lifetimes. The red giant Betelgeuse, for example, was known to be emitting about 10^{-7} solar masses of gas per year, and Harm Habing and colleagues in the Netherlands found on IRAS images that the star is also surrounded by three dust shells out to about 4.5 light years. The dust in these shells appears to have been emitted at up to 50,000 to 100,000 years ago. Likewise, IRAS found that the so-called OH/IR stars,[11] which had been first detected by radio astronomers, are surrounded by dense dust shells apparently emitted by these stars. These old, red OH/IR stars, most of which are variable, have dust shells with temperatures down to about 100 K.

So IRAS detected protostars forming from dust clouds, new stars still surrounded by remnants of their original dust clouds, and dust shells

[11] The OH signifies hydroxyl molecules that were found by radio astronomers because of their maser emission.

emitted by stars near the end of their lives. IRAS also found valuable new data at the galactic level, ranging from that for elliptical and lenticular galaxies, which were found to be very weak infrared emitters, to spiral galaxies like the Milky Way, which are reasonable infrared emitters, to a few exceptional galaxies that produce enormous amounts of infrared radiation.

The spiral galaxy M31 in Andromeda, unlike the Milky Way, was found to be a relatively weak infrared emitter. Its infrared emission at 100 μm was found to come mainly from its central region and from the outer regions of the galaxy, both of which were known to contain dust. In the central region the dust is heated to about 35 K, probably by a large number of red giants, whereas the dust in the outer regions is probably heated by a number of hot OB associations.[12]

In 1966 Halton Arp published his *Atlas of Peculiar Galaxies* in a supplement to the *Astrophysical Journal.* Galaxy number 220 in that catalogue, known as Arp 220, is a bright galaxy that has a most unusual shape. It was thought that it may be two galaxies in the process of merging, then in 1984 Tom Soifer of Caltech and colleagues found, using IRAS data, that it was emitting 80 times more energy in the far infrared than in the optical waveband. As a result, its infrared luminosity was about 100 times that of the Milky Way. Such large amounts of energy were only known to come from quasars. But Arp 220 emits most of its energy in the infrared, whereas quasars emit most of their energy in the X-ray and ultraviolet bands. So is Arp 220 a quasar? Some astronomers suggested that it could be a quasar embedded in a thick cloud of dust, but other astronomers thought that we are simply witnessing an enormous burst of star formation or 'starburst' in the nucleus of Arp 220. The starburst theory was later supported by the detection of intense microwave emission from Arp 220, which is often seen in other star-forming regions.

Further work using archived IRAS data unearthed a number of other ultraluminous infrared galaxies like Arp 220, and Bob Joseph of Imperial College, London, and Tom Soifer suggested that such galaxies may be the result of the interaction or mergers between galaxies. Carol Lonsdale of JPL, Neugebauer and Soifer found that this seems to be the case with many of these ultraluminous galaxies. In comparison Michael Rowan-Robinson and colleagues found that for *normal* IRAS galaxies only about 10% or so seem to be merging. Frank Low then suggested that maybe the merging of galaxies triggers a burst of star formation that then produces a quasar-like infrared galaxy. These ultraluminous, infrared galaxies are now generally called starburst galaxies.

---

[12] OB associations are loose groupings of O- and B-type stars.

Finally, in 1989 and 1990 Rowan-Robinson and colleagues used the Isaac Newton Telescope and the William Herschel Telescope on La Palma to take spectra of faint IRAS sources in 700 square degrees of the sky that had no bright infrared sources. In the process they discovered that one source, catalogued as F10214+4724, had an incredible red-shift of 2.29, compared with the previous record for an IRAS galaxy of 0.4. It was then found to be a radio galaxy, and the emission line spectrum in visible light was found to be similar to that of a Seyfert, suggesting that F10214+4724 is a quasar-like source. But it almost certainly contains a huge amount of dust and gas to produce such a strong infrared signal, so it may be a galaxy in the process of formation rather than a buried quasar. Whichever it is, and either is possible, F10214+4724 was the most luminous object known in the universe at that time, emitting some 10,000 times as much energy as the Milky Way.

14.3 **Exosat**

The year 1983 saw the launch of the European Exosat spacecraft just four months after that of IRAS. Exosat, a medium-sized, X-ray spacecraft had first been proposed as a potential ESRO mission some fourteen years earlier, under the name HELOS (Highly Eccentric Lunar Occultation Satellite). In those days its mission had been to measure the position of X-ray sources by lunar occultation, in the same way that the size of the Crab X-ray source had been measured by a sounding rocket experiment in 1964 (see Chapter 1). It was hoped to launch HELOS in the 1974–75 period, between the launch of Uhuru and the Einstein Observatory. If that had happened it would have fitted very well, as it would have determined source positions more accurately than Uhuru, whilst not having the X-ray imaging capability of Einstein. But in 1969 the ESRO Council had just approved the Cos-B and GEOS programmes, and had been forced to cancel their UV spacecraft because of lack of money (see Section 12.8), so there was no chance that another new spacecraft would be a approved over the next year or so.[13]

Initial studies into the design of HELOS took place in 1970, but in 1971 the ESRO Council decided to delay their decision on the next scientific satellite programme until 1973. There were at that stage three competing projects, namely the ionospheric spacecraft IMP, a Venus orbiter, and HELOS,

[13] At that time a satellite development programme of a medium-sized satellite typically took four years, so programme approval would have been required in 1970–71, if the planned 1974–75 launch was to have been met.

the first two being joint programmes with NASA. Further design studies of these three competing spacecraft took place in 1972, and then in April 1973, after a great deal of heated discussion in the European scientific community, the ESRO Council approved IMP as a 1974 start, and HELOS as a 1975 start with a launch in 1979. IMP was later to be called ISEE and HELOS was to become Exosat.

Although Exosat had been approved, its projected launch date was now after that of the Einstein Observatory, rather than 3 or 4 years before as originally anticipated. Unfortunately, this delay made the original HELOS design obsolete, as it was far less performant than Einstein. As a result ESA decided to radically update the Exosat mission and allow the spacecraft's mass to grow from the 150 kg originally envisaged, to 300 kg at first, and then to some 510 kg by the time that it was launched. This new spacecraft design included a much more sophisticated payload, but the design changes could not be made overnight. So the start date of the main development phase was delayed until 1977, allowing more time for the preceding outline design phase.

Exosat was eventually launched by an American Delta rocket on 26th May 1983 into a 350 × 190,000 km orbit inclined at 71° to the Earth's equator, which allowed for uninterrupted observations of about 72 hours per orbit. Exosat's payload (see Figure 14.3), which was designed mainly to examine already known sources, consisted of a pair of low energy (LE) imaging telescopes, similar in concept to the large X-ray telescope on Einstein but of shorter focal length, a medium energy detector array (MEDA) of proportional counters, and a gas-scintillation proportional counter (GSPC). This suite of instruments was designed to measure the precise location of sources, the mapping of diffuse, extended sources, broadband spectroscopy, and the study of time-variability of sources, as well as the detection of new sources.

Both of the 1 m focal length LE telescopes were equipped with two detectors, namely a position-sensitive detector (PSD) and a channel multiplier array (CMA), mounted in the focal plane on an exchange mechanism. Both of these LE telescopes was also equipped with a 500 and 1,000 line/mm transmission grating which produced spectral resolutions of the order of $\Delta\lambda/\lambda = 0.01$ for the CMA, for example. The PSD covered the range from 0.2 to 2.0 keV, with an on-axis spatial resolution varying from 3 to 1 arcmin, respectively, whereas the CMA, which operated over the range from 0.04 to 2.0 keV, had an almost constant spatial resolution with increasing energy of about 10 to 15 arcsec. The PSD and CMA each had a field of view of about 2°.

Exosat's MEDA, which was the only part of the payload that could be traced back to the original concept of HELOS, consisted of eight proportional

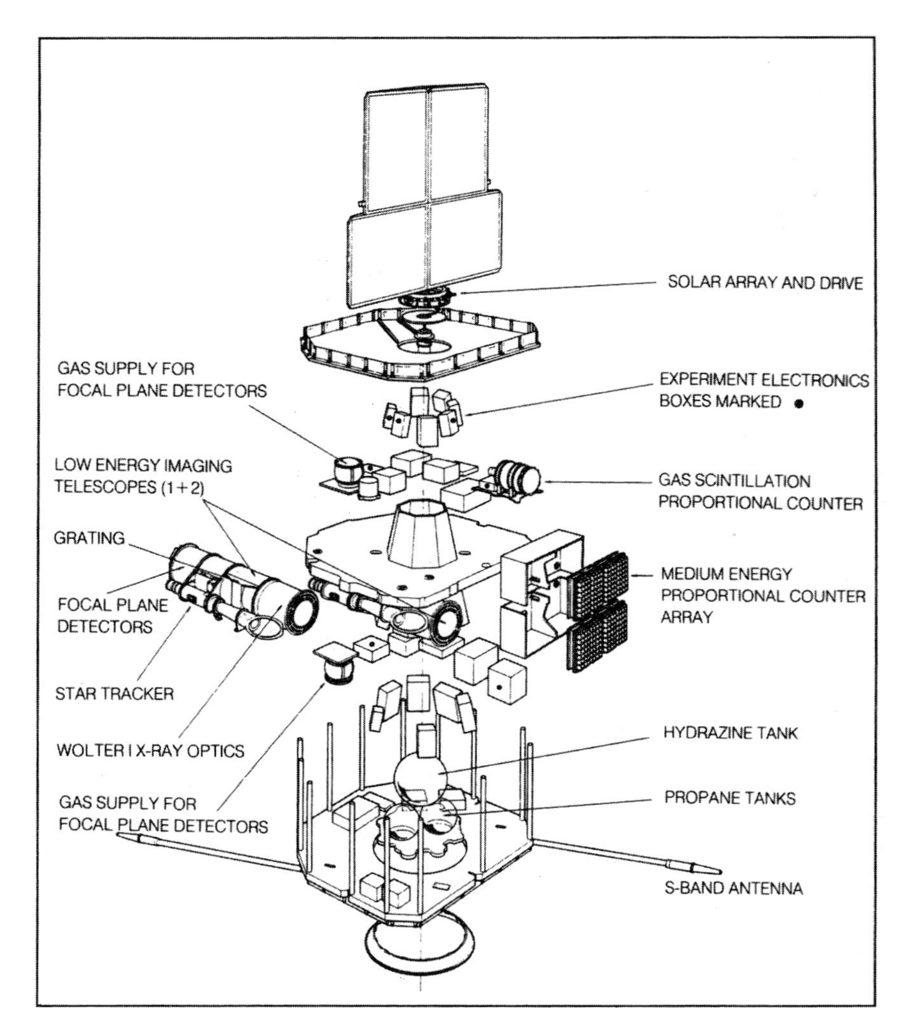

Figure 14.3 An exploded view of the Exosat spacecraft. The two Low Energy Imaging Telescopes were mounted one above the other adjacent to the Medium Energy Proportional Counter Array, otherwise known as MEDA. (Courtesy ESA.)

counters mounted in pairs to form four quadrants. Two mechanisms, each operating on a pair of quadrants, enabled the field of view to be offset by up to 2° so that, for example, one pair of quadrants could observe the desired source, whilst the other pair could measure the background near the source. This background measurement could then be deducted from the source measurement to give a pure source result. The total effective area of the proportional counters was 1,800 cm^2, and they were designed to operate over the energy range from 1.2 to 50 keV. MEDA's field of view was limited by a lead–glass collimator to 45 arcmin, whilst lunar occultation could be used to provide source positions to an accuracy of about 2 arcsec. A time resolution of 10 μs was possible to allow for the study of pulsars and other rapidly varying sources.

Gas-scintillation proportional counters, which were being rapidly developed during the early 1980s, had energy resolutions two or three times better than proportional counters for medium energy X-rays. The Exosat GSPC, a prototype of which had been flown on an Aires sounding rocket, was primarily designed to measure the spectra of relatively strong X-ray sources, and resolve the detailed line structure of the stronger lines in particular. Of specific interest was the structure of the iron emission line complex at 6.7 keV. The GSPC had a similar sized field of view to the MEDA, but a much smaller effective area of about 160 cm^2.

Exosat's initial checkout in orbit indicated that everything was working satisfactorily, except for the sunshade which over-deployed, thus reducing the telescopes' effective area by about 30%. But over the next few months there were a number of other problems. Both PSDs of the two LE imaging telescopes failed, the CMA of telescope 2 also failed, and the mechanism for inserting the grating in telescope 1 failed in the 'out' position. None of these problems could be rectified in orbit so the low energy experiment was seriously compromised. There were also problems with the spacecraft's attitude control system which caused excessive usage of the propane gas supply. Nevertheless, Exosat continued to operate satisfactorily with its remaining experiments past the end of its two year intended mission, until another problem with the attitude control system caused depletion of the remaining propane on 9th April 1986.

The binary AR Lac was known to be an RS CVn system consisting of a G2 solar-type primary and a K0 subgiant secondary, with an orbital period of about two days. It had been first observed in X-rays by HEAO-1, and the Einstein Observatory had been used to monitor it over time, but because of Einstein's low Earth orbit it had only been able to measure AR Lac for about 17% of the binary's orbital period. However, Exosat's highly eccentric orbit allowed continuous monitoring of sources for up to 72 hours, and in July 1984 it became the first X-ray spacecraft to observe AR Lac continuously for one binary period. Both the primary and secondary eclipses were detected in low energy X-rays (see Figure 14.4), with the primary eclipse being particularly deep, and with the secondary eclipse being rather broad and shallow. This indicated that the source of the low energy X-rays, which is gas at a temperature of about 6 million K, must be a bright localised region on or

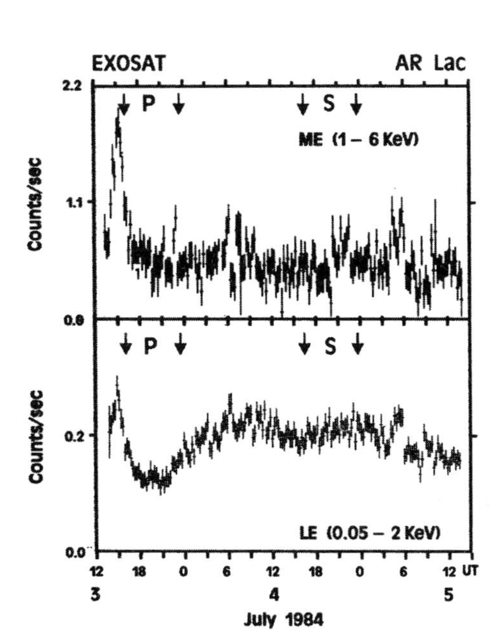

Figure 14.4 Primary (P) and Secondary (S) eclipses for the source AR Lac are seen in the lower graph, which is a plot of the intensity of low energy X-rays versus time as measured by the Exosat spacecraft. No eclipses can be detected for medium energy X-rays in the top graph, however, indicating that the source of these higher energy emissions is larger than that for those at lower energy. (Courtesy ESA.)

near the surface of the G-type star, whereas for the K-type star the source must be extended. No such eclipses could be detected by the medium energy detectors, however. So the 15 million K gas emitting these X-rays must be less localised, and may well surround both stars as a more or less uniform cloud.

Algol, a very well-known eclipsing binary system, consists of a K-type red subgiant star and a B8-type more compact blue star with a binary period of 2.82 days.[14] Exosat observed this binary in 1983 and used the eclipse of the K-type star by its B-type companion to deduce the structure of the corona around the K-type star. Its outer corona was found to have a temperature of about 30 million K, with a lower temperature component of about 6 million K closer to the star. Then on 18th August 1983 Exosat detected a flare at a temperature of 60 million K from the K-type star (see Figure 14.5) that reached a height of about 0.2 stellar radii.

GK Persei or Nova Persei 1901 had been known to exhibit long-duration, irregular optical outbursts of up to 3 magnitudes in amplitude, and it was thought for many years to be a dwarf nova. Then during an optical outburst in 1978 Andrew King and colleagues of Leicester University found, using Ariel 5, that GK Persei was also flaring in X-rays. Another optical outburst

[14] Algol is actually a triple system with the eclipsing binary stars gravitationally linked to a third star with an orbital period of 1.86 years.

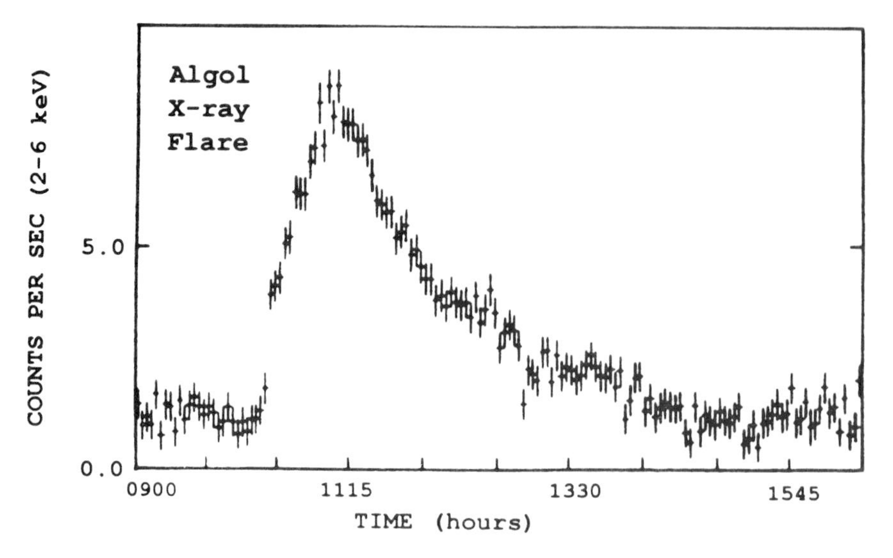

Figure 14.5 The X-ray light curve of Algol showing the flare detected by Exosat on 18th August 1983. (Courtesy ESA.)

was detected in July 1983. So Exosat observations were quickly rescheduled, at Andrew King and Michael Watson's request, to provide eight hours of X-ray data whilst the optical outburst was still under way. This data taken on 8th August showed that GK Persei was pulsing in X-rays with a period of 351 seconds. It was the first detection of such X-ray pulsations in what was thought to be a dwarf nova. There was also a second period of 48 hours detected from Doppler variations in this pulse period, together with evidence of a strong magnetic field, indicating that GK Per is also an intermediate polar.

Tenma or Pegasus, a small Japanese X-ray spacecraft, discovered an X-ray transient on 14th November 1983 which seemed to be a reappearance of the source V0332+53 that had been discovered in June 1973 by the Vela 5B military spacecraft. On this previous occasion it was temporarily the second brightest X-ray source in the sky. Following the Japanese re-discovery, Jaap Davelaar and colleagues then used Exosat to locate V0332+53's position accurately enough for its visible light counterpart to be detected. This proved to be an O- or B-type star with hydrogen emission and helium absorption lines. Exosat also detected pulses from V0332+53 every 4.3 seconds with much more pronounced random flickering on timescales as short as 10 milliseconds. The 4.3 second variation clearly indicated that the source was a neutron star.

Further observations of V0332+53 by Exosat over the next few months showed that the X-ray intensity varied over a period of 34 days due to the binary nature of the source. The binary orbit was found to be highly elliptical, with the transient X-rays occurring only when the neutron star is closest to its companion. These Exosat observations of V0332+53 were crucial as they were the first detection of flickering X-ray emission from a source that was clearly not a black hole. So such flickering cannot be taken as unambiguous evidence of a potential black hole.

As more and more X-ray sources had been detected in the early 1970s it had become evident that the faintest X-ray sources are distributed more or less uniformly across the sky, and so are generally extragalactic in origin, whereas the brightest sources are concentrated in the plane of the Milky Way, with the very brightest sources in the general direction of the galactic centre. These latter are called galactic bulge sources.

Stars in the Milky Way's spiral arms were known to be generally young, whereas those near the centre of the galaxy, in the galactic bulge, were known to be old stars. As massive stars burn up their nuclear fuel very fast, these old galactic bulge sources must consequently have relatively low mass. So, because so many X-ray objects are binaries, it was surmised that the galactic bulge X-ray sources must be binaries with old, low-mass stars. These would be quite unlike the young, massive X-ray binaries in the spiral arms. One of Exosat's targets was to try to aid our understanding of these bulge objects of which virtually nothing was known. Some astronomers suggested, for example, that they may be the progenitors of millisecond pulsars and, if so, this would be a crucial discovery in understanding how pulsars develop.

The first millisecond pulsar, PSR 1937+214, had been discovered in November 1982 using ground-based radio telescopes. Its rapid pulse rate of 1.56 milliseconds was thought to be due to material being deposited on the neutron star over a very long period of time from its binary companion, causing the neutron star to rotate at this rapid rate, but no binary companion could be found. As galactic bulge X-ray sources were also thought to be old binaries, it was thought that maybe the compact source in some of these could show rapid pulsations en-route to becoming a millisecond pulsar. Hence their special interest to X-ray astronomers.

In 1984 and 1985 Michiel van der Klis and colleagues used Exosat to observe one of the brightest galactic bulge X-ray sources called GX5-1,[15] and

[15] GX5-1 was so called because it was at 5° longitude and −1° latitude relative to the galactic centre.

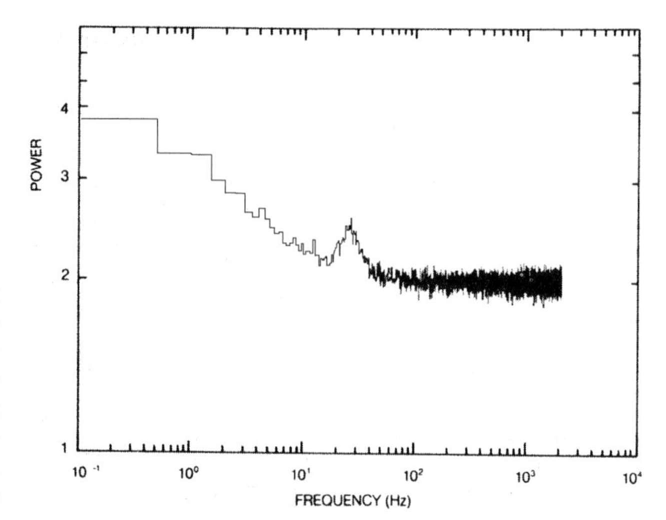

Figure 14.6 The power spectrum of the galactic bulge source GX5-1 as measured by Exosat. Two peaks are clearly seen, one at about 30 Hz and one at the lowest frequency shown of about 0.1 Hz. (Courtesy ESA.)

found that its X-rays showed quasi-periodic oscillations (QPOs), or oscillations whose period was always changing. This discovery caused great excitement amongst the X-ray astronomy community as previously no such oscillations had been found in any galactic bulge sources. The pulsation frequency of GX5-1 showed two peaks (see Figure 14.6), one at very low frequencies, and one at about 30 Hz, but the frequency of both peaks changed randomly with time. Interestingly, however, when the frequency of both peaks was plotted against X-ray intensity of the source, which also varied (see Figure 14.7), the frequency of the higher frequency peak was seen to increase linearly with intensity up to about 40 Hz. But at higher intensity levels the QPOs disappeared completely.

This weird behaviour of GX5-1 presented a major challenge to the theoreticians who had to explain both the linear frequency/intensity relationship and the breakdown of the QPO behaviour at high intensities. The linear relationship was explained as being due to matter being accreted at the poles of the neutron star from the inner edge of the accretion disc whose size changes with time. Because the neutron star is old, its magnetic field is weak, and so the accretion disc extends all the way down to the neutron star's magnetosphere. When there is much material in this accretion disc it compresses the star's magnetosphere, and so the material at the inner edge of the disc will orbit the star more quickly than before. This explains why higher frequencies (or faster rotation of the inner edge of the disc) are associated with higher intensities (or more massive discs). When the accretion disc pressure becomes too great, however, it disrupts the weak magneto-

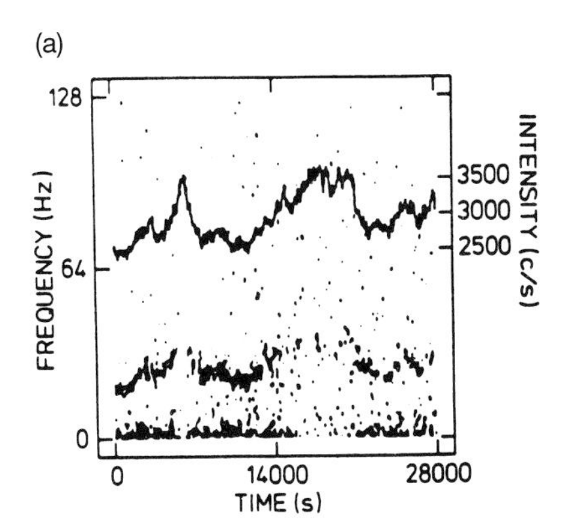

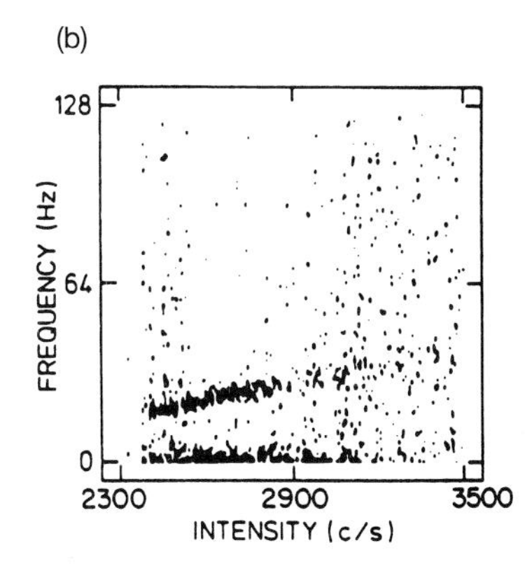

Figure 14.7 (a) The variation in peak frequencies of the QPOs in GX5-1 is shown in the bottom two traces. The X-ray intensity (in counts/s) is shown in the top trace. When the intensity reaches a peak, the QPOs disappear. This is also shown in (b) where peak frequency is plotted against intensity. (Courtesy ESA.)

sphere, the inner edge of the accretion disc becomes chaotic and the QPO behaviour disappears.

Once the QPO behaviour of GX5-1 had been detected, many more such objects were found in the galactic bulge and elsewhere. Sco X-1, Cyg X-2 and the Rapid Burster, in particular, were all found to exhibit quasi-periodic oscillations. Sco X-1 and Cyg X-2 had been known to flicker in X-rays as long ago as the 1960s and, although Bill Priedhorsky and John Middleditch of Los Alamos found QPOs in Sco X-1 (using Exosat), its behaviour was even more confusing than that of GX5-1 as its QPO depended on the form of its X-ray spectrum. The QPO behaviour of the Rapid Burster was discovered using the small Japanese X-ray spacecraft called Hakucho or Cygnus. Remarkably, although the strongest QPOs in the Rapid Burster only occurred during the bursts themselves, QPOs were not observed in all of the strongest bursts! This complex behaviour is not presently understood.

14.4 **Summary**

This set of three medium-sized astronomical observatories operating in the ultraviolet, infrared and X-ray bands that were launched in the late 1970s/early 1980s all had significant European contributions. In fact, the Exosat programme was an ESA programme. Although the Americans were still the leaders in astronomical spacecraft, they continued to seek as much international support as possible for their programmes, as well as supporting ESA and other international programmes with hardware. This period also saw the end of their monopoly in launchers for medium- and large-sized spacecraft with the first launch of the European Ariane launcher in 1979.[16]

These three astronomical observatories of IUE, IRAS and Exosat continued to find new types of astronomical object. For example, IRAS discovered infrared cirrus, starburst galaxies, protostars and dust shells around young stars like Vega and β-Pictoris. It also found dust shells around old, red stars like Betelgeuse and OH/IR stars. Exosat discovered quasi-periodic oscillations in galactic bulge objects, and found that Sco X-1, Cyg X-2 and the Rapid Burster all exhibited QPOs.

IUE enabled the velocities of stellar winds emitted by B-type, Be and Wolf–Rayet stars to be measured. In the latter case the mass outflow seemed to be so great that a Wolf–Rayet star would lose one solar mass of material in only about 25,000 years. Absorption lines were also detected for the first time from a ring nebula surrounding a WR star, which showed that the nebula appeared to be shock-heated, interstellar gas and not gas emitted by the WR star. IUE also helped to analyse Seyfert galaxies and showed that their energy/frequency profile was similar to that of quasars.

Finally, Exosat provided much valuable information on intermediate polars and on stellar flares in binary stars, and showed that all flickering, X-ray sources are not black holes.

---

[16] Although the Japanese also had a launch capability at this time, it was limited to launching small Earth-orbiting spacecraft. It was not until 1985 that their Mu-3SII rocket became available to launch medium-sized spacecraft into low Earth orbit, or small spacecraft into interplanetary space.

Chapter 15 | HIATUS

15.1 COBE

IRAS and Exosat had been launched in 1983, but then there was a long hiatus in both the NASA and ESA astronomy programmes until the launch of Hipparcos[1] and the Cosmic Background Explorer, or COBE, in 1989. This was due to a number of reasons, the most obvious being the Challenger disaster of January 1986, which delayed the launch of a number of spacecraft, especially that of the Hubble Space Telescope (HST, see Chapter 17). But as it happened, NASA did not have many astronomical spacecraft to be launched in the second half of the 1980s anyhow, because of severe budgetary constraints in the 1970s and early 1980s. At the time of publication of the Field Report for the National Research Council in 1982, for example, apart from the HST, the only other approved astronomy programmes were COBE, the Compton Gamma Ray Observatory (then called the GRO), the Extreme Ultra-Violet Explorer (EUVE) and the X-Ray Timing Explorer (XTE). In the event, of these only COBE would be launched before the end of the 1980s. But before going on to explain what COBE was and what its results were, some background to its mission is called for.

---

[1] Hipparcos was an ESA astrometry spacecraft designed to measure the positions, proper motions, parallaxes and intensities of more than 100,000 stars to an unprecedented accuracy. Although Hipparcos was launched in 1989, its main catalogue was not issued until 1997, because of the length of time required to analyse and correct the highly complex raw data. It was only then that the data could be analysed. Because of this, the results are only just starting to appear at the time of writing (in 1998), and so the project is not considered in this book.

In 1937 Theodore Dunham and Walter Adams at the Mount Wilson Observatory discovered cyanogen (CN) in interstellar space. Four years later Andrew McKellar analysed this CN absorption line and concluded that the cyanogen was at a temperature of about 2.3 K. This non-zero temperature was not considered too surprising, however, as it was assumed that the cyanogen was being heated by collisions with electrons or by photon absorption. Then in 1946 Robert Dicke of MIT invented the differential microwave radiometer and used it to scan the sky at about 1 centimetre wavelength, concluding that there was a background radiation in space at a temperature of less than 20 K. Dicke, who was unaware of McKellar's work, thought that the radiation was possibly coming from very distant galaxies, but he did not follow up his observations. In a third parallel, theoretical development, Ralph Alpher and Robert Herman of George Washington University concluded in 1948 that the Big Bang would have produced radiation that should still be observable throughout the universe. They were unaware of McKellar's and Dicke's work, and estimated that the temperature of this background radiation should now be about 5 K, which would mean that its energy peak would be near 2 mm wavelength in the microwave region.

In the early 1950s Tanaka in Japan, Medd and Covington in Canada, and le Roux in France thought that they had found evidence for this background microwave radiation, but their results were not accurate enough to be convincing. Then in 1960 Edward Ohm of Bell Labs was calibrating a horn antenna, to be used for the first trans-Atlantic television broadcast, when he accidentally detected background noise with a temperature of 3.3 $\pm$3 K. Ohm was not an astronomer, however, and so carried on with his engineering calibration, unaware of his potential discovery. Finally, in 1964 Arno Penzias and Robert Wilson, using the same horn antenna at Bell Labs, accidentally found that the whole sky was radiating as a black body at a temperature of 3.5 ± 1.0 K. Unlike Ohm, however, they consulted an astronomer to see if their measurements had any astronomical significance, but that was not until they had spent a year trying to find its origin themselves, without success. The astronomer they consulted was Robert Dicke, who at that time was planning to try to detect the background radiation himself. He immediately recognised what Penzias and Wilson had found.

In observing this microwave background radiation we are looking back in time to a period about 300,000 years after the Big Bang when the universe, which then had a temperature of about 3,000 K, suddenly became transparent.[2] Since then the universe has cooled to this temperature of 3.5 K. Another way of looking at this is to say that the 3,000 K radiation has now been red-shifted to appear to have a temperature of 3.5 K.

The discovery of the microwave background radiation gave the Big Bang theory added credibility, but it also raised a number of interesting questions. For example, is the spectrum of the radiation really that of a black body, and is it isotropic? It was expected that the radiation would not be completely isotropic for two reasons. Firstly, it was difficult to see how galaxies could have formed in an isotropic universe, and secondly, motion of the Earth relative to the background should be detectable as a blue shift in the radiation in the direction of travel, and a red-shift in the opposite direction. This latter type of anisotropy is called a dipole anisotropy.

An anisotropy in the microwave background, of about 1 part in 1,000, was first detected in 1976 by a Princeton high-altitude balloon team led by David Wilkinson, and confirmed the following year by an experiment mounted on a U-2 aircraft by Rich Muller and George Smoot of the University of California at Berkeley. This anisotropy was in the form of a dipole anisotropy, but surprisingly the red and blue shifts appeared to be in the wrong direction if they were just detecting the known rotation of the Sun around the centre of the Milky Way. In fact they had found that the Milky Way itself was moving at a velocity of about 600 km/s relative to the background. This was towards a point, as seen from Earth, in the constellation of Serpens Caput.

In the meantime, in 1974 NASA had issued an Announcement of Opportunity to members of the scientific community for them to propose astronomical missions using small- or medium-sized Explorer spacecraft. This generated 121 proposals, three of which were to investigate the microwave or cosmic background radiation. John Mather (New York), Rainer Weiss (MIT), David Wilkinson (Princeton) and Mike Hauser (Goddard) and their colleagues proposed flying a four experiment spacecraft to measure (i) the anisotropy in the microwave background at wavelengths from 3 to 16 mm, with a sensitivity of one part in about 10,000, (ii) the anisotropy in the far infrared background, with a sensitivity of about 1 part in 100,000 and an angular resolution of about 5°, (iii) the spectral distribution of the microwave

[2] At temperatures in excess of about 3,000 K photons had enough energy to ionise hydrogen atoms, and so were easily absorbed. Once the temperature fell below this level, however, the photons no longer had enough energy to do this, and so could travel more freely through space.

background, and (iv) variations in the infrared background at wavelengths from 8 to 30 microns due to very early galaxies. Samuel Gulkis and colleagues at JPL also suggested measuring the anisotropy in the microwave background on a Scout-launched spacecraft, and a team under the leadership of George Smoot at Berkeley suggested a similar instrument to that of Gulkis, together with a much larger one launched by the Space Shuttle to measure the spectral distribution of the microwave background. The Mather group's spacecraft, which was to have been launched by a Delta, was the most complicated design as it involved cooling most of the detectors down to a liquid helium temperature of just 4 K.

NASA were impressed by the three cosmological spacecraft proposals, but there was far more interest in the twelve proposals for infrared experiments, particularly in the National Academy of Sciences. As a result the USA agreed in 1975 to a joint infrared spacecraft programme with the Netherlands called IRAS (see Section 14.2). In the case of the cosmological spacecraft projects, however, although there was a great deal in common between three proposals in conceptual terms, there were major differences in the detailed designs. As the field was new to space, NASA decided in 1976 to ask specified individuals from all three groups to work together to produce a joint conceptual design. This resulted, in the following year, in the proposed Cosmic Background Explorer (COBE) spacecraft, to be launched into a polar orbit by either a Delta rocket or the Space Shuttle. It would include the following:

(i) A cosmic background mapping instrument to detect the intrinsic[3] anisotropy in the microwave background using a differential microwave radiometer (DMR), with George Smoot as Principal Investigator (PI).[4]
(ii) A far infrared absolute spectrophotometer (FIRAS) to measure the spectrum of the cosmic background radiation to see if it was a black body curve, with John Mather as PI. The sensors would be cooled to about 4 K in a liquid helium Dewar.
(iii) A diffuse infrared background experiment (DIRBE) to detect early infrared galaxies, with Mike Hauser as PI. The sensors would also be contained in the liquid helium Dewar.

[3] That is the anisotropy after that due to the movement of the Earth relative to the background has been eliminated.

[4] The decision as to who would be the PI for this experiment was left open when the COBE proposal was originally submitted on 1st February 1977. George Smoot was appointed as PI in October 1978.

The proposal was approved by NASA in principle provided the spacecraft's cost, excluding launcher and data analysis, did not exceed $30 million, and on this basis the preliminary design phase commenced in late 1977. But the NASA Explorer budget line, against which the COBE project was to be charged, started running into difficulties in the late 1970s because of serious cost over-runs on the IRAS programme. This resulted in the start of COBE's detailed design and development phase being continually delayed and, at one stage, it looked as though COBE would be cancelled. Eventually, however, NASA decided in 1981 to start COBE's detailed design and development phase in the following year, with a launch planned on a Space Shuttle from the Vandenberg Air Force Base on the West Coast at a date to be determined.

By 1980, in spite of many sensitive experiments on board balloons, aircraft and sounding rockets, it had proved impossible to find any anisotropies in the microwave background, other than that due to the movement of the Earth. Then in 1982 George Smoot and his colleagues took advantage of the most recent detector developments, and cooled their DMR detectors down to liquid helium temperatures on both a U-2 and balloon flight. However, even this was not sufficient to detect the intrinsic anisotropy in the microwave background. So George Smoot considered that the DMR design on COBE with its uncooled detectors was not sensitive enough to do its job and, in view of his experience with liquid helium cooled detectors, he pushed for such cooling to be provided for the DMR on COBE. Making significant design changes this late in a project carries with it its own risks, however, and NASA was reluctant to make any such changes, but after many heated exchanges a compromise was agreed. The lowest frequency radiometer (operating at 20 GHz) would be flown as a balloon payload rather than on COBE, the 31 GHz DMR would fly on COBE at 'room temperature', and the two most important DMR's, those operating at 53 and 90 GHz, would be cooled down to about 140 K.

The COBE development programme benefited from IRAS, as both used a liquid helium Dewar and the same design of infrared detectors. So the launch of IRAS in January 1983 and its subsequent successful operation was a great relief to the COBE team. Although it benefited from IRAS, COBE had its own development problems, but by January 1986 its structure had been built and tested, along with much of the spacecraft electronics. So COBE was on schedule to be launched by the Shuttle from Vandenberg in early 1989, its assigned launch date. Then the Space Shuttle Challenger exploded, putting all shuttle launches on indefinite hold.

It was obvious that all plans to launch any space shuttles from the Vandenberg Air Force Base on the West coast would be scrapped following the Challenger disaster, as the US military had decided to return to the use of conventional rockets. So after the suggestion to launch COBE using the European Ariane rocket fell on very deaf ears in NASA, it was decided in September 1986 to use an American Delta launcher from Vandenberg. The main problem was that the mass of the already designed spacecraft had to be reduced by over 50% without delaying the launch date! This difficult mass target was largely met by eliminating the on-board propulsion system that had previously been required to take the spacecraft to its operational altitude of about 900 km from the 300 km orbit of the shuttle, as the Delta could put the spacecraft directly into the required 900 km orbit. This meant that the spacecraft's structure could also be reduced substantially in size. The thermal shield was also reduced in size and made deployable, and the structural elements required to fix the spacecraft to the space shuttle were eliminated. As a side effect this exercise showed how inefficient the space shuttle was in launching (for it) small satellites into a medium altitude polar orbit.

Eventually after further launch delays totalling nine months, COBE was successfully launched by a Delta from the Vandenberg launch site into an almost perfectly circular, polar orbit on 18th November 1989. It was the first NASA astronomical spacecraft to be launched since the Challenger disaster almost three years earlier.

In the meantime, two significant astronomical developments had occurred during COBE's long gestation period. Firstly, in 1981 two teams of astronomers, one led by David Wilkinson at Princeton and the other by Francesco Melchiorri of the University of Florence, announced almost simultaneously that they had detected a quadrupole[5] distribution in the microwave background radiation using balloon-borne instruments. If this was correct it may have been a detection of the intrinsic variability of the background radiation that COBE was designed to measure. A number of experimenters tried to replicate the results but were unable to do so.

The second development was an announcement made in 1987 by a Japanese–American team, headed by Andrew Lange and Paul Richards of the University of California at Berkeley and Toshio Matsumoto of Nagoya University, that the background radiation was not that of a true black body. In particular they detected an excess brightness at 0.7 and 0.5 mm wavelengths using a sounding rocket experiment. This caused consternation amongst theoreticians who then had to explain why the cooling Big Bang

[5] A dipole has one positive or hot pole and one negative or cool pole. A quadrupole has two of each.

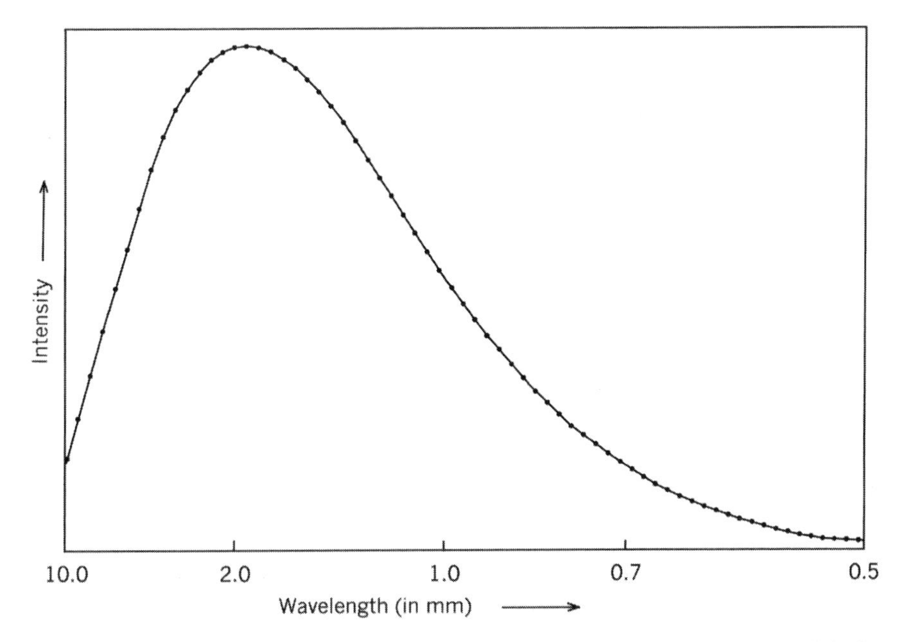

Figure 15.1 COBE's FIRAS data is shown as small circles and the theoretical black body curve is shown for a temperature of 2.73 K. The agreement is excellent.

radiation did not have a black body distribution. A small difference could be explained, but the measured difference was quite large. The Berkeley–Nagoya results even cast doubt on the validity of the Big Bang theory, particularly as the alternative Steady State theory could easily explain the observed effect.

With these two developments at the back of their minds, astronomers watched anxiously as COBE reached its correct orbit and then started to deploy its various appendages. There was a flutter of concern when one of the gyroscopes failed, but the back-up gyro system worked satisfactorily, then four days after launch the Dewar cover was successfully ejected.

The first results from COBE were presented by John Mather, who by then had moved to NASA's Goddard Space Flight Center, to the American Astronomical Society Meeting in January 1990. They were startling (see Figure 15.1), showing as they did an exact fit between the spectrum of the microwave background radiation, as measured by the FIRAS instrument on COBE, and the theoretical curve for a black body at a temperature of 2.73 K. The Berkeley–Nagoya results had been erroneous.

It took two more years before George Smoot of the University of California announced the equally important results from the DMR, which were produced using a whole year's worth of data. Before presenting these results Smoot and his team had had to deduct the effect of the motion of the spacecraft in space, which included the motions of the Earth, Sun and Milky Way. They then had to eliminate known sources of Earth- and Sun-generated interference, and the effect of discrete astronomical sources. Only then were they able to announce that the residual fluctuations from place to place in space, which showed a maximum amplitude of about 3×10^{-5} K, or about one part in 100,000, were greater than the statistical fluctuations caused by errors in measurement. But these preliminary results were not sensitive enough to show individual structures in the universe. They were of enormous importance, however, as they showed, for the first time, that the universe was not perfectly smooth about 300,000 years after the Big Bang, and the magnitude of the fluctuations gave cosmologists something on which to judge their theories.

Finally, the DIRBE instrument showed that the plane of the Milky Way is slightly warped, and that its central bulge is not symmetrical. This instrument also found that the interplanetary dust cloud, which is seen from the Earth as the zodiacal light, is not centred on the Sun, as previously supposed, but on a place in space a few million kilometres away. This is due to the gravitational influence of Jupiter and Saturn.

COBE continued to operate normally until 21st September 1990 when the last of its helium coolant evaporated, and the infrared sensors in the FIRAS instrument started to heat up, terminating that experiment. The DIRBE instrument carried on working a little longer at reduced sensitivity, whilst the DMR, whose detectors were outside the Dewar, continued to operate normally until they were finally switched off in December 1993, just over four years after launch.

15.2 Supernova 1987A

The brightest supernova since 1604 was discovered on 24th February 1987 by Ian Shelton working at the Las Campanas Observatory in Chile. This supernova, now called Supernova (or SN) 1987A, was thought to be a type II supernova. But before going on to describe its discovery and its subsequent space- and ground-based observations, the theory of type II supernovae, as understood before 1987A, needs outlining.

A type II supernova is produced when a star with a main sequence mass in excess of about eight solar masses gets to the end of its lifetime. When hydrogen in the core of a such a heavy star has been converted into helium,[6] the core collapses and the temperature becomes high enough for helium to fuse into carbon and oxygen. This produces even more heat, and when the helium has been used up in the core, the carbon is converted to neon, magnesium and sodium. Successive processes continue, producing ever higher temperatures and heavier elements, until iron is produced in the core. Then there are no further processes possible, and the heat energy is abruptly cut off. At this stage the star consists of shells of gas, with the lightest elements hydrogen and helium in the outer shell, and with shells of ever heavier elements going towards the core of iron.

Once the star is no longer producing heat and radiation, nothing can stop it collapsing and the shells rapidly collapse onto the iron core. There is a limit to this contraction, however, and when the core of the star has been compressed to the density of neutrons in an atomic nucleus, it cannot usually be compressed any more, and rebound occurs, setting up a shock wave which progresses outwards through the star to its surface. Neutrinos are produced when the core collapses, and they move towards the surface much faster than the shock wave. When the shock wave reaches the surface, we see the sudden increase in light which is the visible signal of a type II supernova. As the neutrinos arrive at the surface of the star first, however, and as the neutrinos and light both travel with the speed of light outside of the star, the time difference in receiving their signals on Earth tells us how much longer it took the shock wave to get to the surface of the star from the core. This gives us some idea of the size and condition of the progenitor (i.e. pre-explosion) star.

The theory further predicted that, after the explosion of a type II supernova, there would remain a central very dense neutron star, or possibly a black hole (if the star had been massive enough to overcome the neutron degeneracy pressure during its collapse). Surrounding this neutron star, or black hole, is a rapidly expanding shell of gas which cools at it expands, which has been blown off by the shock wave.

When the iron core collapses, an enormous number of neutrons is released, and these are captured by atomic nuclei in the so-called 'r-process' in which neutrons are added to middleweight nuclei so rapidly that they do not have enough time to decay before another neutron is added. This key process enables elements heavier than iron to be produced.

---

[6] The conversion of hydrogen to helium is the normal energy production process for stars on the main sequence, but only those stars with main sequence masses of more than about 8 solar masses follow the sequence described here, ending up as a type II supernova.

The discovery and investigation of SN 1987A, which was crucial to testing the predictions of this theory of type II supernovae, will now be discussed.

In the early hours of 24th February 1987, Ian Shelton noticed what appeared to be a new bright star in the Large Magellanic Cloud (LMC) on a photographic plate that he had just developed. It was not on the plate that he had exposed the night before at the same Las Campanas Observatory. So he went outside to check, and was astonished to see a new star that could be easily observed with the naked eye. By 0800 hours UT (Universal Time) on 24th February it had brightened to visual magnitude 4.5, making it an easy object to see with the naked eye.

Albert Jones, an amateur astronomer in New Zealand, independently discovered the supernova at about 0900 UT, when he was setting up his telescope for an evening's observation. He immediately telephoned Robert McNaught at the Siding Spring Observatory in Australia, who confirmed the discovery at 1055 UT. They had all seen the nearest supernova to be discovered since Kepler's supernova of 1604, and within hours all the main observatories in the southern hemisphere had been informed. The world astronomical community was buzzing with excitement at the prospect of the first close-up look at a supernova, if 160,000 light years can be said to be 'close-up'.

By good fortune, Robert McNaught had taken some photographs of the LMC on the previous night, but had not developed them. When he did so, he found that his photograph taken at 1040 UT on 23rd February recorded the supernova at magnitude 6.0, or just on the limit of naked-eye visibility. Ian Shelton's first photograph, which had not shown the supernova, had been taken at 0230 UT on 23rd February. Albert Jones had observed the same area with his finderscope at 0920 UT on the 23rd and had not seen anything unusual. So the timing of the supernova's rapid increase in intensity could be deduced as sometime between 0920 and 1040 UT on 23rd February.

As mentioned above, theories of supernova explosions predicted the generation of neutrinos of very high energy a few hours before maximum light output, as they travel through the collapsing star faster than the shock wave which produces the maximum intensity of the supernova. Astronomers did not realise it at the time, but they had recorded the neutrinos from the supernova on 23rd February. Eleven neutrinos had been detected by the Kamiokande II detector in Japan over a period of 13 seconds at 0736 UT on 23rd February, about 3 hours before the supernova had been detected optically, and eight more neutrinos had been detected, simultaneously with the Japanese neutrinos, by the IMB detector in the USA, over a period of 6 seconds.

These neutrinos were the first ever to be detected as the result of a supernova explosion, giving a powerful confirmation of the theory. As neutrinos have no electric charge, and little or no mass, they can travel straight through the Earth, which is why they could be detected in Japan and the United States, when the supernova was below the horizon. The neutrinos that had been detected had come right through the Earth! It was also found that the number of neutrinos detected on Earth was consistent with the theory of the formation of a neutron star in the explosion.

After its initial rapid rise in intensity, SN 1987A continued to gradually brighten from a visual magnitude of 4.5 on 24th February to magnitude 2.8 in mid May, before it began a rapid decline. Astronomers had, in the meantime, examined old photographs of this area of sky, and McNaught had found that the position of a blue supergiant called Sanduleak −69°202, of magnitude 12.2, coincided with that of SN 1987A. Sanduleak −69°202 had been of spectral type B3, with a surface temperature of 16,000 K and a mass of about 15 Suns.[7] This was the first time that the progenitor star had been found for a supernova, but it presented something of a problem as it had been thought that red supergiants, rather than blue ones, developed into supernovae.

IUE spectra taken on 25th February of SN 1987A showed a rapidly expanding envelope at a temperature of about 14,000 K, then on 1st March infrared measurements showed a gas temperature of 5,700 K, with a cloud expanding at 18,000 km/s. By the end of March the gas temperature had fallen to 4,700 K, and the expansion rate had reduced to 10,000 km/s.

Although the peak intensity of SN 1987A was very high, it appeared to be 2 or 3 magnitudes fainter than it should have been. This puzzled astronomers for some time before they realised that it was because the progenitor star was a blue supergiant, rather than a red supergiant. Blue stars are hotter than red stars, and so blue supergiants are smaller than red supergiants of the same intensity, and with a smaller star the explosion is smaller. Interestingly, the time delay on Earth between the arrival of the neutrinos and the visible light of the explosion was only about 2 or 3 hours, confirming that it was a blue supergiant that had exploded, as the time delay for the shock wave to get to the surface of a larger red supergiant would have been about a day.

---

[7] Theory suggests that this star had a main sequence mass of about 20 solar masses, but about 5 solar masses of material had been lost in the subsequent supergiant phase before it exploded.

Theory also predicted that not only are neutrinos produced when the core of the supergiant star collapses, briefly to produce a core temperature of 10,000 million K, but a large amount of nickel 56 is also produced. This nickel 56 is radioactive and decays spontaneously to cobalt 56, which in turn is radioactive, decaying spontaneously to iron 56. In radioactivity not all of the atoms decay at the same time, and the time taken for half of the atoms to decay is called the half-life of the process. The half-lives of the nickel/cobalt and cobalt/iron processes are 6 and 77 days respectively. These radioactive processes heat up the core of the supernova, which is why the light output of the supernova peaked at about 88 days after its initial outburst, again confirming the theory of supernovae.

The decay of cobalt 56 to iron 56 was also expected to produce a great number of very energetic (847, 1238 or 2599 keV) γ-rays, some of which would be converted in collisions in the supernova to less energetic X-rays. It was expected that these γ-rays and X-rays would be detected as the gas shell surrounding the central neutron star got thinner, allowing them to escape into space. The theory was broadly vindicated when, hard (10–30 keV) X-rays were detected coming from 1987A in late June 1987 by the Japanese Ginga spacecraft. This was somewhat earlier than expected, possibly because the radioactive cobalt had been ejected into the higher levels of the expanding debris, and so was observed through less dense layers than originally thought. Alternatively, the debris shell may not be of uniform density, allowing X-rays to escape through the less dense regions. The X-ray intensity peaked in October 1987.

Following the detection of X-rays, 847 and 1238 keV γ-rays were detected by the Solar Maximum spacecraft in August 1987, further confirming the theory of type II supernovae. Then in November 1987 astronomers, using the Kuiper Airborne Observatory, were the first to detect unambiguously a spectral emission line due to a radioactive element in a type II supernova when they clearly detected the line of singly ionised cobalt 56 at about 10.6 microns in the infrared.

A double ring structure, caused by light from the supernova explosion being reflected off an intervening gas cloud, was discovered by Arlin Crotts of the University of Texas on 3rd March 1988, using the 1.0 m telescope at Las Campanas. Subsequent investigations showed that these rings had first been recorded on 13th February 1988 by Michael Rosa using the European Southern Observatory 3.6 m telescope. The two rings indicated from their size that there were sheets of material about 400 and 1,000 light years in front of the supernova.

The remnant of the supernova explosion should be a neutron star which would appear as a pulsar, if the spin axis orientation causes the beams of radiation emitted by the neutron star to be detected on the Earth. It was expected that the best chance of detecting the possible pulsar, through the clouds of gas surrounding it, would be to observe it at far infrared wavelengths, where dust and gas scattering is a minimum. It was assumed that it would be detectable as the gas clouds dispersed.

The pulsar was first apparently detected by John Middleditch of the Las Alamos National Laboratory on 22nd January 1989, when he was analysing data produced four days earlier by Tim Sasseen of the Lawrence Berkeley Laboratory and himself with the 4 m telescope at the Cerro Tololo Inter-American Observatory in Chile. The big surprise was that the pulsar appeared to have a pulsation frequency of 0.508 milliseconds, implying that it was rotating at about 2,000 times per second. If the neutron star was rotating at this rate, its surface velocity must be very close to the velocity of light. The fact that the surface appeared to be rotating at such an incredibly high speed left many astronomers feeling somewhat uncomfortable. What was just as bad, however, was that all previous millisecond pulsars appeared to be old objects, which had probably been spun up by accreting material from a companion star over hundreds or thousands of years, so finding that the fastest pulsar yet discovered was the youngest caused considerable interest. Careful analysis showed that the pulsation rate appeared to vary regularly with a period of 8 hours, and many theories were advanced to explain this.

On 31st January 1989, attempts to measure the pulse rate failed when the pulsar could no longer be detected. Repeated attempts over the next few months also failed, but this was put down to obscuring material from the explosion temporarily blocking the light. Then, in February 1990, it was discovered that the so-called pulsar signal was actually due to radio interference from a television camera used on the telescope. The real pulsar is still awaited.

Chapter 16 | BUSINESS AS USUAL

16.1 Rosat

Astronomical spacecraft were still being designed and built in the late 1980s, even though the immediate prospects of their being launched was remote, as NASA was modifying the space shuttle and its operational procedures following the Challenger disaster. It is not easy to stop design and development work on a spacecraft for a few years, however, and then restart it with the original engineers, as these engineers would have to be otherwise employed in the meantime and may no longer be available. So the general policy was to continue to build spacecraft, albeit at a less hectic pace, and to store them pending the availability of a suitable launch vehicle. Unfortunately, storage is not cheap, as spacecraft have to be kept in a carefully controlled environment, and if they have to be stored for too long some components can deteriorate and need replacing. So the shorter the storage period the better. Rosat (RÖntgenSATellit[1]) was one such spacecraft caught up by this hiatus in American launchings in the late 1980s.

Joachim Trümper of the Max Planck Institute in Garching (near Munich) had originally proposed Rosat as a German national X-ray spacecraft in 1975. But although his proposal to the Federal Minister for Research was met with enthusiasm by other astronomers, the proposed spacecraft appeared to be too expensive for the Germans to undertake as a purely national programme. So an attempt was made to involve other countries. The first to respond positively were the British, who agreed to provide a Wide Field Camera (WFC), operating in the extreme ultraviolet, to be mounted on the side of the main X-ray telescope. Then the Americans

[1] Named after Wilhelm Röntgen, the discoverer of X-rays in 1895.

agreed to provide a shuttle launch free of charge, together with an updated version of the Einstein High Resolution Imager (HRI) as one of the X-ray telescope's detectors.

Uhuru, Ariel 5 and HEAO-1 had undertaken total-sky X-ray surveys, but the more recent HEAO-2 (Einstein) and Exosat spacecraft had been devoted to imaging and spectroscopy of specific X-ray sources. Rosat's mission was to undertake both an all-sky survey and an X-ray imaging mission. HEAO-1's all-sky survey had had a limiting sensitivity of about 3×10^{-3} photons/cm^2/s, and the Einstein spacecraft had observed only 5% of the sky with a sensitivity of about 3×10^{-4} photons/cm^2/s. Rosat's all-sky survey, on the other hand, was to be undertaken at a sensitivity of about 1.5×10^{-4} photons/cm^2/s.

Rosat's X-ray telescope consisted of an 83 cm diameter, 2.25 m focal length, Wolter-type system of four nested cylinders designed and developed by Carl Zeiss. As this was a completely new development in Germany, three small Wolter-type telescopes were flown on sounding rockets to check out their performance in a real space environment. In their focal plane these X-ray telescopes had a prototype Rosat X-ray imaging detector developed and built by Elmar Pfeffermann and Horst Hippmann of the Max Planck Institute. The first sounding rocket flight, which was launched from Woomera, Australia in 1979, was devoted to observing the Puppis A supernova remnant. It was a complete success.

In parallel with these technical developments, the Germans were negotiating with their prospective British and American partners, whilst also trying to find national funds to pay for the their contribution to the project. The negotiations with the Americans were finally completed in 1982, and in the following year the Germans let the spacecraft contract to Dornier Systems in Friedrichshafen, with a launch planned for 1987. Then the Challenger space shuttle exploded in January 1986, and put all spacecraft that were to be launched by the space shuttle on hold.

As the full extent of the Challenger disaster became clear, NASA announced that they could not offer a space shuttle launch for Rosat until 1994 at the earliest. Faced with a delay of at least seven years, it was decided by all three countries to examine the possibility of using an expendable launcher instead of the shuttle. The European Ariane launcher was ruled out as this would involve the Germans paying for the launch, which was out of the question, so the solution of using an American Delta II vehicle, paid for by NASA, was adopted. This change in launcher required the redesign of the Rosat solar arrays and some of the communications antennae to make them foldable, so they could fit into the smaller Delta shroud. Even so a new, larger Delta shroud was required, and further spacecraft testing was needed after hardware modifications, but the spacecraft was able to fly on a Delta II some years before a space shuttle launch was available.

The 2.4 ton Rosat spacecraft was successfully launched by a Delta II on 1st June 1990 into a 580 km sun-synchronous orbit, inclined at 53° to the Earth's equator; the orbital inclination being chosen to facilitate communications with the German ground station. The spacecraft was designed to operate autonomously when out of sight of this ground station at Weilheim, near Munich, with data being received from the spacecraft five or six times per day, for a duration of about eight minutes per transmission. It was planned to spend the first six months after spacecraft check-out undertaking the all-sky survey, to be followed by a one year pointing or observatory phase.

Rosat's design included two almost identical Position Sensitive Proportional Counters (PSPCs) and one HRI detector, all mounted on a carousel in the focal plane of the Wolter telescope, enabling one of the three detectors to be used at a time. The PSPCs, supplied by the Max Planck Institute, each produced images in four X-ray bands between 0.1 and 2.4 keV (or 12 to 0.5 nm, respectively) and each had a field of view of 2° with a ½ arcmin resolution. The HRI, supplied by the Harvard–Smithsonian Center for Astrophysics, operated over the same total X-ray band, but had a ½° field of view with 4 arcsec resolution. The Leicester-University-supplied WFC, which operated in the extreme ultraviolet from 6 to 30 nm, had a field of view of 5° and a resolution of about 2 arcmin. The PSPCs and the WFC operated during the all-sky survey phase, but the HRI was only used during the subsequent observatory phase.

Rosat operated largely without problems for the first few months, then on 25th January 1991, when the all-sky survey was 96% complete, the spacecraft went out of control. Not only did its attitude control system fail, but the attitude control system's fail-safe mode also failed, resulting in the solar arrays no longer being locked on the Sun. As a result the spacecraft had to rely on its batteries for power, but even though the experiments and all non-essential loads on board had been switched off, the batteries were losing charge rapidly. Early in the morning of 26th January, just two orbits after the fault had been discovered, the spacecraft was no longer communicating with the ground, but at its next pass some 90 minutes later ground controllers managed to switch back on the spacecraft's transmitter and also to send commands to activate the back-up attitude control system. The commands worked, and full control of the spacecraft was gradually re-established over the next few orbits. Unfortunately, whilst the spacecraft had been out of control, it had accidentally scanned close to the Sun, and one of the PSPC detectors had been destroyed. In addition, the sensitivity of the WFC had been severely degraded. The redundant PSPC was still fully functional, however, and that continued to operate perfectly until its gas supply ran out 3½ years later in September 1994.

For the first two weeks after its 25th January control failure, the spacecraft had been put into a safe mode whilst engineers tried to establish what had gone wrong. The problem appeared to be associated with a solar storm[2] which had occurred just before the fault, and new software routines were loaded into the spacecraft's memory to minimise the effect of such storms in future. In the meantime, however, the opportunity had passed to complete the all-sky survey, as the parts of the sky not yet covered were no longer visible to the spacecraft. The all-sky survey was completed six months later, however, when those parts of the sky once more came into view.

Rosat had been put into its observatory mode once the above failure analysis had been completed, and for two months it observed about 30 sources per day. Then on 12th May 1991 its Y-axis gyro failed, severely compromising its ability to lock accurately on to a source for a long period. Over the next five months a pointing strategy was developed that did not use the Y-gyro, and the spacecraft operated more or less normally once more until the Z-axis gyro failed just over two years later on 17th November 1993. Eventually, a new pointing strategy was developed using the two remaining gyros, plus the star trackers, magnetometers and Sun sensors, to enable the observational phase to continue well beyond the end of Rosat's design lifetime.

The previous all-sky survey of HEAO-1 had found 842 X-ray sources, and Einstein, Exosat and other spacecraft had increased the number of known X-ray sources to about 6,000 by the time that Rosat was launched. The Rosat all-sky survey increased this number to an incredible 150,000, of which about 35% are stars, 35% are active galaxies and quasars, and about 10% are clusters of galaxies.[3] About 100 supernova remnants were also included in the Rosat catalogue. In addition to the general all-sky survey, Rosat was also used to undertake longer exposures of selected areas of the sky in a so-called

[2] It was known from other spacecraft that increases in high energy particles emitted by the Sun can cause electrical discharges on the surface of spacecraft. These discharges, in turn, often cause problems with spacecraft's electronic units.

[3] These percentages do not add up to 100% as the exact nature of some sources could not be determined.

deep survey. In this it detected sources down to an intensity of about 1×10^{-5} photons/cm^2/s compared with the Einstein faint object limit of 3×10^{-5} photons/cm^2/s.

Prior to Rosat only a handful of Extreme Ultraviolet (EUV) sources were known, because hydrogen clouds in our galaxy absorb energy in this waveband, generally limiting observations to objects only a few hundred light years away. Rosat detected over 700 of these objects, however, of which 384 were published in the first Rosat catalogue. The majority proved to be either white dwarfs, with surface temperatures in excess of 25,000 K, or the coronas, with temperatures of a few million K, surrounding F- to M-type stars. The Vela and Cygnus Loop supernova remnants were also well resolved, and a few Seyfert galaxies and quasars were also identified.

Although this chapter is devoted to astronomy of non-Solar System sources, mention must be made of the startling discovery by Rosat of X-rays from comets. On 27th March 1996 Rosat observed that X-rays, with energies in the range 0.1 to 2.4 keV, were coming from the vicinity of the comet Hyakutake. This was completely unexpected as objects have to be very hot to emit X-rays, and comets were known to be cool objects. The X-rays were not coming from the cometary nucleus, however, but from a cloud about 100,000 km in diameter, centred about 30,000 km sunward of the nucleus. A search of the Rosat archives then revealed three more X-ray emitting comets of which one, called Tsuchiya–Kiuchi, otherwise known as C/1990 N1, had an X-ray emitting cloud about 500,000 km in diameter. X-ray spectroscopy of Hyakutake showed that the X-rays appeared to be produced by the bremsstrahlung process in a gas with a temperature of about 4.6 million K. The detailed mechanisms that produce such a hot gas near the nucleus of a comet are not completely clear, however, but the heating appears to occur where the solar wind meets the cometary coma.

Rosat provided data on the X-ray intensity of stars in young star clusters, and confirmed the effect first noted using the Einstein Observatory that the X-ray intensity of F- or G-type (i.e. solar type) stars reduces significantly with age. Rosat found, for example, that in the very young star cluster NGC 2232, with an estimated age of just 20 million years, the average X-ray brightness of solar type stars is about 10,000 times that of the Sun (with an age of 4.5 billion years), whilst that of solar type stars in the 600 million year old Hyades cluster is about 50 times that of the Sun. As mentioned earlier (see Section 13.3), this effect, of reducing X-ray brightness of F- or G-type stars with age, had been attributed to stellar magnetic fields, which are thought to generate the X-rays, becoming weaker with age as the stars' spin rates slow down. The spin rates of K and M type stars were known to slow down much more slowly with age than those of F- and G-type stars, however. So this explanation was reinforced when Rosat found that the X-ray intensity of K and M type stars in the 600 million year old Hyades cluster is similar to that of such stars in the 40 million year old α Persei group.

If 20 million year old stars are much brighter than the Sun in X-rays, the question obviously arises as to what happens with even younger stars. T Tauri stars are bright, irregular, pre-main sequence stars that are still in the process of formation, so they were an obvious target for Rosat. These stars are generally found in dense interstellar dust clouds, usually near young, main-sequence O- and B-type stars, and are often surrounded by their own low-density gas and dust cloud which produces its own spectral lines.

Thirty-two very young stars were known prior to Rosat in the Chamaeleon dust cloud, for example, of which twenty-five were classical T Tauri stars, and two were a different type of T Tauri star with very few spectral lines. Rosat discovered twenty-five more very young stars in this star forming region, most of which were T Tauri stars with very few spectral lines, indicating that their own surrounding gas and dust shells had been largely dispersed. Surprisingly, Rosat also found a number of isolated T Tauri stars far away from the known star forming regions. It was not clear whether these had formed where they had been found, or whether they had been ejected from the more well-known, larger star forming regions. Nevertheless, T Tauri stars were seen to be strong X-ray sources.

The soft X-ray background had been observed by William Kraushaar and colleagues of the University of Wisconsin in various X-ray bands down to about 0.11 keV in a series of sounding rocket experiments. This data showed that above 2 keV the background is isotropic and, as discussed in Section 13.3, is probably largely due to unresolved quasars. In the 0.45 to 1.2 keV band, however, the central region of the Milky Way is very bright (see Figure 16.1), and three large-scale features could also be seen. The largest is an irregular 110° diameter loop, called Loop I, centred on the Centaurus region of the sky, with the North Polar Spur, which had first been detected by radio astronomers, being the brightest part of the loop. The other two large-scale features are the Eridanus X-ray Enhancement (EXE) near Orion, otherwise called the Orion–Eridanus Superbubble, and the Cygnus Superbubble (see Section 13.2). The X-rays produced by these three large-scale features appear to come from hot gas, probably produced by supernovae and wind from very hot O- and B-type stars. In the 0.11 to 0.19 keV band, the X-ray distribution across the sky is completely different from that at higher energies, with the brightest regions roughly aligned with the galactic poles.

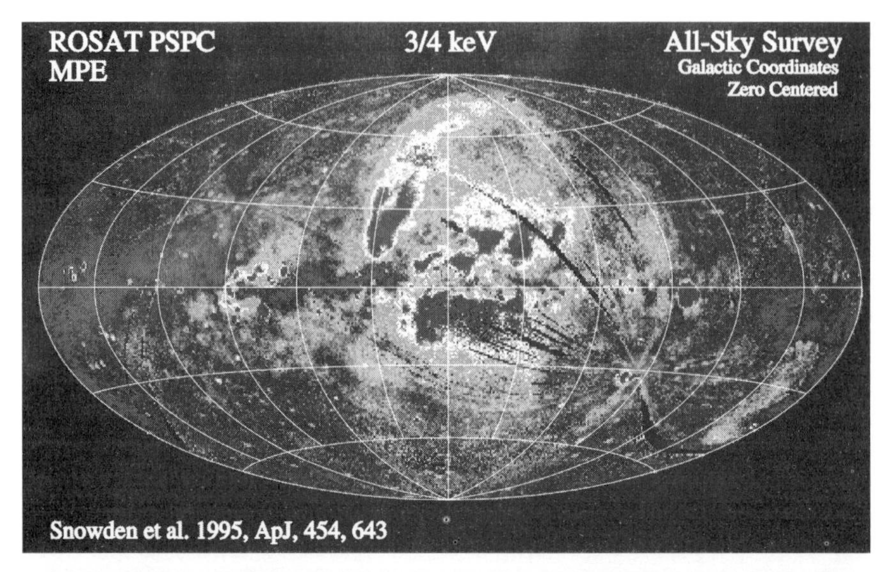

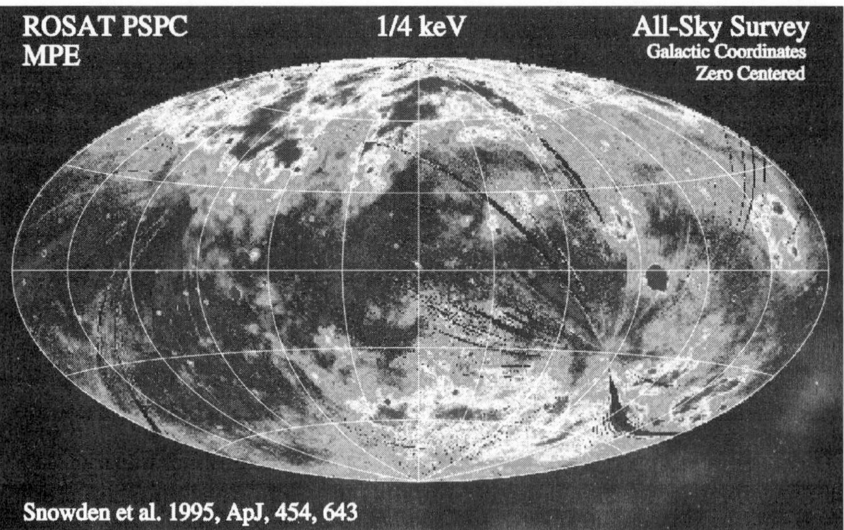

Figure 16.1 These plots show the distribution of X-rays across the sky at different X-ray energies as measured by Rosat. They are similar in structure to those produced by the sounding rocket experiments described in the text but are of a much better resolution. The plots are in galactic coordinates where the galactic plane is at zero latitude and where the galactic centre is at zero longitude. The top plot at 0.75 keV shows Loop I in the upper centre, with the North Polar Spur approximately following the first longitude line to the left of centre. The Cygnus Superbubble is on the galactic equator in the same plot half way from the centre to the left edge. (Courtesy MPE and S.L. Snowden.)

Rosat produced much higher resolution images in these soft X-ray bands below 2.4 keV than was possible with these sounding rocket experiments, and was able to produce the first high resolution image of the Cygnus superbubble, for example. The temperature of the gas in the outer shock front of Loop I was estimated at about 3 million K, and the age of this large loop was estimated to be about 120,000 years. The reason that it appears so large is that it is very close, being only about 600 light years away, and it is relatively old for a supernova remnant.

Another X-ray emitting feature with a diameter of about 20° in Monoceros and Gemini, called the Monogem Ring, had been discovered before Rosat. Rosat showed its temperature to be about 1.5 million K, with an estimated age of about 60,000 years. It appears to be about 1,000 light years away, and a pulsar, PSR 0656+14, which appears to be at about the same distance and age, is not very far away from its centre. If the pulsar is connected with the Monogem ring, this would make it the oldest pulsar/supernova remnant pairing known.

Detailed observations of soft X-rays at about 0.2 keV showed that, although the brightest regions are around the galactic poles, the intensity does not fall to zero in the galactic plane. It should fall to zero in the galactic plane if the soft X-rays are extra-galactic, as there is more than enough hydrogen in the galactic plane to absorb these X-rays. The intensity of soft X-rays near the galactic poles was also found to be higher than would be expected from a simple extrapolation of the data from higher X-ray energies. So there appears to be a source of soft X-rays local to and surrounding the Solar System at a temperature deduced from its X-ray spectrum of about 1 million K. The Sun is known to be in a region where the density of neutral hydrogen is lower than normal, so this 1 million K gas is assumed to be in a local bubble surrounding the Solar System. Calculations suggest that the Sun is near the centre of this bubble, which is estimated to be about 300 light years in diameter, and which was probably created by one or more supernova explosions about 1 million years ago.

One way of checking the contribution to this soft X-ray flux by distant sources is to measure the flux in the direction of known, opaque, neutral hydrogen clouds at known distances. Steven Snowden and colleagues of the Max Planck Institute and David Burrows and Jeff Mendenhall of Pennsylvania State University observed two such clouds in Draco at high galactic latitudes using Rosat. The 0.25 keV X-rays were found to be about half as intense in the direction of the clouds, which are about 1,000 light years away, as from adjacent areas to the clouds, showing that there is a significant soft X-ray contribution coming from more distant regions. Some other galaxies were known to have soft X-ray emitting halos, and so it was concluded that the distant, 0.25 keV X-rays measured by Rosat probably come mainly from a large halo or corona of hot gas surrounding the Milky Way.

The Einstein Observatory had made very long exposures of areas of the sky in the 0.3 to 3.5 keV band that had no known X-ray sources, in an attempt to discover the source of the X-ray background radiation (see Section 13.3). Although only about 15% of its emission could be attributed to known sources, of which 50% were quasars or active galactic nuclei, a statistical analysis suggested that almost all of the remaining background radiation in the range 1 to 3 keV is probably due to unresolved, very distant quasars. Rosat extended this analysis by undertaking an even longer exposure, totalling about eight days, through the so-called Lockman hole, where interstellar matter in our galaxy absorbs no X-rays. The inner part of the image resolved about 250 discrete sources, which were shown to account for at least 75% of the background radiation in that area of sky. Eighty to eighty-five per cent of the energy from these discrete sources was found to come from quasars or active galactic nuclei. So the vast majority if not all of the background radiation at these energies appears to come from discrete sources, most of which are quasars or active galactic nuclei.

16.2 The Compton Gamma Ray Observatory

Gamma ray astronomy had had limited success in the 1970s, due primarily to the limited number of γ-ray photons received per unit detector area in Earth orbit, and the difficulty of building instruments that could accurately detect source locations. Cos-B, for example (see Section 12.8), had detected only two γ-ray photons per hour, and was only able to detect the position of the strong γ-ray source Geminga to within about ½°. So in the mid 1970s there was pressure in America to build a very large γ-ray observatory to follow the three ton HEAO-3 γ-ray and cosmic ray spacecraft that was to be launched in 1979 (see Section 13.5). The new observatory would be more like the size of the original HEAOs, which were to have weighed about 11 tons each before the mission was de-scoped.

The design work on this new Gamma Ray Observatory or GRO, as it was known at the time, started in 1977, and in September of the following year NASA selected its complement of five instruments. The GRO was to be launched by the space shuttle and operate for two years in a 400 km altitude orbit. In 1980 Congress approved a 1981 programme start, but funding

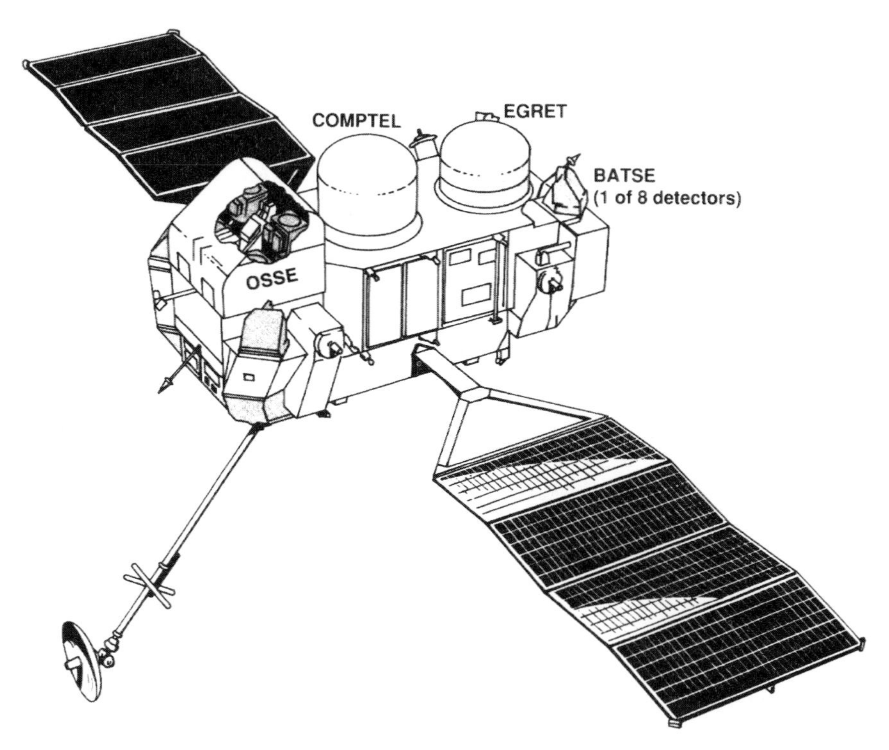

Figure 16.2 The Compton Gamma Ray Observatory in its orbital configuration showing the OSSE, COMPTEL, EGRET and BATSE experiments. The wing-like solar arrays and the boom-mounted communications antenna were deployed whilst the spacecraft was still fitted to the shuttle's remote manipulator arm. (Courtesy *Sky & Telescope*.)

constraints caused NASA to move the launch date from 1987 to June 1988 and delete one of the instruments, the high resolution gamma ray spectrometer. Then in 1986 the launch date was delayed once more because of the Challenger disaster.

The GRO, which was renamed the Compton Gamma Ray Observatory[4], or CGRO, had four instruments on board (see Figure 16.2), each an order of magnitude more sensitive to γ-rays than any previous orbiting γ-ray instruments.[5]

---

[4] Named after the American physicist Arthur Holly Compton, who discovered the Compton effect, in which, when a photon collides with an electrically charged particle, there is a transfer of energy from the photon to the charged particle. The lower energy photon has a resulting lower frequency (or longer wavelength).

[5] As a matter of interest, the lightest instrument on the CGRO, the BATSE, weighed about 1 ton, or about 3.5 times the total mass of the Cos-B spacecraft.

The Burst and Transient Source Experiment (BATSE) was designed to detect γ-ray bursts lasting as little as two milliseconds at energies ranging from 0.03 to 1.9 MeV. To do this it had identical, wide-field instruments placed at each of the eight corners of the spacecraft, so that they could, *in toto*, cover the whole sky. The brightest bursts could be located to about 2° and, if they were also detected by other spacecraft, they could be located by triangulation to within about 1 arcmin or so. A secondary detector, which was attached to each of the eight main BATSE detectors, covered a larger γ-ray range of from 0.015 to 110 MeV. These secondary devices were able to determine γ-ray energies more accurately than the main detector, although they were not as sensitive. When BATSE detected an event it both recorded the data for later transmission to ground and alerted the other three on-board instruments so that they could also observe it.

The Oriented Scintillation Spectrometer Experiment (OSSE) consisted of four steerable detectors, each with a field of view of $3.8° \times 11.4°$, to observe discrete sources over the range from 0.1 to 10 MeV. The detectors rocked back and forth over the target position so that the signal could be unambiguously detected above the general sky background, enabling the source position to be determined to an accuracy of about 10 arcmin or so. Each of the detectors could move backwards and forwards over a 192° arc, and could be reoriented quickly in response to a BATSE detection of a γ-ray burst.

The CGRO's imaging COMPton TELescope (COMPTEL), which was designed to observe γ-rays in the range from 1 to 30 MeV, had a field of view of about 60°, and was able to locate sources to an accuracy of about 30 arcmin. Unlike the OSSE experiment, however, the instrument was fixed and could only be reoriented by moving the whole spacecraft, which took a number of hours.

Finally, the Energetic Gamma Ray Experiment Telescope (EGRET) covered the widest and highest energy range of from 20 to 30,000 MeV, and was able to locate the strongest sources to an accuracy of about 5 arcmin. It was centred on the same area of sky as Comptel, whilst having a slightly smaller field of view. In a way EGRET was similar to Comptel, in so far as it had a wide field of view and a good angular resolution, whilst it complemented Comptel by having an energy range that started about where the latter's finished.

The CGRO was successfully launched by the space shuttle Atlantis on 5th April 1991. It was the heaviest unmanned civilian spacecraft ever launched by NASA, weighing about 17 tons (or 50% more than the Hubble Space Telescope), of which 5.4 tons was instruments. The CGRO was 9 m (30 ft) long and orbited the Earth at the relatively low altitude of 450 km to avoid the Van Allen radiation belts, which would otherwise have caused excessive interference with the experiments. The initial mission plan called for each area of sky to be observed for about two weeks to enable enough γ-rays to be detected, and for a full-sky survey to be completed over the first 15 months. At launch CGRO was designed to have a minimum lifetime of four years, which was double that originally envisaged, with hydrazine thrusters being used to boost the spacecraft back to its nominal orbital altitude of 450 km every year or so.

Once in orbit, the CGRO was lifted out of the shuttle payload bay by the shuttle's remote manipulator arm, and the spacecraft's solar arrays were deployed. Unfortunately, the CGRO's main communications antenna was prevented from deploying by a loose piece of thermal insulation, and it had to be released by astronauts on a spacewalk. When all the spacecraft deployments were complete, the CGRO was released from the remote manipulator arm, leaving the spacecraft on its own in space for the first time. The various instruments were then switched on one by one and checked out, so that on 16th May the CGRO was ready to start its 15 month all-sky survey.

Before the spacecraft had been completely checked out, it had detected γ-rays from the Sun and the BATSE instrument had started to detect γ-ray bursts at about the rate of one per day. Over the next few months there were only minor problems with the CGRO. Then on 18th March 1992, shortly before the first anniversary of its launch, problems with both the prime and back-up tape recorders forced NASA to abandon the recording of data, and communicate with Earth only when the spacecraft was in sight of a Tracking and Data Relay Satellite (TDRS). This resulted in a three month delay to the completion date of the all-sky survey to November 1992.

By the time that the CGRO was launched there were about 500 known radio pulsars, but only two, the Crab and Vela pulsars, were known to pulse in γ-rays. It was important to try to find other γ-ray pulsars in order to refine our models of pulsar behaviour, and so the CGRO was used to try to detect γ-ray pulses from known radio pulsars. The first such detection was made in January 1992 when a γ-ray pulsar (PSR B1509-58) was discovered by BATSE in the constellation of Circinus, in the energy range from 120 to 230 keV. It had a pulsation frequency of 0.15 seconds (see Figure 16.3), which is the same as its radio pulsation frequency. Surprisingly, however, unlike the case of the Crab and Vela pulsars, this new γ-ray pulsar provided only one γ-ray pulse per rotation instead of the usual two. This is presumably because the neutron star is not exactly symmetrical and one of its pulsar beams misses the Earth. The next γ-ray pulsar found by the CGRO, which pulsed at much higher energies of about 100 MeV, also showed the same single

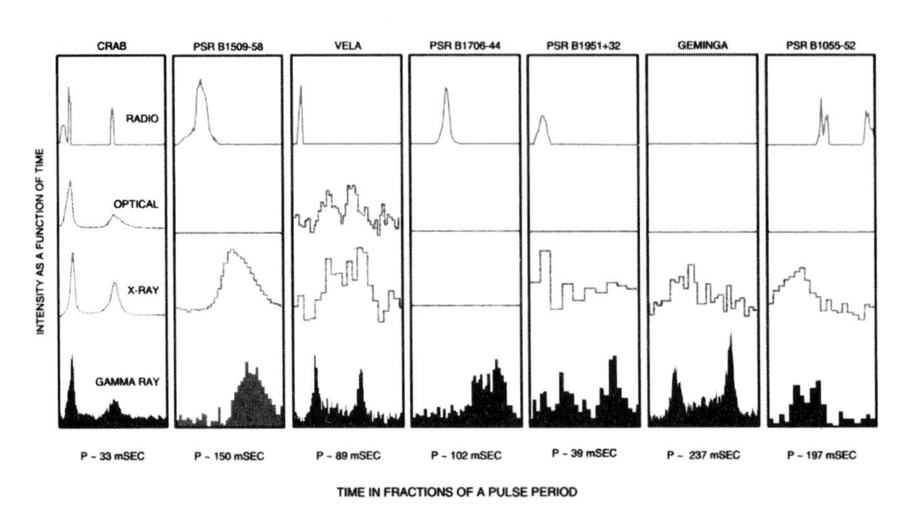

Figure 16.3 Gamma ray pulsars as observed by the CGRO, and their radio, optical and X-ray pulses. Before the CGRO the Crab and Vela pulsars were the only two γ-ray pulsars known. (Courtesy NASA.)

pulse effect. At the time of writing (1998) the CGRO has discovered five new γ-ray pulsars.

Before the CGRO only one quasar, namely 3C 273, was known to emit γ-rays. But the EGRET instrument soon added to the list when the variable quasar 3C 279 was found to be a γ-ray source, emitting more energy in γ-rays than in all other wavebands put together. In fact 3C 279's γ-ray intensity as measured by the CGRO was high enough for it to have been detected by previous γ-ray spacecraft. The fact, that it was not must mean that 3C 279 is variable in γ-rays over a period of a few years at most, indicating that its γ-ray source is relatively small compared with the size of the galaxy. Over the next few months, more γ-ray quasars were discovered, so that as early as April 1992 Neil Gehrels of NASA was able to announce a total of eleven, three of which are so-called BL Lac objects.[6] These γ-ray quasars were found to be very energetic, emitting about 10,000 to 100,000 times the energy in γ-rays that our galaxy emits in all wavebands. To date, the CGRO's EGRET instrument has discovered over 50 quasars that emit energy above 100 MeV, whilst the OSSE instrument has found 17 Seyfert galaxies at lower γ-ray energies.

[6] BL Lac objects are thought to be, like quasars, the nuclei of active galaxies. Unlike quasars, however, they are highly variable over periods of days or months.

The CGRO was also expected to help with the solution to two long standing problems, namely the identification of the intense γ-ray object called Geminga and that of the source of γ-ray bursts (see Section 12.8), as described below.

16.3 **Geminga**

In 1972 the SAS-2 spacecraft found that the second brightest source in the γ-ray sky, which was subsequently called Geminga or 2CG195 + 04, did not correlate with any obvious optical or radio source. Because of this astronomers wondered if Geminga was a completely new type of astronomical object. Unfortunately, however, SAS-2 could only measure Geminga's position to about $1\frac{1}{2}°$, making optical identification difficult. But a few years later Cos-B also observed Geminga and reduced its position uncertainty to about $\frac{1}{2}°$. Then in 1981 an X-ray source was found near the centre of the Cos-B error box by Giovanni Bignami, of the Institute of Cosmic Physics in Milan, and colleagues using the Einstein spacecraft.

The position of the Einstein X-ray source was known to within about 3 arcsec, and this enabled a more thorough search to be made for likely optical counterparts. Hans Bloemen of the Leiden Observatory in the Netherlands found that, although the original 1955 Palomar sky survey plates showed no image in this area, after image processing a 21st magnitude object could be seen about 7 arcsec north-west of the centre of the Einstein X-ray error box. Then in 1983 Patrizia Caraveo and Giovanni Bignami used the Canada–France–Hawaii telescope on Mauna Kea to search the target area. They could not find the Palomar object but found, instead, a 21st magnitude object 4 arcsec south-east of the centre of the X-ray source error box, near the edge of the box. At first it was thought that the two 21st magnitude objects may be Geminga, which had moved at 0.37 arcsec/yr over the 28 years separating the two optical observations. But in 1985 infrared observations using the Multi-Mirror Telescope showed that the Canada–France–Hawaii source is a normal 21st magnitude star, and so cannot be Geminga. At about the same time it was reported that the X-ray source has a periodicity of about 59 seconds, indicating that Geminga is probably a pulsar.

The breakthrough in the search for the optical counterpart of Geminga came in 1987 when Jules Halpern and David Tytler of Columbia University made multicolour CCD images of the area centred on, but larger than, the Einstein error box. They found that one 25th magnitude object stood out from the other sources in the area, as not only was it in the error box but it was unusually blue. In fact it seemed to have the colour, brightness and X-ray spectrum of a neutron star with a surface temperature of about 1 million K located about 2,000 light years away.

The first observations with the CGRO in 1991 caused some surprise, however, as although the EGRET instrument detected Geminga at high energies, the lower energy Comptel instrument did not detect it at all. So Geminga appeared to be bright in X-rays and high energy γ-rays, but not at the intermediate energy of low energy γ-rays. Later that year Jules Halpern of Columbia and Stephen S. Holt of NASA-Goddard found, using Rosat, that Geminga was pulsing in X-rays at a pulse rate of 0.237 seconds. David L. Bertsch of NASA-Goddard and colleagues then observed Geminga with the EGRET instrument on CGRO and found exactly the same pulse rate in high energy γ-rays (see Figure 16.3), showing that Geminga is a pulsar. A gradual slowing down in the pulsation rate was also detected, indicating a pulsar age of about 300,000 years (compared with the 1,000 and 10,000 year ages estimated for the Crab and Vela pulsars, respectively). Finally, in late 1992 Bignami, Caraveo and Sandro Mereghetti found that the optical counterpart of Geminga had moved about 1.5 arcsec in 8 years, indicating that it is very close to us. Geminga appears, after all, to be simply a radio-quiet pulsar whose extreme γ-ray brightness is due mainly to its close proximity to the Earth. Interestingly, because of Geminga's age and distance from Earth, it has been suggested by some astronomers that it may even be the pulsar at the centre of the Local Bubble.

16.4 Gamma Ray Bursts

Four Vela spacecraft had detected about twenty intense Gamma-Ray Bursts (GRBs) over a three year period starting in 1969, with each burst lasting from 0.1 to 30 seconds. Unfortunately, the sources of these bursts, which were all different, could not be pinpointed accurately enough for their optical, radio or X-ray counterparts to be determined. It was confidently expected, however, that once the source positions could be accurately located their counterparts at other wavelengths would be easy to find. As in the case of Geminga, however, this proved not to be so.

The first tentative identification of the source of a gamma-ray burst (or gamma-ray burster) was made following the detection of an intense burst on 5th March 1979 by no less than nine spacecraft.[7] The initial burst, from what

[7] ISEE 3, three Velas, Veneras 11 and 12, the Pioneer Venus Orbiter, the German Helios 2 and the Russian Prognoz 7 spacecraft.

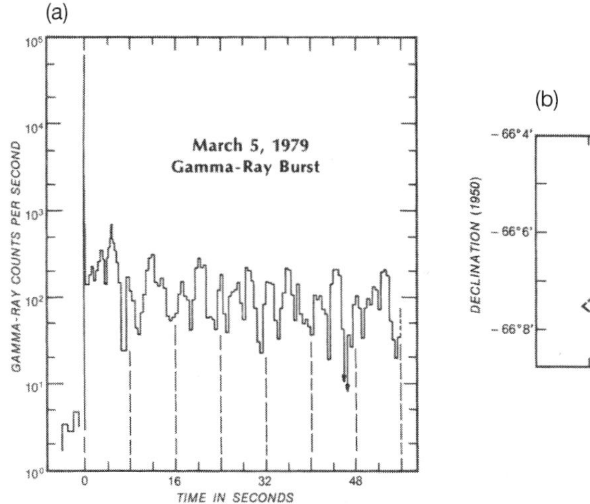

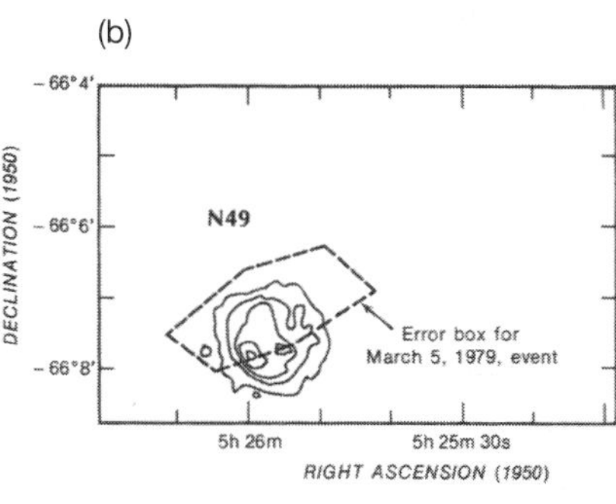

Figure 16.4 (a) The peak and subsequent oscillations in the output of the 5th March γ-ray burst, as recorded by the ISEE-3 spacecraft. (b) The error box for this γ-ray burst is shown in the right-hand plot, as deduced by W.D. Evans of the Los Alamos Scientific Laboratory, and colleagues. It is almost coincident with the X-ray contours of the N49 source in the LMC (as deduced by Helfand and Long) which is also shown on the same plot. (Courtesy T.L. Cline and *Sky & Telescope*.)

is now called SGR 0526-66, lasted only about 0.2 milliseconds, but it was then followed by a series of pulses about 8 seconds apart (see Figure 16.4). The burst, which remained detectable for about three minutes, was much more intense than any other bursts detected up to that time, and had a much softer γ-ray spectrum, suggesting that it may not be a typical burster. Additional but weaker bursts were also detected 0.6, 29 and 50 days after the first event. The observation of the burst by such a large number of spacecraft allowed the source position to be determined to within about 1 arcmin by triangulation. This showed that it appeared to be almost coincident with that of the supernova remnant N49 in the Large Magellanic Cloud, about 170,000 light years away (see Figure 16.4). Unfortunately, if the source was in N49 it must have an energy output of about 10^{44} ergs/s, which appeared unreasonably large. Whilst some astronomers stuck with their N49 identification, others presumed that the alignment with N49 was accidental and favoured a source position only a few hundred light years away, much closer to home. The 0.2 millisecond original burst and subsequent series of pulses suggested to many astronomers that the source of this γ-ray burst, at least, was a neutron star. Unfortunately, no simultaneous event was observed at other wavelengths to aid identification.

There was another problem with the N49 identification that was pointed out by David Helfand of Columbia University. Assuming the identification with N49, then the sources of the previous 100 γ-ray bursts known at that time, all of which had been observed to be less intense than the 5th March event, must be about 3 to 100 times further away, assuming that they were of the same intrinsic intensity of the March event. But Helfand pointed out that there were too few galaxies to support these observed number of events, unless they were more frequent in other galaxies than the Milky Way, where none appeared to have been observed so far. So the bursts must have come from even more distant galaxies, in which case these gamma-ray bursters would have to be even more energetic than the N49 source. As a result of this logic, Helfand favoured the idea that the 5th March burst was from a local source, not from N49.

In 1981 Bradley Schaefer of MIT searched through the photographic plates archived at the Harvard College Observatory, looking for optical counterparts to γ-ray bursts, and found three examples where bright flashes had occurred within the γ-ray error boxes. The optical and γ-ray observations were not simultaneous, however. One optical flash, for example, which had occurred as far back as 17th November 1928, was in the error box of a γ-ray burst of 19th November 1978. Analysis of a series of photographs of that area taken on 17th November 1928 showed that the optical flash probably lasted for only a minute or so, having brightened by at least 12 magnitudes in just a few seconds. Today there is no visible object in the γ-ray error box down to a limiting magnitude of 25, or 22 magnitudes below that detected during the 1928 burst. Whether these optical flashes had any connection with γ-ray bursters was unclear.

By the time that the CGRO was launched in 1991 about 500 γ-ray bursts had been recorded, and their positions were known reasonably accurately for about 200 of them to be plotted on an all-sky map. This showed a more or less uniform distribution across the sky. If the γ-ray bursters had been spread throughout the Milky Way and other galaxies like stars, then we should have observed more sources concentrated in the galactic plane than elsewhere. As this is not so, then bursters must be either very close to us or they must be at galactic distances outside the Milky Way.

All γ-ray bursts are not the same. Some have one brief peak lasting for a tenth of a second or less, followed by a much lower level of γ-rays, whereas other bursts can last for up to about 1,000 seconds and show no initial peak. Three of the 200 bursters with known positions had been observed to emit γ-rays more than once. Unlike the majority of γ-ray bursters, however, these three objects, one of which is SGR 0526-66 discussed above, emit predomi-

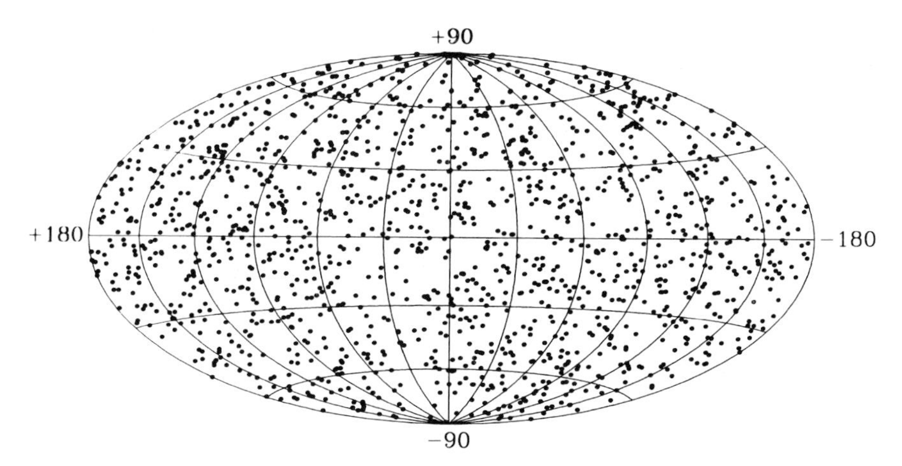

Figure 16.5 Gamma-ray bursts are seen to be uniformly distributed across the celestial sphere, as shown in this diagram in galactic coordinates of 1429 γ-ray bursts detected by the CGRO spacecraft. (Courtesy NASA.)

nantly soft γ-rays, and so are called soft gamma-ray repeaters.[8] Evidence of strong magnetic fields (by Zeeman splitting) or of high gravitational fields (by red-shifting of the spectral lines) had been found in the spectra of about 20% of bursters suggesting that these are neutron stars.

In the first 70 days of operation of the CGRO a total of 55 bursts was detected, and after five months the count was up to 117. The map of these bursts presented by Gerald Fishman of NASA-Marshall in September 1991 confirmed previous data by showing a uniform distribution across the celestial sphere. Later data confirmed this (see Figure 16.5). The CGRO burster distribution was something of a surprise to those astronomers who thought that gamma-ray bursters are close to the Solar System. This was because, as the BATSE detectors on the CGRO are twenty times more sensitive than those on any previous spacecraft, and *if* the previous detectors had only seen local sources, then the more sensitive BATSE detectors should have picked up sources further away in the Milky Way. In that case the BATSE map should show a preferential detection of sources in the galactic plane. As it did not, it was beginning to look as though γ-ray bursters really were distant objects, either in a large halo around the Milky Way or in the more distant

[8] A second of these three soft gamma-ray repeaters was found in 1993 by Shrivinas Kulkarni of ISAS and Dale Frail of NRAO to be coincident, within error, with another supernova remnant called G10.0-0.3. (SGR 0526-66 linked with the supernova remnant N49 being the first.)

universe, unless they were very much weaker than first thought in which case they could still be local events. In the latter case BATSE would still be detecting only those sources in the local region of the Milky Way.

Unfortunately, the BATSE instrument could only measure the position of the weakest γ-ray sources to about 12° and the strongest to about 2°, so triangulation with other spacecraft was required to provide an accurate enough location to enable a search to be made in other wavebands. In April 1996, however, a small Italian–Dutch spacecraft was launched called BeppoSAX,[9] which could do much better in locating γ-ray bursters, as it had both a γ-ray burst detector and two X-ray detectors on board. It was assumed that some γ-ray bursts would simultaneously trigger one of the X-ray detectors, which could measure the source location sufficiently accurately for its optical counterpart to be detected.

On 28th February 1997 BeppoSAX's Gamma-ray Burst Monitor detected a γ-ray burst, now known as GRB 970228, and one of its two X-ray cameras detected a simultaneous X-ray flash. Eight hours later the spacecraft's second X-ray instrument detected a fading X-ray source in the same area (to within 1 arcmin). Within hours John Telting of the Isaac Newton group observed the area with the William Herschel Telescope on La Palma. Comparing the image taken on 28th February with one taken eight days later it was clear that a 21st magnitude star visible on the first image was not visible on the second. Subsequent images from the Keck Observatory and the HST (see Figure 16.6) showed an overlapping 25th magnitude oval that could be the parent galaxy of the gamma-ray burster. The linkage between GRB 970228, the X-ray source, the optical 'star', and the optical galaxy was thought to be reasonably strong, although some astronomers were sceptical and awaited more examples of such linkages before they would accept them as real.

There was no optical counterpart found for the next GRB detected by BeppoSAX on 2nd April, but one was found for the next GRB detected by BeppoSAX on 8th May 1997. An X-ray counterpart for GRB 970508 was detected within hours and optical observers found an object that was still increasing in intensity. Its optical intensity, which peaked two days later at 20th magnitude, was bright enough to enable a spectrum to be taken. This showed lines with a red-shift of 0.84 superimposed on the spectrum of the star by intervening material at a distance of 4 billion light years. This gave the minimum distance to the GRB which clearly showed that it was at cosmological distances, assuming that the GRB and optical source were the same.

---

[9] BeppoSAX is named after Giuseppe 'Beppo' Occhialini, a pioneer in high energy astronomy, and SAX from 'Satellite per Astronomia a raggi X'.

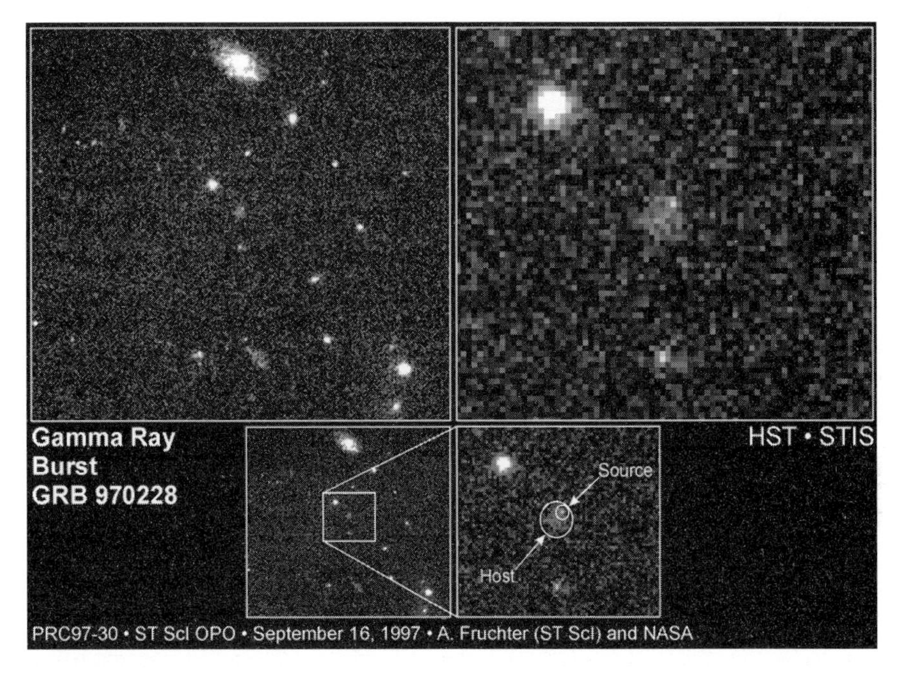

Figure 16.6 The optical counterpart to the γ-ray burst GRB 970228 is clearly shown in the HST image (above right) together with that of its probable host galaxy. (Courtesy A. Fruchter, STScI and NASA.)

At the time of writing (1998) a third[10] optical counterpart has been found to a GRB. This burst, which was detected on 14th December 1997, was found to have an optical counterpart with a red-shift of 3.42, implying a distance of at least 10 billion light years! This in turn implies a γ-ray energy output of at least 3×10^{53} ergs, which is about 100 times the luminous output of a supernova. Theories as to how this energy could be generated are currently still in the formative stage, but most involve black holes or neutron stars.

16.5 The Extreme Ultraviolet Explorer

Back in 1975 the Americans and Russians had met up in space as a 'goodwill gesture' in the so-called Apollo–Soyuz Test Project (ASTP), which involved

[10] An optical counterpart was found to a GRB of 3rd July 1998 with a red-shift of 0.97 after the above had been written.

the mating of the Soyuz 19 and Apollo CSM-111 spacecraft. Amongst other things the Americans took an Extreme UltraViolet (EUV) camera with them, which was sensitive to wavelengths in the band from 5 to 150 nm. At that time it was thought that interstellar hydrogen and helium gas, which absorbs the EUV, would restrict our view in these wavelengths to objects only a few hundred light years away at most, so very few objects would be detected. In the event a total of 30 objects was observed by the Apollo camera including two white dwarfs, called HZ 43 and Fiege 24, the dwarf nova SS Cygni, and the star Proxima Centauri. The white dwarfs were detected as their high surface temperatures are such that they emit strongly in the EUV, and SS Cygni, which was observed during one of its outburst phases, was a known soft X-ray source, so its detection at slightly longer wavelengths was not unexpected. Proxima Centauri is a cool red dwarf, however, and so its surface should not emit much energy in the EUV band, but it is also a flare star, the nearest such star to the Sun, and it was thought that the flares are the source of the EUV radiation. The remaining 26 sources that were detected could not be identified.

Rosat carried the next significant EUV payload into space some fifteen years after the ASTP, and detected over 700 objects in the EUV waveband from 6 to 30 nm. The majority of the objects that could be identified were either white dwarfs or the hot coronas around F- to M-type stars.

NASA had started preliminary design work on their Extreme UltraViolet Explorer (EUVE) spacecraft in the late 1970s, after the successful ASTP project, but it was not until 1984 that the EUVE programme was approved, including a shuttle launch in 1988. As with many other spacecraft already described, the Challenger disaster in 1986 caused the EUVE programme to be radically changed, with the shuttle launcher being replaced by a Delta II rocket.

EUVE was finally launched into a 550 km almost circular orbit, inclined at 28° to the Earth's equator, on 7th June 1992. Weighing 3.4 tons, it had four 40 cm diameter telescopes, three of which, called the survey telescopes, looked out of the side of the spacecraft as it spun on its axis once every 30 minutes. The fourth, the so-called deep survey telescope, pointed along the spin axis. The three survey telescopes were planned to provide a complete sky survey in the first six months after spacecraft checkout in orbit. They were designed to operate in four wavebands between about 7.5 and 75 nm, with a resolution of about 10 arcsec, whilst the axial telescope mapped out a two-degree-wide strip centred on the ecliptic plane with much greater sensitivity. When the six month survey phase was complete, the deep survey telescope was available for spectroscopic observations of individual sources.

When EUVE was in the design phase it was thought that it may not be able to detect any sources outside the Local Bubble of relatively low density gas, which has a diameter of a few hundred light years. The Local Bubble is not spherical, however, and it was known that there was a region virtually devoid of gas for over 1,000 light years in the direction of the star Beta Canis Majoris. There was also a suggestion that in small areas of the sky the interstellar gas may be tenuous enough to allow us to detect other galaxies. It was still something of a surprise, however, when EUVE detected its first extragalactic object, the BL Lac object PKS 2155-304, whilst the spacecraft was still in its initial seven week checkout phase.

During EUVE's subsequent six month survey phase, which lasted from 24th July 1992 to 21st January 1993, it detected 739 sources, about one third of which could not be immediately identified. Of those that could, about 50% were the coronas of F- to M-type stars, and about 35% were the very hot (25,000 to 100,000 K) photospheres of white dwarfs. The remainder were cataclysmic variables, RS CVn stars, active galactic nuclei, supernova remnants and very hot O- and B-type stars.

Rosat had shown that young, rapidly rotating stars are strong X-ray emitters, and EUVE confirmed that this effect also extends into the extreme ultraviolet. RS CVn stars were found to be strong emitters in the extreme ultraviolet by EUVE, due to active regions on their surface and to intense stellar flares. The polar systems VV Puppis, UX Fornacis and RE 1938-461 were also found to be sources of extreme ultraviolet energy. In fact the spectrum and the variation of the extreme ultraviolet radiation during eclipse of VV Puppis enabled Paula Szkody and colleagues of the University of Washington to estimate that the energy is being emitted from a 100 km diameter, 300,000 K temperature region on the surface of the white dwarf. Previous X-ray data had shown that this white dwarf's magnetic field is about 30 million gauss, and it is this extremely strong field that channels the material from the white dwarf's companion onto such a small area of the white dwarf's surface.

It was of no great surprise to find that white dwarfs are strong EUV emitters, as their surface temperatures are such that their energy peak is at these wavelengths, but the white dwarfs' spectra provided a great surprise. It had been expected that the very strong gravitational field of a white dwarf would cause the heavier elements to sink well below the surface, but the EUVE spacecraft showed that this is often not the case with heavy elements still on the surface in surprising concentrations. This is thought to be because the radiation pressure is strong enough within the star to force the heavy elements back to the surface.

As mentioned above, EUVE detected one BL Lac object during its in-orbit checkout phase. The spacecraft then detected four more extragalactic objects over the next three months, none of which had been detected by Rosat. By the end of its second year in orbit, EUVE had increased the number of extragalactic objects to 22, which were a mixture of active galactic nuclei, BL Lac objects and Seyfert galaxies. These detections confirmed that there are tunnels in the EUV-absorbing gas, which enable us to see extragalactic objects in some areas of the sky. Similarly, the observed number and intensity of BL Lac objects indicated that these objects themselves are relatively transparent to the extreme ultraviolet, whereas active galactic nuclei seem to have a fair amount of EUV-absorbing material, thus restricting their visibility in this waveband.

16.6 The Infrared Space Observatory

The study phase of the ESA Infrared Space Observatory (ISO) was started in March 1983, just two months after the launch of IRAS its American/Dutch/British predecessor. Superficially, ISO was to look like a scaled-up version of IRAS, with both spacecraft being constructed around a large cryostat filled with liquid helium, and with both having a similarly shaped sunshade, but there the similarity largely disappeared. IRAS, as the first infrared spacecraft, was mainly designed to undertake an all-sky survey to detect infrared sources, whereas ISO's main task was to examine known sources in more detail.

The ISO payload consisted of a 60 cm diameter, Ritchey–Chrétien reflecting telescope enclosed in a cryostat which was cooled with 2,300 litres of superfluid helium at a temperature of 1.8 K. The telescope's field of view of 20 arcmin was shared between four instruments located in its focal plane, each of which had a field of view of about 3 arcmin. Momentum wheels were designed to point and stabilise the spacecraft, and two star trackers were designed to measure where it was pointing, within specified relative and absolute pointing errors of 2.7 and 12 arcsec respectively (2σ, half cone angle). Its planned 24 hour orbit of about $1{,}000 \times 70{,}000$ km, inclined at 5° to the Earth's equator, was chosen to minimise the spacecraft's exposure to the Earth's Van Allen radiation belts.[11] The orbit enabled scientific data to be down-loaded in real time to two ground stations, thus

[11] This orbit had a much higher apogee than that of IRAS, taking ISO much further away from the Earth's bright infrared background.

eliminating the need for on-board data storage. The ground stations were an ESA station at Villafranca in Spain, and a NASA station at Goldstone in California which was funded by the USA and Japan in return for guaranteed access to ISO. Neither of these ground stations could cover the spacecraft as it rapidly swung past perigee, however, but this did not matter as it was then too close to the Earth, and would suffer from problems caused by the Earth's infrared emissions and by the Van Allen belts. As a result it was planned to close down the instruments for this part of the orbit.

ISO's four payload instruments consisted of one imager, one photopolarimeter and two spectrometers; one spectrometer for the short wavelength infrared and one for longer infrared wavelengths. The imaging camera, called ISOCAM (for ISO CAMera), had two channels operating in different wavebands, each equipped with a 32×32 element infrared array detector. Both the long (5 to 17 micron) and short (2.5 to 5.5 micron) waveband channels had insertable lenses that provided image scales of from 1.5 to 12 arcsec per pixel, together with various filters to restrict the observed waveband further.

The photopolarimeter, called ISOPHOT, had three sub-elements, called PHT-P, PHT-C and PHT-S, for different types of observations. PHT-P was a multi-aperture photometer that covered the waveband from 3 to 110 microns with 14 filters. PHT-C was a simple long-waveband camera with two detector arrays, one, a 3×3 array, covering the 30 to 100 micron range, with the second, a 2×2 array, covered the 100 to 240 micron range. The PHT-S grating spectrophotometer operated over the 2.5 to 12 micron band using two linear detector arrays.

The Short Wavelength Spectrometer (SWS) operated between 2.4 and 45 microns with a spectral resolution ($\lambda/\Delta\lambda$) of about 1,000, which could be increased to about 30,000 in the 15 to 40 micron range by using a Fabry–Perot interferometer. The Long Wavelength Spectrometer (LWS) covered the range from 45 to 200 microns with a spectral resolution of about 200. Again that could be increased selectively up to about 10,000 using a Fabry–Perot interferometer.

Only one of the four instruments was selected to be prime at any one time, but other instruments could be operated in a secondary mode, with a reduced data rate, looking at an adjacent area of sky. Observations were also made routinely with ISOPHOT whilst the spacecraft was slewing from one object to another, thus making full use of the observing time available.

ISO was launched by an Ariane 4 rocket on 17th November 1995 into an interim 500×71,600 km orbit, which was gradually changed to the required 1,000×70,600 km orbit using the spacecraft's hydrazine thrusters. Then on 27th November the cryostat cover was successfully ejected, and about 30 minutes later the first star was detected, confirming that the telescope had an uninterrupted view of space. A long performance verification and calibration phase followed, and on 4th February 1996 the spacecraft started routine observations, undertaking about 50 observations per day.

As anticipated, ISO's scientific instruments had to be switched off around perigee when the spacecraft encountered the Van Allen radiation belts, but the resulting science window of 16 h 40 m per 24 h orbit was some 40 m more than expected. Cosmic ray hits caused problems, however, reducing the sensitivities of the ISOPHOT, SWS and LWS instruments to below initial predictions, although the reductions were not serious. Then on 30th May 1996 ISO accidentally spent about two minutes looking at the Earth, when the spacecraft was out of contact with ground stations around perigee, causing its detectors to heat up from below 4 K to about 10 K. Fortunately the automatic on-board software corrected this pointing error, allowing the detectors to cool down to their previous temperature, with only a small loss of helium coolant.

In spite of these set-backs ISO performed remarkably well, not only having 40 minutes more observing time per orbit than expected, but also having a relative pointing error of 0.5 arcsec (2σ, half-angle) and absolute pointing error (blind pointing; 2σ, half angle) of 5 arcsec, which were both substantially better than specified. In addition, ISO used its helium coolant at a slower rate than anticipated, in spite of the problems in May 1996. This allowed its operational phase to be extended by 30% to 8th April 1998 when it finally ran out of helium. Originally, astronomers had had to choose between observing the centre of the Milky Way or the star forming region of Orion, because of spacecraft viewing constraints. They had chosen the former, but with the 30% extension they were now able to observe the Orion region as well.

Molecules had first been observed in the interstellar medium by Theodore Dunham in 1937 when he detected absorption lines due to methylidyne (CH) and cyanogen (CN) at wavelengths of about 400 nm. The lines were very sharp, suggesting that the gas was very cool. Further developments had to await the development of radio telescopes, but in 1963 Weinreb and Barrett discovered the hydroxyl molecule (OH) in the interstellar medium, and in 1968 water (H_2O) and ammonia (NH_3) molecules were also found. Since then, using radio telescopes and ultraviolet and infrared spacecraft-mounted detectors, there has been a veritable deluge of discoveries of ever more complex interstellar molecules.

Theoreticians had had a problem for some time in trying to explain how collapsing protostellar clouds could cool fast enough as they contracted to

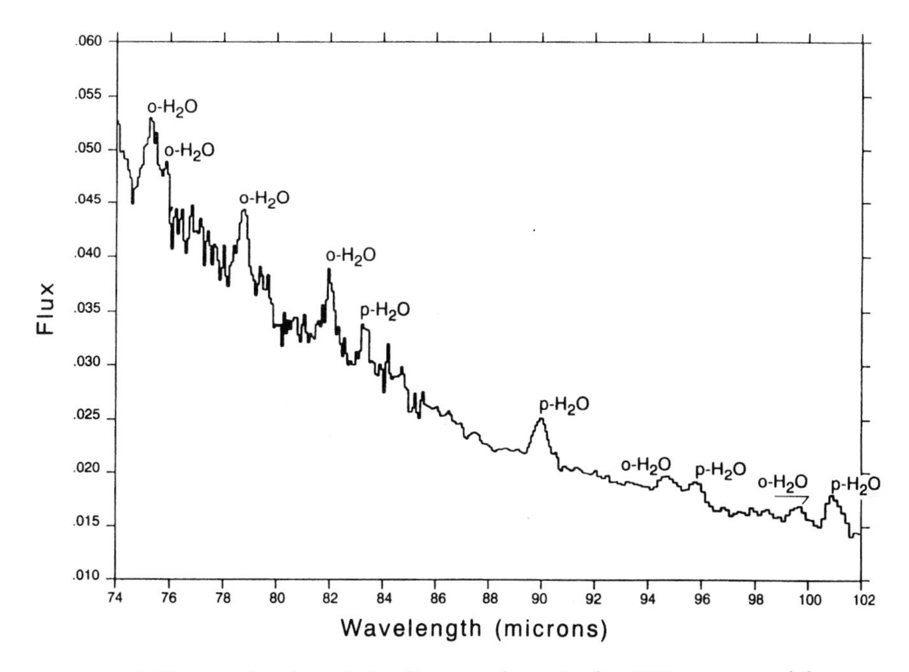

Figure 16.7 Water molecule emission lines are shown in this LWS spectrum of the gas cloud surrounding the M-type red giant W Hydrae. Notations o- and p- refer to different orientations of the hydrogen nuclei. (Courtesy ESA/ISO/LWS.)

enable them to become dense enough to form stars. It had been suggested that water molecules could help to provide such a cooling mechanism, so it was with some relief that ISO found water molecules in the gaseous envelopes surrounding newly formed O- and B-type stars. ISO also found carbon dioxide and carbon monoxide in these clouds, with the carbon dioxide often in the form of ice and with carbon monoxide in the gaseous state.

To most astronomers' surprise ISO found that water molecules are commonplace in our galaxy. Not only were they found in clouds surrounding newly formed stars, but they were also found in the spectra of stars near the end of their lives (see Figure 16.7).[12]

[12] Although this chapter does not discuss Solar System objects, mention should be made that water, carbon monoxide, carbon dioxide, hydrocarbon and nitrile molecules were also detected by ISO in Titan's atmosphere. In addition, ISO provided the best high resolution infrared spectra of Saturn received to date, showing detailed information on molecular hydrogen (H_2), deuterated molecular hydrogen (HD), phosphine (PH_3) and ammonia in the planet's atmosphere.

The C II line (produced by singly ionised carbon) was detected by ISO at a wavelength of 156 microns in different types of objects including H II regions, infrared cirrus, planetary nebulae and colliding galaxies. These observations were important as they will help us to understand the cooling process under way in these objects. The first detection of the 17 micron line of molecular hydrogen in external galaxies will also help astronomers to trace the concentrations of this, the most abundant element in the universe, in these galaxies far more accurately than previously possible.

Colliding galaxies such as the Antenna galaxies (NGC 4038 and 4039) were thought to have large star forming regions in large, dense clouds of dust and gas, triggered by collision shock waves. New stars had already been observed in these regions in visible light, and these areas were first observed in detail in the 2.5 to 17 micron band using ISO's ISOCAM instrument to show significant heating of the gas and dust clouds. Star forming regions of non-colliding galaxies were also clearly seen by ISOCAM. In the case of the Whirlpool Galaxy (M51), for example, the nucleus region was seen to be very warm compared with the remainder of the galaxy (see Figure 16.8). There were also a few warm patches in the spiral arms where star formation is taking place, but otherwise the arms were cool.

A number of potential star-forming regions had been observed in the Milky Way using the JCMT and IRAM ground-based telescopes, but the temperatures of these dark clouds could not be clearly established as their peak emissions were below the minimum wavelength cut-offs of ground-based telescopes. Knowledge of these temperatures was important if astronomers were to understand the early phases of cloud collapse and subsequent star formation. ISO enabled such temperatures to be determined as it operated at shorter wavelengths than the JCMT and IRAM. A two solar mass dark cloud called Lynds 1689, for example, was found to have a temperature of just 13 K, indicating that it was in the very early stages of cloud collapse.

ISO was used to image both the northern and southern Hubble Deep Field (see next chapter). In the southern field, for example, about thirty-five galaxies were found at a wavelength of 7 microns and about twenty-five at 15 microns. One of these infrared sources was found to have no equivalent at optical wavelengths, suggesting that the galaxy may be undergoing a period of rapid star formation. ISO also observed a 3×3 arcmin region of the Lockman hole at 7 microns finding fifteen sources, two of which had no optical counterpart, again indicating that the sources are very red, possibly because of a large amount of star formation.

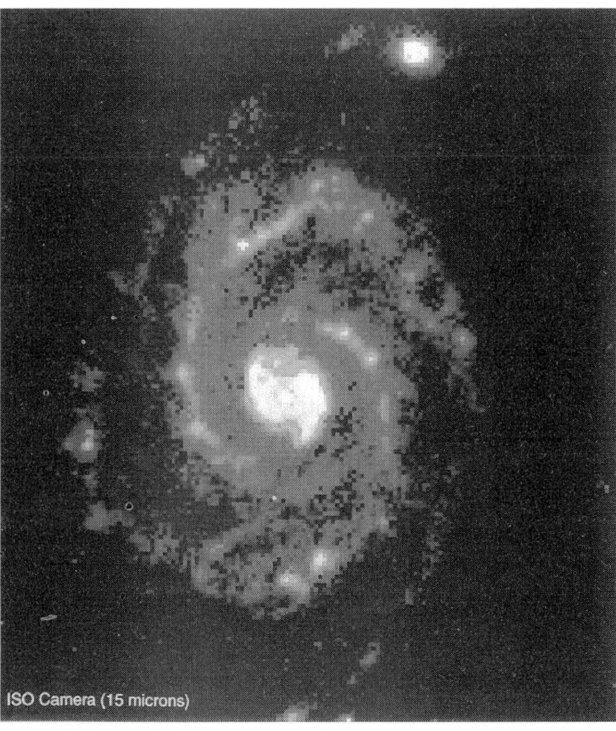

Figure 16.8 Hot spots due to star formation are seen on either side of the nucleus and in places in the spiral arms of the Whirlpool Galaxy (M51). The nucleus of its companion galaxy (at the top of the image) is also bright in the infrared due to star formation. This ISOCAM image was taken at 15 microns. (Courtesy ESA, 96.03.006-001.)

The above are just a glimpse of the sort of results obtained so far (as of 1998) from ISO. Now that the spacecraft is no longer operational, the project has entered a $3\frac{1}{2}$ year post-operations phase where the ISO calibration and data processing will be refined. Further scientific results will then be generated and only after that will we be able to assess the full value of ISO.

Chapter 17 | THE HUBBLE SPACE TELESCOPE

17.1 Pre-Launch History

Way back in 1946 the American astronomer Lyman Spitzer produced a classified report for RAND outlining possible scientific programmes that could be undertaken using space satellites containing optical telescopes of up to 200 to 600 inches (5.1 to 15.2 m) diameter. To say that this was ambitious was something of an understatement, as this report was written before the 200 inch ground-based Palomar telescope had been completed and some eleven years before the launch of Sputnik. At that time the Americans were generally sceptical about the idea of launching even modest sized spacecraft, so the idea of orbiting such large optical telescopes was shelved.

The American attitude to the possibility of building space satellites gradually changed in the 1950s (see Chapter 2), however, and in 1958, the year after Sputnik, Lloyd Berkner, chairman of the Space Science Board of the National Academy of Sciences, solicited suggestions for space-based projects to follow those undertaken in the International Geophysical Year. The two hundred proposals received were then used to aid NASA and its advisory committees to define a space-based observatory programme. In addition to the on-going Explorer series of small spacecraft, it was agreed to undertake a solar programme using small observatory spacecraft, and to undertake a programme devoted to deeper space using large Orbital Astronomical Observatories (OAOs).

The next questions to be resolved about the OAOs were what wavebands should be used? and how large should the spacecraft be? It was thought best to concentrate on those wavebands that could not be observed from the Earth's surface and, as ultraviolet detector technology was the most developed at that time, it was decided to concentrate the OAOs on ultraviolet astronomy. As far as size was concerned, astronomers generally wanted to launch a series of small spacecraft as proof-of-concept observatories to be followed by larger spacecraft later, but NASA wanted to launch large, prestigious observatories immediately to try to up-stage the Soviets. NASA won the argument, and the result was a series of four OAOs, each weighing about two tons, which were launched between 1966 and 1972, of which two were failures (see Section 11.3). OAO-3 or Copernicus, which was the last of the series, was to demonstrate that it was possible to point a large spacecraft to within about 0.1 arcsec and control it to about 0.02 arcsec, which gave confidence that a larger spacecraft could be controlled accurately.

Whilst the OAO spacecraft were still in the design and development phase, Boeing received a contract from NASA to undertake the outline design of a Large Space Telescope (LST) with a mirror diameter of 3.0 m (120"). This diameter was chosen as it was the largest that could be carried by a Saturn launcher, which was the launcher being considered at that time. At this stage it was anticipated that the telescope would be attached to a manned orbiting space station, as such a station was seen by many in NASA as the next major programme after Apollo. Then in 1965 about 100 scientists and engineers met under the aegis of the Space Science Board and NASA at Woods Hole in Massachusetts to discuss possible future NASA programmes in space science. At this meeting the LST was strongly supported by the working party on Optical Astronomy from space, and a more permanent working party was set up as an Ad Hoc Committee of the Space Science Board to examine the technical problems involved in designing such a spacecraft, and to lobby members of the astronomy community to give their support. The main mission of the LST was expected to be in the field of cosmology where it should enable the distances of more distant galaxies to be measured, and so determine the rate at which the expansion of the Universe is changing with time. This should enable the age of the Universe to be deduced, and its future expansion rate to be determined. In particular, it was hoped to answer the question as to whether the Universe will continue to expand for ever, or whether it will eventually stop expanding and contract to a 'Big Crunch'.

The National Academy of Sciences Ad Hoc Committee on the LST held its first meeting under Lyman Spitzer's chairmanship in April 1966, with Nancy Roman as one of the NASA participants. One of the problems faced by this committee was how to persuade a generally sceptical astronomical community that it was worth spending hundreds of millions of dollars on one LST when such a sum could buy many, larger ground-based telescopes. Nancy Roman, as head of astronomy at NASA HQ, had been facing similar problems since she joined NASA in 1959 to set up a NASA astronomy programme. The fact was that money would not have been transferred from space telescopes to ground observatories, even if the space astronomy programme had been scrapped. But many astronomers did not understand or believe this. So to help offset these concerns the Ad Hoc Committee, in their final report published in 1969, recommended that not only should the LST be built, but the number of large ground-based optical telescopes should be at least doubled. This would also avoid astronomers asking for access to the LST when ground-based facilities could do their particular job just as well, or even better. The report also recommended building smaller space telescopes to gain experience before starting to build the 3.0 m, diffraction-limited[1] LST. Such an approach would also get astronomers used to using optical space telescopes, and build up a user community hungry for the LST when it became available.

Two key questions had to be answered by LST protagonists in the late 1960s. What size should the intermediate space telescope be? and should the LST be man-tended? The Copernicus telescope was 0.8 m in diameter, and NASA already had outline designs and costings for a 1.5 m diameter telescope. But in 1969 NASA-Goddard, who were responsible in NASA for space science, had inherited a machine for grinding and polishing mirrors of up to 2.0 m diameter, a 2.0 m diameter telescope was already under consideration for the proposed stellar Apollo Telescope Mount, and Perkin-Elmer had produced an outline design for a 2.0 m space telescope. So in 1970 a 2.0 m telescope was generally foreseen in NASA as an interim space telescope that could be launched within six years, followed by a 3.0 m LST in 1980.

At this stage in 1970 the baseline design of LST assumed a Titan III launch, rather than the original Saturn, but NASA was already starting to consider a launch on the Space Shuttle. In fact in 1969 Grumman had undertaken a study for NASA comparing the costs of a Titan III and of a shuttle launched LST and had concluded that the shuttle option would be about 30% cheaper. In addition, the use of the space shuttle would enable the LST to be repaired and up-dated in orbit with new instruments over its lifetime.

The possible use of the space shuttle, which at that time had not been approved, raised the question of the rôle of astronauts in servicing and operating the LST. Should the LST be a manned facility linked to an orbiting space station, with astronauts replacing film canisters, etc.? or should it be

[1] The term 'diffraction-limited' means that the primary mirror should be sufficiently close to its theoretical curvature that the optical performance of the telescope is limited by its aperture, rather than the quality of its primary mirror.

a stand-alone facility with electronic image detectors like an ordinary spacecraft, just using the shuttle for periodic updating and repair? Should the compartment containing the scientific instruments be pressurised so that film replacement, and repair or replacement of the instruments could take place in a shirt-sleeve environment? Astronomers were in no doubt that they did not want the LST to be used as an excuse for a permanent manned presence in space, as the presence of astronauts could affect the LST's stability, contaminate the optics, and risk damaging it, especially if the astronauts had to work on the LST from the outside using bulky space-suits. As an absolute maximum, astronomers wanted the astronauts to just update and repair the facility.

President Nixon had been faced with a number of major decisions on how the American space programme should develop when he took office in January 1969. Thomas Paine, who was appointed NASA administrator in March 1969, pushed strongly for a rapid expansion of the NASA budget to be spent mainly on large, manned space stations, lunar bases, and the development of a space shuttle. In this he was strongly supported by the US vice president Spiro Agnew who favoured sending men to Mars. Ranged against Paine and Agnew, however, were powerful forces which included the Townes Committee,[2] that had been set up by Nixon to advise on the way forward in Space, the President's Science Advisory Committee, and the Secretary of the Air Force, Robert Seamans, who had previously been a deputy administrator of NASA.

The proposed manned Mars mission (with a landing in the early 1980s) was the first to be dropped, but by January 1970 no decision had been taken on either the space station or the space shuttle, and the NASA budget had been cut once again. Paine resigned in September 1970, as he was exasperated by the lack of a political decision on the future direction of NASA and by the apparent lack of interest in the agency. Matters continued to drift until late 1971, when a decision was finally made to build the space shuttle. This decision, which was announced in January 1972, was to have a major impact on all NASA programmes, because part of the financial justification for building the shuttle involved the phasing out of expendable launch vehicles, whose designs were generally ten or twenty years old, and launching

---

[2] The Townes Committee had been set up in 1968, under the chairmanship of Charles Townes, Nobel Laureate and chairman of the Space Science Board and of NASA's Space Technology Advisory Committee. In the event the Townes Committee recommended a continuation of the existing $6 bn/yr effort including a strong programme of *unmanned* planetary probes.

new spacecraft with the shuttle. This meant that these new spacecraft had to be man-rated for safety reasons, which increased the cost, although such cost increases had not been clearly understood or quantified at that time.

The 1972 Greenstein report for the National Academy of Sciences was written after the successful launch of OAO-2, but after the failures of OAO-1 and -B. So the only observatory spacecraft devoted to ultraviolet studies that had been approved, and had not yet been launched, was OAO-3 or Copernicus, which had as its main instrument an 0.8m diameter UV telescope.

Greenstein was unhappy at the lack of any follow-on UV spacecraft after OAO-3, and so his report considered the case for a Large Space Telescope (LST) of 3.0m diameter, covering the UV, visible and near-IR wavebands. Because Greenstein suspected that such an LST was unlikely to fly until the mid-1980s at the earliest, the report speculated on a possible interim programme. This was to both fill the observational gap for UV studies after OAO-3, and to provide some confidence in the design concepts that such an LST would have to use. It was suggested that this interim programme could include either a replacement for OAO-B or a smaller ultraviolet instrument on a Small Astronomy Satellite, plus a 1.5 m diameter telescope covering the UV, visible and near-IR wavebands as a forerunner to the LST.

Greenstein estimated that the total cost of a programme leading to the launch of a 3.0 m diameter LST in the early 1980s would be about $1.0 bn spread over ten years. This would include the launch of an intermediate LST, probably of 1.5 m aperture, but would exclude operations costs. Because the Greenstein committee recognised that it was unlikely that such a sum would be made available in the foreseeable future, they proposed only the interim programme mentioned above.

The Greenstein committee had been asked to prioritise their recommendations, and had been asked by the Bureau of the Budget to reject the idea of building an LST. NASA had not pressed their case for the LST to the Greenstein committee, however, because of Greenstein's known uneasiness about its potential cost. As a result, and in spite of pleading by George Field[3] and Donald Morton, who were members of the committee, the LST was only included in the Greenstein report as a long-term goal.

In June 1974 the House Appropriations Subcommittee refused to back the LST programme because of its low priority in the Greenstein report, but by early 1974 the National Academy of Sciences' Space Science Board had

[3] Interestingly, Field was to chair the next (1982) report after the Greenstein report for the National Academy of Sciences.

concluded that the Greenstein report, which had been published only two years earlier, had already been overtaken by events, and they commissioned a new study on programme priorities. Not only had new discoveries been made, which had changed the situation, but the approval of the Space Shuttle in 1972 had meant that most long-term space programmes had to be re-evaluated, allowing for a Space Shuttle launch with the possibility of in-orbit servicing. John Bahcall and Lyman Spitzer took this opportunity provided by the Space Science Board to push the case for the LST, and in 1974 persuaded each of the 23 members of the Greenstein committee to back the statement that "In our view, Large Space Telescope has the leading priority among future new space astronomy instruments". In addition, Greenstein stated[4] in 1974, when referring to his report, "that had we not had in mind budget limitations, [and] the at that time unsolved technological problems, and had we fully realised the wide range of discovery that we [have] had even in the last three years, we would not have taken quite so 'conservative' an attitude. Astronomers felt then and feel now that the LST is the ultimate optical telescope and that together with a well-balanced, ground-based program, it will open up new vistas for the human mind to contemplate". So what, two years earlier, had been lukewarm support for the LST, had changed into a glowing endorsement. How much of this change was due to technical progress and how much to lobbying is unclear, but both were clearly involved. As a result of this change of attitude by the scientific community, as represented by the Greenstein committee, Congress agreed in August 1974 to fund the Phase B studies of the LST, but required NASA to find international collaborators who would contribute substantially to the overall costs of the project. In fact, in February 1974 possible ESRO involvement in the LST had already been discussed between NASA and ESRO, and by the time that the Congressional edict was issued in August ESRO was already considering the provision of one of the LST instruments. Although this would not meet the requirement to reduce the American costs substantially, at least it was a start, and ESRO agreed to consider providing other pieces of hardware.

Greenstein had suggested in his committee's 1972 report that an interim space telescope programme be considered whilst waiting for the 3.0 m LST in the longer term. In the event, the end-of-life of the UV spacecraft OAO-3 (Copernicus) was extended to 1980, and before the Greenstein report was issued, NASA had agreed in 1971 to participate in the international IUE programme (see Section 14.1), to provide a further UV satellite capability. So by

[4] Congressional Record, 93rd Congress, 2nd session, 2 July, 1974, H.22088-9.

1974 there was already a working relationship between NASA and ESRO on a space-based ultraviolet telescope programme.

In May 1972 the Marshall Space Flight Center was chosen as the lead centre in NASA for the LST because, even with the approval of the Space Shuttle programme at the beginning of the year, it was having to run down its size significantly following the completion of the Apollo programme. Although Marshall's expertise was in the field of large rockets, it had already been chosen as the lead NASA centre for the HEAO programme. The Goddard Space Flight Center, on the other hand, who were generally responsible within NASA for space science programmes, were overloaded with work and could not take on another large programme at this stage. Goddard was made responsible to Marshall for the scientific instruments.

At this stage Marshall estimated that the LST would cost between $570 and $715 million in 1972 dollars ($650 million in 1972 dollars is equivalent to about $2.6 billion in 1998 dollars) covering the design, development and manufacture of an intermediate and final LST. This set alarm bells ringing in NASA HQ,[5] resulting in James Fletcher, the NASA administrator setting a LST target programme cost of $300 million up to the end of its first year in space.[6]

The LST design was divided by NASA into three sections, the Optical Telescope Assembly (OTA), the Support Systems Module (SSM) and the Scientific Instruments (SIs). The OTA was the basic optical telescope (primary and secondary mirrors and tube, but excluding the detectors), the SSM was the module providing power, telemetry, data handling, attitude

---

[5] NASA HQ should not really have been surprised about these price estimates, as the four-spacecraft OAO programme had cost $490 million in real year dollars, or about $2.2 billion in 1998 dollars, for example. There was a great deal of similarity in design between the four OAO spacecraft, so the cost of the first OAO would probably have been about 35% to 40%, rather than 25% of the total programme cost. This gives a cost for the first OAO of about $825m in 1998 dollars. It does not seem unreasonable that the 11½ ton LST, plus an intermediate LST, should cost about $2.6bn, which is only about three times the cost of the first 2 ton OAO. If anything the estimate looks on the low side.

[6] At this time NASA were in the middle of the Mars Viking programme, whose costs were getting out of hand. The original estimate had been $364 million, but the total programme was to cost about $1.0 billion by its completion. So if the LST programme could be undertaken for $300m it would be remarkably cheap compared with Viking, leading again to the feeling that the $300m was largely an artificial price for the LST. In fact it was not really a price, but rather a figure that Fletcher thought he could persuade Congress to accept.

and orbit control and other 'housekeeping' functions, and the Scientific Instruments were the photometers, spectrographs, and other detector assemblies mounted at the base of the OTA.

There were only two credible contractors for the OTA, but there were a number of possible contractors for either the total LST contract or the SSM, if the OTA and SSM were contracted separately. There would be a problem if NASA were to contract for a prime LST contractor responsible for both the SSM and OTA, as all the potential bidders for the prime contract would be chasing just the two potential OTA contractors. Because of this NASA decided not to appoint a Prime Contractor, but to let the contracts for the SSM and OTA separately, and control the interface between the SSM and OTA themselves. Controlling such an interface is not as easy as it sounds as, unless the interface is relatively simple and stable, it is likely that the contractors on either side of the interface will have to make frequent changes. These would then require the contractor on the other side of the interface to make changes also, leaving NASA to pick up the bill.[7]

In order to keep the initial costs down, Marshall suggested in April 1973 bringing the LST back to the Earth every now and again to update it during its planned fifteen year lifetime and to refurbish it, thus avoiding the costs of making it refurbishable in orbit. Although this solution may cost more in the long run, it reduced the all-important initial costs as required by Fletcher. NASA also scrapped the more cautious approach of building a prototype spacecraft followed by a flight spacecraft, and decided to adopt a proto-flight philosophy instead. In this the prototype spacecraft is tested, modified and uprated to become the flight spacecraft. It has the advantage in only requiring one complete spacecraft to be built instead of two, but the disadvantage that any problems found with the prototype have to be put right on what will eventually become flight hardware. After modification the spacecraft has to be tested again, which also produces risks that the two model approach does not entail, but it had the overriding advantage to

[7] The alternative of running a competition for the OTA contractor, and then running a competition for the Prime Contractor, with the Prime Contractor being made responsible for the chosen OTA contractor, does not seem to have been seriously considered. This would have had the advantage that the Prime has to trade off the changes on either side of the SSM/OTA interface and pay for them, assuming that the contract is appropriately worded. In the event, part of the system engineering, including the optimisation of this interface, was performed by Lockheed, the SSM contractor, and part by Perkin–Elmer, the OTA contractor, but Perkin–Elmer was not made responsible to Lockheed.

NASA of saving money. Finally, NASA decided in 1973 that an interim LST was no longer required, as the risk of jumping immediately to a 3 m telescope was then thought to be reasonable, now that a space shuttle launch had been agreed, which would allow refurbishment of the LST if there were problems with it in orbit. In addition a 'Big Bird' military surveillance satellite had already been launched that allegedly used similar technology to the LST.

These measures as a whole reduced Marshall's 1973 estimate for the LST, up to the end of its first year in orbit, to about $290 to $345 million in 1973 dollars (excluding the shuttle launch costs), which was very close to Fletcher's requirement. Two years later, however, further budgetary problems forced NASA to reduce the mirror diameter of the LST from 3.0 m to 2.4 m, saving an estimated $61m in 1975 dollars. Later in 1975 the Large Space Telescope's name was changed to Space Telescope, so as to avoid giving Congress and the tax payers the impression that NASA were being greedy in asking for anything 'large' during a time of financial stringency in the USA. Some astronomers were concerned that this change in name signalled that NASA were eventually going to cut its diameter even more to 1.8 m, which George Low, NASA's deputy administrator, assured them was not the case.

The Space Telescope programme was finally approved by Congress in 1977, with estimated costs to NASA in the range $425 to $475 million (in 1978 dollars)[8] including the cost of operations up to the end of the first year in orbit, and with a launch date of December 1983. In addition ESA would contribute about 15% of the total programme costs by supplying one of the scientific instruments, namely the Faint Object Camera, the solar arrays, and some staff for the Space Telescope Science Institute, which was to be responsible for managing the Space Telescope in orbit. In return, European astronomers were guaranteed at least 15% of the observing time with the Space Telescope.

The phase C/D or design and development phase of the Space Telescope project was started in late 1977 with contracts let to Lockheed for the SSM and Perkin–Elmer for the OTA. At about the same time the following scientific instruments were chosen for the Space Telescope's first flight:

- Wide Field/Planetary Camera with James Westphal of Caltech as PI
- Faint Object Spectrograph, Richard Harms of the University of California, PI

---

[8] This is an average of about $290 million in 1972 dollars, as required by Fletcher in 1972, or about $1.2 billion in 1998 dollars.

- High Speed Photometer, Robert Bless, University of Wisconsin, PI
- High Resolution Spectrograph, John Brandt, Goddard Space Flight Center, PI
- Faint Object Camera supplied by ESA

In addition, one of the three fine guidance sensors, which were to be used to detect where the telescope was pointing, would be used for high accuracy astrometry.[9]

The Space Telescope team at Marshall started phase C/D with a high degree of confidence that the programme could be completed within the agreed budget of $425 to $475 million (1978 dollars), as almost half of this budget had been reserved to cover technical problems. Marshall had reached this position by using a so-called 'low cost' approach, which assumed that failed units, including scientific instruments, could be readily repaired or replaced in orbit by the space shuttle as and when required.[10] Because of this the Space Telescope and its subsystems did not have to be as rigorously tested on the ground as for a normal spacecraft. Its structure could also be designed much more conservatively, again reducing the level of risk and the level of ground testing required.

Within a year or two of the start of phase C/D, however, this whole philosophy began to unravel as technical problems began to appear, and it became clear that, unless the design was improved, the shuttle would have to make significantly more repair trips than originally envisaged. So Marshall reluctantly agreed to return to a more conservative design approach with more testing, consequently extending the programme duration and increasing its costs. As a result, only two years into the programme about 50% of the cost contingency had been used up and a potential launch slip of 6 to 12 months had been flagged. A year later, at the end of 1980, with costs almost out of control, Marshall was forced to reduce the number of orbital replacement units from 124 to about 15, whilst NASA administrator Robert Frosch had accepted a new cost estimate of $700 to $750 million at 1981 rates ($500 to $540 million at 1978 rates), with a launch date of January 1985. This change had only been accepted in NASA after a thorough review of the options available, as some senior NASA personnel thought that with

---

[9] Work in the subsequent development programme showed that all three sensors were needed to locate enough guide stars when observing in a sparse region of the sky.

[10] At this stage NASA were assuming that the Shuttle would fly much more frequently than finally proved to be the case, with turn-round times between flights taking days rather than weeks or months.

these costs and schedule the Space Telescope Project may well be cancelled by Congress. In fact at one stage Bob O'Dell, NASA's Space Telescope Project Scientist, had been so concerned about possible project cancellation that he had even suggested deleting two of the spacecraft's scientific instruments to keep the project alive. Fortunately that was not necessary.

NASA hoped that their major problems were now behind them after they had simplified the programme and increased the budget. But that was not to be, as at the end of 1982 Perkin–Elmer admitted that their part of the Space Telescope programme was in serious trouble. The Marshall project team had been aware throughout 1982 that Perkin–Elmer were having technical problems, but these rapidly escalated in the second half of the year such that by December Perkin–Elmer said that they needed about another $50 million if they were to keep to their delivery schedule for the Optical Telescope Assembly. This caused Samuel Keller, NASA's deputy associate administrator in the office of Space Science, to investigate the Space Telescope programme, and in February 1983 he concluded that the programme as a whole was running at least six months late and would probably need an extra $100 million to complete.

The following month an independent NASA review team headed by James Welch concluded that technical problems on the Space Telescope programme were being compounded by poor management and by a poor management structure for the whole project. In particular, Lockheed, who acted as Marshall's system engineering advisor, had no direct authority over Perkin–Elmer or the scientific instrument contractors. Unfortunately Marshall had not been able to undertake the system engineering task themselves, as the number of personnel that they had been allowed to employ on the Space Telescope programme had been strictly limited by the Department of Defense[11] via NASA HQ.

Following further extensive discussions with Congress in 1983 the Space Telescope budget was increased to $1,175 million at 1983 rates ($770 million at 1978 rates), or an increase over the original budget of about 70%, after taking inflation into account. At the same time ESA re-estimated their expected cost-to-completion to be about 130 MAU (somewhat less than $130 million) at various price levels, or about 85 MAU at 1976 prices, which

[11] The Department of Defense had been concerned because elements of the Space Telescope designs were based on military contract work, and they wanted to limit the possibility of secure information leaking out. The DoD's concern was further enhanced when ESA and their contractors became part of the programme, simply because they were not American.

was an increase of about 40% above their original budget after inflation. ESA's problems were more technical than organisational in nature, chief of them being with the Photon Detector Assembly for the Faint Object Camera. The supplier of the basic detector was having trouble producing tubes of acceptable quality, and there were also problems caused by high voltage breakdown in the glass interface at one end.

Launch was set by NASA in 1983 for the second half of 1986. At about the same time the name of the telescope was changed to the Hubble Space Telescope, or HST, after the astronomer Edwin Hubble who had discovered the expansion of the universe. The new launch date meant, unfortunately, that the HST would not be available to support Voyager 2's closest approach to Uranus in January 1986, or to observe Halley's comet during its perihelion passage the following month.

Because of the long list of problems in the American part of the programme, NASA HQ decided in 1983 to take a much more active rôle in the control of the Space Telescope programme in future. NASA administrator James Beggs appointed a new programme manager, James Welch, and NASA contracted the BDM Corporation to undertake systems engineering. With a much larger budget, NASA also decided to make the programme less risky, by partially reversing some of the earlier decisions which had been taken to reduce the cost. For example, the spares holding was increased, and the number of units that could be replaced in orbit was also increased. NASA also decided to undertake all repair and upgrading of the telescope in orbit, rather than bring it back to Earth, because of the risk to the telescope in bringing it back to Earth and launching it again every five years as originally foreseen. Later in the programme this new concept was endorsed because of the perceived risk to the astronauts of landing the space shuttle with nine tons of hardware in its payload bay.

The next three years after the 1983 crisis were more like those of a normal spacecraft programme, with Lockheed working 24 hours/day, 7 days/week for part of 1985 to try to maintain the HST schedule. Unfortunately, problems caused the system level thermal vacuum test[12] to be delayed from October 1985 to 1986, and in January 1986 the launch date had to be slipped by two months to 27th October of that year. Then the Space Shuttle Challenger exploded, putting the launch of the HST on indefinite hold.

Sceptics or realists, depending on your point of view, had doubted that the Hubble Space Telescope could have been launched in October 1986 had

---

[12] In this test the whole spacecraft is put in a vacuum chamber and operated under simulated orbit conditions.

the Challenger not exploded, as the system level thermal vacuum test had not started until May 1986. Any but the most trivial problems discovered with the spacecraft in this test would almost certainly have resulted in a launch delay. In the event the test, which took two weeks longer than expected, showed that the HST had a number of problems which could now be addressed in 'slow time' because of the indefinite launch delay caused by the Challenger accident.

Indications from the shuttle programme in late 1986 gave a new launch date for the HST of November 1988, but the shuttle took longer than expected to return to operational status and the HST was not launched for a further eighteen months. In this time, both the design of the HST and the system for operating it, once it was in orbit, were significantly improved.

17.2 The Spacecraft and its Instruments

The Hubble Space Telescope that was launched in April 1990 (see Figures 17.1 and 17.2) was one of the most sophisticated and costly scientific instruments ever built. It was an f/24 Ritchey–Chrétien telescope with a 2.4 m diameter primary mirror that in itself weighed almost a ton. The whole HST in the launch configuration (i.e. without its solar arrays, aperture door and antennas deployed) was 13.1 m long and 4.3 m in diameter and weighed 11.6 tons. It was to be placed in a 610 km altitude orbit inclined at 28.8°, have a lifetime of 15 years, and be refurbished every three years or so by astronauts using the space shuttle. At launch it had cost about $2,000 million to NASA and $350 million to ESA at 1998 rates (or about $500m and $70m at 1972 rates).[13] This is a similar total to that estimated by Marshall in 1972, but that was for a prototype and a flight 3.0 metre space telescope, plus an intermediate size flight instrument, whereas the $2,400 million actually spent was for only the flight 2.4 metre instrument.

The best ground-based telescopes in 1990 could achieve a resolution in the visible waveband of about 0.5 arcsec on the best seeing nights, with an average resolution of about 1.0 arcsec for about 2,000 hours per year. The

[13] In addition the launch of the HST cost about $300m and in-orbit operations cost about $250m per year at 1998 rates. Assuming refurbishment and updating every three years at a cost of $350m per launch and $150m per hardware set, gives a 15 year total cost for the HST programme of an incredible $8,400m at 1998 rates (made up of $2,350m for original hardware + $300m for original launch + 15 years of operations at $250 m/yr + 4 refurbishments at $500m each).

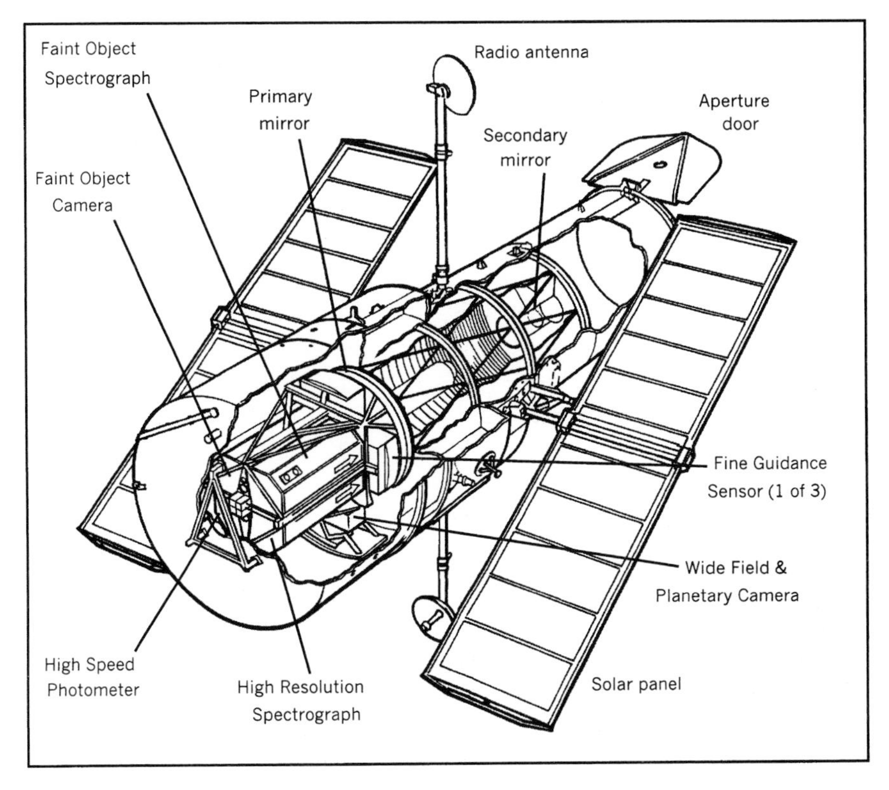

Figure 17.1 Cut-away view of the HST in its orbital configuration, showing the deployed solar array panels, radio antennae and aperture door. The experiment packages are shown in the compartment behind the primary mirror. (Courtesy ESA.)

HST was expected to have a resolution of about 0.1 arcsec in the visible, and about 0.015 arcsec at 121.6 nm (Lyman-α) in the ultraviolet, and be available for observations for about 7,000 hours per year. In fact, the HST was expected to operate eventually over the ultraviolet, visible and infrared wavebands from about 110 nm up to about 1 mm; a far wider band than possible with any Earth-based telescope, although there were no infrared detectors in HST's first set of scientific instruments.

There is no point in having an optical system of very high spatial resolution unless it can be kept stable to a higher accuracy during an exposure. The HST was no exception to this with a requirement that it could slew and point to within 0.01 arcsec of a desired object, and remain within 0.007 arcsec of that position for 24 hours. With such a high resolution and pointing specification it was expected that the telescope would be able to detect stars of visual magnitude 27 in a 4 hour exposure and of magnitude 29 after 24 hours.

Figure 17.2 The Hubble Space Telescope during testing. (Courtesy Lockheed Missiles and Space Company, Inc.)

This compared with a magnitude limit of about 25 for the 5 m Palomar telescope, although by the time that the HST was launched the 10 m Keck was almost complete, which was expected to go down to about magnitude 27.

The Wide Field and Planetary Camera (WF/PC) was the only dedicated instrument lying radial, not parallel, to the HST's optical axis, and the only one observing light from the centre of the telescope's field of view. The WF/PC was developed at JPL and consisted of eight CCDs arranged in two groups of four, one group operating in the so-called wide field mode at f/12.9, and the other in planetary mode at f/30. The first group covered, in total, a square field of view of 2.7×2.7 arcmin, with 1,600×1,600 pixels each covering 0.1 arcsec. The second, planetary group covered in total a field of view of 69×69 arcsec, with each of the 1,600×1,600 pixels covering 0.043 arcsec. The CCDs, which were cooled to about 180 K to increase their signal to noise ratio, were coated with a phosphor material called coronene to improve their response in the UV and enable a spectral range of 115 nm to 1.1 microns to be covered.

ESA's contribution to the HST's set of instruments was the Faint Object Camera (FOC) built by Dornier Systems of Germany, which had both an f/48 and an f/96 camera system. The f/48 system had a field of view 22 arcsec square, with various filters and prisms, and a grating to provide spectra. The f/96 system had various filters, prisms and polarisers, plus a coronagraph system to enable a faint object to be detected just 1 arcsec away from another object with a magnitude difference of about 17. In the very high resolution mode, to be used in the ultraviolet, the f/96 system became an f/288 system and covered a field of view of just 4 arcsec with a pixel size of 0.007 arcsec. The FOC detectors, built by British Aerospace, consisted of high efficiency photomultipliers coupled to television cameras that could, as a system, detect individual photons of light. The full 1,024×1,024 pixel resolution could be produced at 8 bits of digitisation, or a reduced 512×512 pixels could be produced at 16 bits.

For imaging applications the WF/PC provided the largest field of view, but the FOC provided the highest angular resolution at all wavelengths. The WF/PC was more sensitive than the FOC to red light, but the FOC was far more efficient than the WF/PC in the ultraviolet, where the FOC could take full advantage of the very high resolution expected from the HST.

Digicons, which were used as the detectors for both the Faint Object Spectrograph (FOS) and the High Resolution Spectrograph (HRS), consist of linear arrays of diodes to measure intensities at various wavelengths in the spectra. The FOS was designed by Martin Marietta to provide a spectral resolution of 200 for faint sources of magnitude 22 to 26 and of 1,300 for magnitudes of 19 to 22, using either a 'blue' digicon operating between

115 and 500 nm or a 'red' digicon between 180 and 850 nm. The HRS made by Ball Aerospace was designed to produce medium and high resolution spectra in the ultraviolet for bright objects, ranging from a spectral resolution of 2,000 at magnitude 19, through 20,000 at 16, to 100,000 at 14. The 2,000 mode operated from 110 to 170 nm, giving an average spectral resolution of 0.07 nm, with the other two modes operating from 110 to 320 nm, giving average spectral resolutions of 0.01 and 0.002 nm, respectively.

Finally, of the dedicated instruments, the High Speed Photometer (HSP) of the University of Wisconsin was by far the simplest and cheapest, containing no moving parts. Nevertheless, its specification was still impressive; it being required to measure fluctuations in brightness of objects down to a magnitude of 24, with a time resolution of 0.016 milliseconds, and a photometric accuracy of 0.2%, over the spectral range from 120 to 800 nm. The detectors covering the ultraviolet and visible bands were image dissectors, which are magnetically focused photomultipliers. There was also a standard photomultiplier that covered the red end of the spectrum.

As already mentioned, one of the three Fine Guidance Sensors (FGS) was to be used to perform astrometry, measuring the relative position of two stars in the visual magnitude range from 4 to 18 within 0.003 arcsec. This was about four times more accurate than for Earth-based astrometric measurements. The brightness of the stars were also to be measured to an absolute accuracy of about 1% between 510 and 690 nm.

Mention has already been made of problems with producing the Photon Detector Assembly for the Faint Object Camera, but this was by no means the only instrument to have had development problems. For example in 1984, during thermal vacuum testing of the Wide Field and Planetary Camera, it was found that the sensitivity was not uniform across its CCD arrays. That in itself would not necessarily have been a problem, as the different pixel sensitivities could have been mapped across the CCDs, and corrected for by subsequent computer analysis. Unfortunately, the effect seemed to be variable with time in a random manner. It was then realised, however, that the sensitivity in blue light was being affected by the exposure history of each pixel. So that, if a bright object had been imaged on the CCD array, later images taken with that CCD would, for a time, show enhanced sensitivity where the bright image had previously fallen, producing a ghost-like image of the previous object. After a detailed investigation of the effect, which is called 'quantum efficiency hysteresis', it was found that it could be solved by illuminating the back of the CCD array from time to time with ultraviolet light. This resulted in the suggestion that the effect could be eliminated in orbit by pointing the telescope at the Sun, but with the aperture cover closed and with a filter fitted in a small hole in that cover to just let through a small amount of the Sun's ultraviolet light. Not surprisingly, the idea filled most people with horror in case the telescope should be accidentally pointed towards the Sun without the aperture cover being properly closed. Even if it was closed, a damaged filter could let through sufficient sunlight to ruin the telescope. In the end a much less risky solution was adopted of illuminating the back of the CCDs with ultraviolet light from the Sun, using a light pipe fitted through the side of the telescope. This solution also had the beneficial side effect of raising the ultraviolet sensitivity of the CCDs compared with that of their new state.

Other problems were found with the WF/PC during development. For example, the graphite–epoxy material used in its construction was found to take a long time to outgas in vacuum, releasing water vapour that could, in orbit, condense on cold optics whilst the telescope was in eclipse. The instrument also seemed to suffer from contamination of the edges of its optical window by epoxy cement used nearby.

The Fine Guidance Sensors had been a continuous source of concern during the HST development programme. Although they had been through one major redesign at the start of the programme, they were still in trouble in 1983 and required the injection of an extra $15m of funds to help to solve their problems.

Both the High Resolution Spectrograph and the High Speed Photometer had their development problems, but the Faint Object Spectrograph caused much larger headaches. For example, problems with its heat pipes, which were designed to cool its detectors, were only solved in 1988 when the aluminium pipes were replaced by stainless steel units. But the most serious problem with the FOS, which was due to the red digicon losing sensitivity, was finally solved by changing the faulty detector[14] for a new one in 1987. So the time allowed by the long launch delay was well used in improving both the main telescope and its instruments.

17.3 Launch and Post-Launch Checkout

On 10th April 1990, some thirteen years after the start of the design and development phase, the Hubble Space Telescope was ready to be launched by the Space Shuttle Discovery. But four minutes before lift-off a fault with

[14] This was the second time that the red digicon had had to be replaced. The first time was in 1984 when both the red and blue digicons had had to be replaced.

a power supply unit caused the launch to be cancelled. Two weeks later, on 24th April, NASA were ready to try again. This time the count-down got to within 31 seconds of launch, when a fault with a valve in the liquid oxygen line to the shuttle's fuel tank caused another unscheduled hold. On this occasion, however, the valve responded to further commands, and so a few minutes late, at 8.34 am local time, the shuttle finally roared into the sky with its precious payload.

Originally it had been planned to launch the HST into a 590 km high orbit, but this height had been increased to 610 km because the HST was now to be launched during solar maximum. At such times the Earth's atmosphere is denser than average, and this would cause the spacecraft's altitude to decrease faster than originally expected. In the event the shuttle managed to achieve a 614 km high orbit, which was the highest flown to date by any shuttle.

Over the next few days the HST's solar arrays were unfurled, the high gain antennas deployed, the HST released from the shuttle, and the aperture door opened. Unfortunately, a number of relatively minor problems caused a gradual slippage of the in-orbit commissioning, so that the first attempt to focus the telescope, by moving the secondary mirror, did not take place until some days later than planned on 15th May. The Fine Guidance Sensors were used to obtain an approximate focus, and on 20th May the WF/PC was used to produce the first HST image. This was of an area in the loose, open star cluster NGC 3532 which contained a double star with a 1 arcsec separation. The double was clearly resolved, even though the two stars were not clearly defined points of light.

NASA were delighted, and at the subsequent press conference explained that the images would look much clearer when the telescope was properly focused. Not everyone was convinced, however, and on the next day Roger Lynds of the NOAO, and a member of the WF/PC team, suggested that the image showed that the HST had spherical aberration. Independently, Chris Burrows of ESA came to a similar conclusion, concluding that the image showed coma, which could be eliminated by moving the secondary mirror, and spherical aberration, which could probably be eliminated by using the actuators behind the primary mirror to change its shape slightly.

There was a possibility that the image problem was due to the WF/PC, but this idea was disproved on 17th June when the first images taken by the FOC showed exactly the same effect.[15] Two days later it became clear that,

---

[15] ESA were somewhat unfair in their initial publicity as they compared the first light image of the FOC with a relatively poor (2 arcsec) image taken with a ground-based telescope. Ground-based telescopes can achieve significantly better resolution than this.

if the effect was due to spherical aberration, it was too severe for the actuators behind the primary mirror to correct. Jon Holtzman of the WF/PC team was then tasked to calculate what the images from the telescope would look like as the secondary mirror was moved through the focus, assuming various amounts of spherical aberration. A focusing sweep was undertaken a few days later with the HST, and the structure of the images compared with Holtzman's calculations. The match was virtually perfect for 0.5 wave rms of aberration, indicating that the HST had severe spherical aberration caused by the primary mirror.

As a result, NASA immediately set up a review board under Lew Allen, the director of JPL, to examine what had happened during the primary mirror programme and why. The board soon concluded that an error had occurred during the testing of the primary mirror at Perkin–Elmer in 1981, which had not been picked up later because the telescope had not been optically tested as a complete system. In retrospect there had even been indications that something was wrong during this test of the primary mirror, but these had been erroneously dismissed. The idea of a complete optical system test had been rejected at the start of the programme as being too expensive on what was supposed to be a minimum cost programme. It was also considered that such a test could well compromise the overall cleanliness of the telescope, so the test was left out of the programme.

In the optical waveband the HST was originally expected to focus 70% of the light from a star into a radius of 0.1 arcsec, whereas the measured figure in orbit was about 15% within that radius, and 70% within a radius of 0.7 arcsec. So the images were seven times larger linearly than expected, and within the 0.1 arcsec radius they were a factor of five dimmer. This would obviously severely compromise the HST's imaging performance, but it would cause fewer problems with its spectrometers, although exposure times would have to be significantly larger to compensate. Not all was lost, however, as computer processing of the images, in a process known as deconvolution, could partially offset the effect of spherical aberration for relatively bright objects, and so significantly improve the clarity of the images. Starting on 3rd August NASA took a number of images with the HST and, after deconvolution, published them to show the, by now, highly critical public that the HST was still capable of first class science. The problem was that it was not producing two-billion-dollars-worth of science,[16] or anything like it.

[16] The money spent up to May/June 1990, when the problem was first discovered, was about \$2.7bn in 1998 rates, or about \$2.0bn at 1990 rates.

17.4 Scientific Results – The First Three Years[17]

It had been hoped that the HST may be able to detect planets around nearby stars using the coronagraph system in the Faint Object Camera to blot out the stars' overwhelming light. Unfortunately, the HST's optical defect precluded use of this coronagraph, so such planets could not be detected. The telescope could still be used, however, to examine the gas and dust clouds around young stars like β-Pictoris and Vega which had been detected by IRAS, and which may or may not be associated with young planets. In 1991, Albert Boggess of the Goddard Space Flight Center and colleagues used the High Resolution Spectrograph to examine the ultraviolet spectrum from β-Pictoris, and showed that gas from the cloud was spiralling in towards the star at velocities of about 50 km/s. In addition, the spectrum was seen to vary over timescales as short as a month, which was thought to be due to clumps of matter in the disc rotating around the star. The fact that there seemed to be detectable clumps of gas was encouraging to astronomers who anticipated that such clumps could eventually produce planets, even if planets could not be detected by the HST because of its optical problem.

In 1843, the star η-Carinae had flared up to briefly become the second brightest star in the sky, and observations many years later showed that the star was surrounded by a small nebula. Then in 1968 Gerry Neugebauer and Jim Westphal at Caltech found that η-Carinae is the second brightest star in the sky at the infrared wavelength of 20 μm. This is due to the star heating up a surrounding dust cloud to a temperature of about 250 K. It was later shown that this cloud is still gradually expanding and that its ultraviolet spectrum is typical of a cloud of gas and dust ejected by a star near the end of its lifetime. This cloud or nebula, which is peanut-shaped with its long axis oriented north-west/south-east, was thought to be a thin dusty shell of matter, as it had a sharp edge and did not get progressively brighter towards the centre.

The HST added to this emerging picture of η-Carinae and its surrounding nebula by producing the first high resolution image of the nebula with its WF/PC. This resolved a completely new feature in the object, showing two highly constrained jets of material shooting out from the nebula perpendicular to its long axis. In one of the jets there appeared to be a series of par-

---

[17] The HST was also used to observe planetary cloud movements, particularly on Saturn and Jupiter, and to observe comets. Although the planetary data was of interest to planetary meteorologists, I have not considered it above as this section is devoted to the HST's main results.

allel standing waves, although it was not clear whether these were real or an artefact produced by the computer enhancement process.[18] To resolve this question astronomers would have to wait for the HST optical defect to be corrected in the first servicing mission (described below).

The first spectrum obtained by the HST was produced by the High Resolution Spectrograph in October 1990 of the 4th magnitude star χ-Lupi, which was known to be a chemically peculiar star with a relatively normal temperature of about 10,000 K. This star was known to have not only an abnormally high concentration of mercury, platinum and gold, but the mercury appeared to be in the form of mercury 204, its heaviest stable isotope. High resolution ultraviolet spectra were required to help to understand this peculiar star and, in particular, to see if there were other isotopes of mercury present.

David Leckrone of the Goddard Space Flight Center and colleagues had first used the HRS in October 1990 to observe the 194.2 nm line of singly ionised mercury in χ-Lupi to see if the mercury was all isotope 204. Unfortunately, this first attempt to produce a spectrum was only partially successful, as although the HRS produced a spectrum using its 2 arcsec aperture, a computer problem prevented any high resolution observations with its ¼ arcsec aperture. A second attempt which was made a few months later was completely successful, however, showing detailed structure of the 194.2 nm line which had never been resolved before. This showed that 99% of the mercury in the line-forming region of χ-Lupi is mercury 204, compared with about 7% in normal stellar atmospheres.

The cause of χ-Lupi's high concentration of mercury 204 is not understood. Maybe the star's photosphere is layered, producing a very high concentration of mercury 204, platinum and gold for some reason in the thin layer of photosphere that we are observing, or maybe there are strange nuclear reactions taking place. Maybe the reason is connected in some unknown way with χ-Lupi's small binary companion. At present we do not know.

Jay Bookbinder, Frederick Walter and Alexander Brown and colleagues used the HRS to observe flare stars during their flaring activity and during the immediate aftermath to detect changes in their spectra. In the case of the Sun, which has very much smaller flares than flare stars, the material

[18] The image had been computer enhanced by 180 passes through a deconvolution algorithm, which some astronomers thought may have produced the wave effect. Other astronomers disagreed, however, as they thought that they could detect very faint waves in the unprocessed image.

Figure 17.3 An early HST image of the ring of material around the supernova SN 1987A. The expanding cloud of gas from the supernova explosion is at the centre of the ring. The two sources on either side of the ring are background stars. (Courtesy STScI and NASA.)

may fall back to the Sun *after* a flare, but the HST found that, in the case of the flare star AD Leonis, material was falling towards the star *during* the flare. This effect was seen in the lines of triply ionised carbon near 155 nm and triply ionised silicon near 140 nm. These lines can only be produced by gases at temperatures of about 100,000 K,[19] indicating that they are emitted by material in the star's transition region.

On 23rd August 1990 the Faint Object Camera produced a 28 minute exposure of SN 1987A (see Section 15.2) in the light of doubly ionised oxygen at 500.7 nm, showing that the supernova was surrounded by a clear oval structure about 1.6 arcsec across at its widest extent (see Figure 17.3). It was thought that this structure was a ring of material, inclined at 43° to

[19] The minimum temperature of formation of Si IV is about 60,000 K and of C IV is about 110,000 K.

our line of sight, that had been ejected by the central star some time ago. It had apparently been made to glow by ultraviolet light emitted by the supernova which had reached the ring 240 days after the explosion. Calculations showed that the really interesting time should occur sometime between 1996 and 2002 when the material from the supernova explosion hits the material of this ring.[20]

This luminous ring surrounding 1987A enabled a more accurate estimate to be made of the distance to the LMC, in which the supernova had been found, which had been previously estimated as 161,000 ± 18,000 light years. The non-circularity of the ring had enabled its inclination to our line-of-sight to be determined, and analysing the data from the IUE spacecraft gave the difference in time taken for the light to reach the Earth from the nearer and further edges. Given the geometry, the distance to the LMC was then calculated as 169,000 ± 8,000 light years.

The Milky Way's globular clusters are about 12 billion years old and at that age all their heavy white and blue main sequence stars should have died long ago, if all the stars had been formed at the same time, leaving just cool red and orange stars today. It had been known for some time, however, that some of the Milky Way's globular clusters contain bright, blue and presumably massive stars, called blue stragglers, but their origin was unclear.

It had been suggested that these blue stragglers may be the result of mass transfer between binary stars or of their complete merger. This theory received support in 1990 when Mario Mateo of the Carnegie Institution of Washington and his colleagues discovered three blue stragglers in very short period, eclipsing binary systems in the globular cluster NGC 5466. The very short periods indicated that the stars in these binaries are very close together and so mass transfer between them would be relatively easy. At about the same time Michel Aurière of the Pic du Midi Observatory and colleagues found five blue stragglers within 0.2 light years of the centre of the globular cluster NGC 6397. Stars are very densely packed in the central regions of globular clusters and, as a result, collisions are expected to be relatively frequent there, so the discovery of these five blue stragglers also helped to support the merger theory, but more such data was required.

Francesco Paresce of the Space Telescope Science Institute (STScI) and colleagues used the HST's Faint Object Camera in 1990 to image the central region of the bright globular cluster 47 Tucanae. They then filtered out the light from the dominant red giant stars and found twenty-one blue stragglers in the core region, again supporting the merger theory. It had been thought

[20] At the time of writing (1998) there are indications that this impact has just started.

that all stars in globular clusters were of the same age, but if blue stragglers had been produced by stellar mergers or by mass transfer between binaries, some of the globular cluster stars are being rejuvenated, even if they are strictly the same age as their red giant companions.

In 1991 Jon Holtzman of the Lowell Observatory and colleagues used the WF/PC to image the central region of the giant elliptical galaxy NGC 1275 at the heart of the Perseus cluster of galaxies, and found what appeared to be about fifty blue, very bright globular clusters only about 300 million years old. This surprising discovery could, according to Holtzman, be due to a recent merger between two spiral galaxies which could have formed NGC 1275. The galaxy's irregular shape gives some credibility to this merger theory, but elliptical galaxies have up to 100 times as many globular clusters compared with spirals, leading some astronomers to doubt the theory. Just before Holtzman and his team made their announcement, however, Keith Ashman of the STScI and Stephen Zepf of the University of Durham showed theoretically how merging galaxies could stimulate massive gas clouds to produce globular clusters. It had been thought that globular clusters were the remains of an ancient era of galaxy formation, but if the evidence of the clusters in NGC 1275 has been correctly interpreted this is not necessarily so.

As mentioned above, the Milky Way's globular clusters were known to be very old and to have a very high density of stars in their central regions. In view of this it was thought that many of the stars in a globular cluster may by now have gradually spiralled in to the centre to create a massive black hole there. One globular cluster that may contain a black hole at its centre was thought to be M15, but the HST images showed no high intensity source at its centre that would have been produced by matter spiralling in to a black hole. Instead, when the images of the many red giants in M15 had been filtered out, its central 0.8 light year diameter was found to contain thousands of faint stars.[21]

Some time ago it had been thought that the radio galaxy M87 (or 3C 274) may harbour a supermassive black hole at its centre (see Section 13.3), so the high resolution optical images of the central region of this galaxy was awaited with interest from the HST. Images of M87 taken with the WF/PC in the near infrared showed that the very bright central core is still unresolved, so that it must be less than 6½ light years in diameter. Tod Lauer and colleagues analysed these images and showed in 1992 that the intensity

[21] If the stars were this dense near to the Sun, there would be over 100,000 stars closer to the Earth than Proxima Centauri, which is the Sun's nearest stellar neighbour.

falls off with distance from the bright core, as expected if the latter contains a black hole with a mass of 2.6 billion Suns.

The light from distant quasars passes through many cool hydrogen clouds on its way to us, and each of these clouds absorbs light from the quasar. In the laboratory the wavelength of the Lyman-α line of hydrogen is 121.6 nm. So if we are observing a very distant quasar whose light traverses a hydrogen cloud of red-shift 2.5, say, the cloud will produce a Lyman-α absorption line in the light of the quasar at a wavelength of $121.6\times(1+2.5) = 425.6$ nm. If there are a number of hydrogen clouds between the quasar and Earth, the Lyman-α absorption line for each cloud will be at a different wavelength, depending on each cloud's red-shift. This produces what is now called the 'Lyman-α forest' of absorption lines.

Because of atmospheric absorption, Earth-based observations of hydrogen clouds using their Lyman-α absorption lines are limited to a minimum wavelength of about 320 nm, which corresponds to a Lyman-α red-shift of about 1.7. These ground-based analyses had shown that the number of absorbing clouds declined steadily from the furthest detected down to those with a red-shift of about 2, implying that very few such clouds would be present in the present universe if this trend continued to zero red-shift. The HST could observe the Lyman-α forest for the near universe, however, as it could observe ultraviolet spectra down to the Lyman-α rest wavelength of 121.6 nm. Two groups of astronomers set out to measure the number of hydrogen clouds in the near universe, one group led by Ray Weymann of the Carnegie Observatories using the HST's High Resolution Spectrograph, and the other led by John Bahcall of Princeton University using the Faint Object Spectrograph. Both groups found appreciably more local Lyman-α absorption than expected (see Figure 17.4), and their results indicated that the density of hydrogen clouds in the near universe is similar to that at a red-shift of about 2.

It is important to know the amount of deuterium present in the universe, as explained in Section 12.1, as this should indicate if the universe is open, flat or closed. A deuterium abundance of about 15 parts per million (by mass) had been detected in the 1970s by observing the deuterium absorption in the ultraviolet spectrum of β-Centauri. This deuterium abundance figure, which was not very accurately known, indicated an open universe, but as it has crucial implications astronomers wanted to have a more accurate estimate, which could be produced using the HST.

Jeffrey Linsky of the University of Colorado and colleagues re-estimated the deuterium cosmic abundance using the HST High Resolution Spectrograph. This time they chose to observe the star Capella as it had

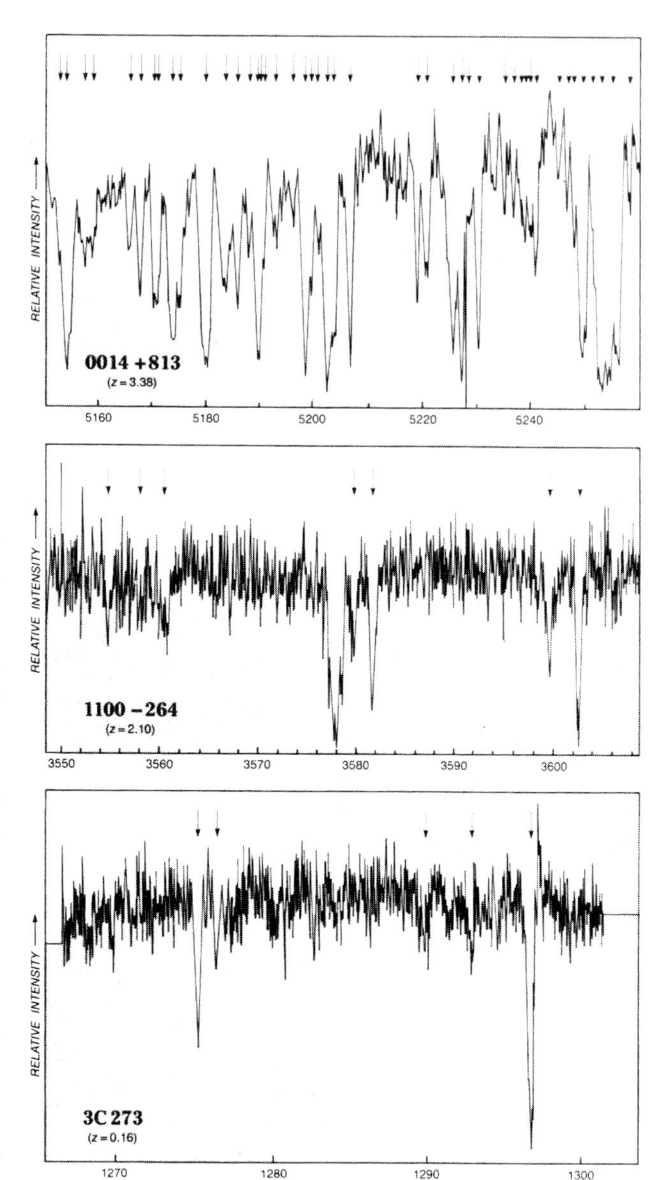

Figure 17.4 These plots show the spectra of three quasars at red-shifts (z) ranging from 3.38 (top) to 0.16 (bottom). The top two graphs were obtained from ground-based observatories and the bottom one using the HST's High Resolution Spectrograph. Although the density of Lyman-α absorption lines per unit wavelength (when normalised for red-shift) has reduced significantly between the top two spectra, there is no significant difference between the second two, showing that the density of hydrogen absorption clouds has not changed appreciably in the recent universe. (Courtesy S. Morris and *Sky & Telescope*.)

only one interstellar cloud between it and the Earth, making the data analysis simpler than for β-Centauri. The deuterium to hydrogen ratio of 15 parts per million ($\pm$10%) deduced from these new measurements confirmed previous suspicions that there is not enough ordinary matter to close the universe. There may be enough dark matter to close the universe, however, but this new deuterium abundance figure implied that there would have to be about ten times as much dark matter as ordinary matter in the universe for it to be closed. Otherwise the universe will continue to expand for ever.

17.5 Hardware Problems and the First Servicing Mission

By far the most serious problem with the HST discovered prior to its first servicing mission of December 1993 was the spherical aberration of its primary mirror, but this was by no means the only important problem experienced with this complex spacecraft. In fact almost immediately after the spacecraft had been set adrift from the space shuttle, it was noticed that the HST shuddered every time it crossed the Earth's day/night terminator. The cause was eventually tracked down to the 12 metre long solar array wings which flapped in response to the thermal shock caused by crossing this terminator. The worst vibration, which had a frequency of 0.1 Hz, caused the spacecraft to be outside of its stability specification of 0.007 arcsec for about seven or eight minutes almost every time it crossed the terminator, which happened twice on every 97 minute orbit. Eventually, NASA managed to reduce the effect of these vibrations by modifying the on-board software to make the rotation speed of the HST's reaction wheels vary every time that vibrations were detected. This software largely solved the problem, but it used valuable on-board computer memory, so it was planned to replace the solar arrays on the first servicing mission with suitably modified arrays.

Gyroscopes are notoriously unreliable, which is why the HST carried six, rather than just the three required to measure the spacecraft's orientation in three axes. These six were packaged in three pairs, each with their own electronic control unit. The first gyro failed at the end of 1990, the second in 1991, and the third in 1992, but fortunately one of each pair was still operational in 1993, allowing full attitude control of the spacecraft. Analysis of the various faults indicated that, to fully reconfigure this subsystem, the astronauts would have to fit two new pairs of gyros and associated electronic control units, together with more robust fuses.

Other failures included two of the six memory modules in the HST's main computer, a low voltage power supply in the Goddard High Resolution Spectrograph, and the solar array control electronics unit (although the back-up unit was still operational). There were also problems with both magnetometers, a high voltage image intensifier in the Faint

Object Camera's f/48 channel, and worn bearings in one of the Fine Guidance Sensors.

This catalogue of problems with the HST gives the impression that the spacecraft was very unreliable. But in many cases the problem could be bypassed by using a redundant unit that was on-board to cover for such eventualities, according to standard spacecraft design practice. In fact, with the exception of the spherical aberration, the in-orbit failure rate of the HST units as a whole was very close to that predicted.

The pressing question in 1990, after the spherical aberration had been discovered, was what could be done to correct it? NASA ruled out the idea of returning the HST to Earth for repair, and it was impossible to replace the primary mirror in orbit, so what could be done? It had been decided before launch to design an improved WF/PC camera to replace the one on the HST at the first servicing mission. This new camera's optics could be modified to correct for the primary mirror's spherical aberration for this camera.

As far as the other instruments were concerned, an investigation panel, which had been set up to review the various options, submitted its report in late October 1990 to Riccardo Giacconi, the director of the STScI. The main recommendation was that the most lightly used instrument, the High Speed Photometer, should be replaced at the first servicing mission by a unit, now called COSTAR,[22] that would intercept and correct the beams of light *en route* to the remaining three axially located instruments,[23] namely the Faint Object Camera, the Goddard High Resolution Spectrograph and the Faint Object Spectrograph.

COSTAR consisted basically of five pairs of mirrors, two pairs for the FOC, two for the FOS, and just one pair for the Goddard HRS as its two apertures were very close together. The first mirror of each pair intercepted light reflected from the telescope's secondary mirror. It then reflected the light on to the second of the pair of mirrors, which corrected the spherical aberration in the process of reflecting it on to the appropriate instrument. These mirrors were all mounted on deployable fingers attached to an optical bench which would be deployed once COSTAR had been fitted to the HST in place of the High Speed Photometer. The optical alignment of these mirrors was critical, so they had motors to allow them to be individually adjusted in orbit.

---

22 COSTAR stands for 'Corrective Optics Space Telescope Axial Replacement'.

23 The Fine Guidance Sensors, which are mounted radially, would still have to operate with no corrective optics until the second servicing mission in 1997.

It was estimated that COSTAR should enable 60% of a star's light to fall inside the central 0.1 arcsec of its image in visible light, compared with the pre-COSTAR figure of 15% and the specification of 70%. Unfortunately, the extra reflections involved in COSTAR reduce the light reaching the instruments by about 20% in the visible and about 40% in the ultraviolet. The fields of view of the instruments were also reduced, so that the f/48 channel of the FOC became an f/75 channel, for example, and the FOC's coronagraph remained useless. Nevertheless, the overall performance of the FOC, FOS and Goddard HRS would be significantly improved.

The new WF/PC, called imaginatively the WF/PC-2, no longer had a switchable system between four wide field and four 'planetary' CCD mosaics. Instead, to save money, three wide field CCDs were arranged in an 'L' shape, and one smaller, high-resolution CCD was fitted in the fourth quadrant. Although there were fewer CCDs than before, their performance was better in both the visible and ultraviolet wavebands, allowing shorter exposures. COSTAR would not correct the light beam *en route* to the WF/PC-2, however, as the WF/PC-2 was, like its predecessor, to be mounted radially, so the WF/PC-2 design was modified to make the correction itself.

NASA had initially intended to replace two of the existing HST instruments by two so-called 'second generation' instruments, named NICMOS[24] and STIS,[25] during an in-orbit servicing mission, as well as replacing the WF/PC by WF/PC-2 (with no corrective optics originally, of course). NASA had not decided initially in which order these instruments would be flown, however, leaving that decision to rest on circumstances. In 1991, when the outline COSTAR design was being produced, the first in-orbit servicing period was expected to take place in late 1993 or early 1994. So it was at this time, with just two or three years to go to launch, that NASA had to decide on whether to fly COSTAR on that mission, and when to install the WF/PC-2, NICMOS and STIS instruments. After much agonising, NASA finally decided in December 1991 to install both COSTAR and the WF/PC-2 at the first servicing mission, but only on condition that COSTAR did not significantly delay that mission, and to fly NICMOS and STIS later. Unfortunately, the HST budget could not be increased, so the funds for COSTAR, and for including corrective optics in NICMOS and STIS, all had to come from the budget for the second generation instruments. Consequently, the designs of NICMOS and STIS were radically simplified to provide money for COSTAR and, assuming that all this could be achieved within the budget, it was

[24] Near Infrared Camera and Multi-Object Spectrometer. This was the first infrared instrument to fly on the HST. [25] Space Telescope Imaging Spectrograph.

planned to fly NICMOS and STIS on the second servicing mission expected in 1997.[26]

In the event Ball Aerospace completed COSTAR in the remarkably short period of just 28 months, and it was ready on time to be launched on the first servicing mission, which started on 2nd December 1993.[27] Then during the course of five epic spacewalks, lasting about six hours each, astronauts from the Space Shuttle Endeavour installed COSTAR and the WF/PC-2, and replaced all the faulty units, including the solar arrays and gyros. After the servicing mission, which was a complete success, the COSTAR and the WF/PC-2 units produced stunning images, free at last from the effects of spherical aberration. In addition, the replacement solar arrays caused no more oscillation problems.

17.6 The Second Three Years

There are numerous clouds of gas and dust spread across the constellation of Orion. These are generally very cold, but in some regions they are heated up or even ionised by very hot stars. For example, the very bright, colourful Orion Nebula is visible in the sword of Orion because the gas in this region is being ionised by the very bright, blue star called θ^1C Orionis. This O-type star, which is one of the four trapezium stars at the centre of the nebula (see Figure 17.5), emits more ionising radiation than the hundreds of other stars in the nebula put together, making it the dominant star.

Astronomers had known for some time that the Orion Nebula, which contains a number of very young stars, is probably still forming stars from the plentiful amounts of gas and dust that it contains. In 1967, for example, Eric Becklin and Gerry Neugebauer had discovered in the nebula an object, now called the BN object, that had a temperature of only 600 K. At first it was thought that it was a protostar, but radiation was soon found coming from hydrogen atoms at a temperature of 10,000 K, indicating that the BN object is a cloud of warm dust surrounding a hot, young star. Later, George

---

[26] In fact NICMOS and STIS were installed in the HST during the second servicing mission in February 1977, along with an FGS with corrective optics. But this mission, and the subsequent HST results, fall beyond the scope of this book (as explained in the Preface).

[27] The total cost of the first servicing mission was an incredible $692 million, consisting of $429m shuttle and operational costs, $251m NASA HST hardware costs, and $12m ESA solar array costs.

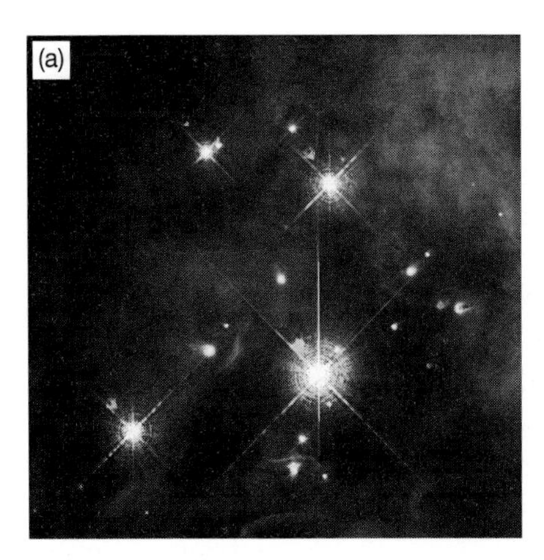

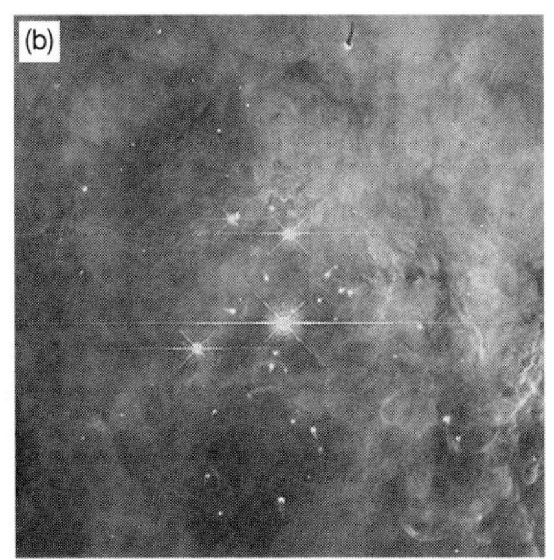

Figure 17.5 (a) The four young trapezium stars in the Orion Nebula as seen by the WF/PC-2 are strong emitters of ultraviolet radiation, but θ^1C Orionis, the star to the lower right, outshines all the others. This has resulted in the proplyds, which are clouds of dust and gas surrounding very young stars, in this region forming tails that generally point away from θ^1C Orionis. More of these proplyds are shown in (b), also from the WF/PC-2, which shows a larger region centred on the trapezium. (Courtesy D. Johnstone (CITA), STScI and NASA.)

Herbig of the University of Hawaii showed that the density of stars in Orion's Trapezium Cluster at the centre of the nebula is about 10,000 times that in the vicinity of the Sun. He also found that most of these stars were still contracting, not having yet reached the Main Sequence, so the cluster is a dense stellar nursery.

It was not surprising, therefore, that the Orion Nebula was one of the first objects imaged by the original WF/PC, enabling John Stauffer of the Harvard–Smithsonian Center for Astrophysics to find even more young stars. But the images of this complex region were severely compromised by HST's faulty primary mirror, and so it was with special expectation that new ones were taken with the replacement WF/PC. Robert O'Dell of Rice University, the HST Project Scientist from 1972 to 1983, produced the first of these images just three weeks after the replacement camera had been installed. They clearly showed a number of stars surrounded by discs of dusty material, which O'Dell called 'proplyds' (for *pro*to*pl*anetar*y* *d*iscs). Some proplyds were expected, but the sheer number was a big surprise, with over 30% of the stars imaged being surrounded by proplyds of one sort or another (see Figure 17.5). Mark McCaughrean of the Max Planck Institute and John Stauffer then showed that these are low mass, pre Main Sequence stars by observing them in the near infrared.

The structure of the individual proplyds seen in the WF/PC's high resolution images proved to be very interesting. They are basically small clouds of dust and gas whose outer parts, which face the star θ^1C Orionis, are being ionised by its intense ultraviolet radiation, in the same way that the gas in the Orion Nebula is being ionised. Those proplyds nearest to θ^1C Orionis have a bright cusp on the side facing this star, and have a comet-like tail, created by the strong stellar wind, trailing in the opposite direction.

Although dust clouds around young stars had been seen elsewhere before, these proplyd images were the first seen at high resolution. Whether the proplyd dust and gas clouds were really protoplanetary[28] discs was unclear, however. It was possible that some of the proplyds may have started to form planets, but the proplyds were estimated to be only about 1 million years old, and so, if planets have started to form, they would only be in the very earliest stages of formation. In addition, to detect planets at such a distance of about 1,500 light years would require probably a two orders-of-magnitude increase in resolution, which is currently not available, and it would be even more difficult to detect small protoplanets. These proplyds are being continuously eroded by radiation from θ^1C Orionis, and it is not clear, if and when planets start to form, whether they will become large enough before all the material has been driven away by the star.

In August 1994 Robert O'Dell and Kerry Handron of Rice University imaged part of the planetary nebula NGC 7293, otherwise known as the Helix Nebula, with the WF/PC-2 and found hundreds of bright, arc-like objects pointing towards the central white dwarf star. Behind the arc-like heads were comet-like streamers making a number of what are called 'cometary knots' (see Figure 17.6). These cometary knots were thought to be clouds of neutral hydrogen, with the arc-like heads being ionised by the white dwarf's ultraviolet light. Unlike the proplyds of the Orion Nebula, however, none of these cometary knots seemed to have a central star. Shortly after their discovery, John Meaburn of the University of Manchester and colleagues used the Anglo-Australian Telescope to measure their recession velocity from the nebula's centre, and found a figure of 10 km/s, compared with the nebula's overall expansion velocity of 24 km/s. So the knot material seems to have been ejected by the central star when it was in its red giant phase, before it ejected the planetary nebula. The latter is now overtaking the knots, causing erosion and the comet-like tails.

One of the most spectacular images produced by the WF/PC-2 camera was that of the Eagle Nebula, otherwise known as M16, taken in April

---

[28] A protoplanet is a planet in the process of formation.

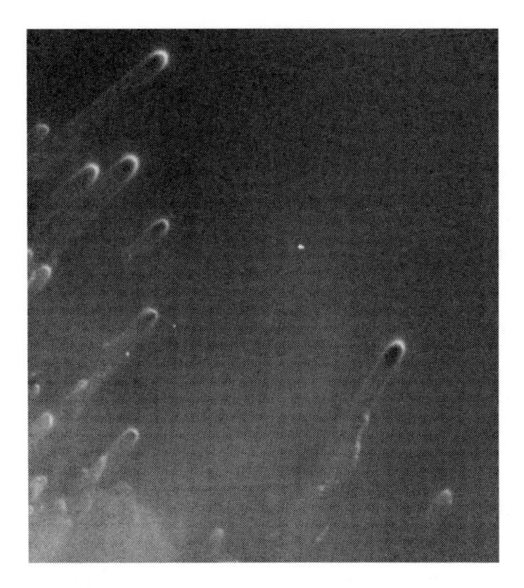

Figure 17.6 Cometary knots in the Helix Nebula have tails pointing away from the central white dwarf (out of picture) in this WF/PC-2 image. Although these knots look superficially similar to proplyds, they are quite different as they have no central star. The knots appear to be formed from clouds of material ejected by the white dwarf when it was in its red giant phase. (Courtesy R. O'Dell, K. Handron (Rice University), STScI and NASA.)

1995 (see Figure 17.7a). The central region of this nebula, which is 7,000 light years away, consists of fingers of a dark molecular cloud protruding into a bright region of ionised hydrogen in which there are a number of hot young stars. Originally, this whole region was one dark molecular cloud, parts of which started condensing to form the young stars that we see today. Ultraviolet radiation from these young stars is now causing the hydrogen to glow, and is gradually eroding away the dark cloud. It is thus an ideal region to study how ultraviolet radiation interacts with a molecular cloud.

The HST image (see Figure 17.7b) showed that there were small protuberances at the edges of the large dark fingers of dense molecular gas and dust. Some of these small protuberances, or Evaporating Gaseous Globules (EGGs) as they became known, appeared to contain embryonic stars. However, near infrared images taken with the 3.6 m telescope of the ESO about a year later showed that this association of young stars with EGGs was incorrect. In fact, in only one case was a star associated with an EGG. All the other stars thought to be associated with EGGs were shown to be background objects. These EGGs, now called 'preprotostellar regions', are dense regions of the nebula which are contracting under their own gravity whilst being illuminated with strong ultraviolet light. Stars should eventually form in them, but the stellar masses will be limited by the evaporation of these regions by the ultraviolet radiation which has already, in some cases, cut off the regions from the main molecular cloud.

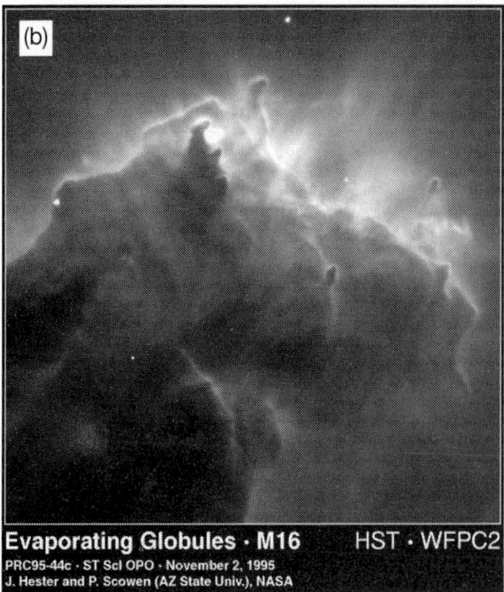

Figure 17.7 The spectacular Eagle Nebula (a) consists of long dark fingers of dense molecular gas and dust which is being irradiated by intense ultraviolet light. This light is gradually eroding the fingers, leaving small protuberances of material (b) which may eventually condense to form stars. (Courtesy J. Hester and P. Scowen (Arizona State University), STScI and NASA.)

The Crab Nebula has a centrally located pulsar embedded in an inner shell of material which is emitting synchrotron radiation in its intense magnetic field. This 'synchrotron bubble' appears to be surrounded by an outer shell of dense material which is seen as a complex network of bright wisps ejected during the supernova explosion about 950 years ago. A few years ago this outer shell was seen to contain large amounts of dust which seem to have survived or possibly to have been produced in the supernova explosion.

Jeff Hester and Paul Scowen of Arizona State University used the WF/PC-2 to image the Crab Nebula at an unprecedented resolution. Surprisingly, their images showed a number of new features in what was already a thoroughly researched object. A halo was found in the synchrotron nebula around the direction of one of the pulsar's X-ray polar jets, with the halo varying its structure radically over a timescale of a few weeks. In addition, a small knot of bright emission, which kept changing its location, was found very close to the pulsar along the direction of the other X-ray polar jet. Both the polar halo and the bright knot were thought to be probably caused by interactions between the pulsar's wind of charged particles, which is emitted along its polar axis, and the gas of the nebula.

The HST's images of the filamentary structure of the Crab Nebula showed that it was much more complex than previously thought. These filaments, which are heated to fluorescence by the intense ultraviolet radiation of the synchrotron nebula, were often found to have cool dense, dusty cores surrounded by hotter, more highly ionised regions. It had been thought that the synchrotron nebula was constrained to stay relatively close to the pulsar by this outer filamentary region, but the HST images indicated that the synchrotron nebula is pushing its way outwards through the filamentary region, causing the filaments to be elongated and often fragmented.

We left η-Carinae with the news that the WF/PC-1 image showed two highly constrained jets of material shooting out of the nebula perpendicular to its north-west/south-east long axis, and this raised the question as to which was the star's polar axis, the long axis or the jet axis?

In 1992 John Hillier of the University of Munich and David Allen of the Anglo-Australian Observatory used the Anglo-Australian Telescope to measure the spectra of the two bulbous shells that had been ejected along the nebula's north-west/south-east axis. Lines of helium and nickel showed that the incomplete spherical shell of the south-eastern lobe is expanding towards us at 660 km/s, whilst the partly filled north-western lobe is moving away from us. At this rate the nebula would have been expanding for about 150 years, which implies that it was emitted during η-Carinae's 1843 outburst.

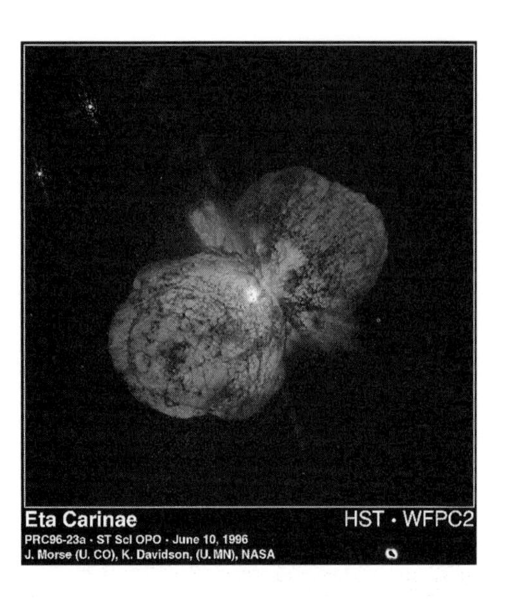

Figure 17.8 The star η-Carinae exploded in 1843 to produce the two large lobes seen in this image. A less powerful explosion occurred about fifty years later to produce the jet pointing to the top left. It is thought that the central star, which has survived these two explosions (as well as many smaller events), may be a binary, as a 5½ year period has been detected in its optical and infrared emission. (Courtesy J. Morse (University of Colorado), K. Davidson (University of Minnesota), STScI and NASA.)

Shortly after the first servicing mission had been completed, Jeff Hester used the WF/PC-2 to image the Carina Nebula (see Figure 17.8) and found that the standing wave patterns on the previous computer-enhanced WF/PC-1 image was largely an artefact of the computer processing. It was also concluded that the jets are in the star's equatorial plane, with the two bulbous shells or lobes expanding along the star's polar axis pointing north-west/south-east.

Recent observations of the Carina Nebula at other wavelengths have shown that the nebula is continuing to develop in surprising ways. For example, over the second half of 1992 Michael Corcoran and colleagues found, using Rosat, that the Carina Nebula's X-ray intensity doubled. Also, between mid-1992 and early-1994 Robert Duncan of the Anglo-Australian National Facility, Stephen White of the University of Maryland, and their colleagues found that there was a threefold increase in its radio intensity. Then in 1996 Augusto Damineli of the University of São Paulo found a 5½ year period in the optical and infrared intensity of η-Carinae and in the intensity of certain spectral lines. Damineli attributed this to pulsations in the star's photosphere, although more recent analysis tends to support the idea that η-Carinae is a binary.

Finally, in 1996 Kris Davidson of the University of Minnesota and Jon Morse of the University of Colorado announced that they had compared WF/PC-2 images of the Carina Nebula taken in April 1994 and September 1995 and, even in such a short space of time, the expansion of the nebula

Figure 17.9 Images of SN 1987A taken after the new WF/PC had been fitted to the HST surprised astronomers by resolving two large rings of material, in addition to the smaller ring seen previously. The material in the larger rings seems to have been emitted when the original star was a red giant. (Courtesy C. Burrows, STScI and NASA.)

could be clearly detected. The main lobes were confirmed as expanding at about 700 km/s, showing that they were the result of the explosion in 1843. But the perpendicular jets of material seem to have been in motion for only about 100 years, which suggests that they are the result of a smaller explosion noticed in the 1890s.

It will be recalled that the WF/PC-1 had shown a ring of material around SN 1987A (see Figure 17.3 earlier). Immediately after the new WF/PC had been fitted, a new image was taken and this showed (see Figure 17.9) not only the original ring, but also two larger, fainter rings which appeared to intersect each other as seen from Earth. The origin of these fainter rings is still unclear, but spectral analysis shows that they were emitted when the progenitor star was a red giant, before the material of the smaller ring was emitted. The best guess is that these two large rings are the outer edge of an hourglass-shaped shell, but that is still only a hypothesis.

Interestingly, the WF/PC-2's image of the planetary nebula MyCn 18 (see Figure 17.10) showed a similar, albeit more complex structure than for that of SN 1987A. Otherwise known as the Hourglass Nebula, this planetary nebula seems to be the result of a fast stellar wind expanding through a cloud of previously ejected matter. Because the original cloud was probably denser at the star's equator, this would have constrained the fast stellar wind there, producing the narrow, hourglass waist, whereas at the star's poles the wind has apparently escaped freely. The various large rings in the hourglass's

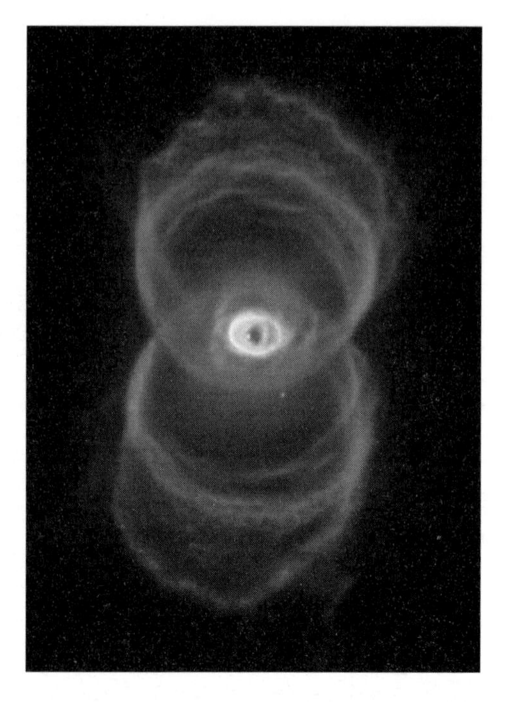

Figure 17.10 The ring structure of the Hourglass Nebula, otherwise known as MyCn 18, seen in this image looks like a more complex version of the double ring structure of SN 1987A. Irregular emissions of material are thought to be responsible for the various rings. The non-central white dwarf is clearly seen inside the bright, central elliptically shaped structure. (Courtesy R. Sahai and J. Trauger (JPL), STScI and NASA.)

walls are thought to be due to mass being ejected in spurts during the star's red giant phase. The HST imaged the 'central' white dwarf for the first time, which was unusually found to be off-centre. It is not clear why this is so, but Raghvendra Sahai and John Trauger of JPL, who produced the image, suspect that there is also an unseen companion.

The WF/PC-2 camera has produced many stunning and interesting images since its installation as part of the first servicing mission, of which the above are just a few. Particularly impressive are the images of planetary nebulae, like the Helix and Hourglass Nebulae discussed above, showing the complex patters of gas emitted by stars towards the end of their lives.

The WF/PC has also been used to observe a number of colliding galaxies, probably the most spectacular example being the Antenna Galaxies NGC 4038 and 4039 (see Figure 17.11). The Antenna Galaxies had been imaged by Bradley Whitmore of the STScI and Francois Schweizer of the Carnegie Institution of Washington using the original WF/PC. Although the resolution of these WF/PC-1 images was poor, these astronomers were able to detect about 700 blue compact star clusters, generally within large regions of ionised hydrogen, together with a dozen or so red star clusters. The blue clusters, which are less than 50 million years old, appeared to be about the same size of the Milky Way's globular clusters. They appeared to

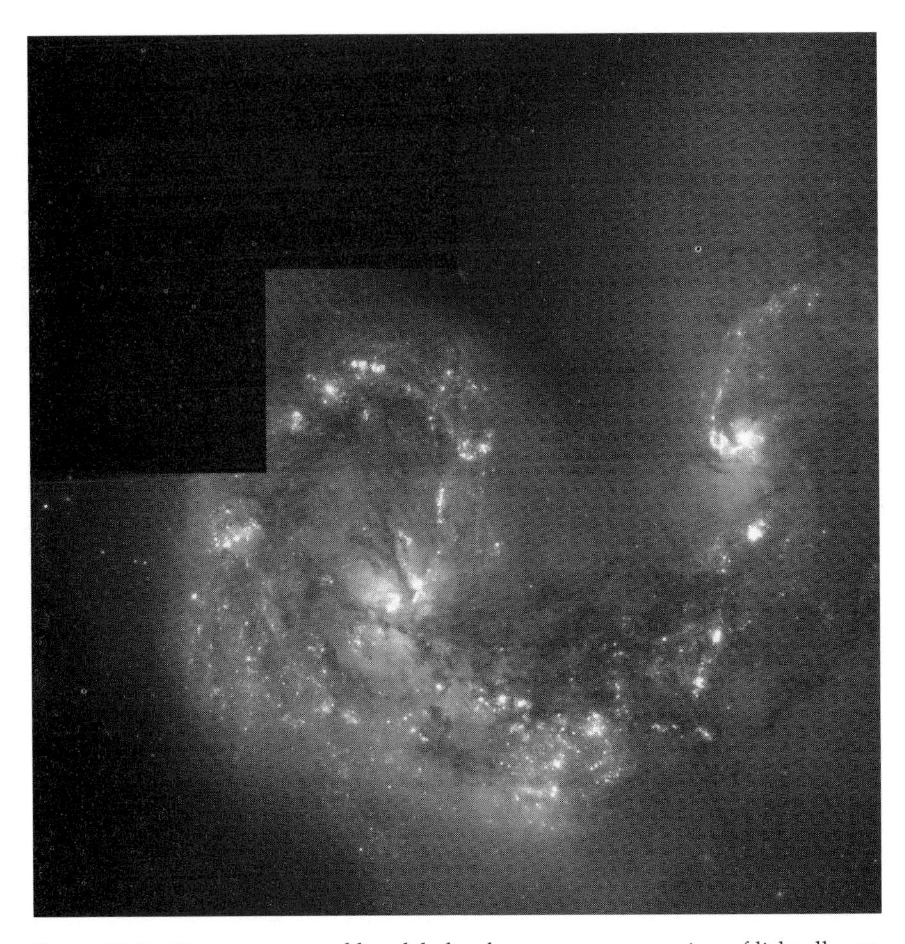

Figure 17.11 Numerous young, blue globular clusters are seen as points of light all over this image of the Antenna Galaxies which are still in the process of colliding. The cores of the two galaxies are left of centre and near the top right; most of the other objects in this image are blue globulars. (Courtesy B. Whitmore, STScI and NASA.)

have been formed in the collision process between the two galaxies, like the case of NGC 1275 described in Section 17.4, although the Antenna merger is more recent so their globulars are younger. The structure of the galactic merger was much more clearly seen using the WF/PC-2 camera, which resolved over 1,000 of these blue globular clusters.

Another example of colliding galaxies is the Cartwheel Galaxy, which appears to be the result of a head-on collision of a spiral galaxy and another, smaller galaxy about 200 million years ago. Kirk Borne's image (see Figure 17.12), released in January 1995, shows the yellow, dusty spiral galaxy at the

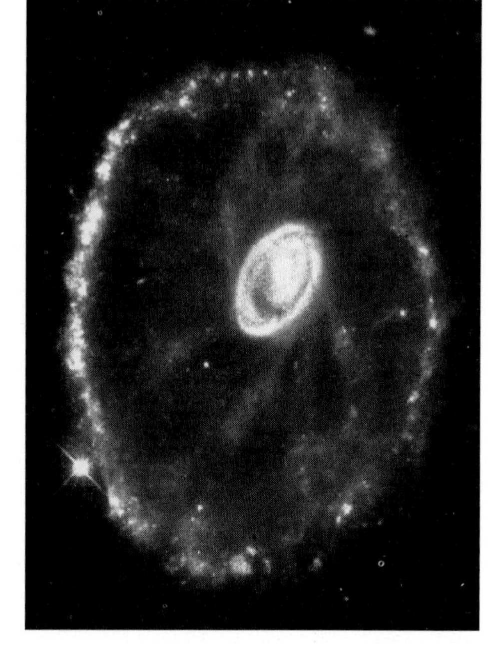

Figure 17.12 The yellow dusty spiral galaxy near the centre of this image of the Cartwheel Galaxy has apparently been forced to eject material during its head-on collision with another galaxy. This ejected material forms the large outer ring seen in this image which consists of gas, dust and giant clusters of hot, massive, blue stars. (Courtesy K. Borne, STScI and NASA.)

centre surrounded by blue spokes leading to a ring of intense, blue star formation. This ring was apparently created by material that was originally in the spiral galaxy but has now been flung out, like ripples on a pond, by the collision. Blue stars are also forming in the core galaxy and its immediate surround, but the star formation rate there is nothing like as intense as that in the blue outer ring.

Earlier work had indicated that there may be a supermassive black hole at the centre of M87 (or Virgo A), so it was with interest that the centre of the galaxy was imaged by the WF/PC-2. Holland Ford of Johns Hopkins University, Richard Harms of the Applied Research Corporation and colleagues reported in 1994 that the image showed, much to everyone's surprise, that the central region of this elliptical galaxy exhibited spiral structure. The Faint Object Spectrograph also showed that the gas at a radius of 60 light years has a velocity of 550 km/s, implying a central mass of about 3 billion solar masses, consistent with Tod Lauer's earlier estimate (see Section 17.4). This supports the contention that M87 has a supermassive black hole at its centre.

In January 1995 John Bahcall, Sofia Kirhakos of the Institute for Advanced Study, and Donald Schneider of Pennsylvania State University presented data to an American Astronomical Society meeting, showing that

11 out of the 15 low red-shift quasars that they had imaged using the WF/PC-2 had apparently no host galaxies. This astonished the astronomical community as it had been thought that quasars are the nuclei of galaxies in which a supermassive black hole is accreting material from the surrounding galaxy. Ground-based images of the closest quasars had previously shown what appeared to be faint galaxies around many of these quasars, and yet the WF/PC-2 with its superior resolving power seemed to disprove this interpretation. Further work in 1995 on the WF/PC-2 images by Kim McLeod of the Harvard Center for Astrophysics and George Rieke of the University of Arizona showed that Bahcall, Kirhakos and Schneider's conclusions had been incorrect, however. The host galaxies *were* there, but they were so faint that they had been obscured by the intense quasar emission. McLeod and Rieke were also able to confirm the presence of these quasar host galaxies using near infrared images taken with the 2.2 metre Steward Observatory reflector on Kitt Peak. So the astronomical community could relax once again.

Although it was now clear that quasars are the nuclei of certain types of active galaxies, the HST images also indicated that the quasar mechanism may not be the same in all cases. Some quasars are seen at the centre of otherwise normal spiral and elliptical galaxies, but other images showed that quasars are also associated with colliding or merging galaxies. So at this stage the cause of quasar activity is still somewhat uncertain.

It had been known for some time, from an analysis of the dynamics of stars within galaxies, and of galaxies within galaxy clusters, that there appeared to be more mass in the universe than that visible in stars, gas and dust. This extra mass is assumed to be in the form of so-called dark matter. Then in 1986 Roger Lynds of the Kitt Peak National Observatory and Vahe Petrosian of Stanford University discovered giant luminous arcs in the galaxy clusters 2242-02 and Abell 370, which were thought to be due to the gravity of the foreground clusters distorting images of galaxies behind the clusters. These, and later examples of this so-called 'gravitational lensing', allowed an estimate to be made of the mass of the foreground galaxy clusters and any other intervening objects.

The WF/PC-2 showed some spectacular examples of gravitational lensing by galaxy clusters. For example, the image of the cluster Abell 2218, which has a red-shift of 0.18, was found to be lensing galaxies of red-shift 2.5. The HST also provided a very clear image of the galaxy cluster CL0024 + 1654, which is imaging a galaxy cluster about twice as far away. Analysis of this latter image enabled the mass of CL0024 + 1654 to be determined as about 200 times the mass of its visible galaxies. About 10% of the cluster's total mass appeared to be gas and dust, so the remaining 90% must be dark matter.

Robert Williams, director of the STScI, decided in early 1995 to use the WF/PC-2 to find out what the universe looked like billions of years ago. To do this he decided to point the HST towards a part of the sky in which there were no stars, and to observe that 'blank field' for as long as possible. In the event the WF/PC-2 took a total of 342 exposures in four wavebands centred on 300, 450, 606 and 814 nm over the period from 18th to 28th December 1995. The resulting Hubble Deep Field image showed over 2,000 galaxies down to a magnitude of 30 in the red region. Neutral hydrogen in intergalactic space absorbs ultraviolet light from distant galaxies below the so-called Lyman wavelength limit, and as this Lyman limit is red-shifted it makes galaxies appear redder the further they are away, so much so that the most distant galaxies could only be detected at 814 nm. Using this simple technique of comparing the redness of galaxies enabled their red-shifts to be estimated, even when they were too dim to have their spectra recorded. On this basis, of those objects with a magnitude of at least 28, 367 galaxies were found to have a red-shift of between 0 and 1, 512 between 1 and 2, 135 between 2 and 3, and so on, with 4 galaxies having a red-shift greater than 6; these latter four being observed when the universe was only a few billion years old. Interestingly, this Hubble Deep Field image showed that about 30% to 40% of distant galaxies are irregular in shape, compared with only a few per cent today. Clearly, a more detailed analysis of this and other deep field images will help substantially in our understanding of galaxy evolution.

When the Hubble Space Telescope was launched, it was widely expected that it would enable the age of the universe to be determined much more accurately than before, by enabling astronomers to determine the so-called 'Hubble constant' H_0. In 1929, when Edwin Hubble showed that the recession velocity of galaxies increased with increasing distance from us, he calculated a value for this rate of expansion of $H_0 = 500$ km/s/Mpc.[29] Since then recession velocities have been determined at greater and greater distances so that, at the time of the HST launch, the true value of H_0 was thought to be somewhere in the range from about 50 to 100 km/s/Mpc. If the universe had been expanding at these rates for ever, these latter values of H_0 would imply an age for the universe of 20 to 10 billion years, respectively. But the globular clusters in the Milky Way appear to have an age of at least 13 billion years, so the 10 billion year figure appears to be wrong, implying that H_0

[29] Mpc stands for megaparsec, or 3.26 million light years.

must be less than 100 km/s/Mpc. The universe has a finite density, however, tending to slow down its expansion, so the universe would have expanded faster in the past than today, producing a *lower* age for a given current value of H_0, implying that H_0 must be much lower than 100 km/s/Mpc. Before going further to discuss this situation, however, the density parameter Ω_0 needs defining.

The density parameter Ω_0 is defined as the ratio of the actual density of the universe to the critical density (see Section 12.1). So for a universe that will expand for ever, i.e. an open universe, $\Omega_0 < 1.0$, for a flat universe $\Omega_0 = 1.0$, and for a closed universe $\Omega_0 > 1.0$.

If $H_0 = 50$ km/s/Mpc, the age of the universe would be about 20 billion years for a universe with $\Omega_0 \approx 0$, about 17 billion years if $\Omega_0 = 0.2$, and only about 13 billion years for a flat universe (i.e. with $\Omega_0 = 1.0$). If H_0 is higher, these ages are lower. So if H_0 is somewhere between 50 and 100 km/s/Mpc, and if the estimated ages of globular clusters are correct at 13 billion years minimum, then the universe cannot be closed (as Ω_0 cannot be greater than 1.0).

To determine H_0 we need to estimate the distance of a galaxy or cluster of galaxies, as their recession velocities can be easily deduced from the red-shift of their spectral lines. Many years ago Henrietta Leavitt of the Harvard College Observatory showed that the period of oscillation of Cepheid variable stars increases as their absolute intensities or magnitudes increase. Ejnar Hertzsprung then used this period/absolute magnitude relationship to determine the distance of Cepheid variables from their measured periods and observed magnitudes. A little later Edwin Hubble was able to detect Cepheid variables in galaxies for the first time, and so was able to use Hertzsprung's technique to estimate galactic distances. In fact, it was distances determined by this method that enabled Hubble to discover his relationship between distance and red-shift, mentioned above, and to make his first estimates of the value of H_0. Since then, the distances of more and more distant galaxies have been determined, and it was hoped that the Hubble Space Telescope would continue to expand our knowledge of galactic distances, and hence enable H_0 to be determined more accurately.

In 1992 Sandage, Tammann, Macchetto and colleagues used the HST to examine 27 Cepheid variables in the galaxy IC 4182, which lies about 16 million light years from Earth, in an attempt to improve the estimate of H_0. A type Ia supernova had exploded in IC 4182 in 1937, and by accurately measuring the distance of this galaxy using Cepheids, Sandage and colleagues could determine the *absolute* magnitude of the 1937 supernova. This in turn allowed the distance of further galaxies to be determined by measuring the *apparent* magnitude of their type Ia supernovae, assuming that all type Ia supernovae have the same absolute magnitude. The value of H_0 determined using this technique was 45 ± 15 km/s/Mpc.

After the installation of the WF/PC-2 it became possible to observe, for the first time, Cepheids in galaxies in the Virgo cluster. In late 1994, Wendy Freedman of the Carnegie Observatories and colleagues observed twenty Cepheids in the Virgo cluster galaxy M100, and concluded that this galaxy is about 56 ± 6 million light years away. Knowing its recession velocity, and after correcting for the Milky Way's known movement in local space enabled these astronomers to estimate H_0 as 80 ± 17 km/s/Mpc. Shortly afterwards H_0 was estimated by Nial Tanvir of Cambridge University as 69 ± 8 km/s/Mpc from Cepheids in galaxy M96, and by 1996 the most likely value of H_0 using Cepheids in ten galaxies was calculated to be about 72 km/s/Mpc. This gave a maximum age for the universe (i.e. if $\Omega_0 \approx 0$) of about 14 billion years, with a more likely value of less than this (because Ω_0 is not zero). In particular if $H_0 = 72$ km/s/Mpc and $\Omega_0 = 1.0$, the age of the universe is about 9 billion years. In these cases the universe appears to be about the same age as the Milky Way's globular clusters (if $\Omega_0 \approx 0$), or possibly less, which it clearly cannot be, depending on the actual value of Ω_0.

According to the Big Bang theory, if Ω_0 is between 0.1 and 10 today, the initial value of Ω_0 just after the Big Bang was within 10^{-59} of 1.0. Many theorists have concluded, therefore, that as such a tiny deviation from 1.0 is unlikely, Ω_0 must be exactly equal to 1.0. Present best observational estimates of Ω_0, including both visible and dark matter, lie in the range from 0.1 and 1.0, so it seems plausible from both a theoretical and direct observational point of view that $\Omega_0 = 1.0$. There is something wrong, however, with an age of the universe of about 9 billion years, as that is appreciably less than the estimated age of the Milky Way's globular clusters. There is, however, a way out of this problem.

Years ago Einstein invented a cosmological constant Λ to stop the universe expanding or contracting in his cosmological theory, as he thought at that time that the actual universe was doing neither. In 1931, however, he dropped his cosmological constant when it became apparent that the universe was actually expanding. Einstein's cosmological constant implied that the vacuum of space has a finite energy density, and, if this is reintroduced into the Big Bang theory, the equality $\Omega_0 = 1.0$ is modified to read $\Omega_0 + \Omega_\Lambda = 1.0$. If Ω_Λ is non zero, for a given value of Ω_0 the age of the universe becomes larger. So, for example, if $H_0 = 72$ km/s/Mpc and $\Omega_0 = 0.2$, say, the age of the universe increases from about 11 billion years if $\Omega_\Lambda = 0$ to 15 billion years if $\Omega_0 + \Omega_\Lambda = 1.0$. Observational astronomers are trying to still provide better estimates of H_0 and Ω_0 using the HST and other observatories.

17.7 Concluding Remarks

Whilst this book has been in preparation the Hubble Space Telescope, BeppoSAX and a number of other spacecraft mentioned in the text have been producing more and more interesting results. In addition, the full Hipparcos archive has been published, and a number of new spacecraft have been launched to the planets or to undertake remote observations of the universe. It is too early, at the moment, to judge the full significance of their results, but it is clear that we are living in a very interesting time as far as astronomy is concerned. There is every reason to believe that in ten years time, say, some of the problems mentioned in this book will have been solved, but a number of new ones will have taken their place. So, no doubt, the old adage that 'The more we know, the more we know we don't know' will be valid for many more years to come.

Appendix

Table 1A. *Main Characteristics of the Inner Planets and the Moon*

	Mercury	Venus	Earth	Moon	Mars
Mean distance from Sun (10^6 km)	58	108	150		228
Ditto (AU)	0.39	0.72	1.00		1.52
Orbital period (days)	88	225	365	27	687
Ditto (years)	0.24	0.62	1.00	0.08	1.88
Equatorial diameter (km)	4,880	12,100	12,760	3,480	6,790
Ditto (Earth = 1)	0.38	0.95	1.00	0.27	0.53
Mass (Earth = 1)	0.06	0.81	1.00	0.012	0.11
Mean density (g/cm^3)	5.44	5.25	5.52	3.34	3.90
Spin period	58.7 days	243.0 days	23.9 hrs	27.3 days	24.6 hrs
Inclination of equator to orbit	0°	177°	23°	6°	24°
Solid surface?	Yes	Yes	Yes	Yes	Yes
Ring system?	No	No	No	No	No
No. of known satellites	0	0	1		2

Table 1B. *Main characteristics of the outer planets*

	Jupiter	Saturn	Uranus	Neptune	Pluto
Mean distance from Sun (10^6 km)	778	1,427	2,870	4,497	5,910
Ditto (AU)	5.2	9.5	19.2	30.1	39.5
Orbital period (years)	11.86	29.5	84.0	164.8	248.0
Equatorial diameter (km)	142,800	120,400	51,100	49,520	2,320
Ditto (Earth = 1)	11.21	9.45	4.00	3.88	0.18
Mass (Earth = 1)	318	95	14.5	17.1	0.002
Mean density (g/cm^3)	1.33	0.69	1.28	1.64	2.13
Spin period[a]	9.93 hrs	10.66 hrs	17.24 hrs	16.11 hrs	6.38 days
Inclination of equator to orbit	3°	27°	98°	29°	122°
Solid surface?	No	No	No	No	Yes
Ring system?	Yes	Yes	Yes	Yes	No
No. of known satellites	16	18	15	8	1

Notes:

[a] The figures for Jupiter, Saturn, Uranus and Neptune are the internal rotation periods deduced from the variability of their radio signals.

Table 2. *Energy balance for the gas giants*

	Jupiter	Saturn	Uranus	Neptune
Mean solar distance (AU)	5.203	9.539	19.181	30.058
∴ Incident solar energy[a] (W/m^2)[b] I	50.50	15.02	3.72	1.51
Average reflectivity of planet R	0.34	0.34	0.30	0.29
∴ Solar energy absorbed by planet[c] (W/m^2)	8.29	2.46	0.65	0.27
∴ Average expected temperature (K)	109.5	82.4	58.2	46.6
Effective temp. deduced from Voyager (K)	124.4	95.0	59.1	59.3
∴ Emitted energy deduced from Voyager (W/m^2)	13.73	4.47	0.68	0.70
∴ Energy balance = Emitted/ Absorbed energy	1.66 ± 0.10	1.82 ± 0.11	1.05 ± 0.09	2.60 ± 0.29
Internally generated energy[d] (W/m^2)	5.44 ± 0.43	2.01 ± 0.14	0.03 ± 0.04	0.43 ± 0.05

Notes:

[a] On planar surface perpendicular to the solar radiation at the mean solar distance of the planet.

[b] As a comparison, this is 1,367 W/m^2 for the Earth.

[c] This is the solar energy absorbed, on average, per unit surface area *of planet*. It equals $\frac{1}{4} I \times (1 - R)$.

[d] Equals emitted minus absorbed energy.

Table 3. *magnetic fields of the gas giants compared with that of the Earth*

	Earth	Jupiter	Saturn	Uranus	Neptune
Equatorial radius (km)	6,380	71,400	60,200	25,550	24,760
Inclination of equator to orbit	23.4°	3.1°	26.7°	97.9°	28.8°
Angle between magnetic and spin axes	11.5°	9.6°	0°	58.6°	47.0°
Dipole offset (km)	460	7,000	2,400	7,700	14,000
Ditto (radii)	0.07	0.10	0.04	0.30	0.55
Magnetic field at equator[a] (gauss)	0.31	4.28	0.22	0.25	0.15
Dipole moment (Earth = 1)	1	19,000	540	50	28
Av. sunward distance of magnetopause (planetary radii)	10.4	70	22	20	27

Notes:
[a] These magnetic field values are at the surface of the Earth or at the cloud tops of the other planets.

Glossary

active galaxy A galaxy which emits appreciably more energy than that produced by its stars. Seyfert galaxies and quasars are types of active galaxies.

adaptive optics A method of improving the quality of the image produced by a ground-based optical telescope by using a small, rapidly deformable, secondary mirror to correct for atmospheric turbulence.

albedo Reflectivity of an astronomical object. The units used are either on the scale 0 to 1 or 0% to 100%.

anorthosite An igneous rock consisting mainly of plagioclase feldspar, with much smaller amounts of pyroxene and olivine.

apoapsis The point in the path of an orbiting body at which it is furthest from the primary

apogee The point in the orbit of the Moon or Earth-orbiting spacecraft where it is furthest away from the Earth.

astrometry The measurement of the position and apparent motions of celestial objects.

atomic number The number of protons in an atomic nucleus. So hydrogen has an atomic number of one, helium two, and so on.

basalt A volcanic igneous rock consisting mainly of pyroxene and plagioclase, with smaller quantities of ilmenite.

Be-type stars B-type stars with hydrogen emission lines superimposed on the normal spectrum.

black body A body that absorbs and re-emits all the radiation incident on it.

black body radiation The thermal radiation emitted by a black body. The wavelength where the emission is greatest reduces with increasing temperature. So for the Sun, with a surface temperature of about 6,000 K, the maximum is in the visible band, but for the gas of the Sun's corona, which is at a temperature of a few million K, the maximum is in the X-ray band. (It is no coincidence, of course, that the maximum thermal emission of the Sun's surface is in the visible band, as our eyes have evolved so that they are most efficient where our light-giving star produces the most energy.)

BL Lac object A highly variable, quasar-like object that it thought to be the nucleus of an active galaxy; the variability being due to the fact that we are observing the active galactic nucleus down one of its active jets.

bow shock The boundary around a planet's magnetosphere where the solar wind is deflected.

breccia A rock made of broken fragments held in a fine-grained matrix. They are commonly the result of impacts of, for example, meteorites on the Moon.
bremsstrahlung A process in which a photon is emitted when an electron is decelerated by passing close to an atomic nucleus.
caldera A large volcanic crater created by the collapse of the surface and/or by an explosive eruption.
cataclysmic variable (CV) A star that exhibits sudden, recurrent increases in brightness, e.g. a nova, recurrent nova, dwarf nova or flare star.
chasma A steep-sided canyon.
chromosphere The region of the Sun above the photosphere. The temperature of the chromosphere falls from about 6,000 K at the top of the photosphere to 4,500 K about 500 km above the photosphere, before increasing to 20,000 K at the top of the chromosphere.
column mass The mass of atmosphere over unit area of surface.
Compton effect A process in which a photon collides with an electron, resulting in the photon losing energy (and so increasing its wavelength).
conjunction An alignment of three bodies. So, for example, the planet Mercury can be in a line with the Earth and Sun, and be either between the Earth and Sun, in an arrangement described as an *inferior conjunction*, or on the other side of the Sun to the Earth, described as a *superior conjunction*.
corona The outermost visible region of the Sun that is seen during total solar eclipses. It extends for a number of solar radii from the photosphere and has a temperature in excess of one million K.
coronagraph An instrument containing an occulting disc that creates an artificial eclipse of a bright source like the Sun, and allows observers to detect faint objects close to the bright source.
cosmic rays Very energetic elementary particles travelling through space. When they are above the Earth's atmosphere they are called *primary cosmic rays* or *cosmic ray primaries*. After collisions with the Earth's upper atmosphere they produce showers of *secondary cosmic rays*.
critical density The density of the universe that will just cause its expansion to cease at an infinite time in the future.
crust The outermost solid layer of a planet or moon, usually consisting of rock and/or ice. It is the outermost part of the lithosphere.
dwarf nova A star that shows sudden increases in brightness over periods up to about one year. It consists of a white dwarf surrounded by an accretion disc, which is fed with material emitted from a main sequence star. The first dwarf nova to be detected was U Geminorum in the nineteenth century.
ecliptic The plane of the Earth's orbit around the Sun.
emissivity The ratio of the power per unit area radiated from a surface to that radiated from a black body at the same temperature.
filaments Long features that appear dark when seen against the Sun's disc, but

appear bright, and are called prominences, when seen at the limb during total solar eclipses. They are masses of cool dense gas.

first ionisation potential The minimum energy required to remove the first electron from an atom or molecule to infinity.

flare A small, bright structure in the Sun's chromosphere which generates an explosive release of energy lasting for a few minutes.

flare star Also known as a UV Ceti-type star, a flare star is a dwarf M-type star that shows an increase in intensity over the order of minutes. This is thought to be due to a flare, like those on the Sun, although such stellar flares are much more energetic than solar flares.

Fourier analysis The analysis of periodic data into its various constituent frequencies.

Fraunhofer lines The dark lines in the solar spectrum which give information about the elements in the Sun. Each element is characterised by lines of given wavelengths. The detailed structure of these lines gives information on the environment (e.g. pressure, temperature, magnetic field, etc.) to which the element is subjected.

gabbro Granular igneous rocks.

gamma rays Electromagnetic radiation with wavelengths shorter than about 0.01 nm (or with $E>0.1$ MeV).

geomagnetic index A measure of the effect of solar activity on the Earth's magnetosphere. It is based on measuring variations in the ground-level magnetic field.

helioseismology The study of the interior of the Sun by an analysis of its natural oscillation modes.

hydrogen-α (or Hα) The most prominent line in the visible part of the solar spectrum. It is produced when an electron in its second excited state falls to its first excited state. At a wavelength of 656.3 nm it is the first line in the Balmer series.

ilmenite An iron titanium oxide, $FeTiO_3$.

infrared waveband From about 0.7 to 300 microns.

intermediate polar A cataclysmic variable in which the compact source is a magnetic white dwarf in a binary system with a red giant. Material from the red giant is fed on to the poles of the white dwarf via an accretion disc.

inverse Compton effect A process in which an electron collides with a photon and transfers some of its energy to the photon.

ionisation Gases are said to be ionised when their atoms or molecules lose electrons by collisions with ionising radiation in the form of photons or elementary particles. The higher the energy of the photons (i.e. the shorter the wavelength) the more likely are they to ionise a gas.

ionosphere A region of a planetary atmosphere where the atoms and molecules are ionised. The Earth's ionosphere has four layers or regions, namely the D region from an altitude of 50 to 90 km, the E region from 90 to 160 km, the F_1 region from 160 to 230 km, and the F_2 region from 230 km to about 600 km. These different regions reflect radio waves of different wavelengths.

Lagrangian points Points in the orbital plane of two objects that orbit around their common centre of mass, where a particle of negligible mass can remain in equilibrium. There are five Lagrangian points for two bodies in circular orbits, three of which are unstable to small perturbations. The other two, which are 60° in front of and behind the less massive body, and in the same orbit, are stable.

lithosphere The rigid outer layer of a planet or moon, including the crust and part of the upper mantle.

Low Mass X-ray Binary (LMXB) A binary where the compact source is a neutron star or black hole, and where the other star is a post B-type main sequence star.

Lyman-a A line at 121.6 nm which is produced when an electron in the first excited state of a hydrogen atom decays to its ground state.

magnetic inversion line The line where the photospheric line-of-sight magnetic field is zero.

magnetopause The boundary of the magnetosphere.

magnetohydro-dynamic theory The theory of a magnetised fluid in an electric and magnetic field.

magnetosphere The region around a planet in which its magnetic field is constrained by the solar wind.

main sequence stars Stars on a band on the Hertzsprung–Russell diagram running from very luminous, hot, white, O-type stars through B, A, F, G and K stars of progressively lower luminosities and temperatures to low intensity, cool, red, M-type stars.

mantle The layer of a planet or moon between the crust and the core.

mare (pl. maria) The extensive dark areas on the Moon that were originally thought to be seas of water. They are now known to be seas of solidified lava.

maser In the maser process an incoming microwave photon stimulates an atom or molecule in a high energy state to emit energy identical to that of the incoming photon. If there are enough atoms or molecules initially in the same excited state, a large number of identical transitions are stimulated, leading to a burst of monochromatic energy. At optical wavelengths this is called a laser process. MASER comes from Microwave Amplification by the Stimulated Emission of Radiation.

Massive X-Ray Binary (MXRB) A binary where the compact source is a neutron star or black hole, and where the other star is an O- or B-type.

meteoroid A piece of rock or dust in space that can burn up in the Earth's atmosphere as a meteor, or fall to Earth as a meteorite.

neutron star The stellar remnant of stars with masses of from 1.4 to 3.0 solar masses, when they have stopped producing energy by thermonuclear fusion. A neutron star consists almost entirely of neutrons, and is prevented from further collapse by neutron degeneracy pressure, which is a quantum mechanical effect analogous to electron degeneracy pressure in white dwarfs.

occultation The passage of one astronomical object in front of another so that an observer can no longer see the more distant object.

OH source An astronomical source emitting microwave radiation characteristic of the hydroxyl molecule (OH), usually by maser emission.

olivine A magnesium or iron silicate, $(Mg,Fe)_2SiO_4$.

P Cygni profile A spectrum with blue-shifted absorption lines and red-shifted emission lines due to an expanding disc or shell of material around a star.

periapsis The point in the path of an orbiting body at which it is nearest to the primary.

perigee The point in the orbit of the Moon or Earth-orbiting spacecraft where it is nearest to the Earth.

photosphere The visible surface of the Sun at a temperature of about 6,000 K. Sunspots and faculae are in the photosphere.

plages Bright areas in the chromosphere associated with sunspots.

plagioclase A feldspar of sodium or calcium aluminium silicate.

plasma An ionised gas consisting of electrons and ionised atoms.

plate tectonics The mechanism on the Earth's surface where structural plates move together or apart, thus producing continental drift.

polar A cataclysmic variable in which the compact source is a highly magnetic white dwarf in a binary system with a red giant. Material from the red giant star is fed directly on to the white dwarf at its poles, rather than via an accretion disc.

prominences Solar filaments seen at the edge of the Sun (see 'filaments').

protostar A star in the earliest stage of formation, before the onset of nuclear reactions in its core.

pyroxene A calcium or sodium and magnesium, iron or aluminium silicate, i.e. $ABSi_2O_6$, where A is Ca or Na, and B is Mg, Fe or Al.

quasar A very bright extragalactic object with a high red-shift. It is generally thought to be the most luminous type of nucleus of an active galaxy.

regolith The top layer of a solid planetary or similar body (i.e. the Moon) covering the crust and composed of dust and loose fragments of rock.

rhyolite A volcanic igneous rock of relatively high silica content. Lighter in colour and more viscous than basalt.

scintillation counter A detector using crystals that emit flashes of light when irradiated by γ-rays. Each flash of light is usually detected by a photomultiplier.

Seyfert galaxy A type of active galaxy with a brilliant point-like nucleus and inconspicuous spiral arms. It has a broad emission-line spectrum.

shield volcanoes Gently sloping volcanoes built up from successive lava flows from a single vent.

spherical aberration An optical aberration in which rays of light from the edge of a lens or mirror are brought to a different focus from those on the optic axis.

spicules Small, radial jet-like structures at the top of the Sun's chromosphere.

star spectral types See 'main sequence stars'.

sublimation The act of a material changing from the solid to the vapour phase without going through an intermediate liquid phase. So, for example, at certain temperatures and pressures ice can evaporate without passing through the water phase.

supernova-type I The explosion of a white dwarf in a binary whose mass has exceeded the Chandrasekhar limit of 1.4 solar masses. Prior to the explosion the white dwarf has been gaining mass from its red giant binary companion.

supernova-type II An explosion produced when a main sequence star of more than eight solar masses stops producing energy by thermonuclear fusion.

synchrotron radiation Radiation produced by energetic electrons as they spiral in a magnetic field.

synodic period The mean interval for a planet between successive inferior or superior conjunctions with the Earth and Sun.

tomography The method of producing cross-sectional images.

transition region The region on the Sun between the chromosphere and the corona in which the temperature increases from about 20,000 to 1,000,000 K in a very short distance.

trojans Small bodies at the $\pm 60°$ Lagrangian points and co-orbital with a planet or one of its satellites.

ultraviolet waveband From about 9 to 380 nm

Van Allen radiation belts Two ring-shaped belts around the Earth of electrically charged particles trapped by the Earth's magnetic field. The inner belt extends from about 2,000 to 5,000 km altitude, and the outer belt from about 13,000 to 19,000 km.

visible light Covers the waveband from about 380 nm (blue) to about 780 nm (red).

Wolf–Rayet stars Very hot, O-type stars producing a spectrum with strong, broad emission lines on a continuum background.

white dwarf The stellar remnant of stars with a maximum mass of about 1.4 solar masses when they stop producing energy by thermonuclear fusion. A white dwarf consists of atomic nuclei and electrons that have been completely stripped from atoms. White dwarfs are prevented from further collapse by so-called electron degeneracy pressure, which is a quantum mechanical effect.

X-rays Photons of very short wavelengths. The X-ray waveband is from about 0.01 to 9 nm (100 to 0.1 keV). Those X-rays at the short wavelength (high energy) end are called hard X-rays, and those at the long wavelength (low energy) end are called soft X-rays.

Zeeman effect The splitting of spectral lines into a number of components when the source is in a magnetic field.

Bibliography

This bibliography includes the general sources used in the preparation of this book, which can be consulted if the reader requires more detail than I have been able to give in this overview. It only includes the main sources, however, ignoring many detailed articles and papers that were used to check and complete some of the details. The book publishers mentioned below are generally those for copies available in the UK. In other countries the publishers may differ. Earlier and/or later editions are sometimes available.

Books

General

Audouze, J., and Israël, G. (eds.), *The Cambridge Atlas of Astronomy*, Cambridge University Press, Third edition, 1994, ISBN 0-521-43438-6.

Baker, D., *Spaceflight and Rocketry: A Chronology*, Facts on File, 1996, ISBN 0-8160-1853–7.

Bonnet, R.M, and Manno, V., *International Cooperation in Space: The Example of the European Space Agency*, Harvard University Press, 1994. ISBN 0-674-45835-4.

Bromberg, J.L., *NASA and the Space Industry*, Johns Hopkins University Press, 1999, ISBN 0-8018-6050-4.

Burrows, W.E., *This New Ocean: The Story of the First Space Age*, Random House, 1998, ISBN 0-679-44521-8.

Canizares, C.R., Becker, F., and Culhane, J.L. (eds.), *U.S.–European Collaboration in Space Science*, National Academy Press, Washington, 1998, ISBN 0-309-05984-4.

Cornell, J., and Gorenstein, P. (eds.), *Astronomy from Space: Sputnik to Space Telescope*, MIT Press, 1985, ISBN 0-262-53061-9.

Friedman, H., *The Astronomer's Universe: Stars, Galaxies, and Cosmos*, W.W. Norton, 1990, ISBN 0-393-02818-6.

Harford, J., *Korelev: How one Man Masterminded the Soviet Drive to Beat America to the Moon*, Wiley, 1997, ISBN 0-471-14853-9.

Harland, D.M., *The Space Shuttle: Roles, Missions and Accomplishments*, Wiley, 1998, ISBN 0-471-98138-9.

Heppenheimer, T.A., *Countdown: A History of Space Flight*, Wiley, 1997, ISBN 0-471-14439-8.

Huntley, J.D., *The Birth of NASA: The Diary of T. Keith Glennan*, NASA SP-4105, 1993, ISBN 0-16-041936-0.

King-Hele, D.G., *et al.*, *The R.A.E. Table of Earth Satellites 1957–1989*, RAE Farnborough, Fourth edition, 1990, ISBN 0-9516542-0-9.

Krige, J., and Russo, A., *Europe in Space 1960–1973*, ESA SP-1172, 1994, ISBN 92-9092-125-0.

Logsdon, J.M., (ed.) *Exploring the Unknown: Selected Documents in the History of the US Civil Space Program: Vol. I. Organizing for Exploration*, NASA SP-4407, 1995.

Logsdon, J.M. (ed.) *Exploring the Unknown: Selected Documents in the History of the US Civil Space Program: Vol. II. External Relationships*, NASA SP-4407, 1996, ISBN 0-16-048899-0

Massey, H., and Robins, M.O., *History of British Space Science*, Cambridge University Press, 1986, ISBN 0-521-30783-X.

McDougall, W.A., *The Heavens and the Earth: A Political History of the Space Age*, Johns Hopkins University Press, 1997, ISBN 0-8018-5748-1.

Murdin, P., and L., *Supernovae*, Cambridge University Press, 1985, ISBN 0-521-30038-X.

Robinson, J., *The End of the American Century: Hidden Agendas of the Cold War*, Hutchinson, 1992, ISBN 0-09-177065-3.

Tucker, W., and Tucker, K., *The Cosmic Inquirers: Modern Telescopes and Their Makers*, Harvard University Press, 1986, ISBN 0-674-17435-6.

Pre-Spacecraft Era

De Vorkin, D.H., *Science with a Vengeance: How the Military Created the US Space Sciences after World War II*, Springer-Verlag, 1992, ISBN 0-387-94137-1.

Hall, A. (ed.), *Man in Space. Vol. 1, The First Small Step*, Peterson, 1974.

Piszkiewicz, D., *Wernher Von Braun: The Man Who Sold the Moon*, Praeger, 1998, ISBN 0-275-96217-2.

Rosen, M.W., *The Viking Rocket Story*, Faber & Faber, 1956.

Spangenburg, R., and Moser, D.K., *Wernher Von Braun: Space Visionary and Rocket Engineer*, Facts on File, 1995, ISBN 0-8160-2924-5.

The Solar System

Beatty, J.K., and Chaikin, A. (eds.), *The New Solar System*, Cambridge University Press, Third edition, 1990, ISBN 0-521-36162-1

Briggs, G.A., and Taylor, F.W., *The Cambridge Photographic Atlas of the Planets*, Cambridge University Press, 1982, ISBN 0-521-23976-1.

Chaikin, A., *A Man on the Moon: The Voyages of the Apollo Astronauts*, Penguin, 1995, ISBN 0-14-024146-9.

Calder, N., *Giotto to the Comets*, Presswork, London, 1992, ISBN 0-9520115-0-6

Golub, L., and Pasachoff, J.M., *The Solar Corona*, Cambridge University Press, 1997, ISBN 0-521-48082-5.

Greeley, R., *Planetary Landscapes*, Chapman and Hall, Revised edition, 1987, ISBN 0-04-551081-4.

Hufbauer, K., *Exploring the Sun: Solar Science since Galileo*, Johns Hopkins University Press, 1991, ISBN 0-8018-4098-8.

Kaufmann, W.J., *Planets and Moons*, W.H.Freeman, 1979, ISBN 0-7167-1040-4.

Kippenhahn, R., *Discovering the Secrets of the Sun*, Wiley, 1994, ISBN 0-471-94363-0.

Lang, K.R., and Whitney, C.A., *Wanderers in Space: Exploration and Discovery in the Solar System*, Cambridge University Press, 1991, ISBN 0-521-42252-3.

Littman, M., *Planets Beyond: Discovering the Outer Solar System*, Wiley, 1990, ISBN 0-471-51053-X.

Miner, E.D., *Uranus: The Planet, Rings and Satellites*, Wiley, Second edition, 1998, ISBN 0-471-97398-X.

Moore, P., and Hunt, G., *The Atlas of the Solar System*, Mitchell Beazley, 1983, ISBN 0-85533-468-1.

Phillips, K.J.H., *Guide to the Sun*, Cambridge University Press, 1995, ISBN 0-521-39788-X.

Reeves, R., *The Superpower Space Race: An Explosive Rivalry through the Solar System*, Plenum, 1994, ISBN 0-306-44768-1.

Surkov, Y., *Exploration of the Terrestrial Planets from Spacecraft: Instrumentation, Investigation, Interpretation*, Wiley, Second edition, 1997, ISBN 0-471-96429-8.

Tatarewicz, J.N., *Space Technology and Planetary Astronomy*, Indiana University Press, 1990, ISBN 0-253-35655-5.

Spacecraft Observatories (X-Ray etc.)

Aschenbach, B., Hahn, H-M., and Trümper, J.,: Jenkner, H., (trans.), *The Invisible Sky: Rosat and the Age of X-Ray Astronomy*, Copernicus, 1998, ISBN 0-387-94928-3.

Chaisson, E.J., *The Hubble Wars: Astrophysics Meets Astropolitics in the Two-Billion-Dollar Struggle over the Hubble Space Telescope*, Harvard University Press, 1998, ISBN 0-674-41255-9.

Charles, P.A., and Seward, F.D., *Exploring the X-ray Universe*, Cambridge University Press, 1995, ISBN 0-521-43712-1.

Clark, D.H., *The Quest for SS 433*, Adam Hilger, 1986, ISBN 0-85274-828-0.

Culhane, J.L., and Sanford, P.W., *X-ray Astronomy*, Faber & Faber, 1981, ISBN 0-571-11550-0.

Davies, J.K., *Astronomy from Space: The Design and Operation of Orbiting Observatories*, Wiley, 1997, ISBN 0-471-96018-7.

Elvis, M. (ed.), *Imaging X-ray Astronomy: A Decade of Einstein Observatory Achievements*, Cambridge University Press, 1990, ISBN 0-521-38105-3.

Fischer, D., and Duerbeck, H.: Jenkner, H., (transl.), *Hubble Revisited: New Images from the Discovery Machine*, Copernicus, 1998, ISBN 0-387-98551-4.

Henbest, N., and Marten, M., *The New Astronomy*, Cambridge University Press, 1983, ISBN 0-521-25683-6

Hirsch, R., *Glimpsing an Invisible Universe: The Emergence of X-ray Astronomy*, Cambridge University Press, 1983, ISBN 0-521-31232-9.

Kondo, Y. (ed.), *Exploring the Universe with the IUE Satellite*, Kluwer, 1989, ISBN 90-277-2380-X.

Kondo, Y. (ed.), *Observatories in Earth Orbit and Beyond*, Kluwer, 1990, ISBN 0-7923-1133-7.

Labuhn, F., and Lüst, R. (eds.), *New Techniques in Space Astronomy*, Reidel, 1971, ISBN 90-277-0202-0.

Mather, J.C., and Boslough, J., *The Very First Light: The True Inside Story of the Scientific Journey Back to the Dawn of the Universe*, Penguin, 1998, ISBN 0-14-027220-8.

Petersen, C.C., and Brandt, J.C., *Hubble Vision: Astronomy with the Hubble Space Telescope*, Cambridge

University Press, 1995, ISBN 0-521-49643-8.

Rowan-Robinson, M., *Ripples in the Cosmos*, W.H.Freeman, 1993, ISBN 0-7167-4503-8.

Smith, R.W., *The Space Telescope: A Study of NASA, Science, Technology and Politics*, Cambridge University Press, 1993, ISBN 0-521-45769-6.

Smoot, G., and Davidson, K., *Wrinkles in Time: The Imprint of Creation*, Abacus, 1995, ISBN 0-349-10602-9.

Strong, K.T., Saba, J.L.R., Haisch, B.M., and Schmelz, J.T. (eds.), *The Many Faces of the Sun: A Summary of the Results from NASA's Solar Maximum Mission*, Springer, 1999, ISBN 0-387-98481-X.

NASA History Series
(in date order)

General

NASA SP-4101, 1966, Rosholt, R.L., *An Administrative History of NASA, 1958–1963.*

NASA SP-133, 1967, Corliss, W.R., *Scientific Satellites.*

NASA SP-4202, 1970, Green, C.McL., and Lomask, M., *Vanguard: A History.*

NASA SP-4401, 1971, Corliss, W.R., *NASA Sounding Rockets.*

NASA SP-4211, 1980, Newell, H.E., Jr., *Beyond the Atmosphere: Early Years in Space Science.*

NASA SP-4403, 1981, Anderson, F.W., Jr., *Orders of Magnitude: A History of NACA and NASA 1915–1980.*

NASA EP-177, 1981, Chipman, E.G., *et al., A Meeting with the Universe: Science Discoveries from the Space Program.*

NASA SP-4102, 1982, Levine, A.S., *Managing NASA in the Apollo Era.*

NASA SP-4405, 1985, Roland, A., *A Spacefaring People: Perspectives on Early Spaceflight.*

NASA SP-4301, 1985, Rosenthal, A., *Venture into Space: Early Years of the Goddard Space Flight Center.*

NASA SP-4012, rep. ed. 1988, Van Nimmen, J., Bruno, L.C., and Rosholt, R.L., *NASA Historical Data Book, Vol. I, NASA Resources, 1958–1968.*

NASA SP-4012, 1988, Ezell, L.N., *NASA Historical Data Book, Vol. II, Programs and Projects, 1958–1968.*

NASA SP-4012, 1988, Ezell, L.N., *NASA Historical Data Book, Vol. III, Programs and Projects, 1969–1978.*

NASA SP-4406, 1989, Bilstein, R.E., *Orders of Magnitude: A History of the NACA and NASA, 1915–1990.*

NASA SP-4012, 1994, Gawdiak, I.Y., *NASA Historical Data Book, Vol. IV, NASA Resources 1969–1978.*

NASA SP-4311, 1997, Wallace, H.D. Jr., *Wallops Station and the Creation of an American Space Program.*

Spacecraft and Missions

NASA SP-57, 1965, *NASA Orbiting Solar Observatory OSO 1: A Project Summary.*

NASA SP-59, 1965, Jet Propulsion Laboratory, *Mariner-Venus 1962, Final Project Report.*

NASA SP-190, 1967, *Mariner Venus 1967.*

NASA SP-184, 1969, Surveyor Program

Office, *Surveyor Program Results.*

NASA-214, 1969, NASA Manned Spacecraft Center, *Apollo 11 Preliminary Science Report.*

NASA-235, 1970, NASA Manned Spacecraft Center, *Apollo 12 Preliminary Science Report.*

NASA-272, 1971, NASA Manned Spacecraft Center, *Apollo 14 Preliminary Science Report.*

NASA-289, 1972, NASA Manned Spacecraft Center, *Apollo 15 Preliminary Science Report.*

NASA-315, 1972, NASA Manned Spacecraft Center, *Apollo 16 Preliminary Science Report.*

NASA-330, 1973, NASA Lyndon B. Johnson Space Center, *Apollo 17 Preliminary Science Report.*

NASA SP-329, 1974, *Mars as Viewed by Mariner 9.*

NASA SP-337, 1974, Hartman, W.K., and Raper, O., *The New Mars: Discoveries of Mariner 9.*

NASA SP-350, 1975, Cortright, E.M. (ed.), *Apollo Expeditions to the Moon.*

NASA SP-404, 1975, Lundquist, C.A. (ed.), *Skylab's Astronomy and Space Sciences.*

NASA SP-396, 1977, Fimmel, R.O., Swindell, W., and Burgess, E., *Pioneer Odyssey*

NASA SP-4210, 1977, Hall, R.C., *Lunar Impact: A History of Project Ranger.*

NASA SP-424, 1978, Dunne, J.A., and Burgess, E., *The Voyage of Mariner 10.*

NASA SP-425, 1978, *The Martian Landscape.*

NASA SP-402, 1979, Eddy, J.A., *A New Sun: The Solar Results from Skylab.*

NASA SP-4205, 1979, Brooks, C.G., Grimwood, J.M., and Swenson, L.S., *Chariots for Apollo: A History of Manned Lunar Spacecraft.*

NASA SP-439, 1980, Morrison, D., and Samz, J., *Voyage to Jupiter.*

NASA SP-441, 1980, Carr, M.H., *et al.*, *Viking Orbiter Views of Mars.*

NASA SP-446, 1980, Fimmel, R.O., *et al.*, *Pioneer: First to Jupiter, Saturn and Beyond.*

NASA SP-451, 1982, Morrison, D., *Voyages to Saturn.*

NASA SP-461, 1983, Fimmel, R.O., *et al.*, *Pioneer Venus.*

NASA SP-466, 1984, Tucker, W.H., *The Star Splitters: The High Energy Astronomy Observatories.*

NASA SP-4212, 1984, Ezell, E.C., and Ezell, L.N., *On Mars: Exploration of the Red Planet, 1958–1978.*

NASA SP-487, 1987, Lord, D.R., *Spacelab – An International Success Story.*

NASA SP-4214, 1989, Compton, W.D., *Where No Man Has Gone Before: A History of Apollo Lunar Exploration Missions.*

NASA SP-520, 1995, Roth, L.E., and Wall, S.D., *The Face of Venus: The Magellan Radar-Mapping Mission.*

NASA SP-4218, 1996, Butrica, A.J., *To See the Unseen: A History of Planetary Radar Astronomy.*

ESA History Series

ESA HSR-2, Oct. 1992, Russo, A., *ESRO's First Scientific Satellite Programme 1961–1966.*

ESA HSR-3, Nov. 1992, Russo, A., *Choosing ESRO's First Scientific Satellites.*

ESA HSR-6, Mar. 1993, Russo, A., *The definition of a scientific policy: ESRO's satellite programme in 1969–1973.*

ESA HSR-8, May 1993, Krige, J., *Europe in Space: The Auger Years (1959–1967).*

ESA HSR-20, Sept. 1997, Russo, A., *The Definition of ESA's Scientific Programme for the 1980s.*

ESA HSR-24, May 1999, Russo, A., *ESA's Scientific Programme Towards the Turn of the Century,* ISBN 92-9092-531-0.

ESA HSR-Special, July 1993, Russo, A. (ed.), *Science Beyond the Atmosphere: The History of Space Research in Europe.*

ESA BR-142, May 1999, Wilson, A., *ESA Achievements: More than Thirty Years of Pioneering Space Activities,* ISBN 92-9092-629-5.

ESA BR-147, 1999, Calder, N. (ed.), *Success Story; 30 Discoveries from ESA's Science Missions in Space,* ISBN 92-9092-610-4.

Units

I have tried to use a set of units which would be acceptable to both amateur and professional astronomers. Some of these people normally use SI units, but some feel more familiar with cgs or imperial units, so my selection has been a compromise. The units used are as follows:

Type of unit	Unit	Abbreviation	Equivalent
Distance	megaparsec	Mpc	10^6 pc
	parsec	pc	3.26 light years or 3.1×10^{16} m
	light years		6.3×10^4 AU or 9.5×10^{15} m
	Astronomical Unit	AU	150 million km
	kilometre	km	0.62 miles (1 mile ≡ 1.6 km)
	metre	m	39.4 inches
	centimetre	cm	10^{-2} m (1 inch ≡ 2.54 cm)
	millimetre	mm	10^{-3} m
	micron	μm	10^{-6} m
	nanometre	nm	10^{-9} m or 10 Ångströms
Angle	arc minute	arcmin	1/60 degree
	arc second	arcsec	1/3600 degree
Mass	kilograms	kg	2.2 lb (1,000 kg ≡ 1 metric ton)
	grams	g	
Density	grams per cubic centimetre	g/cm^3	10^3 kg/m^3
Pressure	bar		1.01 atmosphere or 1.0×10^5 Pa
Temperature	kelvin	K	°C + 273
Energy	electron volts	eV	1.6×10^{-19} J
	erg		10^{-7} J
Frequency	gigahertz	GHz	10^9 Hz
	megahertz	MHz	10^6 Hz
Magnetic field	gauss		10^{-4} Tesla or 10^5 gamma

Abbreviations[1]

ABMA	Army Ballistic Missile Agency
AFBMD	Air Force Ballistic Missile Division
AFCRL	Air Force Cambridge Research Laboratories
AGN	Active Galactic Nucleus
ALSEP	Apollo Lunar Surface Experimental Package
ANS	Astronomical Netherlands Satellite
APL	Applied Physics Laboratory
ARPA	Advanced Research Projects Agency (of the DoD)
AS & E	American Science and Engineering Inc.
ASTP	Apollo–Soyuz Test Project
AU	Accounting Unit (of ESRO/ESA)
BAC	British Aircraft Corporation
Caltech	California Institute of Technology
CERN	Conseil Européen pour la Recherche Nucléair
CGRO	Compton Gamma Ray Observatory
CIA	Central Intelligence Agency
CME	Coronal Mass Ejection
CNES	Centre National d'Etudes Spatiales (French Space Agency)
COBE	Cosmic Background Explorer
COSPAR	Committee on Space Research (of the International Council of Scientific Unions)
CTIO	Cerro Tololo Inter-American Observatory
DISCO	Dual Spectral Irradiance and Solar Constant Orbiter
DoD	Department of Defense
DSN	Deep Space Network
EEC	European Economic Community (the Common Market)
ELDO	European Launcher Development Organisation
ESA	European Space Agency
ESO	European Southern Observatory
ESRO	European Space Research Organisation
Estec	European Space Technology Centre
EUV	Extreme Ultraviolet
EUVE	Extreme Ultraviolet Explorer

[1] Excluding the abbreviations for spacecraft instruments, which are explained in the text, and for star and other catalogues.

FIP	First Ionisation Potential
FY	Financial or Fiscal Year
GDS	Great Dark Spot (on Neptune)
GE	General Electric
GEOS	Geosynchronous Scientific Satellite
GRB	Gamma Ray Burst (or Burster)
GRIST	Grazing Incidence Solar Telescope
GRO	Gamma Ray Observatory (Compton)
GRS	Great Red Spot (on Jupiter)
HEAO	High Energy Astronomy Observatory
HELOS	Highly Eccentric Lunar Occultation Satellite
HST	Hubble Space Telescope
IACG	Inter-Agency Consultative Group
IAF	International Astronomical Federation
IAU	International Astronomical Union
ICBM	Inter-Continental Ballistic Missile
ICE	International Cometary Explorer
ICSU	International Council of Scientific Unions
IDC	Image Data Compression
IGY	International Geophysical Year
IMP	Interplanetary Monitoring Platform
IR	Infrared
IRAM	Institut d'Astronomie Millimétrique
IRAS	Infrared Astronomical Satellite
IRBM	Intermediate Range Ballistic Missile
ISAS	Institute of Space and Astronautical Science (Japan)
ISEE	International Sun-Earth Explorer
ISO	Infrared Space Observatory
ISPM	International Solar Polar Mission
ISTP	International Solar–Terrestrial Physics (programme)
IUE	International Ultraviolet Explorer
IUS	Interim or Inertial Upper Stage
JCMT	James Clerk Maxwell Telescope
JFET	Junction Field Effect Transistor
JPL	Jet Propulsion Laboratory
KAO	Kuiper Airborne Observatory
LAS	Large Astronomical Satellite
LMC	Large Megellanic Cloud
LPR Project	Long Playing Rocket Project (later called Vanguard)
LST	Large Space Telescope
MAU	Million Accounting Units (ESA's currency)
MHD	Magnetohydrodynamic
MIT	Massachusetts Institute of Technology
MOSFET	Metal Oxide Semiconductor Field Effect Transistor

MOU	Memorandum of Understanding
MSS	Medium Sensitivity Survey
NACA	National Advisory Committee for Aeronautics
NASA	National Aeronautics and Space Administration
NOAO	National Optical Astronomy Observatory
NRAO	National Radio Astronomy Observatory (in the USA)
NRL	Naval Research Laboratory
NSSDC	National Space Science Data Center
OAO	Orbiting Astronomical Observatory
OGO	Orbiting Geophysical Observatory
OMB	Office of Management and Budget
OSO	Orbiting Solar Observatory
PI	Principal Investigator
PSAC	President's Science Advisory Committee
PVO	Pioneer Venus Orbiter
QPO	Quasi-Periodic Oscillation
RAE	Radio Astronomy Explorer
RAL	Rutherford Appleton Laboratory
Project RAND	Research ANd Development
ROSAT	Röntgensatellit
RTG	Radioisotope Thermoelectric Generator
SAR	Synthetic Aperture Radar
SAS	Small Astronomy Satellite
SGR	Soft Gamma Ray Repeater
SMC	Small Megellanic Cloud
SNR	Supernova Remnant
SOHO	Solar and Heliospheric Observatory
SSB	Space Science Board (of the National Academy of Sciences)
STScI	Space Telescope Science Institute
SIRTF	Space Infrared Telescope Facility
STSP	Solar–Terrestrial Science Programme
TCP	Technological Capabilities Panel
TD	Thor Delta
TDRS	Tracking and Data Relay Satellite
TOPS	Thermoelectric Outer Planet Spacecraft
TOS	Transfer Orbit Stage
UGO	Unidentified Gamma-Ray Object
UT	Universal Time (or Greenwich Mean Time, GMT)
UV	Ultraviolet
UVAS	Ultraviolet Astronomical Satellite
VLA	Very Large Array (radio telescope)
WR	Wolf–Rayet (stars)
XTE	X-Ray Timing Explorer

Name Index

Spacecraft Index*

Main references are in bold type

* Sometimes spacecraft programmes are listed separately from the individual spacecraft of which they are composed. In this case the programme reference usually covers the early planning phase.

Subject Index

Main references are in bold type